CW00522466

# Sir Charles Frank OBE, FRS:

## An eightieth birthday tribute

# Sir Charles Frank, OBE, FRS:

## An eightieth birthday tribute

Edited by

R G Chambers, J E Enderby, A Keller, A R Lang and J W Steeds

*H H Wills Physics Laboratory, University of Bristol*

Adam Hilger
Bristol, Philadelphia and New York

*British Library Cataloguing in Publication Data*

Sir Charles Frank, OBE, FRS: an Eightieth birthday tribute
1. Crystals
I. Chambers, R. G. (Robert Guy) 1924–
548

ISBN 0-7503-0100-7

*Library of Congress Cataloging-in-Publication Data are available*

The frontispiece photograph is reproduced courtesy of Gordon Kelsey, University of Bristol Arts Faculty Photographic Unit.

Published under the Adam Hilger imprint by IOP Publishing Ltd
Techno House, Redcliffe Way, Bristol BS1 6NX, England
335 East 45th Street, New York, NY 10017-3483, USA
US Editorial Office: 1411 Walnut Street, Suite 200, Philadelphia, PA 19102

Typeset by Mathematical Composition Setters Limited, Salisbury, UK.
Printed in Great Britain by Galliard (Printers) Ltd, Great Yarmouth

# Contents

# Foreword

**Sir Nevill Mott**

It is a pleasure to write a foreword to a volume celebrating the eightieth birthday of Sir Charles Frank. I first met him during the war, when he was working with R V Jones, who tells the story of his exploits in his contribution to this volume. I doubt if I knew what he was doing, but I must have sensed the quality of the man sufficiently to know that I wanted Charles and Maita to come to Bristol on the long awaited day when we could return to our universities and to our beloved physics. Probably, too, we talked about his distinguished research before the war in Oxford and in Berlin. In any case, he came in 1946 and there he stayed to become director of the laboratory in 1969 until his retirement in 1976, Vice-president of the Royal Society and honoured by a multitude of medals and honorary doctorates.

Arthur Tyndall was director of the laboratory at that time; I held the chair of theoretical physics and we worked closely together. Tyndall was immensely ambitious for his laboratory and proud of his young men; we were supported by an outstanding Vice-Chancellor, Philip Morris. It was a period of optimism. Science (we felt) had won the war, and could win the peace. We had lost some of our outstanding pre-war team (Herbert Skinner, Walter Heitler, Herbert Fröhlich, Ronald Gurney and Harry Jones), but we still had Cecil Powell who built up the team that led to the pi-meson and a Nobel prize, and on the solid state side we appointed, among others, Frank, Jack Mitchell and Dirk Polder and a little later for three years we had with us Jacques Friedel. One thing we wanted to do was to hold summer schools for industrial scientists, and introduce them to concepts in electronics introduced before or during the war, or just afterwards, such as the transistor and dislocations. It is from one of these summer schools that I derive my most vivid recollections of Charles at that time.

I think it was I who introduced him to the problem of crystal growth and also to the dislocation as introduced by G I Taylor in his model of work-hardening—though I cannot be sure of this. I had published a paper before the war with Frank Nabarro on a dislocation theory of age hardening. In any case Charles took to these subjects with enthusiasm. Nabarro and I had not gone beyond the edge dislocation, but Charles soon read up the work of Burgers and others on the more general concepts, saw that the screw dislocation was essential for the explanation of crystal growth and—at one of our summer schools—expounded how his ideas predicted growth spirals. And then, someone in the audience got up and said he had seen such a thing, and showed us a picture. Or at any rate that is how I remember it.

I do not want in this introduction to catalogue his later achievements, except perhaps the Frank–Read source which finally made the dislocation theory of plastic flow fully credible. Bristol was centre of dislocation activity at that time—with Jack Mitchell 'decorating' dislocations in silver halides and Charles 'decorating', almost accidentally, growth steps on metal surfaces. Was the concept of 'decorating' a Bristol invention? I would like to think so.

Charles has been a leader of Bristol physics since he came to us, and Maita and Charles in their home in Coombe Dingle a centre for friendship within the university that all of us remember with gratitude.

# Preface

Friends and colleagues of Sir Charles Frank offer him this collection of essays as a tribute to his pioneering work in many fields. The diversity of topics discussed in the following pages bears witness to the wide scope of Sir Charles' scientific activities and influence: the remarkable extent to which current understanding is based upon ideas first put forward by him becomes abundantly apparent.

Imaginative analysis, logical and penetrating, often drawing strongly upon basically simple geometric concepts, is a characteristic running through the research of Sir Charles Frank. His method has produced results elegant, durable and widely applicable. Countless workers, spreading across a range of disciplines and all the continents, are indebted to Sir Charles for inspiration, instruction and helpful advice given unstintingly over many years.

The editors wish to record the enthusiasm with which contributors prepared their birthday tributes presented in this volume, and to express warm thanks to all authors.

Sir Charles Frank delivering his Award Address after receiving the first triennial Crystal Growth Award of the American Association for Crystal Growth, July 1978, Washington, DC. (Report in *J. Crystal Growth* **46** (1979) 591–4.)

# List of Contributors

**K H G Ashbee**, Department of Materials Science and Engineering, The University of Tennessee, Knoxville, TN 37996-2200, USA

**P Bennema**, RIM Laboratory of Solid State Chemistry, University of Nijmegen, Toernooiveld, 6525 ED NIJMEGEN, The Netherlands

**M Berry**, FRS, H H Wills Physics Laboratory, Tyndall Avenue, Bristol BS8 1TL, UK

**J W Cahn**, Metallurgy Division, Materials Science and Engineering Laboratory, National Institute of Standards and Technology, Gaithersburg, MD 20899, USA

**S Chandrasekhar**, FRS, Raman Research Institute, Bangalore 560080, India

**T Evans**, FRS, J J Thomson Physical Laboratory, University of Reading, Whiteknights, Reading RG6 2AF, UK

**A J Forty**, University of Stirling, Stirling FK9 4LA, UK

**J Friedel**, For. Mem. RS, Laboratoire de Physique des Solides Associé au CNRS, Université Paris-Sud, Bâtiment 510, 91405 Orsay, France

**C A Handwerker**, Metallurgy Division, Materials Science and Engineering Laboratory, National Institute of Standards and Technology, Gaithersburg, MD 20899, USA

**D T J Hurle**, Royal Signals and Radar Establishment, St Andrews Road, Malvern, Worcs WR14 3PS, UK

**R V Jones**, FRS, Aberdeen, UK

**A Keller**, FRS, H H Wills Physics Laboratory, Tyndall Avenue, Bristol BS8 1TL, UK

**M Kléman**, Laboratoire de Physique des Solides Associé au CNRS, Université Paris-Sud, Bâtiment 510, 91405 Orsay, France

**M R Mackley**, Department of Chemical Engineering, University of Cambridge, Pembroke Street, Cambridge CB2 3RA, UK

**F R N Nabarro**, FRS, Condensed Matter Physics Research Group, University of the Witwatersrand, PO Wits 2050, Johannesburg, Republic of South Africa

**J F Nye**, FRS, H H Wills Physics Laboratory, Tyndall Avenue, Bristol BS8 1TL, UK

**J P Poirier**, Département des Géomatériaux, Institut de Physique du Globe de Paris, 4 Place Jussieu, 75252 Paris Cedex 05, France

**R C Pond**, Department of Materials Science and Engineering, The University of Liverpool, PO Box 147, Liverpool L69 3BX, UK

**L I Ponomarev**, I V Kurchatov Atomic Energy Institute, Moscow 123182, USSR

**S K Runcorn**, FRS, Department of Physics, University of Newcastle-Upon-Tyne, Newcastle-Upon-Tyne, NE1 7RV, UK

**J E Taylor**, Department of Mathematics, Rutgers University, New Brunswick, NJ 08854, USA

**D C Tozer**, Department of Physics, University of Newcastle-Upon-Tyne, Newcastle-Upon-Tyne, NE1 7RV, UK

**J H van der Merwe**, Department of Mathematics, Applied Mathematics and Astronomy, University of South Africa, Pretoria, Republic of South Africa

**I M Ward**, FRS, Interdisciplinary Research Centre in Polymer Science Technology, University of Leeds, Leeds LS2 9JT, UK

# Auld Acquaintance

R V Jones

English education in the 1920s was both highly competitive and heavily biased towards the classics. Some of us then at school, though, were drawn towards science by good teachers, and we were further fascinated by hearing Oliver Lodge speak on 'the wireless' about the solar-system atom of Rutherford and Bohr, and James Jeans talk about the stars in their courses. As early as 1923 we might even have heard Rutherford himself giving his Presidential Address to the British Association in the BBC's first national broadcast; and by the late '20s news of the wave-like nature of the electron and of the expansion of the universe was trickling into our schools.

As we approached Sixth Form, we were excited to realise that if we proved good enough we might be able to go on to university studies in one or other of the sciences, and perhaps even spend our lives in such a privileged pursuit. But would we be good enough?

The acid test of that would be to win a scholarship to a university, with Oxford and Cambridge at the top of the list. Without such a scholarship few of us could afford to go. Years afterwards, C F Powell remarked to me that, so far from this being a hindrance, it was almost an advantage because of the incentive it provided to work towards a qualification that would enable us to rise above the humdrum level that we found around us. Little wonder, then, that competition for the few scholarships in science was intense. Typically, for ten open awards there would be around a hundred candidates, each picked by his school as deserving a chance. Candidates would then go to spend four days or so in the college of their choice, and sit the scholarship examination set by a group of colleges, morning and afternoon for the whole time of their stay. Questions were intended to probe for promise rather than received knowledge, '*SPES* rather than *RES*'.

## 1 OXFORD

Such, then, was the December scholarship examination in 1928 for the group of Oxford colleges that included Balliol, St John's and Lincoln. Balliol hall was the appointed venue for the written papers and a hundred or so boys from schools all over England gathered outside the hall in the sharp December morning. A long flight of stone steps, steep and wide, led up to the door of the hall: on these were assembled the earliest arrivals, keen to get to their seats as soon as the door opened so as to gain maximum time on the papers, not unlike the rush at the Harrods' sale. I was not even on the lowest step, but at the top I could see a bespectacled cherubic figure leaning against the very door as though to make sure of being first in.

I did not know who the cherub was, but evidently he won a scholarship for I saw him again in the following autumn, coming away from Blackwell's bookshop reverently nursing a large and pristine textbook of chemistry. I was inclined to think 'What a swot!' But there was something about his youthful expression that suggested happy enthusiasm. It was, of course, the time of the Great Depression of 1929, and any of us who had won a scholarship could therefore count himself doubly fortunate that year but the cherub had an entranced look, different from any of my contemporaries.

A few weeks later I saw him again, this time on the river as cox of one of the Lincoln torpids. I did not learn his name and college for another two years, for he was reading chemistry while I was reading physics, and we were in different colleges. The one common meeting point for science undergraduates was the Junior Scientific Club, and it was there that I heard him introduce himself as F C Frank of Lincoln, where he was a scholar and a pupil of N V Sidgwick. He was making a good-natured criticism of the Club's policy, in which he bantered that it only succeeded in attracting respectable attendances because it bribed its members with an issue of sweets during its lectures.

We came into closer direct acquaintance in 1932, after we had both graduated and were free to live in unlicensed lodgings. (Undergraduates in those days had to live in lodgings licensed by the Proctors and subject to strict regulations that, among other things, and perhaps mindful of the Reeve's tale, specified that no undergraduate should sleep on the same floor as his landlord's daughter.) Unlicensed lodgings were cheaper, and although even as postgraduates we were still liable to be 'progged' if we entered a public house, we could live where we pleased. My lodgings were at 10 St Michael's Street where one of Charles Frank's chemist friends also lived, while his were at 8 Museum Road which he shared with two other chemists. One of these was a Wadham man whom I knew well and who was repeatedly vulnerable to the practical jokes which his dig-mates, sometimes aided by me, played on him. These jokes, and above all the 5th November 'Rag' of 1932, brought Charles Frank and me closely together.

## 2 GUY FAWKES NIGHT, 1932

It was traditional that there should be a 'Rag' in the streets of Oxford in the evening of Guy Fawkes Day, in which hosts of undergraduates sallied from their colleges in a show of high spirits. Cornmarket, The High, Carfax, St Giles' and the Broad were densely packed, and although police were drafted in from outside Oxford they never had sufficient numbers to keep complete control. In fact, the height of undergraduate ambition was to return to college with a policeman's helmet. Exuberance had increased in the years after World War I when the ex-service undergraduates were finding relief in high spirits after their appalling experiences in the trenches, and although these spirits were cooling by the time we went up in 1929, I myself saw two cars set on fire that year.

Despite such boorishness, though, I have long been grateful for the experience, for it taught me how in a crowd normally intelligent and law-abiding people can react on one another to behave in an increasingly ridiculous manner and carry themselves away in a riot. For the 1932 Rag Charles Frank had the idea that a few of us should out-do the raggers with their fireworks by introducing chemical warfare. He was in his first year of research, and was working on dipole moments. His allotted compound was p-brom-benzyl bromide, which turned out to be a virulent lachrymator, and he proposed to make enough of it and seal it into glass phials so that several of us could be armed with a few phials apiece. We were to distribute ourselves at strategic points, including the entrances to the public lavatories (likely to be in much demand because of pre-rag drinking), and to all drop our phials at the same moment. Charles made the compound and sealed all the phials himself, and in the process he so upset his respiratory system that it took years to recover. We met in his lodgings, where he gave us the phials, and we went off to mix in the crowd. When zero hour came, I was disappointed to find that the crowd at my assigned location in Cornmarket was rather thin, and even more so when I dropped a phial and shattered its contents on the roadway without apparent effect. I then dropped another, and another, until my supply was exhausted. Where were the tears that Charles had expected? The explanation, we realised only later, was that the evening was cold and the lachrymator failed to vaporise—until the hitherto thin crowd was reinforced by a surge up Cornmarket, when the liquid was picked up on the soles of shoes, and was thereby warmed. Eyes began to stream. The police became alarmed. Ambulances arrived. Their crews shouted 'Mustard gas! Clear out!', and undergraduates ran for their colleges. But there was no escape, for there was still plenty of the lachrymator on shoes and, when these were further warmed by firesides, Oxford wept. The London papers had headlines such as 'Gas Attack at Oxford', and not only was that particular 5th November Rag quelled, but it seemed to break the whole tradition, for things were very much quieter in succeeding years.

## 3 POSTGRADUATE LIFE

After the Rag episode Charles' path and mine crossed with increasing frequency, particularly when most of our contemporaries left Oxford in 1933 and only a few of us were successful in getting research studentships that would enable us to stay. I worked in the Clarendon, conveniently near Charles' 'digs', while mine were immediately opposite the Union, of which he was an active member. Gradually we discovered our common interests, both inside and beyond science, with Charles trying to extend mine to some others which he himself was pursuing. These included the formation of a society to make films, one of his proposed subjects being the then-new bypasses around Oxford. He also took up ju-jitsu, and after each of his training sessions he would drop in to my lodgings to practise a newly-learnt 'hold' on me.

He had wandering eyes and hands that would pick up any new gadget in the room, and investigate how it worked. In the process he could become so absorbed that he would drop out of the conversation if others were present; and I can still hear the gruff 'Oh!' that told us that he had started to dismantle the gadget and either that he could not put it together again or, worse, that his dismantling had unexpectedly released a hidden but essential spring that was now flying across the room, necessitating a long search for its recovery. Since my 'gadgets' now included pistols and revolvers, it was advisable to hide them if there was a chance that Charles would call. Brusque of manner when anyone was talking nonsense, he would occasionally surprise us by attending a religious service, for example in Christ Church Cathedral, which he did—he explained to us—to reassure himself that he was right in not accepting religious dogma.

From time to time he revealed humorous ingenuity. He won an Association of Scientific Workers competition for an advertisement to sell water as a patent medicine. After dressing up the properties of water in recondite but snappy scientific jargon likely to impress the public, it concluded with a footnote 'Himmelwasser's Calcined Protium *now* contains the rare and costly substance Calcined Deuterium'—this was in 1933 when heavy water had just been discovered. Many years later, when there was a correspondence in *The Times* deploring the increasing use of 'four-letter' words in polite literature, he contributed a letter consisting entirely of words of four letters, all of them in unexceptionable taste.

As a chemist, he conformed to the practice of the chemistry school that included a year of research after three years of undergraduate study before honours were finally awarded. For this year he worked in the Dyson Perrins Laboratory on dipole moments of molecules. After that, he worked in the engineering laboratory on dielectrics with the electrical engineer Willis Jackson. I still have the reprint that he gave me of his letter to *Nature* of 11 February 1936 headed 'Electrical Evidence on Calcite Imperfection'. This may have been the first earnest of his lifelong interest in defects in crystals, and it argued that the large polarisation discovered by Joffe in conduction through calcite could be explained by calcium ion vacancies in the crystal lattice.

## 4 BERLIN 1936–8

Charles travelled more widely than most of us, and his suitcase began to collect an impressive array of hotel and travel labels. It was therefore no surprise when he chose to spend two years with Peter Debye in the Kaiser Wilhelm Institut für Physik in Berlin. He left Oxford in 1936, not yet having taken his DPhil. I had been awarded mine in 1934, feeling driven to take it in the minimum time of two years because the qualification of a doctorate seemed the only hope of getting a further grant to continue research. In retrospect I envied Charles the detachment with which he could delay his doctorate and thereby ignore the constraint of short-term researches which could produce quick results, for his longer-term approach enabled him to ground himself more thoroughly in the basics of his subject, whose emphasis was changing from chemistry to physics, and where he was acquiring mathematical techniques. He approached mathematics 'from the top down' as he once put it to me, because when he found that he needed the results of such branches as Fourier analysis he then learned the underlying techniques.

By the time that he left for Berlin in 1936, I was already caught up in the problems of detecting aircraft for air defence; and before we parted I suggested that he might 'keep his eyes open' for any relevant developments in Germany. This was in no sense an officially inspired request—just a personal hunch on my part that with the Nazis in power a war was likely sooner or later, and that both sides would draw on the resources of science.

While he was still in Germany I was moved from Oxford into the Air Ministry in London, and among the tasks that came my way was to look at the few reports that came in from our intelligence service. Although it might have been expected that there would be plenty to discover of technical interest in the obvious German preparations for war, MI6 was producing very little. Almost the only report of any substance in the three months or so that I spent in the Air Ministry was about some curious activity involving a new tower on the Brocken in the Harz Mountains.

## 5 THE BROCKEN

I wondered how I might alert Charles in Berlin, in the hope that he might take an interest. Fortunately I knew of the Brocken Spectre from atmospheric optics, and so I wrote to him suggesting that if he could visit the Brocken he might be able to give me a better description than that available in textbooks. Immediately realising that I must be after something, he promptly burnt the letter and paid a visit to the Brocken. What he found was a tower, ostensibly for a television transmitter but reported by MI6 as possibly intended for paralysing internal combustion engines; he also found an array of poles with pear-shaped objects at their tops rather like Belisha beacons, the purpose of which we never discovered.

A month or two later Charles returned home to Ipswich, and wrote telling me that he did not believe in ghosts but that it would be interesting to meet. It happened that I was due to visit the radar research station at Bawdsey which was near Ipswich, and so a meeting was easy; Charles told me what he had seen, supplemented by a postcard photograph of the television tower. I took it into the Air Ministry the following morning, where it transpired that between us Charles and I had produced a better and prompter description of what was actually happening on the Brocken than the MI6 agent who had been briefed to follow up the original report. This was at the peak of the Munich crisis in September 1938, and in the aftermath Charles and I went our appointed ways, he to the Colloid Science Laboratory in Cambridge and I back to the Admiralty Research Laboratory at Teddington.

## 6 SCIENTIFIC INTELLIGENCE

Although we heard nothing more, it seems that the Brocken episode did not go unnoticed. The Tizard Committee for the Scientific Survey of Air Defence had been asking the question: since we ourselves were successfully developing radar, were the Germans doing so too? The inability of MI6 to answer this question, and others of a technical nature, or to produce even pointers to the answers, led the Tizard Committee to propose early in 1939 that a scientist should be allowed to sit with the intelligence organisation to see why information was so scarce, and to suggest how the organisation might be strengthened in the fields of science and technology. The Committee had in mind a scientist of sufficient seniority and eminence to command respect—their ideal candidate may well have been Thomas Merton, a friend of Tizard (and, incidentally, also of Lindemann) who was outstanding for his skill and ingenuity in instrumentation and who had, I believe, worked with intelligence on such aids as secret inks in the First World War.

However, although the proposal was accepted by the intelligence authorities, the Treasury refused to sanction it, on the grounds that it was unnecessary. Surely our scientists could meet their German counterparts at international conferences, et cetera, and so keep themselves aware of the lines of scientific thought in Germany that might have military applications.

It was in the face of this refusal by the Treasury that the Secretary of the Tizard Committee, A E Woodward-Nutt, recalled the interest that I had already shown in intelligence and suggested that I might be brought back from the Admiralty, which would not involve creating a new post requiring Treasury sanction. The Brocken episode may well have been a factor in his suggestion. Finding that I would be enthusiastically interested he persuaded his Director in the Air Ministry to move me back there for work with intelligence.

With unconscious prescience we fixed 1 September 1939 as the date on which I should start with intelligence, and in *Most Secret War* I have described what happened over the next few months, with Hitler's 'Secret Weapon' speech and the 'Phoney War'. It was clear that the scope of scientific intelligence would be beyond the efforts of a single scientist to satisfy, and I had little difficulty in convincing Tizard and the Air Ministry that I should need help. So, who would I suggest should join me? Obviously Charles Frank, both for his science and for his knowledge of German and Germany. I was authorised by the Air Ministry to approach him, which I promptly did; but the proposal was vetoed by the Deputy Director of Scientific Research at the Admiralty, who argued that all that was needed was for the intelligence services to pass their reports to the Directorates of Scientific Research, who could deal with them without the intervention of scientists actually working in the intelligence organisations. In *The Paperclip Conspiracy* the author, Tom Bower, quotes the Minute of 29 February 1940 by the Deputy Director, J Buckingham, objecting to the proposal on the ground that 'there was not enough work to justify the employment of two people'. So I had to continue single-handed, while Charles went to the Chemical Warfare Experimental Establishment at Porton, where he worked with Oliver Gatty on smoke generators to hide factories and other key objectives from aerial bombing. In one of the trials while Charles was present, a generator exploded and Gatty, who was outstanding among the younger Oxford chemists, was killed.

## 7 MAITA

In the meantime both Charles and I had married. I remember him telling me—while it was still a possibility in February 1940 that he would join me—how he had met Maita. At the end of November 1939 Russia had made its monstrous attack on Finland, and in the respite of the Phoney War there was a feeling that over and above our own efforts against Germany we should do all we could to support the Finns. Fund-raising ventures to help Finland were therefore instituted, and Charles became prominent in the Cambridge effort, which was to be launched by a public meeting. The Fund committee thought it would be a good idea if one of the speakers were to be a Finn who could say something about the country and the situation that it now faced. The suggestion, though, was almost stillborn because no member of the committee knew of a Finn; but they had the bright idea of going to the Cambridge police and asking whether they had any Finns on their register of aliens. The police looked through their records and found two, both female. One was fat and forty, the other young and pretty, but had forgotten her Finnish. Would she do? The committee decided she would. Charles agreed 'and so', he told me 'I'm going to marry her'. Happily the year of this *Festschrift* has also seen their golden wedding.

Events in the summer of 1940, with Dunkirk, the Battle of Britain and the onset of the Blitz, were to prove just how short-sighted the Admiralty view about Scientific Intelligence had been. Indeed it had become so vital that—thanks to Bletchley—I, a single and lowly scientific officer, was able (occasionally at first but later frequently) to identify the target selected by the German bombers some hours ahead of their attack, with all that this could mean for our nightfighters, guns and civil defence services. Since I myself was in London, and alone had the expertise, I was now able to argue that unless there was at least one other to whom I could impart it, a single bomb could extinguish the service I had been providing. So in October 1940 it was at last agreed that someone should be authorised to join me, and I insisted that he should be Charles Frank.

## 8 THE BLITZ

Charles joined me shortly before the Coventry Raid of 14/15 November 1940, and I have described in *Most Secret War* his introductions both to the world of secret intelligence and to the rigours of the London Blitz. His first visit with me to Bletchley on 28 November ended with us both in the casualty ward of St Alban's Hospital because we had gone through the windscreen of my car, which was a write-off, when I hit a lorry stranded without lights on the main A5 road as we tried to get back into London in the middle of the nightly bombing. The following night, 29/30 November, will remain with us as long as our memories last, for with our wives we were in Richmond, no more than a mile from the aiming point of the main nightly attack. One bomb missed us by about eight yards, and there were another twenty or so within 200 yards. Again, *Most Secret War* has told the story.

Besides the powerful help which he provided in general, the first of Charles' unique contributions that I can recall was the solution of a Bletchley puzzle. German meteorological codes were studied there by an astrophysicist of international repute, who sent us some strings of figures which had baffled him and which he described to us as having no relation to any known meteorological data. Within an hour or so Charles succeeded in showing that they were no more recondite than reports of height, bearing and distance of meteorological balloons probing wind variation with height.

One of our happiest, if at times frustrating, pursuits was to pore over the photographs taken by our reconnaissance aircraft in the search for evidence of new German technological developments. During 1940 and 1941 these included especially the transmitters and antennas for radio navigational systems such as the beams: then as the bombing threat from the Luftwaffe declined and our own air offensive grew from 1941 onwards, we became primarily concerned with German radar; and anticipating the threat of German retaliation in the later phases of the war, our emphasis then moved to the V-1 and V-2 weapons, while all the time we had to watch for possible developments in the nuclear field.

## 9 GERMAN RADAR

Charles was a most acute observer. In January 1941 he spotted that the width of a shadow of a small suspected structure near Cherbourg had changed in the interval between successive photographs in a reconnaissance run. This suggested that the structure had rotated in the interval; and although the change in the width of the shadow was only about a tenth of a millimetre, and hardly more than the resolution limit of the photograph, it decisively led to our first photographic identification of a German radar station.

In the autumn of the same year a further acute observation by him led to our identifying the still smaller Würzburg 53 centimetre radar used for controlling anti-aircraft fire and nightfighters. Studying photographs of the coastal radar station at Cap d'Antifer, a few miles north of Le Havre, he followed a track that led away from the main radar station towards a villa perched on a cliff-top. Most observers would have dismissed the track as simply leading to the villa but Charles noted that the track stopped in a loop some thirty or forty yards short, and near the apex of the loop was a small 'blob'. I can remember him stubbing at it with his finger and saying 'What's that?' It could well have been a mere dust spot on the print, and we had to check that it appeared on more than one photograph. Even a subsequent photograph taken at lower level proved to be so indistinct and indecisive that our expert interpreter at Medmenham, the Photographic Intelligence Centre, warned us that the object might be no more than 'a piece of garden statuary'. But subsequent pictures taken at very low level showed that it was indeed the radar equipment for which we had been searching for months. The discovery led to the very successful parachute raid at Bruneval on 27/28 February 1942 which decisively clinched the Government decision to establish the Parachute Regiment.

With his knowledge of German, Charles used to read the transcripts of conversations between prisoners-of-war that our interrogators had overheard, and for me it was one of the unforgettable moments of the war when one afternoon in March 1943 he looked across to me from his desk and said 'It looks as though we shall have to take those rockets seriously!' What he had just read was a remark by one captured German general to another in which he had said that something must have gone wrong with the rockets because he knew that they were imprisoned somewhere near London and yet they had heard no explosions, even though he had been told at least a year before in Germany that the rockets would be ready. This was not the first information that we had received about rockets, but the fact that a German general whose technical knowledge had already commanded our respect could make such a comment suggested that there was substance to the other reports about rockets that we had been receiving.

Charles obviously enjoyed his fluency in German, which one of our colleagues hoped to engage to persuade him into a gentler way of dealing with doors. A door separated the room which he and I shared from the one next door in which two other colleagues worked. Charles often moved between the rooms, and the door invariably banged abruptly after his passage. One of our neighbours therefore put a large notice on the door saying 'Please close the door quietly!' But it had absolutely no effect. So his fellow sufferer said 'Perhaps he will take notice of it if we put it in German!' And so they made a new notice saying 'Schliessen die Tur *leise*, bitte!' The next time Charles went through the door he did indeed take notice. He stopped, and grunted 'You've left out the Umlaut on Tür', and the door continued to bang as noisily as ever.

## 10 THE BOMBER OFFENSIVE

At about the same time in 1943 there came to a head the great controversy concerning whether or not Bomber Command should be allowed to drop packets of resonant metal strips to confound German radar. The argument against using this countermeasure was that the Germans might then adopt it and in turn use it against our own radar; ultimately the argument was resolved when we succeeded in showing that by its aid we could do much more damage to the German defences than they could do to ours and the strips (known to us as 'Window' and the Americans as 'Chaff') were first used on 24/25 July 1943. Charles and I spent several nights at the Interception Station at Kingsdown in Kent listening to the radio telephonic orders and replies passing between the German ground controllers and their nightfighters in order to gauge as thoroughly as possible the extent to which the controllers were upset by the clouds of spurious echoes appearing on their radar screens. Of the first few nights, the official Luftwaffe assessment stated 'The technical success of this action must be designated as complete... By this means the enemy has delivered the long-awaited blow against our decimetre radar sets both on land and in the air'.

The reason why we were paying such close attention to the reactions of the German defences was that the exchange of thrust and counterthrust was likely to quicken on the introduction of Window, and we needed to react at once to every change that would inevitably be made in the defences. We listened with some amusement as one bewildered German radar officer exclaimed 'The British bombers are multiplying themselves' as echo after echo appeared on his screen, while another had fixed his attention on the echo from a packet of Window and with increasing exasperation ordered it to waggle its wings, 'Rolf-Lisa machen!', which would have made the echo wax and wane if it had indeed come from the nightfighter he was trying to control.

After the shock of the first few nights, though, the German defences of course reacted: they gave up their rigid control of individual nightfighters each associated with its own ground station, and instead assembled their nightfighters to hunt in packs to attack our main bomber stream, whose track could be reported by ground observers using no more sophisticated aid than their ears. Each of the five main Fighter Divisions had its own pack which under the Divisional Controller could hunt over a wide area; and so that the pack could recognise its particular controller in the flurry of orders coming over the radio, each controller had his own code-name. The four that I can remember were Kakadu, Leander, Möbelwagen, and Prima Donna, whose names were shortly adopted by Charles for the hens which he and Maita had started to keep at their home in Golders Green in the hope of eggs to alleviate wartime rationing.

## 11 THE FELDBERG

Besides our concern with such devices as Window to protect our bombers we were also involved in countering the efforts of the German defences to defeat the radio aids to navigation that Bomber Command had been forced to employ. One of these was the Gee system, in which pulses, sent out simultaneously from three ground stations, could be picked up by a bomber which could then establish where it was from the time intervals between the three pulses as it received them. Obvious countermeasures to the system were to jam or to transmit false pulses on the radio channel on which the Gee pulses were transmitted; and in the autumn of 1944 the Germans brought into operation a powerful new pulse transmitter which threatened considerable trouble to Bomber Command. Fortunately an enterprising young RAF Signals Officer had succeeded in getting a line on the transmitter by using the Gee system in reverse: receivers at two of the Gee transmitting stations had picked up the German pulses, and by measuring the time interval between the reception of the pulses at two of these stations he determined that the offending transmitter must lie on one branch of a hyperbola about the two stations as foci, and such hyperbolae were already charted on the Gee maps of Germany. This information was sent back to us in London and, scanning along the relevant hyperbola, Charles saw that it ran close to the Feldberg near Frankfurt, which we already knew to be the site of a television tower, sister to the one he had seen on the Brocken. So we recommended that the tower should be attacked: within a few days it was destroyed and the trouble ceased. A court of enquiry ensued in Germany as to how we could have identified the tower so quickly. Charles and I sometimes wondered who the bright signals officer was who had so neatly and effectively 'turned the tables' on the jammer, but in the maelstrom of war no information about his identity came our way. One of the first happy results of Charles' move to Bristol after the war, when wartime experiences were being exchanged, was the discovery that the same signals officer had also himself come to the H H Wills Laboratory—he was Peter Fowler.

## 12 THE V-WEAPONS

At the same time as the Feldberg affair we were also in the middle of the German retaliation campaign with the V-weapons. The first V-1 had fallen at Bethnal Green early on 13 June 1944, and a trial V-2 rocket fired the same day from Peenemünde had gone astray and fallen in Sweden. When some of our officers examined the wreckage they reported that one of the pumps injecting fuel into the combustion chamber had the curious feature of being lubricated by the liquid that it was pumping. When they, as puzzled engineers, reported this to Charles it immediately reminded him of the physics textbooks of our schooldays which told of the Claude process of making liquid air, where Claude had introduced this form of lubrication for his pump because normal lubricants would have frozen solid. The evidence was decisive in pointing to liquid oxygen as a key fuel, and this enabled us to sort out the many conflicting and confusing intelligence reports that had been coming in over the past year.

Incidentally, when the alarm over the V-2 had started in the spring of 1943 there was much debate among British experts as to whether the Germans could possibly have made a rocket with a range of 200 miles, some arguing that it would be impossible. I can remember Charles doing the simple calculation of the minimum velocity needed, and hence the minimum energy for a given mass; then he went to the standard physical tables for heats of reaction, and found that there were several combinations of materials, such as liquid oxygen and alcohol, that would contain more than sufficient energy for a given mass. So the impossibility argument could be dismissed.

We were able to make an accurate assessment of the V-2 before it was fired against London, which showed that its maximum range was just over 200 miles. With the rapid advance of the Allied armies through Belgium in August 1944 it looked as though they might sweep the Germans out of range—so much so that, with the V-1 already defeated, Duncan Sandys on 7 September held a press conference to celebrate the end of London's ordeal. He and others had overlooked, though, our assessment of the maximum range; and although nearly all Belgium had been liberated, Holland was still under the Nazis, and there was a small region around the Hague in their hands that was still within the maximum range. In the early evening of the day after the press conference Charles and I were in our office when we heard a bang at about twenty to seven. Simultaneously we said 'That's the first one!' And Charles went on to say that he had heard two explosions and not one.

It turned out that this was the first observation of the 'double bang' which could often be heard if a rocket had fallen to the west of the observer. The first bang was due to the downcoming bow-wave created by the supersonic missile as it passed overhead and the second was from the explosion when it struck the ground. At the moment of impact, for a missile descending at 45° with a bow-wave making an angle of 15° with the trajectory, and hence 30° with the ground, the downcoming front of the bow-wave is just half the distance from the observer that the impact point is. Roughly, if one heard a double bang the missile had fallen to the west of you, at a westward distance of one mile for every $2\frac{1}{2}$ seconds between the bangs. If it had fallen to the east, you only heard the single bang of the explosion, since the bow-wave to the west of the missile's trajectory had gone upwards at about 60°, and hence far over your head.

## 13 FUTURE DEVELOPMENTS

Charles and I were surprised to find how little our experts knew about the theory of long-range rockets, and so he set out the theory in one of our reports. The American Eighth Army Air Force asked us to write an article for their magazine on possible future developments, and it is worth quoting from the concluding paragraphs. The words are mine, but the figures are mainly Charles':

> There can be no doubt that with the A4 the rocket has come to stay for a long time, if only for its non-military applications: in no other way can we get free of the earth's atmosphere, with all that this freedom may mean to astrophysical studies. The attainment of the upper atmosphere will in itself be a major factor in experimental meteorology, and sooner or later someone will seriously try to reach the moon—and succeed. Military applications are bound to be made, whatever the limits imposed by treaties, and we should do well to keep an eye on the possibilities. If we were to allow ourselves more liberty of conjecture, we might consider using atomic fuels to drive an exhaust of hydrogen molecules, or perhaps lighter particles, giving an entirely different order of performance.
>
> It is an often stated requirement that a weapon of war should have a probable error comparable with its radius of destruction, so that a few shots would ensure the obliteration of the target. Practical weapons seldom approach this ideal, although in the future it may be attainable through homing devices. With a very long range rocket we may have to accept errors, and it may be easier to increase the radius of destruction by the use of new types of explosive based on the fission of the uranium atomic nucleus. If such an explosive becomes practicable, it will probably have a radius of destruction of the order of miles, and on this account alone it might best be carried in some unmanned projectile, of which the rocket would be a particularly suitable type by virtue of its relative immunity from interception and of its potentially better accuracy at long ranges compared with pilotless aircraft. Speculation of this kind is fascinating, but can well wait for a paper at a later time when it is nearer realization.
>
> Reviewing therefore what we have seen to be reasonable extrapolation from present practice, a two-stage rocket of about 150 tons starting weight could deliver a 1 ton warhead to nearly 3,000 miles range, with a probable error of 10 miles in range and 3 miles in line. This might be a feasible weapon for delivering a uranium bomb, should such a bomb become practicable. It would be almost hopeless to counter by attacks on the ground organization, because the increased range would allow an almost unlimited choice of firing site, while the trajectory could be so varied that the firing point could not be deduced with sufficient accuracy for countermeasures. Production would probably take place underground. At the moment such a rocket could not be intercepted, but by the time it becomes a serious possibility it may itself be a target for smaller defence rockets fitted with predictors and homing devices: but these would depend upon adequate warning, and the defences might also be saturated by a salvo of long range rockets.

With reference to the possibility of defence by homing missiles and its counter of saturation by firing rockets in salvoes, we may claim to have anticipated the key problem of the Strategic Defence Initiative which lay some 40 years ahead.

The mention in the foregoing passage of the prospect of a uranium bomb was nearly a year ahead of the successful explosion at Alamogordo, and it caused many eyebrows to rise. We had been well informed about the prospect partly because another of our activities was to watch for parallel atomic developments in Germany and we shared the rare privilege of looking after Niels Bohr when he came to England after his escape from Denmark via Sweden. Kept secluded from public notice, he was only too glad to have us spend our free hours with him, and he gave us some memorable tutorials on fission and the 'liquid drop' theory that he had put forward with John Wheeler.

As the pressure of the war eased off we began to think about the main fields in which physics might be expected to advance after the war. Besides nuclear physics and radio astronomy, I can recall Charles pointing out that the mechanism of separation of chromosomes at mitosis would be worthy of study, especially as regards the nature of the separating force, while I had made an elementary estimate that indicated that whatever the form a genetic code might have the 'bits' must be of the order of molecular dimensions. Francis Crick who had then joined us has said that his own interest in the field was stimulated by our discussions. Advances in the understanding of the solid state also seemed likely, and Charles gave me several tutorials on his pre-war subject of defects in crystals: one elegant point, which I hope that my memory has not unduly scrambled, was that the vapour pressure of holes entering a crystal at a given temperature from a vacuum was equal to that of atoms evaporating from the crystal into the vacuum at the same temperature. In his descriptions of crystal structure and the nature of imperfections I was repeatedly struck by his outstanding ability to visualise things in three dimensions. I also admired his mastery of what G P Thomson, following H G N Moseley, called the characteristic British approach to physics through 'pictorial thinking'.

## 14 BRISTOL

How long we would have continued in intelligence or defence science after the war is problematic. We had much enjoyed working together, and we should have liked to continue; but with the reorganisation of scientific intelligence on the recommendation of P M S Blackett, with the unhappy results that *Most Secret War* described, we regretfully started to look for other opportunities. I knew that that Charles wished to return to academic life, and so I recommended him to Nevill Mott, whom I had come to know well during the war because of his work with the Army Operational

Research Group at Petersham and because he often came in for an evening to our nearby flat in Richmond. On 21 January 1946 he wrote to me from Bristol: 'I have been looking into the possibility of Frank coming here and I have very little doubt that we could fix him up...The appointment would be for work on dielectrics, ions in solution and all that field'.

A few months later I myself was offered the chair of natural philosophy at Aberdeen, and arrangements were finalised for Charles to join Nevill Mott in Bristol, where it was a truly exciting time with Cecil Powell discovering the pi- and mu-mesons. One of Charles' earliest impacts was to suggest to Powell to use these prefixes, rather than alpha- and beta-, in case another meson might be discovered that might prove an earlier member of the family. Of his many subsequent contributions, particularly to solid state physics and to the well-being of the H H Wills Laboratory, other authors in this *Festschrift* will be writing with deeper knowledge than I myself possess. I did see, though, the birth of the spiral dislocation—and can remember an exultant Charles sending me a message when he received the first photographic confirmation, saying that it was like our discovery of the first German radar station. And whenever we met he had something new to expound: our wartime interest in continental drift recurred in his dynamics of the flow of basalt away from the ocean ridges and his theory of island arcs, while his friendship with John Fisher aroused an interest in the development of cancers, and his travels in France led him to espouse the possibility that Joan of Arc had been saved from burning and had survived to old age. Firmer of his possibilities were that comets might be snowballs and that dislocations could be subject to traffic jams. At our most recent meeting in October 1988 I was treated to a tutorial on the prospects for muon-induced fusion. Charles, indeed, is a true Clerk of Oxenforde: 'gladly wolde he lerne and gladly teche', and many of us have accordingly benefitted.

Our friendship has lasted nearly sixty years. In the hectic days—and nights—of war we shared some acute physical dangers: in these Charles was imperturbable. He was equally so in those wartime Whitehall crises which *Most Secret War* described, and where his command of basic science was the firmest of anchors. Kindness towards juniors and directness towards contemporaries have been prominent among his attributes, along with constant good temper, good humour and good balance, and determined courage in facing physical disability.

When I was last in Bristol he pointed out to me that, in the portrait with which his colleagues and pupils so signally honoured him on his retirement as head of the Laboratory, he holds in his hands a copy of *Most Secret War*, in which the intensest of our shared experiences are to be found. It reminded me of Aeschylus wishing to be remembered not for his plays but for the fact that he fought at the Battle of Marathon. With such a mark of Charles' own regard for our long association, I delight in the opportunity to join in this salute to him on his eightieth birthday, with deepest gratitude for his unflagging friendship.

## REFERENCES

Bower T 1987 *The Paperclip Conspiracy* (London: Michael Joseph)
Jones R V 1978 *Most Secret War* (London: Hamish Hamilton)

# The History of One Hypothesis

L I Ponomarev

Half of published scientific papers are never cited in subsequent publications. From the other half about 80% are cited only once. In ten years after the papers have been published only 20% of them can be cited in the future, in fifteen years—only 5%; by this time all the rest have become obsolete. This is truth as it is and this is the penalty for the fast development of science today. However, a small number (<1%) of papers continue to be cited after 20, 30 and even 40 years. These are classical papers.

In issue no 4068 of the magazine *Nature*, dated 18 October 1947, a short paper entitled 'Hypothetical alternative energy sources for the "second meson" events' by Charles Frank was published (see figure 1). Today, 40 years later, the paper is cited more often than 90% of all the papers published in the last year. The secret of its longevity is the same as that of man's longevity, namely, to maintain stability they should change with time. It seems timely now to trace the life of Charles Frank's fruitful hypothesis and to comprehend the results available now and based on it.

The year this paper appeared in *Nature* I had reached the age of ten and was a pupil in a provincial school in the south of Russia. Certainly, I did not suspect the existence of the whole system of sophisticated knowledge of contemporary physics. Ten years later I became a student of the Physical Faculty of the Moscow State University, but still would not have guessed any relation between my future life and the paper by Charles Frank. Even in 1967 I had not read the paper, but I had heard something about it from my teacher, Professor S S Gerstein. Only ten years later, when I found myself in the CERN library, did I at last read this paper, and

# HYPOTHETICAL ALTERNATIVE ENERGY SOURCES FOR THE 'SECOND MESON' EVENTS

By Dr. F. C. FRANK, O.B.E.
H. H. Wills Physical Laboratory, University of Bristol

IN the paper by Lattes, Occhialini and Powell*, it is shown that cosmic ray mesons coming to the end of their track in a photographic emulsion rather frequently give rise to a secondary meson of kinetic energy about 4 MeV., and no other visible particle. This observation appears to demand for its interpretation the existence of two sorts of meson, the source of the observed kinetic energy of the secondary residing in the mass difference between the two. In view of the importance of this conclusion, we have tried to consider all reasonably imaginable processes to explain the observations without introducing a new elementary particle; processes, that is, in which the energy is derived from the material of the photographic emulsion instead of from the meson. Reasons are given below for the confident rejection of all such processes.

It is concluded that there is no energy to be gained, from elements present in substantial quantity in the photographic emulsion, from the following processes:

(a) Induced β- or K-capture processes: on grounds of known mass defects for the lighter nuclei, and systematic principles for the heavier nuclei present. There should be energy (though not enough to explain the observations) to gain from such processes from $K^{40}$, $Rb^{87}$, $Lu^{176}$, $Os^{187}$ and from one each of the pairs $In^{113}$, $Cd^{113}$, $Sn^{115}$, $In^{115}$, $Te^{123}$, $Sb^{123}$; none of these is present in substantial quantity. It is improbable that any natural nucleus possesses a more stable adjacent isobar not discovered in Nature.

(b) Induced emission of single nucleons: on grounds of known mass defect for the lighter nuclei and systematic principles for heavier ones; such processes, from naturally existing nuclei, always absorb energy.

(c) Induced α-emission: excluded for the lighter nuclei by known mass-defects. Not excluded for the heavier nuclei, since mass defects are not known with sufficient accuracy, and by the Geiger–Nuttall relation the life-times of low-energy α-emitters are too great for detection. However, with no heavier nuclei than silver present, energies greater than about 2 MeV. are excluded by this consideration, because they would lead to spontaneous decay at a detectable rate. Moreover, the Coulomb repulsion from a medium-heavy nucleus would assuredly give the α-particle sufficient energy to make a visible track.

(d) Induced fission: once again Coulomb repulsion would assuredly produce a visible track from the fission fragments, and it is inherently improbable that a meson could activate the process, say in silver, the only element which comes in question from the point of view of availability of energy.

(e) Processes with change of charge by 2, for example, of the type

$$At^{A}_{Z} + Y^{-} \rightarrow At^{A}_{Z-2} + Y^{+},$$

where $Y^{-}$ and $Y^{+}$ denote negative and positive mesons. It is to be expected that this process, if it can occur, would yield energy from about half the nuclei which possess stable isobars of charge diminished by 2. In the case of $Zn^{64}$ going to $Ni^{64}$, measured mass defects indicate an energy yield of $9 \cdot 8 \pm 3$ MeV. The conditions for this process are not satisfied by any nucleus present in substantial quantity in the photographic emulsion, namely, by isotopes of hydrogen, boron, carbon, nitrogen, oxygen, silver, bromine, iodine and sulphur; of these only $S^{36}_{16}$, the rarest (0·016 per cent) of the isotopes of sulphur, which is itself present only in traces, has a known stable isobar $A^{36}_{18}$. That is in the wrong direction, requiring the improbable converse process starting with a positive meson.

(f) Induced decay of naturally persisting nuclear isomers: this would imply the existence of isomers with life-times about $10^{12}$ times as great, and stored energies about ten times as great, as any yet known—a very improbable combination.

This seems to exhaust the possibilities among processes of nuclear degradation. On the other hand, it needs more consideration to exclude the possibility of processes of nuclear build-up, in which a proton is added to some other nucleus: such processes are almost all exothermic. The possibility of such processes arises from the special properties of the combination of proton and negative meson, which may be called a mesonic hydrogen atom, or an excited neutron, at choice.

A meson stopping in the emulsion loses kinetic energy from, say, 100,000 eV. down to 2,000 eV. in a fraction of a micron. This may occur either in silver bromide, or in gelatine, with comparable probability. In the latter case it is very likely to be in the neighbourhood of a proton, of which there are about twice as many as all other nuclei together in the gelatine. It should then, at first, enter a hydrogen-like orbit about the proton. The resulting compact neutral atom (radius, in the 'Coulombic ground-state', about 1/200 of the Bohr radius, thus $2 \cdot 6 \times 10^{-11}$ cm., and binding energy about 200 times that of a hydrogen atom, thus 2,700 eV.) should appear, from distances exceeding $10^{-10}$ cm., like a slow neutron. It should be able to pass freely through the electronic clouds of other atoms, and approach close to nuclei, without Coulomb repulsion. Indeed, the polarization of the mesonic atom in the field of a second nucleus should lead to an initial attraction.

The most probable next step would appear to be the capture of the meson by the nucleus of higher charge, and repulsion of the proton. There should, however, be a non-vanishing probability that the proton is captured, and the meson expelled. In one case at least, namely, that in which the second nucleus is a deuteron, the probability of the alternative reaction should be high. For by analogy with the hydrogen molecule-ion, $H_2^-$, the combination of deuteron, proton and meson should be stable in respect of Coulombic interactions with an internuclear separation of $10^{-8}/200 = 5 \times 10^{-11}$ cm. At this distance, with a very low Gamow barrier, the two nuclei should very readily unite, the meson acquiring kinetic energy by what may be described as 'internal conversion'. Thus, formally, we should have the reaction:

$$H^1_1(e^-) + Y^- \rightarrow H^1_1(Y^-) + e^- + 2{,}700 \text{ eV.}$$
$$H^1_1(Y^-) + D^2_1 \rightarrow H^1_1D^2_1(Y^-) + 500 \text{ eV.}$$
$$H^1_1D^2_1(Y^-) \rightarrow He^3_2 + Y^- + 5{\cdot}46 \text{ MeV.}$$

It is highly improbable that the small amount of deuterium present in the normal emulsion could account for the observed phenomena. The corresponding process with a second proton (forming a deuteron, since $He^2$ is not stable) requires the emission of a positron and neutrino which, between them, would take most of the 1·43 MeV. available, which is in any case insufficient. With heavier nuclei the details of interaction are more complicated, since, among other considerations, there will only be binding in the molecule-ion while it is in an excited state.

The mesonic hydrogen atom will be attracted to a second nucleus by virtue of polarization. Some energy can be dissipated to form the molecular combination, by Auger effect, which will cease to operate when the nuclei and meson are well within the K-shell of electrons. The combination dissociates when the meson falls by a radiative transition into a level in which it is concentrated around the nucleus of higher charge. The time taken for this radiation should be of the order $(hc/e^2)(\lambda/r)^2(1/\omega)$, which is about $10^6Z^{-2}$ periods of the emitted radiation (as in hydrogen-like spectra) whether we are dealing with mesons or electrons. The characteristic frequency should be 200 times that of the corresponding electronic spectrum, that is, say, $10^{18}Z^2$, and the frequency of molecular vibration $(200)^{3/2}$ times that of electronically bound hydrogen compounds, that is, about $3 \times 10^{17}$. Hence, we estimate the life-time of the bound state as about $10^{-10}Z^{-4}$ sec., or about $3 \times 10^7Z^{-4}$ periods of oscillation of the mesonic molecule-ion.

The Gamow barrier to be penetrated by the proton extends from the internuclear separation in the excited molecule-ion, $r_1$, to the distance at which the nuclei make contact, $r_0$. The meson still provides some screening within this distance, and we take the potential energy to be $E = (Z - 1)e^2/r$. We estimate $r_1$ as that distance at which the polarizing action of the second nucleus, on the mesonic hydrogen, regarded as linear with field strength, produces an equivalent displacement of the meson equal to its Bohr radius, $a_Y = 2 \cdot 6 \times 10^{-11}$ cm. Thus, $r_1 = 3a_Y\sqrt{Z/2} = 5 \cdot 5 \times 10^{-11}Z^{1/2}$ cm.: $r_0$ we take as $1 \cdot 45 \times 10^{-13}(A^{1/3} + 1)$ cm., where $A$ is the mass number of the second nucleus. Then, if we write $\lambda^* = h/\sqrt{2M(E - E_0)}$ (where $M$ is the mass of the proton) and neglect the initial kinetic energy $E_0$, the transparency of the barrier to a proton wave should be of the order

$$\exp\left\{-4\pi\int_{r_0}^{r_1}(1/\lambda^*)dr\right\}$$
$$= \exp\left\{-\frac{8\pi e\sqrt{2M(Z-1)}}{h}(r_1^{1/2} - r_0^{1/2})\right\}$$
$$= \exp\left\{-3 \cdot 33\sqrt{Z-1}\,(7 \cdot 4Z^{1/4} - 0 \cdot 4\sqrt{A^{1/3} + 1})\right\},$$

which is about $10^{-27}$ for $B^{11}_5$. This, subject to considerable error, for example, from the uncertainty of $r_1$, expresses the probability that the proton will penetrate the barrier in one attempt. Multiplying by the number of oscillations before dissociation, we find the probability of penetration before dissociation to be $10^{-21}$ in this case. This result may well be in error by several powers of 10, but it appears that if the model is at all relevant, the chance of penetration is insignificant for all nuclei beyond the deuteron. The same calculation for the latter, neglecting the inner screening and so writing 1 in place of $\sqrt{Z-1}$, gives a penetration probability of $10^{-9}$ per vibration, so that penetration may occur in about $10^{-8}$ sec.

Specific forces of the meson have been disregarded. There is an obvious chance that it will suffer nuclear capture or destruction before the two nuclei can interact, the probability of which is reduced by the fact that the meson is in excited states throughout the period significant for this process. If, on the other hand, the mesonic forces lead to some closer union of proton and meson than we have considered (differing from a neutron), it might well react with any nucleus.

The energy available from proton adhesion is small (1·96 and 0·51 MeV. respectively) for the 'saturated' nuclei $C^{12}_6$ and $O^{16}_8$. For nearly all other nuclei it is 5 MeV. or more. The next most common nucleus in the gelatine is $N^{14}_7$, yielding 7·3 MeV., which would account for the observations if the product, $O^{15}_8$, was formed with an excitation of about 3·5 MeV. Boron was also present in these emulsions, and could yield more than enough energy by proton adhesion, but in this case the most likely reactions liberate α-particles.

This process can in any case be rejected statistically. For a total of 380 mesons observed to stop in the emulsion, four secondary mesons have been observed, with at least 500 μ of secondary track in the 50-μ thickness of the emulsion, and mutually consistent in energy yield. Geometrical probability indicates that about thirty times as many events should have occurred, in which the secondary meson passed out of the emulsion earlier: it could then either go undetected or be indistinguishable from a proton. On the other hand, one should not fail to detect a projected length of 50 μ of the end of a meson track. Hence the four observed secondary mesons are representative of about 120 ± 60 (the limits expressing probable error) whereas the 380 mesons observed to stop should represent not more than about 900 actually stopping. (For randomly directed straight tracks ending at random depths in an infinite layer of thickness $H$, the statistical proportion of all tracks which end in the layer which have horizontally projected ranges within the layer exceeding $R$ is $(\sqrt{R^2/H^2 + 1} - R/H)$. For $R/H = 1$, this is 0·414. For $R/H > 1$, it is well approximated by $H/2R$, but for $R/H \gg 1$, track curvature will reduce the proportion below this estimate.)

Thus it appears probable that the production of a 4 MeV. secondary meson occurs for 13 ± 7 per cent of the mesons stopping in the emulsion. Presumably only about half the mesons are negative, and only about half of these can form mesonic hydrogen atoms (since H comprises 40 atomic per cent of the whole emulsion): if we suppose the interaction with a second proton is weak, so that the mesonic hydrogen is most likely, and about equally likely, to be attracted to carbon, nitrogen or oxygen (1 in 8 of which nuclei, in the gelatine, is $N^{14}$), we have an expectation of about 3 per cent if *every* encounter of a mesonic hydrogen atom with an $N^{14}$ nucleus leads to capture of the proton, and on every occasion there is 'internal conversion' of the liberated energy, producing a secondary meson of 4 MeV. Unconsidered factors are mostly unfavourable, so that even on the small number of observations it is statistically improbable that they represent this process.

A further consideration is that the simple theory leads us to expect a larger number of easily observable 2 MeV.—and possibly of ½ MeV.—secondary mesons deriving their energy from proton adhesion to $C^{12}_6$ and $O^{16}_8$: though it is a simple and plausible addition to the theory to suppose that these reactions are relatively 'forbidden'.

It should be added that such processes may be of importance in other circumstances; if it is correct to suppose that mesons can survive for, say, $10^{-8}$ sec. in hydrogen-like orbits about protons, then there is a finite chance that a meson can induce a nuclear build-up reaction, causing the attachment of a proton to a deuteron. However, with two or more cosmic ray mesons in the field, the whole of our observational knowledge about mesons requires re-examination.

I am indebted for a number of discussions to Dr. H. Fröhlich and Messrs. Lattes, Occhialini and Powell.

*Note added in proof.* Later observations enable the 'corrected observed' proportion of mesons producing secondaries to be refined from 13 ± 7 per cent to 12·8 ± 2·5 per cent (117 ± 20 in 917 ± 70), thus increasing the confidence with which the hypothetical alternative process can be rejected.

* See *Nature*, **160**, 453 and 486 (1947).

**Figure 1** The paper by F C Frank in *Nature* **160** 525–7 (1947), where the idea of muon-catalysed fusion was considered for the first time. Reproduced by permission of Macmillan Magazines Ltd.

I was fascinated by the strength and creative resources of the human intellect.

The history of muon catalysis † is already forty years old, even older than that of thermonuclear fusion.

In 1947 Lattes, Occhialini and Powell discovered the $\pi$-meson and its decay to a muon and a neutrino: $\pi^{\pm} \to \mu^{\pm} + \nu_{\mu}$. But at that time physicists were not yet ready to accept the existence of two types of meson and, hence, were searching intensively for alternative explanations of the observed facts. In particular, possible sources for the 'secondary meson' energy, other than the natural mass difference between pion and muon were sought. Among many hypotheses analysed, Sir Charles Frank considered the muon catalysis chain (see figure 1, where one should substitute $Y^- \to \mu^-$),

$$\mu^- \to p\mu \to d\mu \to pd\mu \to {}^3He + \mu^- + 5.4 \text{ MeV} \tag{1}$$

which could imitate the decay process $\pi^- \to \mu^-$, provided the isotope exchange $p\mu \to d\mu$ rate is high enough at the natural deuterium concentration (0.015%) in hydrogen. It turned out that to explain the experiment this hypothesis was not necessary. However, it was in this paper the notion of the muon catalysis cycle (1) in a ($H_2 + D_2$) mixture was first introduced to scientific use.

Analogously to the Columbus egg, the simplicity of and seeming evidence for the muon catalysis hypothesis, expounded in Charles Frank's papers, should not deceive anybody. In fact, in 1947 the muon had already been known for ten years, and the mesic atom existence was also predicted, but nobody had made the conclusive step of considering the possibility of mesic molecule formation and, moreover, the consequences of this hypothesis, i.e. the complete chain of muon catalysis reactions.

Similarly to any fruitful hypothesis, this one prompted the ways for its verification and further development.

In 1948 A D Sakharov noticed that after $dd\mu$ mesic molecule formation in the muon catalysis cycle the nuclei of such a molecule undergo fusion immediately and suggested the use of the phenomenon for the production of energy and neutrons (see figure 2).

In 1953–7 the first theoretical calculations (Ya B Zeldovich, A D Sakharov, J D Jackson, S S Gerstein) of various muon catalysis characteristics appeared: the rates of $pp\mu$ and $pd\mu$ mesic molecule formation, of the isotope exchange $\mu p \to d\mu$, the sticking coefficient $\omega_s$ for the reactions $dd\mu \to \mu{}^3He + n$ and $dt\mu \to \mu{}^4He + n$, etc, were all calculated. In particular, it was already then understood correctly that the mesic molecules $pp\mu$ and

† Other details of the history and essence of muon catalysis can be found in papers by Zeldovich and Gerstein (1960), Gerstein and Ponomarev (1975), Ponomarev and Fiorentini (1987), and Ponomarev (1990).

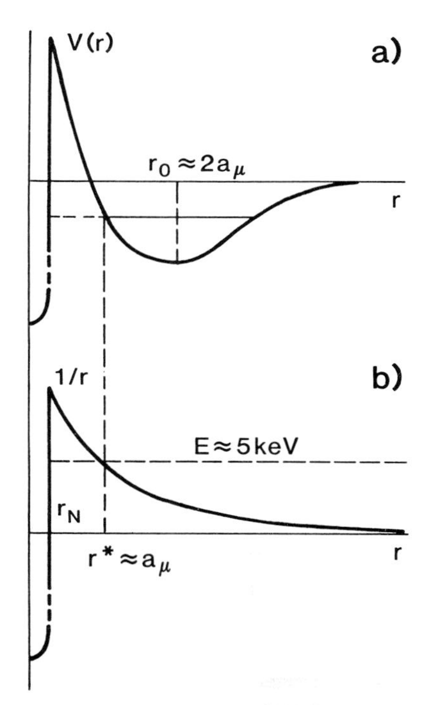


**Figure 2** The scheme of Coulomb barrier penetration in the nuclear fusion reaction in the mesic molecule (*a*) and in the nuclear collision 'in flight' (*b*). The nuclei undergo fusion at small distances $r < r^*$. Because of muon screening the classical turning point $r^*$ in the mesic molecule is the same as for the nuclei collision at energy $E \approx 5$ keV.

pd$\mu$ are formed in reactions of the type

$$\mathrm{d}\mu + \mathrm{H}_2 \rightarrow [\,(\mathrm{pd}\mu)\mathrm{pe}]^+ + \mathrm{e} \tag{2}$$

i.e. the mesic molecule pd$\mu$ (more accurately, the mesic molecular ion $(\mathrm{pd}\mu)^+$) formed becomes the 'nucleus' of the usual 'molecule' (of the mesic molecular complex $[(\mathrm{pd}\mu)^+\mathrm{pe}]^+$), the released energy being carried away by the conversion electron. At the same time the role of the mesic molecule formation reaction was realised. Indeed, the fusion rate for the 'in-flight' reaction, e.g.,

$$\mathrm{d}\mu + \mathrm{p} \rightarrow {}^3\mathrm{He} + \mu$$

is suppressed by a factor of $10^6$ compared to the fusion rate in the mesic molecule

$$\mathrm{pd}\mu \rightarrow {}^3\mathrm{He} + \mu^-.$$

However, for almost ten more years nobody observed the phenomenon.

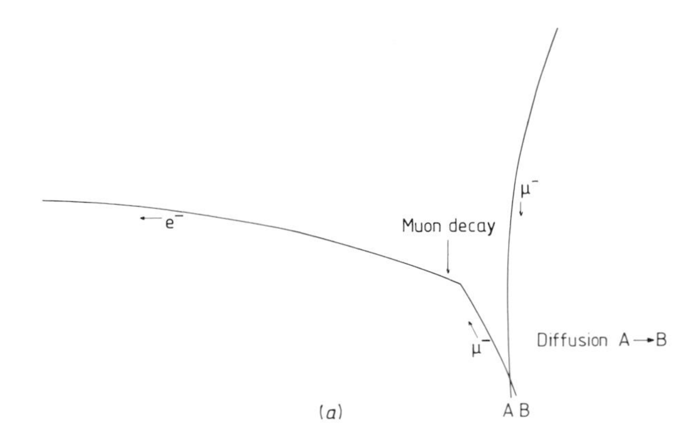


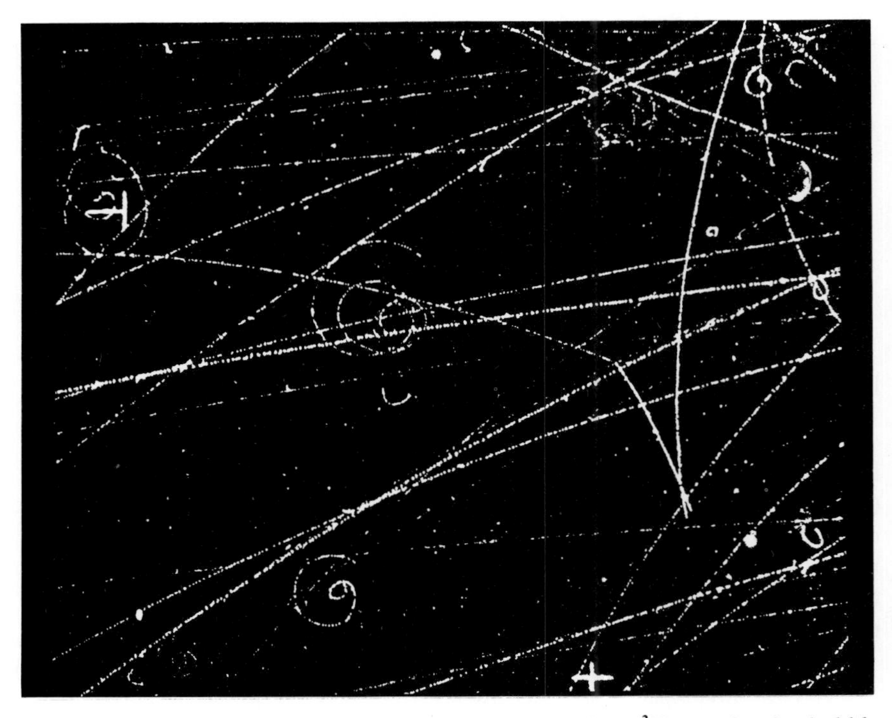

**Figure 3** (*a*, *b*) The muon-catalysed fusion events $pd\mu \rightarrow {}^3He + \mu$ in the bubble chamber observed by Alvarez's group at Berkeley. (*a*) From Alvarez *et al* (1957), reproduced by permission of Lawrence Berkeley Laboratory. (*b*) A double catalysis event; from Alvarez L W 'Recent developments in particle physics', Nobel Lecture, 11 December 1968. Reproduced by permission of the Nobel Foundation.

On Christmas eve 1956, L Alvarez and his collaborators at Berkeley observed the muon catalysis cycle in the bubble chamber invented by him (see figure 3). And only one month later a pioneering and thorough theoretical paper by Jackson (1957) appeared†, where he was first to

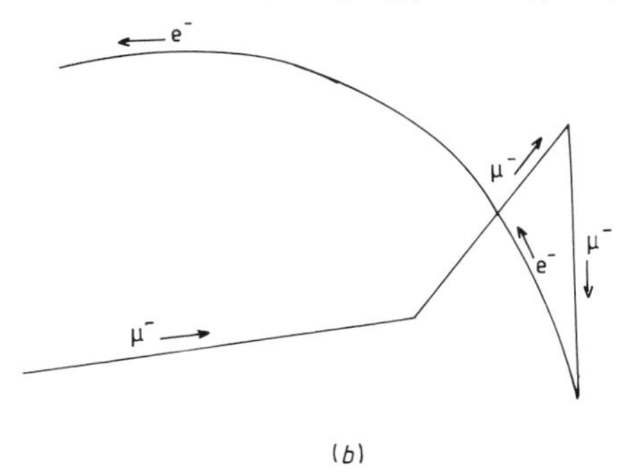


(*b*)

**Figure 3** (*cont.*)

† This paper is unique. Deliberately using an incorrect calculation method, the author, nevertheless, calculated three principal quantities (the $dt\mu$-molecule formation rate, $\lambda_{dt\mu} \sim 10^8\ s^{-1}$, the coefficient of muon sticking to helium, $\omega_s \sim 10^{-2}$, and the energy spent on $\mu^-$ production, $c_\mu \sim 5\ GeV/\mu^-$) which prove to be rather close to modern values (see table 1). He also estimated the maximum number of muon catalysis cycles $X_c \sim 100$ (also in agreement with contemporary data).

consider the muon catalysis cycle in the deuterium–tritium mixture (see figure 4). The explosion of enthusiasm which followed is rarely generated by a scientific paper: muon catalysis was discussed everywhere, even in the evening newspapers (see figure 5).

From this moment on the study of mesic atomic processes started all over the world: L Lederman *et al* and Doede, Fetkovich *et al* in the USA, Ashmore *et al* in Great Britain, Dzhelepov *et al* in Russia, etc. One should mention especially the contribution of S S Gerstein to the theoretical development of muon catalysis problems. The survey which was published by him with Ya B Zeldovich in 1960 still remains the best introduction to the muon-catalysed fusion problem.

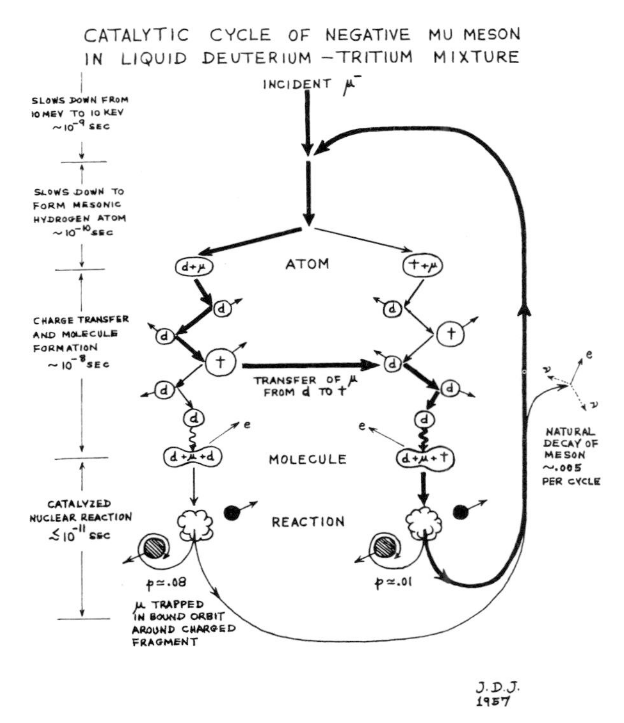


**Figure 4** The muon catalysis fusion cycle in the deuterium–tritium mixture, according to Jackson (1957). Reproduced by permission.

During five years (1957–62) due to the efforts of the groups mentioned it became clear that it is impossible to realise a large number of muon catalysis cycles either in the $H_2 + D_2$ mixture, or in pure deuterium. Therefore, there was no hope of using the phenomenon under these conditions for energy production. At the same time, rather elaborate theoretical estimates for the rate of mesic molecule formation in the deuterium–tritium

($D_2 + T_2$) mixture $\lambda_{dt\mu}$ gave a value of the order of $10^4$ $s^{-1}$ and thus ruled out any hope for practical yield in this case also.

Over the next five years (1962–7) interest in mesic atomic processes and muon catalysis in the hydrogen isotope mixture was supported mainly by the needs of weak interaction physics. In particular, it is necessary to take these processes into account in experimental studies of $\mu^-$ capture in

**Atomic Energy Produced By New, Simpler Method**

**Coast Scientists Achieve Reaction Without Uranium or Intense Heat—Practical Use Hinges on Further Tests**

Special to The New York Times.

MONTEREY, Calif., Dec. 28—A third and revolutionary way to produce a nuclear reaction was described here today. It does not involve uranium, as in the fission reaction, or million-degree heat, as in the fusion reaction.

The new process is called "catalyzed nuclear reaction." It was discovered accidentally a few weeks ago during routine work with the huge atom-smashing bevatron at the University of California radiation laboratory.

A team of twelve scientists from the university explained the process to the American Physical Society here. The team was headed by Dr. Luis W. Alvarez, assistant director of the laboratory.

Curiously enough, it was made not at the laboratory at Livermore, where scientists are attempting to control thermonuclear reaction for practical uses, but at the Berkeley laboratory, which is devoted to fundamental research.

Thus far, the new reaction is little more than a laboratory curiosity, the scientists said. The energy it produced came from the fusion of a few hydrogen atoms, they explained, and was scarcely enough to register on highly sensitive measuring instruments.

The process has no commercial value now, though it suggests possible industrial uses of immeasurable importance. It may, scientists said, point a way toward taming the intense heat of the hydrogen bomb to make it useful for peacetime purposes.

Others in the University of California group were Dr. Hugh Bradner, Dr. Frank S. Crawford Jr., Dr. John A. Crawford, Dr. Paul Falk-Vairant, Dr. Myron L. Good, Dr. J. Don Gow, Dr. Arthur H. Rosenfeld, Dr. Frank Solmitz, Dr. M. Lynn Stevenson, Dr. Harold K. Ticho and Dr. Robert D. Tripp.

One method of obtaining nuclear reaction—the so-called "fission reaction" employed in the atom bomb—relies on the bombardment of atomic nuclei with other atomic particles.

The other—the "thermonuclear reaction" of stars and the modern hydrogen bomb—depends upon the union or fusion of two light atomic nuclei to form one heavy nucleus at temperatures of about 1,000,000 degrees.

The type described today employs a medium-weight atomic particle (known as a negative mu-meson) as a catalyst to make a hydrogen nucleus fuse with a deuterium (heavy hydrogen) nucleus. This fusion occurs at low temperatures.

One result is the formation of helium—a variety known as helium-3. Another is the release of prodigious amounts of energy, calculated at about 5,400,000 electron volts for each reaction.

The mu-meson, which triggers this change of elements, is not used up as a catalyst, but remains free to bring together other nuclei of hydrogen and deuterium, and form more helium-3 and produce more energy.

**Catalyst Short-Lived**

But the catalyst is extremely short-lived, Dr. Alvarez noted, and thus limits the process. The mu-meson has a life of approximately one-millionth of one second, a period sufficient to let it catalyze no more than one or two fusions before it perishes.

In commenting on the future of the new reaction, Dr. Alvarez said:

"If this is to become of practical importance, we would have to find a different catalyzing particle which has properties similar to the mu-meson but has a lifetime of at least ten or twenty minutes."

Such a particle would permit millions of energy-producing reactions and, it may be presumed, the release of enough energy to operate electric generators, motors and other heavy equipment.

In this connection, Dr. Alvarez—who recently traveled through the Soviet Union and visited scientific laboratories there—observed:

"It is interesting that Russian scientists have reported evidence that such a particle does exist in cosmic rays."

The announcement of the discovery of the "catalyzed nuclear reaction" was made simultaneously by the Atomic Energy Commission in Washington. The commission provides financial support for the fundamental research at the Berkeley Atomic Laboratory.

**Figure 5** *The New York Times*, 29 December 1956, presenting the muon-catalysed fusion phenomenon to its readers.

hydrogen ($p\mu \to n + \nu_\mu$) and deuterium ($d\mu \to 2n + \nu_\mu$). In all the world only two experimental groups continued to work in this field: in the USSR at Dubna (V P Dzhelepov, P F Ermolov *et al*) and in Italy at Bologna (E Zavattini, A Bertin, A Vitale *et al*). Practically all theoretical investigations were carried out at Dubna at that time. In 1964–6 studying the muon catalysis process in gaseous deuterium at a temperature $T \sim 240$ K, Dzhelepov *et al* (1966) at Dubna measured the $dd\mu$ mesic molecule formation rate (figure 6). It proved to be unusually high:

$$\lambda_{dd\mu} = (0.75 \pm 0.11)\ 10^6\ \mathrm{s}^{-1}$$

i.e. approximately ten times higher than that measured in liquid deuterium

$$\lambda_{dd\mu} = (0.076 \pm 0.015)\ 10^6\ \mathrm{s}^{-1} \qquad \text{(Fetkovich } et\ al\ 1960)$$
$$\lambda_{dd\mu} = (0.103 \pm 0.004)\ 10^6\ \mathrm{s}^{-1} \qquad \text{(Doede 1963).}$$

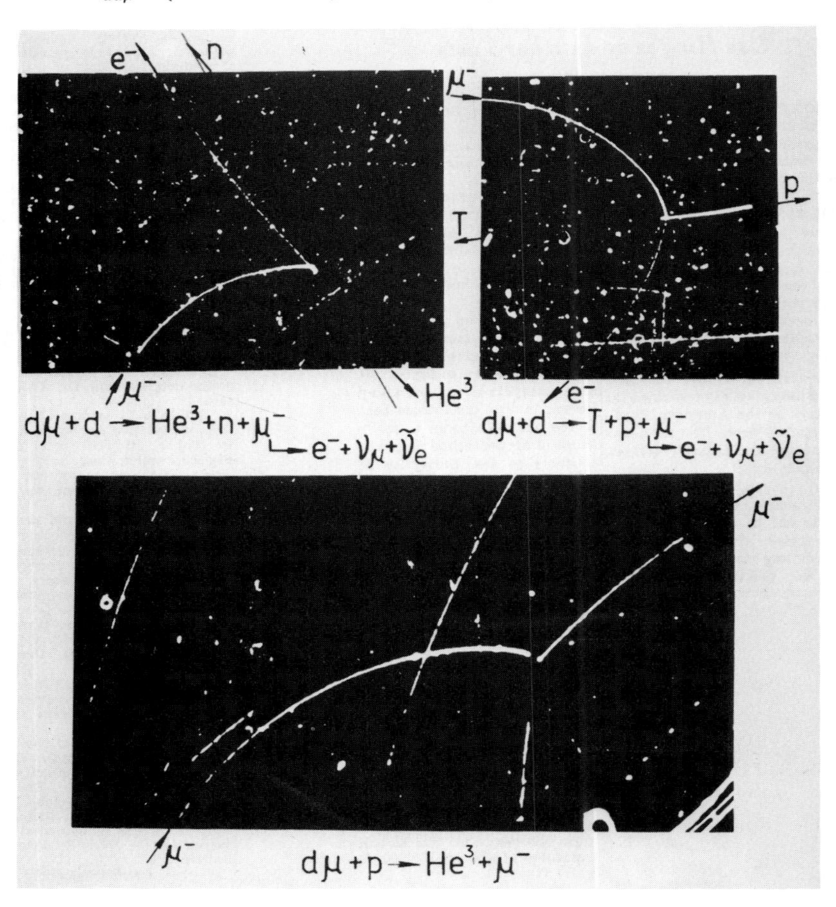


**Figure 6** The muon-catalysed fusion events $dd\mu \to {}^3\mathrm{He} + n + \mu$ and $dd\mu \to t + p + \mu$ observed by Dzhelepov *et al* (1966) in Dubna.

The result obtained was so improbable that some experimentalists considered it to be mistaken (and some of them even spoke about it out loud).

Really, the result was rather difficult to believe: the temperature in the diffusion chamber exceeds the liquid deuterium temperature by only 220 K. This means that the kinetic energy of the d$\mu$ mesic atom ($\varepsilon_{d\mu} \cong \frac{3}{2}kT$) in the diffusion chamber differs by only 0.03 eV from its energy in liquid deuterium ($\sim$0.003 eV). In comparison with the characteristic scale of mesic atom and mesic molecule energies this difference is negligible: the d$\mu$ atom ground-state binding energy equals, e.g., 2500 eV, the dd$\mu$ mesic molecule ground-state binding energy is 325 eV, and the isotopic difference between the energies of p$\mu$ and d$\mu$ mesic atom ground states is equal to about 137 eV. Hence, it proved rather difficult to believe that a change in d$\mu$ mesic atom kinetic energy of only 0.03 eV can lead to a tenfold increase in the rate of dd$\mu$ mesic molecule formation in the reaction

$$\mathrm{d}\mu + \mathrm{D}_2 \rightarrow [(\mathrm{dd}\mu)\mathrm{de}]^+ + \mathrm{e}. \tag{3}$$

To explain somehow the observed fact the experimentalists recalled the argument by Ya B Zeldovich (1954), which was actually incorporated in his first paper on the subject. In contrast to process (2), where the pd$\mu$ mesic molecule is formed by the intense dipole E1 transition from the continuum s-wave state of the d$\mu$ + p system (with total angular momentum $J = 0$) to the pd$\mu$ mesic molecule p-wave state (total angular momentum $J = 1$), the dd$\mu$ mesic molecule can be formed in reaction (3) in a state with total angular momentum $J = 0$ by a monopole E0 transition. The energy of this state ($\sim$30 eV) exceeds the electron ionisation energy in reaction (3) by a factor of only two. For this reason the conversion coefficient for reaction (3) should be extremely high and, in addition, one can, in principle, expect resonance effects.

Just when these hot discussions were taking place, in the Laboratory of Nuclear Problems of JINR in Dubna E A Vesman, a postgraduate student from the University of Tartu (Estonia), was passing his course. His professor S S Gerstein participated actively not only in the discussions but in the experimental work as well. Gerstein ten years before, in 1958, had received his PhD degree working under L D Landau; his thesis was entitled 'Nuclear reactions, involving $\mu$-mesons in hydrogen' (Gerstein 1958). In it he calculated, in particular, the mesic molecule energy levels. It followed from these calculations that the dd$\mu$ mesic molecule had four stable states with rotational ($J$) and vibrational ($v$) quantum numbers ($Jv$) = (00), (01), (10), (20). However, in addition, Gerstein suspected the existence of a fifth dd$\mu$ molecule state with quantum numbers ($J = v = 1$) (see figure 7) and estimated its binding energy: $|\varepsilon_{11}| \leqslant 7$ eV. He did his best; to calculate this value more accurately was impossible at that time. The reason was that neither powerful computers were available nor had effective algorithms to solve the Coulomb three-body problem been created.

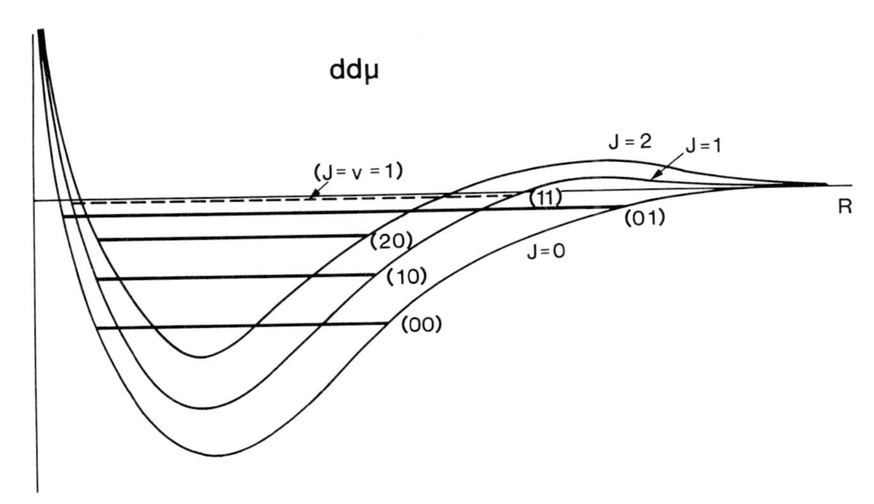


**Figure 7** The scheme of mesic molecule dd$\mu$ rotational–vibrational ($Jv$) states.

Nevertheless, Gerstein told Vesman about his conjecture and soon afterwards the latter suggested his hypothesis for mesic molecule resonance formation. Its essence is the following (Vesman 1967). Assuming the existence of a loosely bound state of dd$\mu$ mesic molecules ($J = v = 1$) with binding energy $|\varepsilon_{11}| \cong 2$ eV, one can consider, instead of reaction (3), another possible reaction

$$d\mu + D_2 \rightarrow [(dd\mu)dee]^*_{\nu K} \tag{4}$$

where the released binding energy of a mesic molecule is transferred not to the conversion electron (this is impossible, since its ionisation energy is $\approx 15$ eV), but to the excitation of the vibrational–rotational degrees of freedom ($\nu K$) of the mesic molecular complex $[(dd\mu)dee]^*_{\nu K}$. Since the value of the vibrational quantum of this complex is $\cong 0.34$ eV, at $|\varepsilon_{11}| \cong 2$ eV the latter is formed in the vibrational state $\nu = 7$ (see figure 8; according to the first version by Vesman, in the state $\nu = 6$).

It is evident that such a process should be of a resonance character, its rate being dependent on the temperature of the medium, i.e. on the average kinetic energy of d$\mu$ mesic atoms. In particular, the maximum cross section for dd$\mu$ molecule formation is observed when the resonance condition is fulfilled:

$$\varepsilon_0 + |\varepsilon_{11}| = \Delta E_\nu \tag{5}$$

where $\Delta E_\nu$ is the energy difference between the initial ($D_2$ molecule) and the final ([(dd$\mu$)dee] complex) states, and $\varepsilon_0$ is the kinetic energy of the relative d$\mu$ + $D_2$ motion (see figure 8). When the temperature $T$ of deuterium is varied, the fraction of mesic atoms with the resonant kinetic energy $\varepsilon_0$ changes, and thus the observed rate $\lambda_{dd\mu}$ of $dd\mu$ mesic molecule resonant formation (3) changes. In particular, if the mesic atom kinetic energies obey the Maxwell distribution

$$f(\varepsilon, T) = 2(\varepsilon/\pi)^{1/2} T^{-3/2} \exp(-\varepsilon/T) \quad (6)$$

then the observed dependence of the rate of process (4) on temperature reproduces the dependence (6) at the resonance energy $\varepsilon_0$:

$$\lambda_{dd\mu}(T) \sim f(\varepsilon_0, T) \sim \sqrt{\varepsilon_0}\, T^{-3/2} \exp(-\varepsilon_0/T).$$

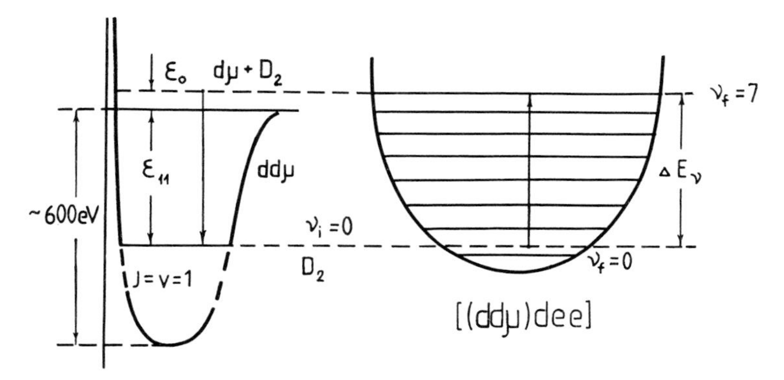


**Figure 8** The scheme of $dd\mu$ mesic molecule resonant formation in reaction (4). The resonance condition is $\varepsilon_0 + |\varepsilon_{11}| = \Delta E_\nu$.

The first estimate of the rate $\lambda_{dd\mu}$ of reaction (4) performed by Vesman in 1967 (see figure 9) explained qualitatively the results of Dubna experiments, as well as their difference from earlier results. However, to confirm (or to reject) Vesman's hypothesis, one should firstly prove the existence (or non-existence) of a loosely bound ($J = v = 1$) state of the $dd\mu$ mesic molecule, and secondly develop a comprehensive theory to calculate the rate $\lambda_{dd\mu}$. The first problem was solved only ten years later, and the second one required twenty years. (Figure 10 displays the status of theory and experiment today.)

For quantitative comparison between theory and experiment it is necessary to calculate the energy $\varepsilon_{11}$ of the state ($J = v = 1$) with an accuracy of about 0.001 eV (or better), which is comparable with the kinetic energy of the mesic atom thermal motion at about 10 K. This is a very high accuracy: for comparison remember that the mesic atom ground-state energy $E_\mu \cong -m_\mu e^4/2\hbar^2 = -m_\mu/m_e$ Ry $\approx -2500$ eV, while the energy of the $dd\mu$ mesic molecule ground state ($J = v = 0$) is $\varepsilon_{00} = -325$ eV. Thus, the necessary absolute accuracy, 0.001 eV, is equivalent to a relative accuracy of around $10^{-6}$, which had never before been achieved in the three-body problem even for the ground state (it is seen from figure 6 that the level ($J = v = 1$) is the fifth excited state of the dd$\mu$ three-body system). The problem is particularly complicated, since the value of $\varepsilon_{11} \cong -2$ eV $\sim 10^{-3} E_\mu$ is extremely small and is close to the boundary of the continuum.

Nevertheless, in Dubna in the period 1965–9 a general approach to this problem (the so-called adiabatic representation in the three-body problem) was developed. The essence of the idea is the decomposition of the wave-function of the three-body system over the basis set of the solutions of the quantum mechanical two-centre problem (Vinitsky and Ponomarev 1982). First, algorithms were developed to solve the two-centre problem for muon motion in the Coulomb field of two fixed nuclei (T P Puzynina, N F Truskova, L I Ponomarev), and after this the method for solution of the resulting system of ordinary differential equations (the so-called, continuous analogue of the Newton method) was developed (I V Puzynin). In these studies we were essentially developing the early works by M Born, D Bates, A Dalgarno, S S Gerstein and others.

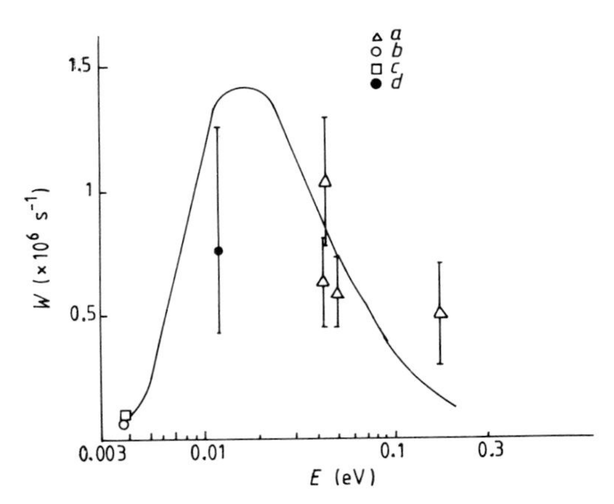


**Figure 9** The dd$\mu$ mesic molecule formation rate plotted against the average d$\mu$ atom kinetic energy. The experimental data are from (*a*) Dzhelepov *et al* (1966), (*b*) Fetkovich *et al* (1960), (*c*) Doede (1963), (*d*) Alvarez *et al* (1957); the curve shows the result of theoretical calculation by Vesman (1967).

This activity resulted in the first calculation of the dd$\mu$ molecule ($J = v = 1$) state energy, performed in Dubna in 1973. This value ($\varepsilon_{11} = -0.7$ eV) was still far from the modern one, but it followed rigorously from the calculation that the true level is more deeply bound, and it took four years more to achieve adequate accuracy. It was a difficult but rather intriguing problem†. Nevertheless, in the summer of 1977 this work was, in general, finished and we could state with confidence that the hypothesis of Vesman was relevant. We presented the first quantitative calculations of $\lambda_{dd\mu}$ rates which could be compared with experiment.

Understanding the importance and the necessity of such a comparison we inspired the experimentalists to perform new measurements of the temperature dependence of $\lambda_{dd\mu}(T)$. In the summer of 1977 these measurements were completed and we had an opportunity to analyse them before they were published (two years later, Bystritskii *et al* 1979). These measurements evidenced uniquely both the validity of Vesman's hypothesis and the correctness of our calculations.

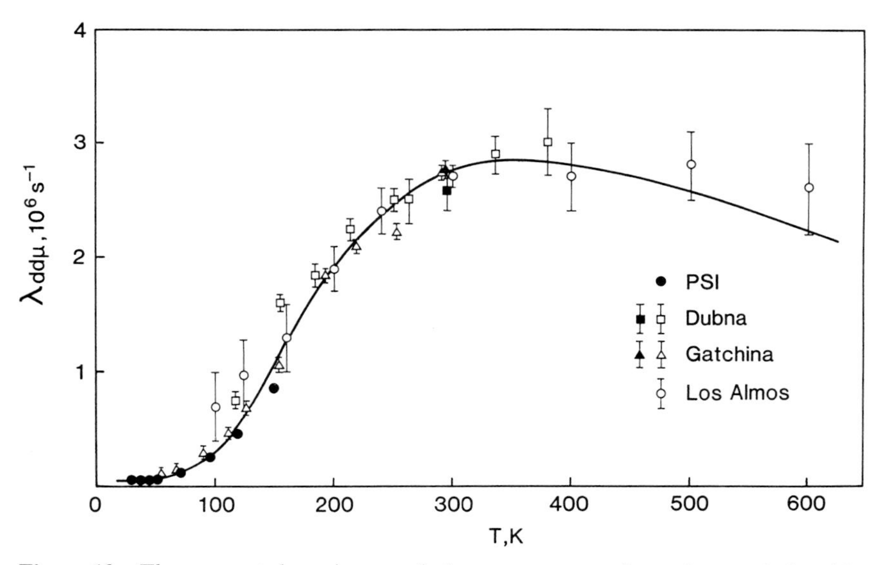


**Figure 10** The present-day picture of the temperature dependence of the dd$\mu$ mesic molecule formation rate $\lambda_{dd\mu}(T)$. The experimental data are from Bystritskii *et al* (1979), Balin *et al* (1984), Jones *et al* (1986) and Zmeskal *et al* (1990); the theoretical result shown by the curve is from Menshikov *et al* (1987).

Naturally, it was not the end of the story: we continued to improve the accuracy of our value of $\varepsilon_{11}$ for ten more years until 1987. At present, mostly due to efforts of scientists from the USSR, USA and Japan, this value has been calculated with enormous accuracy (to $10^{-5}$ eV) and is

† To have sufficient computer time we often had to work at night. I remember clearly one of them, the night of 10/11 December 1976. Early in the morning, we had a telephone call from Vinitsky's home: he was informed that his wife was in labour. In ten minutes we were at his home, in half an hour his wife was in hospital, and at midday he became the father of two charming daughters.

(Alexander and Monkhorst 1988)

$$\varepsilon_{11} = -1.97487 \text{ eV}.$$

In addition, about ten corrections to this value have been calculated (in most part by Bakalov *et al*). Only after these calculations and the development of new methods to calculate $\lambda_{\mathrm{dd}\mu}$ (L I Men'shikov, M P Faifman, M Leon, J Cohen), has one managed to obtain the picture in figure 10 in 1986, almost 20 years after Vesman's hypothesis appeared.

In parallel with calculations of mesic molecule energy levels, in Dubna the methods for calculation of various mesic atomic processes were developed (A V Matveenko, M P Faifman, L I Ponomarev) and in 1973 we could satisfactorily calculate the cross sections of practically all interesting processes and, in particular, of the elastic scattering

$$\mathrm{t}\mu + \mathrm{d} \rightarrow \mathrm{t}\mu + \mathrm{d}. \tag{7}$$

Once, in spring 1974, early in the morning (after a night without sleep) M P Faifman phoned me and told me excitedly that the cross section of the reaction (7) displayed a narrow resonance in the scattering state with angular momentum $J=1$ at collision energy $\varepsilon \cong 0.72$ eV (see figure 11). Though I didn't take it very seriously, this was good news. The reason was that all the calculations in progress at that time were carried out in the so-called two-level approximation, i.e. in the expression for the three-body system wavefunction only the two first terms of the decomposition over the two-centre basis set were retained. Hence the result obtained was certainly an intermediate one and should definitely have changed after the effective potentials in the equations were improved.

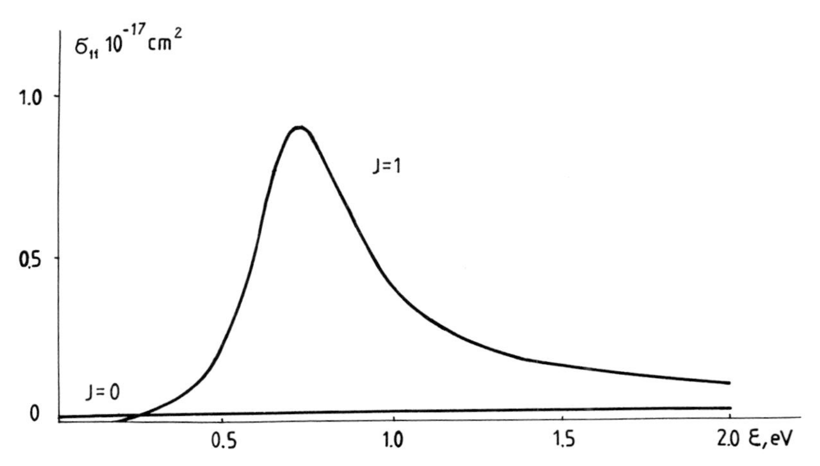


**Figure 11** The energy dependence of the s-wave ($J=0$) and p-wave ($J=1$) elastic $\mathrm{t}\mu + \mathrm{d}$ scattering cross section calculated in the two-level approximation (Matveenko *et al* 1975).

But I still continued to ponder over this result and in December 1974 I came to the conclusion that in the dt$\mu$ mesic molecule a loosely bound state ($J = v = 1$) should exist with a binding energy $|\varepsilon_{11}| = 0.6$ eV. I appealed to the analogy which existed between the energy level systems of dd$\mu$ and dt$\mu$ mesic molecules (see table 1), where present-day values for the binding energies $|\varepsilon_{Jv}|$ of mesic molecule ($Jv$) states are presented.

It can be seen from table 1 that the binding energy $|\varepsilon_{01}|$ of the excited (01) state of dt$\mu$ is rather close (only 1 eV less) to the value of $|\varepsilon_{01}|$ for the dd$\mu$ molecule. The first calculations gave the value $\varepsilon_{11} \approx -0.7$ eV for the energy of the ($J = v = 1$) state of the dd$\mu$ mesic molecule. Vesman's hypothesis required the value $\varepsilon_{11} \approx -2$ eV. Provided that this was true, one concluded that improving the accuracy of the energy of the ($J = v = 1$) state, one could find it more deeply bound by 1.3 eV.

**Table 1** The binding energies of dd$\mu$ and dt$\mu$ molecules.

| | (00) | (01) | (10) | (11) | (20) |
|---|---|---|---|---|---|
| dd$\mu$ | 325.074 | 35.844 | 226.682 | 1.97482 | 86.434 |
| dt$\mu$ | 319.140 | 34.834 | 232.472 | 0.66017 | 101.416 |

Further, according to Bethe and Peierls, the wavefunction of the quasistationary state in the t$\mu$ + d system (a p-wave resonance in the scattering cross section with energy $\varepsilon = 0.72$ eV; see figure 11) is similar to that of the stationary state with energy $\varepsilon_{11} = -\varepsilon$. Hence, after the improvement of the potentials in the corresponding system of equations the resonance energy should also decrease by approximately 1.3 eV, i.e. the quasistationary state in the dt$\mu$ system should become a stationary state with energy $\varepsilon_{11} \cong 0.7 - 1.3 = -0.6$ eV.

If this is true, then the resonance reaction

$$\mathrm{t}\mu + \mathrm{D}_2 \rightarrow [(\mathrm{dt}\mu)\mathrm{dee}]^*_{\nu K} \tag{8}$$

is possible, which is analogous to reaction (4), the only difference being that the mesic molecular complex [(dt$\mu$)dee] is formed in the vibrational state $\nu = 2$ rather than $\nu = 7$ for the complex [(dd$\mu$)dee]. As a consequence, one should expect the process (8) rate to be rather high (estimates gave about $10^8$–$10^9$ s$^{-1}$).

Such a conclusion needed serious checking. This was completed by our group only three years later, in the summer of 1977, and confirmed the rough estimates of the energy $\varepsilon_{11}$ and of the process (8) rate (Vinitsky *et al* 1978).

These theoretical predictions were announced first at the VIIth International Conference on Elementary Particles and Nuclear Structure at the end of August 1977. Soon after this V P Dzhelepov organised the experimental group in Dubna headed by the first class experimentalist V G Zinov. They managed to observe the process of muon catalysis in the deuterium–tritium mixture at the beginning of the summer of 1979 (Bystritskii *et al* 1981).

In this experiment the tritium fraction in the $D_2 + T_2$ mixture was very low (~3%), hence, even at the high rate of process (8), the neutron yield from fusion in the mesic molecule

$$\mathrm{dt}\mu \rightarrow {}^4\mathrm{He} + \mathrm{n} + \mu^- \tag{9a}$$

was expected to be low. But, in any case, neutrons should appear provided the reaction (8) rate is high ($\geqslant 10^8\,\mathrm{s}^{-1}$), in contrast to earlier predictions ($\sim 10^4\,\mathrm{s}^{-1}$).

In the evening of 4 June 1979 a decisive run on muon catalysis in the $D_2 + T_2$ mixture started in Dubna. I was at home this time and all night I was waiting for news from the experimentalists. They called me only at 5 am and told me briefly: 'Go to bed, we see neutrons'. The run lasted for three weeks and during this period it was established that the rate $\lambda_{\mathrm{dt}\mu} > 10^8\,\mathrm{s}^{-1}$, and the rate of isotope exchange $\lambda_{\mathrm{dt}}$ in the process

$$\mathrm{d}\mu + \mathrm{t} \rightarrow \mathrm{t}\mu + \mathrm{d} \tag{10}$$

was $\lambda_{\mathrm{dt}} = (2.8 \pm 0.4) \times 10^8\,\mathrm{s}^{-1}$. Both these rates exceed by more than 100 times the rate $\lambda_0 = 0.46 \times 10^6\,\mathrm{s}^{-1}$ of muon decay

$$\mu^- \rightarrow \mathrm{e} + \nu_\mu + \bar{\nu}_\mathrm{e}. \tag{11}$$

Due to this fact a muon can catalyse more than 100 fusions in process (9*a*) during its lifetime, the only condition being that the probability $\omega_\mathrm{s}$ of the muon sticking to helium in the reaction

$$\mathrm{dt}\mu \rightarrow \mu{}^4\mathrm{He} + \mathrm{n} \tag{9b}$$

be small. And this is in fact the case: as early as 1957 this quantity was estimated by Ya B Zel'dovich and J D Jackson and found to be $\omega_\mathrm{s} \sim 10^{-2}$.

Were the muon sticking in reaction (9*b*) equal to zero, the number of muon catalysis cycles would be equal to the ratio of the cycle rate $\Lambda_\mathrm{c}$ (see figure 12) to the muon decay rate $\lambda_0$ (11):

$$X_\mathrm{c} \cong \Lambda_\mathrm{c}/\lambda_0. \tag{12a}$$

In the opposite case, when the cycle rate is extremely high ($\Lambda_\mathrm{c} \rightarrow \infty$), the number of cycles is

$$X_\mathrm{c} \cong \omega_\mathrm{s}^{-1}. \tag{12b}$$

In reality, the number of cycles is defined by the expression

$$X_\mathrm{c} = \frac{\Lambda_\mathrm{c}}{\lambda_0 + \omega_\mathrm{s}\Lambda_\mathrm{c}} \tag{12c}$$

from which both (12*a*) and (12*b*) follow in the limits $\omega_\mathrm{s} \rightarrow 0$ and $\Lambda_\mathrm{c} \rightarrow \infty$.

The expression for $\Lambda_c$ can be obtained from the system of equations describing the muon catalysis kinetics in the $(D_2 + T_2)$ mixture (see figure 12):

$$\Lambda_c^{-1} = (\lambda_c\varphi)^{-1} \cong \frac{1}{\lambda_{dt\mu}C_d\varphi} + \frac{C_d}{\lambda_{dt}C_t\varphi} \qquad (13)$$

where $\varphi = N/N_0$ is the reduced mixture density, $N_0 = 4.25 \times 10^{22}\ \text{cm}^{-3}$ is the liquid hydrogen density, and $C_d$ and $C_t$ are deuterium and tritium fractions in the $(D_2 + T_2)$ mixture, with $C_d + C_t = 1$.

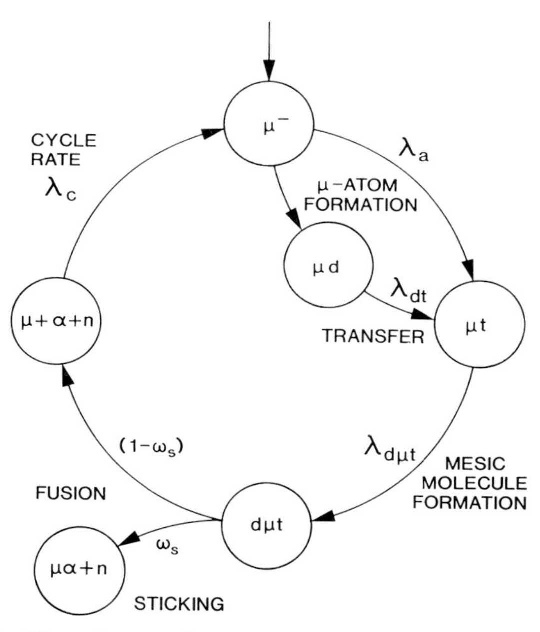


**Figure 12** The scheme of muon-catalysed fusion processes in the deuterium–tritium mixture.

One can see easily that at small $\varphi$ and $C_t$ (if $\lambda_{dt\mu} \geqslant 10^8\ \text{s}^{-1}$ and $\lambda_{dt\mu} \sim \lambda_{dt}$), the first term in (13) can be neglected. This means that, measuring the number of cycles $X_c$, one can determine $\lambda_{dt}$ from (12*c*), but not $\lambda_{dt\mu}$. In addition, at $\varphi \approx 0.05$ and $C_t \approx 0.03$ (the conditions of the Dubna experiment), the cycle rate was $\Lambda_c \cong 10^5\ \text{s}^{-1}$, i.e. $\Lambda_c/\lambda_0 \sim 0.2$, and only one in five muons can realise at least one cycle of muon catalysis.

These were the facts established in the experiment of the Dubna group in the summer of 1979. Shortly afterwards they were reported by V P Dzhelepov at the International Conference in Vancouver in August 1979 and stimulated the group led by S E Jones to begin to experiment, the first stage being finished at Los Alamos, USA, in 1983.

In contrast to the Dubna experiment, in this experiment muon catalysis was observed in the $D_2 + T_2$ mixture at a density $\varphi = 0.6$ and tritium fraction $C_t \cong 0.5$. From (12*c*) and (13) one can easily estimate that about 100 fusions per muon should be expected in this case. The experimentalists at Los Alamos did see just this.

Subsequent experiments were also carried out on the liquid deuterium–tritium mixture by two groups: at Los Alamos and at Institute of Nuclear Research, Switzerland (SIN, now Paul Sherrer Institute), which gave the following results:

$X_c = 150 \pm 20$ (Collaboration Los Alamos–Idaho–Provo)
$X_c = 114 \pm 10$ (Collaboration PSI Villigen–Wien–Munich–Berkeley–Los Alamos)

Long before these direct measurements and soon after the prediction of high efficiency for muon catalysis in the deuterium–tritium mixture Yu V Petrov, a well-known expert in reactor physics from the Leningrad Nuclear Physics Institute asked himself in the winter of 1977/78 whether one could use fusion neutrons from reaction (9) for nuclear energy and neutron production. His conclusion was quite definite: a muon catalytic hybrid reactor exploiting the muon catalysis phenomenon could be envisaged (see figure 13), which could produce nuclear fuel for four thermal atomic power stations equal to it in power (Petrov 1980).

Though the validity of this conclusion is still questioned, without doubt, it has 'catalysed' the development of muon catalysis studies in the world. In my view all the efforts to resolve the controversy at present are premature. In fact, these will be the problems of the next century, when the level of technology and the scale of priorities on which the future fate of muon catalysis depends will be quite different. More fruitful activities at present are to find the conditions for the optimal procession of the muon catalysis cycle and to investigate all details of the physics of this interesting and beautiful phenomenon.

Several such peculiar features of the phenomenon became clear at once, when one accepted the mechanism of resonance mesic molecule formation. First of all, due to the interaction of muon and nucleus spins, the d$\mu$ and t$\mu$ atom ground states are split into two sublevels (see figure 14). This means, in its turn, that the scheme of the resonance reaction (4) should be specified in the following way:

$$(\mathrm{d}\mu)_F + \mathrm{D}_2 \rightarrow [(\mathrm{dt}\mu)\mathrm{dee}]^*_{\nu K}. \tag{14}$$

Obviously, the resonance condition (5) will be different for various initial states $F$ of the d$\mu$ mesic atom and, hence, the temperature dependences of mesic molecule dd$\mu$ formation rates $\lambda_F(T)$ will differ for these states.

This phenomenon was actually found by the Vienna–SIN collaboration (Kammel *et al* 1982), and the most recent results obtained at the accelerator of the Paul Sherrer Institute are presented in figure 15.

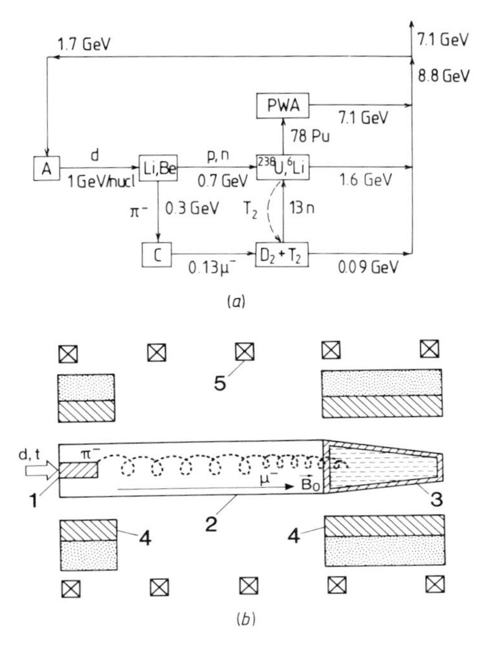


**Figure 13** (*a*) The conceptual scheme of a muon-catalytic hybrid reactor (MCHR) according to Petrov (1980): A, accelerator; Li, Be, target for pion production; C, converter $\pi^- \to \mu^-$; $D_2 + T_2$, synthesiser; $^{238}U$, $^{6}Li$, blanket; PWR, nuclear power reactor. (*b*) The design of the MCHR: 1, target for pion production; 2, converter; 3, $\mu$CF target with deuterium–tritium mixture; 4, blanket.

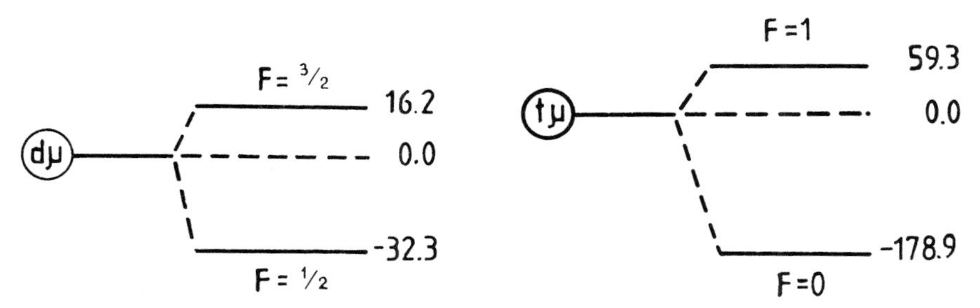


**Figure 14** The hyperfine splitting of the ground states of d$\mu$ and t$\mu$ atoms. The energy levels are given in meV.

In the same manner the conditions for resonance (5) will differ when t$\mu$ atoms collide with $D_2$ and DT molecules, because the vibrational quanta for these two molecules differ essentially. This effect was also observed in experiments by Jones *et al* (1983) at Los Alamos.

The modern 'full' picture of the mechanism of the resonance reaction (9) is much more complicated than the initial hypothesis by E A Vesman about reaction (4). First of all, the present-day calculations give for the dt$\mu$ mesic molecule ($J = v = 1$) state energy (Alexander and Monkhorst 1988)

$$\varepsilon_{11} = -0.660\ 17\ \text{eV} \tag{15}$$

i.e. $|\varepsilon_{11}| > \Delta E_\nu$ at $\nu = 2$ for reaction (9), while for reaction (4) at $\nu = 7$ the inverse inequality is valid, $|\varepsilon_{11}| < \Delta E_\nu$. In the case of dd$\mu$ mesic molecule formation the negative resonance defect $\Delta\varepsilon = |\varepsilon_{11}| - \Delta E_\nu < 0$ is compensated by the kinetic energy $\varepsilon_0$ of the d$\mu$ + $D_2$ collision (see figure 7). On the other hand, for the case of dt$\mu$ molecule formation according to scheme (9) the positive resonance defect $\Delta\varepsilon > 0$ is carried away by the third particle in reactions of the type (see figure 16)

$$t\mu + D_2 + D_2 \rightarrow [(dt\mu)dee]_{\nu K} + D_2'. \tag{16}$$

The comprehensive theory of such processes has not been developed up to now.

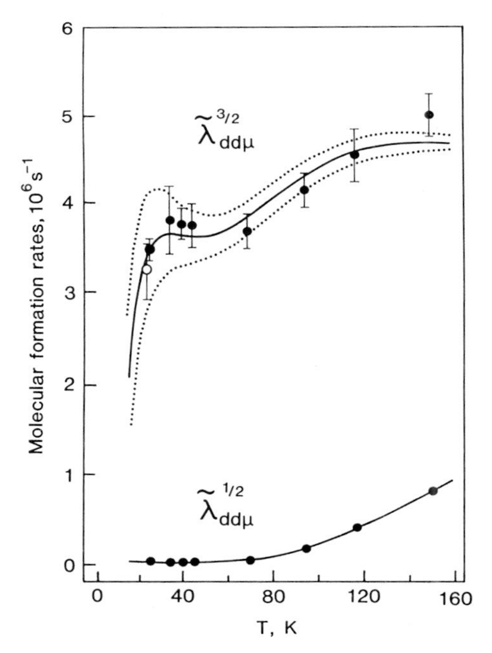


**Figure 15** The temperature dependence of the dd$\mu$ mesic molecule formation rate for different initial hyperfine structure states ($F = \frac{3}{2}, \frac{1}{2}$) of the d$\mu$ atom. The experimental data are from Zmeskal *et al* (1990), and the theoretical result shown by the curve is from Faifman (1988).

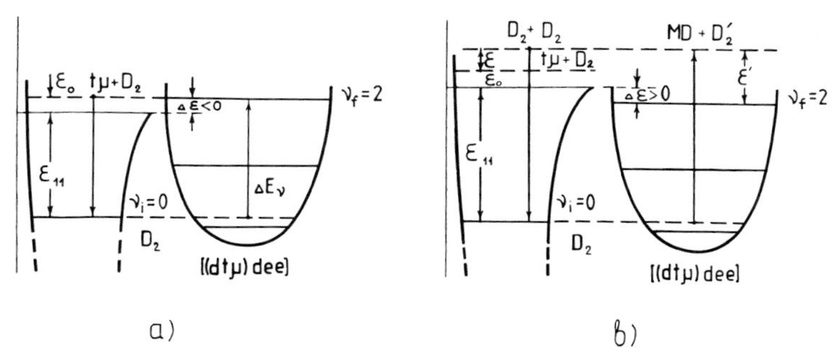


**Figure 16** The scheme of dt$\mu$ mesic molecule resonance formation ($a$) in the pair collisions (8) with the negative resonance defect $\Delta\varepsilon = |\varepsilon_{11}| - \Delta E_\nu < 0$, and quasiresonant formation ($b$) in triple collisions (16) at positive resonance defect $\Delta\varepsilon > 0$.

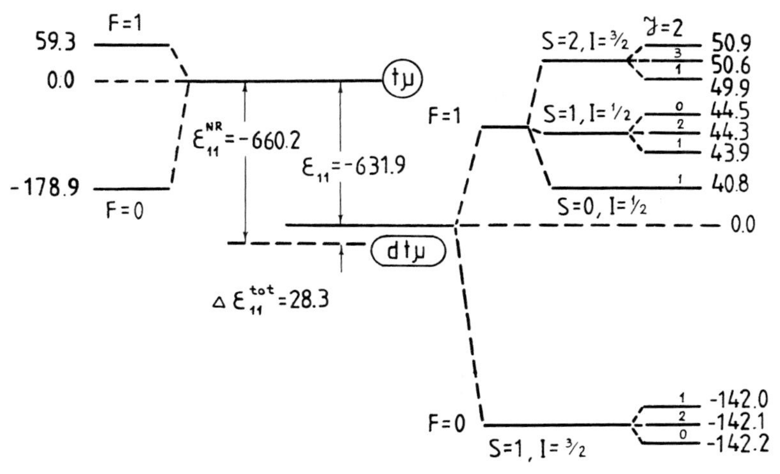


**Figure 17** The schemes of the hyperfine splitting for the ground state of the t$\mu$ atom and the ($J = v = 1$) state of the dt$\mu$ mesic molecule: $F$ is total atomic spin, $S$ is total mesic molecule spin, $I$ is total spin of nuclei and $J = S + I$ is the total angular momentum of the system. Energy splitting (meV) is according to Bakalov and Korobov (1989).

In reality, the picture of dt$\mu$ molecule resonance formation is much more complicated, because due to spin–spin and spin–orbital interaction the energy of the dt$\mu$ molecule ($J = v = 1$) state is split into ten sublevels (see figure 17), and in correspondence with it, the number of possible resonance conditions (5) is increased.

The rotational excitations $K'$ of the complexes [(dt$\mu$)dee] and rotational states $K$ of $D_2$ molecules being taken into account, the problem to describe quantitatively the reaction

$$(t\mu)_F + [D_2]_{\nu K} \rightarrow [(dt\mu)^{J\nu}_{IS}\ dee]_{\nu' K'} \tag{17}$$

(here $S$ is the dt$\mu$ molecule total spin and $I$ is the total spin of the nuclei) becomes rather involved and its complete solution is a matter for the future.

In the same way, an adequate description for the process of the muon sticking to helium is expected, as reaction (9*b*) is followed by the whole complex of muon stripping processes in collisions of $(\mu He)^+_{nl}$ mesic atoms in state ($nl$) with hydrogen isotope nuclei, e.g.

$$(\mu He)^+_{nl} + d \longrightarrow \begin{cases} (\mu He)^+_{n'l'} + d \\ He^{2+} + (d\mu)_{n'l'} \\ He^{2+} + d + \mu^- . \end{cases} \tag{18}$$

(For further information see, e.g., Markushin (1988).)

Figure 18 represents a comparison of theoretical and experimental results on the probability of muon sticking to helium $\omega_s$. Though complete agreement has not yet been achieved, it is seen from this figure that $\omega_s \leqslant 0.5 \times 10^{-2}$, i.e. the theoretical limit for the number of cycles $X_c$ in the deuterium–tritium mixture is $X_c^{max} = 1/\omega_s > 200$.

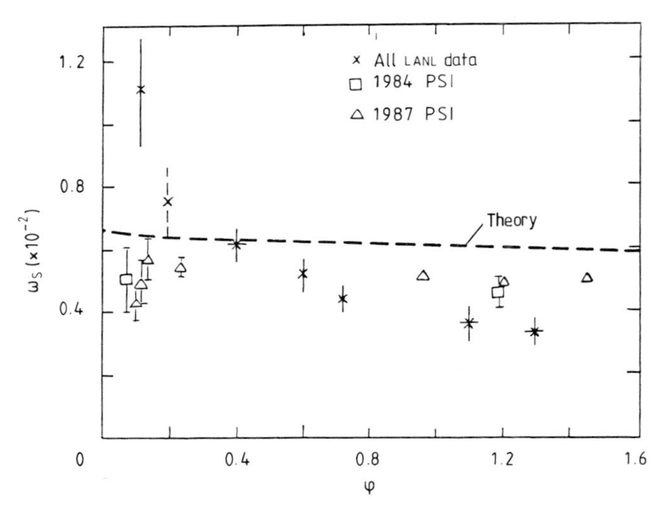


**Figure 18** The dependence of sticking probability $\omega_s$ on the density of deuterium–tritium mixture. Experimental data are from LANL and PSI and the present-day theoretical prediction is shown by the broken curve (Petitjean 1988).

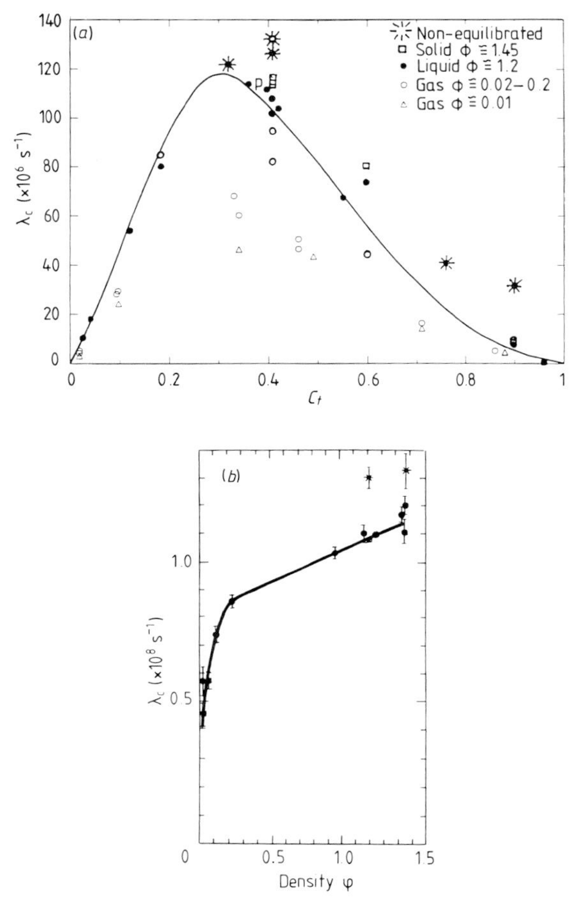


**Figure 19** The muon-catalysed fusion cycle rate: (*a*) the dependence on tritium concentration $C_t$, (*b*) the dependence on density $\varphi$. The experimental data are from the PSI Villigen–Wien–Munich–Berkeley–Los Alamos Collaboration.

Up to now the muon catalysis cycle rate $\lambda_c$ has been also measured experimentally in the $D_2 + T_2$ mixture. This rate is a rather complicated function of the temperature $T$ and the mixture density $\varphi$ and also of the tritium fraction $C_t$ (see figures 19(*a*), (*b*)). To extract from these data the values of many quantities which characterise the muon catalysis cycles ($\lambda_{dt\mu}$, $\lambda_{dt}$, $\omega_s$, $\lambda_{dd\mu}$, etc), one should consider in detail the kinetics of this process.

The kinetics of the muon catalysis process in the hydrogen isotope mixture $H_2 + D_2 + T_2$ (see figure 20) is rather complicated and is, in essence, rather similar to that of neutron deceleration in the nuclear reactor, when the neutron resonance capture in $^{238}$U, fission processes in $^{235}$U, the absorption in admixtures, etc, are all taken into account. To describe it as a whole one should know the cross sections of many mesic atomic and mesic molecular processes; in particular, of mesic molecule formation (2)–(4), (14), (17), of the isotope exchange process (10), of the stripping process (18) and of many elastic scattering processes, e.g. (most of them have now been calculated in the papers by Melezhik *et al*):

$$\begin{aligned} &\mathrm{d}\mu + \mathrm{p} \rightarrow \mathrm{d}\mu + \mathrm{p} \\ &\mathrm{t}\mu + \mathrm{d} \rightarrow \mathrm{t}\mu + \mathrm{d}, \text{ etc} \end{aligned} \tag{19}$$

spin-flip processes

$$\mathrm{d}\mu(\uparrow\uparrow) + \mathrm{d} \rightarrow \mathrm{d}\mu(\uparrow\downarrow) + \mathrm{d} \tag{20}$$

cascade transitions in mesic molecules, e.g.,

$$[(\mathrm{d}\mathrm{t}\mu)_{Jv}\mathrm{dee}] \rightarrow [(\mathrm{d}\mathrm{t}\mu)_{J'v'}\mathrm{de}]^{+} + \mathrm{e} \tag{21}$$

and mesic atom acceleration in inelastic collisions

$$(\mathrm{d}\mu)_{nl} + \mathrm{d} \rightarrow (\mathrm{t}\mu)_{n'l'} + \mathrm{d}. \tag{22}$$

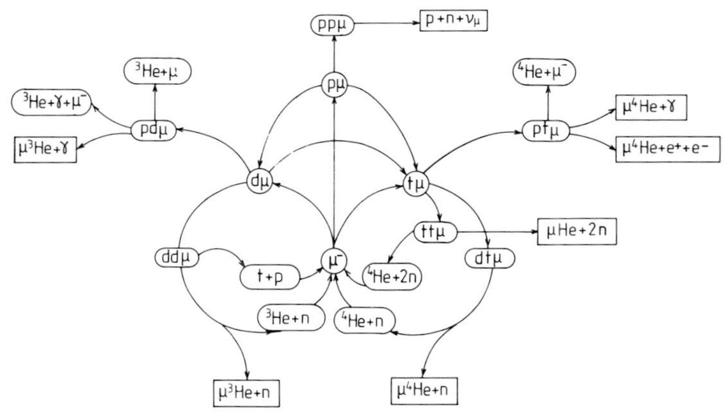


**Figure 20** The muon-catalysed fusion processes in the mixture $H_2 + D_2 + T_2$.

In addition, the fusion rates in mesic molecules are involved:

$$\mathrm{d}\mathrm{t}\mu \longrightarrow \begin{cases} {}^4\mathrm{He} + \mathrm{n} + \mu^{+} \\ \mu\,{}^4\mathrm{He} + \mathrm{n} \end{cases} \tag{23a}$$

$$\mathrm{d}\mathrm{d}\mu \longrightarrow \begin{cases} {}^3\mathrm{He} + \mathrm{n} + \mu^{-} \\ \mu\,{}^3\mathrm{He} + \mathrm{n} \\ \mathrm{t} + \mathrm{p} + \mu^{-} \\ \mathrm{t}\mu + \mathrm{p} \end{cases} \tag{23b}$$

$$\mathrm{pd}\mu \longrightarrow \begin{cases} {}^3\mathrm{He} + \mu^- \\ \mu{}^3\mathrm{He} + \gamma \end{cases} \tag{23c}$$

$$\mathrm{pt}\mu \longrightarrow \begin{cases} {}^4\mathrm{He} + \mu^- \\ \mu{}^4\mathrm{He} + \gamma \\ \mu{}^4\mathrm{He} + \mathrm{e}^+ + \mathrm{e}^- \end{cases} \tag{23d}$$

$$\mathrm{tt}\mu \longrightarrow \begin{cases} {}^4\mathrm{He} + 2\mathrm{n} + \mu^- \\ \mu{}^4\mathrm{He} + 2\mathrm{n}. \end{cases} \tag{23e}$$

In its turn, studying muon catalysis processes, one can extract information about elementary particle, atomic and nuclear physics processes. For example, the structure of nuclear transitions in fusion reactions (23) can be investigated, the data on which are rather scarce at present, particularly, at low energies of nuclear collisions (Bogdanova 1988).

Space limitations do not permit me even to mention all the results obtained during 40 years study of muon catalysis and I have not tried to. (See table 2 for the main characteristics of muon-catalysed fusion in the different mixtures of hydrogen isotopes.)

**Table 2** The main characteristics of muon-catalysed fusion†.

| Mesic molecule | $\omega_s$ | $\lambda_m(s^{-1})$ | $\lambda_f(s^{-1})$ | Energy release (MeV) |
|---|---|---|---|---|
| pd$\mu$ | 0.99 | $5.8\times10^6$ | $2.6\times10^5$ | 5.4 |
| pt$\mu$ | 0.94 | $6.8\times10^6$ | $0.7\times10^5$ | 2.0 |
| dd$\mu$ | 0.12 | $\sim4\times10^6$ | $4.3\times10^3$ | 3.3 |
| dt$\mu$ | $0.43\times10^{-2}$ | $\sim3\times10^8$ | $1.2\times10^{12}$ | 17.6 |
| tt$\mu$ | 0.14 | $3\times10^6$ | $1.5\times10^7$ | 11.3 |

† All the data are presented at $\varphi = 1$ and room temperature. The molecular formation rates $\lambda_m$ are averaged over the inner variables.

But taking advantage of this good opportunity, I have tried to follow the interesting and emotionally coloured history of the evolution of our knowledge about this beautiful phenomenon. I have also tried to give a general picture of the growth of the whole branch of physical research originated by a short paper published by Sir Charles Frank more than forty years ago. Now muon catalysis is studied at about 40 laboratories in 15 countries, every year international conferences on this subject are organised, a special Journal *Muon Catalyzed Fusion* is published in Basel, and respectable experts consider muon catalysis as a serious alternative to other methods of nuclear breeding developed at present: hot fusion, inertial fusion, laser fusion, etc.

It is a great honour for me to present the modern state-of-the-art in this specific field of research to Sir Charles Frank, whose creative initiative stimulated the development of many branches of modern science.

## REFERENCES


Alexander S A and Monkhorst H J 1988 *Phys. Rev.* A **38** 26

Alvarez L W *et al* 1957 *Phys. Rev.* **105** 1127–8

Bakalov D and Korobov V I 1989 *JINR Rapid Commun.* **2** (35) 15–20 (in Russian)

Bakalov D, Melezhik V S, Menshikov L I and Vinitsky S I 1985 *Phys. Lett.* **161B** 5–8

Balin V D, Vorobyov A A, Vorobyov An A, Zalite Yu K, Mayev E M, Medvedev V I, Semenchuk G G and Smirenin Yu V 1984 *Pisma Zh. Eksp. Teor. Fiz.* **40** 318–20 (1984 *JETP Lett.* **40** 112)

Bleser E J, Anderson E W, Lederman L, Meyer S L, Rosen J L, Rothberg J E and Wang I T 1963 *Phys. Rev.* **132** 2679–91

Bogdanova L N 1988 *Muon Catalyzed Fusion* **3** 359–76

Bystritskii V M *et al* 1979 *Zh. Eksp. Teor. Fiz.* **76** 460–9 (1979 *Sov. Phys.–JETP* **49** 232)

Bystritskii V M *et al* 1981 *Zh. Eksp. Teor. Fiz.* **80** 1700–14 (1981 *Sov. Phys.–JETP* **53** 877)

Doede J H 1963 *Phys. Rev.* **132** 1782–99

Dzhelepov V P, Ermolov P F, Moskalev V I and Filchenkov V V 1966 *Zh. Eksp. Teor. Fiz.* **50** 1235–51 (*1966 Sov. Phys.–JETP* **23** 820)

Faifman M P 1988 *Muon Catalyzed Fusion* **2** 247–60

Fetkovich J G, Fields T H, Yodh G B and Derrick M 1960 *Phys. Rev. Lett.* **4** 570–2

Frank F C 1947 *Nature* **160** 525–7

Gerstein S S 1958 *Thesis* P N Lebedev Physical Institute Moscow (in Russian)

Gerstein S S and Ponomarev L I 1975 *Muon Physics* vol 3, eds V Hughes and C S Wu (New York: Academic) p. 141

Jackson J D 1957 *Phys. Rev.* **106** 330–9

Jones S E *et al* 1983 *Phys. Rev. Lett.* **51** 1757–60

Jones S E *et al* 1986 *Phys. Rev. Lett.* **56** 588

Kammel P *et al* 1982 *Phys. Lett.* **112B** 319

Lattes C M G, Occhialini G P S and Powell C F 1947 *Nature* **160** 453–6

Markushin V E 1980 *Muon Catalysed Fusion* **3** 395–420

Matveenko A V, Ponomarev L I and Faifman M P 1975 *Zh. Eksp. Teor. Fiz.* **68** 437–46 (1975 *Sov Phys.–JETP* **41** 212–16)

Melezhik V S, Ponomarev L I and Faifman M P 1983 *Zh. Eksp. Teor. Fiz.* **85** 434–46

Menshikov L I, Ponomarev L I, Strizh T A and Faifman M P 1987 *Zh. Eksp. Teor. Fiz.* **92** 1173–87 (1987 *Sov. Phys.–JETP* **65** 654)

Petitjean C 1988 *Fusion Engng Design* **11** 255–64

Petrov Yu V 1980 *Nature* **285** 466–9

Ponomarev L I 1990 *Contemp. Phys.* **31** 219–45

Ponomarev L I and Fiorentini G 1987 *Muon Catalyzed Fusion* **1** 3–20

Ponomarev L I, Puzynin I V and Puzynina T P 1973 *Zh. Eksp. Teor. Fiz.* **65** 28–34 (1973 *Sov. Phys.–JETP* **38** 14)

Sakharov A D 1948 *Internal report* P N Lebedev Physical Institute Moscow

Vesman E A 1967 *Pisma Zh. Eksp. Teor. Fiz* **5** 113–15 (1967 *JETP Lett.* **5** 91)

Vinitsky S I and Ponomarev L I 1982 *Fiz. Elem. Chast. At. Yad.* **13** 1336–418 (1982 *Sov. Phys.–Particles and Nuclei* **13** 557)

Vinitsky S I, Ponomarev L I, Puzynin I V, Puzynina T P, Somov L N and Faifman M P 1978 *Zh. Eksp. Teor. Fiz.* **74** 849–61 (1978 *Sov. Phys.–JETP* **47** 444)

Zeldovich Ya B 1954 *Dokl. Acad. Nauk SSSR* **95** 493–6

Zeldovich Ya B and Gerstein S S 1960 *Usp. Fiz. Nauk* **71** 581–630 (1961 *Sov. Phys.–Usp.* **3** 593)

Zmeskal J *et al* 1990 *Phys. Rev.* A **42** 1165–77

# Morphology of Crystals: Integration of Principles of Thermodynamics, Statistical Mechanics and Crystallography

**P Bennema**

## 1 INTRODUCTION

### 1.1 Preliminary remarks, aim of this paper

In 1951 the famous paper of Burton, Cabrera and Frank, 'The growth of crystals and the equilibrium structure of their surfaces' was published (Burton *et al* 1951). In this paper the two-dimensional lattice gas or Ising model plays an essential role. The interface between a crystal and vacuum was considered as a lattice gas situated on top of a flat crystal. Within such a lattice gas an order–disorder phase transition occurs. BCF interpreted this order–disorder phase transition as what is nowadays called the roughening transition. This implies that below a certain critical temperature crystal surfaces are flat and above this temperature they are rough. Below the roughening temperature crystals are limited by large areas of two-dimensional solid and vacuum domains. These domains are limited by edges having an edge free energy larger than zero. At the roughening temperature and above it, solid and vacuum domains become one phase and the edges disappear. Consequently, the edge free energy of steps vanishes.

BCF showed that contrary to a two-dimensional lattice gas no phase transitions occur for steps and that steps are always rough. Steps have no fixed boundaries and will move under the influence of a positive or negative driving force for crystallisation without any thermodynamic barrier.

These concepts: order–disorder phase transition (roughening transition) and rough steps are the basic ingredients of the spiral theory of BCF. This theory can be summarised as follows: in order to have growth at low supersaturations below the roughening temperature a step source must be available, because steps cannot be formed by statistical fluctuations. It is Sir Charles Frank who for the first time put forward the idea that a screw dislocation provides a source of steps (Frank 1949). If a screw dislocation emerges at the surface a spiral automatically develops (see also Bennema (1984)).

It will be shown in this paper how the two-dimensional lattice gas which gives an exact solution for the order–disorder (or roughening) temperature can be generalised for real crystals, which may have complex structures. It will be shown how the morphological theory of Hartman and Perdok and the lattice gas theory can be logically integrated. In addition, the limitations of such an integrated approach will be discussed.

In the introduction a short survey of Ising models for the interface crystal–mother face and the concepts used will be presented. In section 2 the results of a theory yielding an exact solution for the order–disorder phase transition will be discussed. In section 3 the basic principles of the Hartman–Perdok theory will be discussed. In section 4 examples of the application of the integrated Hartman–Perdok–Ising theory will be given for organic and inorganic crystals. In the last section, section 5, a generalisation of the old crystallographic theory which showed that crystal faces have to be indexed with three integers (*hkl*) to two new types of crystals, namely modulated and quasi crystals, will be discussed briefly. For these two types of crystals crystal faces have to be indexed with four indices (*hklm*) and six indices (*hklmno*), respectively.

## 1.2 Ising models, crystal surfaces and crystal growth

In order to apply statistical mechanical Ising models to the interface crystal–mother phase this whole interface is partitioned into cells of equal size and shape (see figure 1(*a*)) and the cells can have only two properties, which are solid (s) or fluid (f). (Note that the vacuum cells mentioned above are now generalised to fluid cells.) Within such an interface model the solid on solid (SOS) condition can be introduced (Bennema and Gilmer 1973, Bennema and van der Eerden 1987, Temkin 1966, Gilmer and Bennema 1972, Gilmer 1976, de Haan *et al* 1974, Leamy and Gilmer 1974). This implies that solid cells only can occur on top of solid cells and overhangs are ruled out. Also isolated fluid cells in the solid body and the reverse are ruled out. Looking at the top of an SOS surface it can be seen that each surface configuration is characterised by the height of towers $\alpha_{xy}$ (see figure 1(*b*)). Monte Carlo simulations have been carried out on such surfaces, using as independent variables the quantities

$$\frac{\phi}{kT} = \frac{\phi^{sf} - \frac{1}{2}(\phi^{ss} + \phi^{ff})}{kT} \tag{1}$$

and

$$\beta' = \frac{\Delta\mu}{kT} = \frac{\mu^{f} - \mu^{s}}{kT}. \tag{2}$$

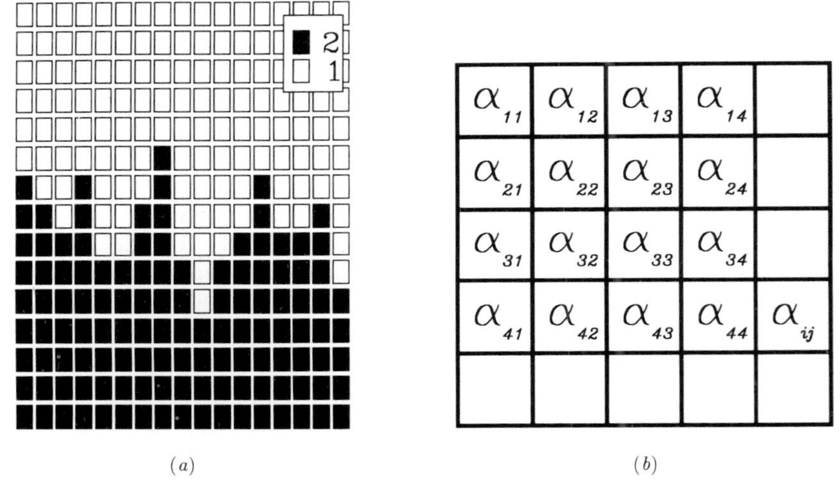


**Figure 1** (*a*) A cut through an SOS 'landscape'. Fluid cells □ (1), solid cells ■ (2). Solid cells only occur on top of other solid cells. This makes the solid phase a connected phase. The same holds automatically for the fluid phase. (*b*) The SOS surface seen from above. Due to the SOS restriction the whole interface can be characterised by a set of 'towers', which are now given by the matrix elements $\alpha_{xy}$. Theoretically the number $\alpha_{xy}$ varies in a discrete way from $-\infty$ to $+\infty$ and characterises this height of towers referenced to a reference plane. The set of numbers $\alpha_{xy}$ characterise a surface configuration.

Here $\phi$ is a generalised bond energy, $k$ is the Boltzmann constant and $T$ is the absolute temperature. $\phi$ represents the energy of formation of a solid–fluid (sf) bond in reference to half a solid–solid (ss) bond and half a fluid–fluid (ff) bond. The bond energies $\phi^{ss}$, $\phi^{ff}$ and $\phi^{sf}$ are supposed to be negative. In the case where the fluid phase is considered as vacuum, we have

$$\frac{\phi}{kT} = -\frac{1}{2}\frac{\phi^{ss}}{kT} \tag{3}$$

which is a positive number. In the following we will assume that $\phi$ is always positive. The quantity $\phi/kT$ defines the equilibrium properties of the surface such as roughness, etc. The quantity $\beta'$ defines the non-equilibrium driving force for crystallisation since $\mu^f$ is the chemical potential of a fluid cell and $\mu^s$ is the chemical potential of a solid cell. The dimensionless driving force $\beta'$ can be interpreted as being proportional to the supersaturation or undercooling.

In figure 2 the results of computer simulation studies are presented for different $\alpha$ values, where $\alpha$ is defined for a simple square lattice as

$$\alpha = 4\phi/kT. \tag{4}$$

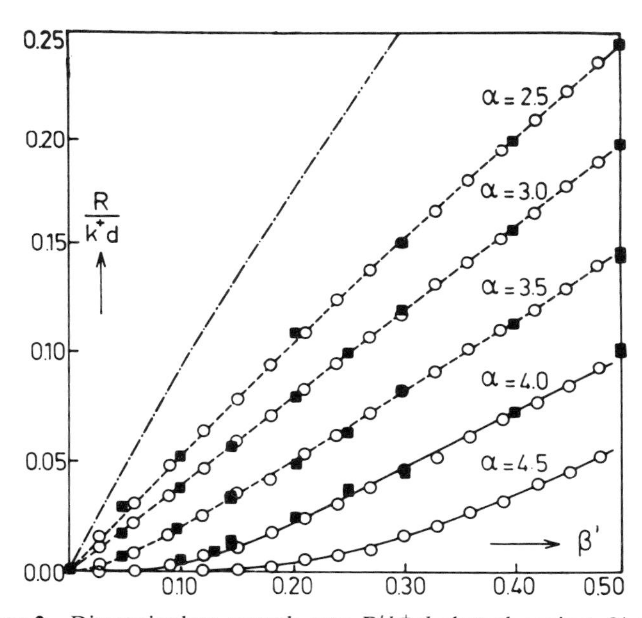


**Figure 2** Dimensionless growth rate $R/k^+d$ plotted against $\beta'$ for $0 < \beta' < 0.5$: ■, results of Gilmer and Bennema (1972); ○, special purpose computer results (De Haan *et al* 1974); full curves, two-dimensional birth and spread nucleation formula; broken curves, empirical relation; uppermost line, maximal rate of growth (of a kinked or rough surface).

It can be seen that the lower the $\alpha$, which means keeping $\phi$ constant, the lower the temperature, and so the lower the growth rate. This is understandable because the lower the temperature the less rough the surface, and so the lower the sticking fraction. At about $\alpha \cong 3.2$ a change from a linear to a non-linear $R$ against $\beta'$ curve occurs. It is now well known that this change marks the so-called roughening transition in an SOS or $XY$ model. Independent of computer simulations the character of this phase transition is now clarified thanks to the work of theoretical physicists like Leamy and Gilmer (1974), van Beijeren (1977), Swendsen (1978), Müller-Krumbhaar (1978), van der Eerden and Knops (1978), and Weeks and Gilmer (1979). The roughening transition is defined as a (dimensionless) temperature

$$\theta^R = \left(\frac{2kT}{\phi}\right)^R \tag{5}$$

such that if the actual (dimensionless) temperature $\theta$ of the surface

$$\theta < \theta^R \quad \text{then} \quad \gamma_{\text{step}} > 0 \tag{6}$$

and if

$$\theta > \theta^R \quad \text{then} \quad \gamma_{\text{step}} = 0 \tag{7}$$

where according to the convention used in this paper $\theta$ is defined as

$$\theta = 2kT/\phi. \tag{8}$$

$\gamma_{\text{step}}$ is defined as the edge free energy of a step. This is an order parameter playing a key role in theories on crystal surfaces and crystal growth. It will be clear that for $\theta < \theta^R$ or, what amounts to the same, $\alpha > \alpha^R$ where $\alpha^R$ is defined as

$$\alpha^R = 4\left(\frac{\phi}{kT}\right)^R \tag{9}$$

the surface remains in essence flat apart from statistical fluctuations, over 'infinitely' long distances. This is because large islands with a height of one atomic layer, or holes, cannot be formed due to the fact that $\gamma_{\text{step}} > 0$. If, however, $\theta > \theta^R$ surfaces roughen up and will lose their crystallographic orientation (*hkl*). No energy price has to be paid for this roughening, since $\gamma_{\text{step}} = 0$.

In case of growth ($\beta' > 0$) if $\theta > \theta^R$ or $\alpha < \alpha^R \cong 3.2$, the crystal face can grow barrierlessly since $\gamma_{\text{step}} = 0$, while if $\theta > \theta^R$ it has to grow by a layer mechanism, i.e. a two-dimensional nucleation mechanism or a spiral growth mechanism because $\gamma_{\text{step}} > 0$. The non-linear curves for $\alpha \geqslant 4$ can be fitted with a non-linear two-dimensional nucleation curve (figure 2 and Bennema (1984), Bennema and Gilmer (1973), Bennema and van der Eerden (1987), Gilmer and Bennema (1972), Gilmer (1976), de Haan *et al* (1974)). If the supersaturation becomes sufficiently high, the size of a two-dimensional nucleus becomes equal to a few atoms (molecules). Then roughening also occurs. This is the so-called kinetical roughening caused by the driving force for crystallisation. Also in this case crystal faces, which at low supersaturations grow as flat faces below their roughening temperature, lose their crystallographic orientation at higher supersaturation. This also depends on the 'strength' of the connected net. The lower this strength, the lower the edge free energy and the lower the supersaturation for which kinetical roughening occurs (Human *et al* 1981, Jetten *et al* 1984, Elwenspoek and van der Eerden 1987, Elwenspoek *et al* 1987).

It is interesting to note that simulation of Ising models developed from 1966 to 1980 and in fact most essential points, such as the roughening transition, were clarified at the end of the seventies (see the survey by van der Eerden *et al* (1978) and Chernov and Lewis (1967), Gilmer and Bennema (1972), Gilmer (1976), de Haan *et al* (1974), Leamy and Gilmer (1974)).

Yet in order to clarify specific theoretical and experimental issues, simulation studies on Ising models are still being carried out. As examples we have the studies of Zhao *et al* (1982) on stress fields around dislocations and on modified Ising models (Zhao 1988, Chen *et al* 1986) and work carried out by Clarke *et al* (1989a,b) on the reconstructed (001) surface of silicon. We note that, in the model of Clarke *et al*, cells and bonds are anisotropic, but the stacking to simulate the (001) reconstructed surface fulfils the space group symmetry $F^4/_d\bar{3}^2/_m$ of the diamond structure. Similar simulations on anisotropic orthorhombic structures have been carried out previously (van Dijk *et al* 1974, van der Eerden *et al* 1977, van der Eerden 1979). We also refer to recent work of Cheng and co-workers on growth and dissolution (Cheng *et al* 1989).

### 1.3 Two-dimensional interface models

The change from a linear so-called continuous crystal growth mechanism obeying a linear $R$ against $\beta'$ law to a non-linear law is a key issue in crystal growth theories and was studied in the pioneering paper of Burton *et al* (1951) and also by Jackson (1958), Mutaftschief (1965) and Voronkov and Chernov (1966). In these papers the interface as presented in figure 1(*a*) was reduced to one or two mixed solid–fluid layers. Such a layer can be considered as a two-dimensional mixed fluid solid crystal, and to such a two-dimensional mixed solid fluid model the theory of Onsager (1944) for an order–disorder phase transition applies. The work of Onsager was another pioneering paper, because it showed that a real order–disorder phase transition occurred. Moreover, Onsager was able to calculate the dimensionless order–disorder phase temperature $\theta^c$ exactly. Although the SOS model is definitely a better model for the interface crystal–mother phase, because it allows for more layers and hence more fluctuations, it can be noted that upon comparing the dimensionless roughening temperature $\theta^R$ and $\theta^c$ for simple interfaces, as given in table 1, it can be justified to assume that these values are close to each other, so that

$$\theta^R \cong \theta^c. \tag{10}$$

The BCC (110) SOS value was taken from the special model introduced by van Beijeren (1977), which can be solved exactly. We note that the character of the two phase transitions, namely a roughening transition in an SOS or SOS-like model and an order–disorder phase transition in a two-dimensional Ising model, is quite different. The transition in the first case is a very smooth phase transition of infinite order. For the second case it is a transition of second order, yet the numerical values of the transition temperatures are rather close. This must be attributed to the fact that the fluctuations of the top layer give the highest contribution to the free energy of a step. As opposed to physicists and chemists, who study the character of phase transitions in detail, in crystal growth we are more interested in the values of $\theta^R$ and not in the precise character of the phase transition.

In the next sections a mathematical approach will be discussed which yields an exact solution for the Ising temperature $\theta^c$. Exact solutions are obtained also if for complex two-dimensional crystals (nets) only first-nearest-neighbour bonds are taken into consideration. Next-nearest-neighbour bonds may be possible, but crossing bonds may not occur. It will be shown below that the theory of order–disorder phase transition according to Onsager (1944) and its generalisation can be integrated with the crystallographical theory of Hartman and Perdok. In addition the relation with the classical Bravais–Friedel–Donay–Harker theory and its recent generalisation for modulated and quasi-crystals will be discussed at the end of this paper.

**Table 1** Comparison of Ising temperatures $\theta^c$ and roughening temperatures for faces of simple structures.

| | $\theta^c$ | $\theta^R$ |
|---|---|---|
| SC(100) | 2.269 18... | 2.56 |
| BCC(110) | 2.269 18... | — |
| BCC(110)SOS | — | 2.885 5390... |
| FCC(100) | 2.269 18... | 2.2 |
| FCC(111) | 3.641 48 | 3.0–4.0 |
| Diamond(111) | 1.518 652 | 1.48 |

## 2 EXACT SOLUTIONS FOR THE DETERMINATION OF THE ISING TEMPERATURES OF TWO-DIMENSIONAL ISING LATTICES AND MEAN FIELD SOLUTION

### 2.1 Introduction

As discussed above, in section 1.2, the interface crystal–mother phase may be considered as an interface with a thickness of one atomic layer, and this can be described as a two-dimensional Ising lattice. The solution of the problem of how to integrate the world of real complex structures and their surfaces and the world of statistical mechanical Ising models (exact and mean field solutions) will be postponed till section 3. In section 2.2 we will discuss the results of an order–disorder phase transition temperature for rectangular two-dimensional crystals (nets) and refer to the paper of Rijpkema *et al* (1982). Till now crystals were mostly considered as structures consisting of cells having the two properties: solid or fluid (or spin up and spin down). Between the cells interaction energies due to 'contacts' between cells occurred. In the following, cells will be reduced to centres of gravity or points and these points will be called cells, atoms, ions or simply points. Contacts between cells will be reduced to covalent-like chemical bonds represented, as usual in chemistry, by strings. The energy between the *i*th and the *j*th points is indicated by a positive number $\phi_{ij}$. We will discuss the exactly soluble problem first. The theory of the exactly soluble model is written in the language of a spin formalism where each point $i$ is in a solid state with the spin variable $s_i = 1$ or in a fluid state with $s_i = -1$.

## 2.2 Exact determination of Ising temperature of complex nets with first nearest neighbours

### 2.2.1 *Bond and spin conventions*

Before discussing the implications of the model of Rijpkema *et al* (1982) we need to discuss the sometimes confusing spin and bond energy conventions used in this paper and others first.

For the time being we assume that we have a surface in contact with vacuum. We want to investigate the process of breaking a bond between the *i*th and *j*th cells, expressed in the language of spins. We write for the energy of a particular two-dimensional surface configuration filled with solid and vacuum cells the energy

$$E = \sum_{\langle ij \rangle} \phi'_{ij} s_i s_j \tag{11}$$

where $s_{i,j} = \pm 1$. The summation is over all $ij$ pairs of points (cells). $\phi'_{ij}$ is a (negative) interaction energy between a solid cell $i$ and a solid cell $j$. Here we use a similar convention as given in equation (3) above. Within the framework of the Ising formalism used before this $\phi'_{ij}$ is not the same as the $\phi^{ss}$ given by equation (3). Now the ground state for the $ij$ bond is a solid–solid interaction and is given by

$$E_{gr} = \phi'_{ij} s_i s_j \tag{12}$$

where $s_i s_j = 1$. The excited state corresponds to

$$E_{ex} = \phi'_{ij} s_i s_j \tag{13}$$

where $s_i s_j = -1$.

The difference is

$$E_{ex} - E_{gr} = \Delta E = -2\phi'_{ij} \tag{14}$$

and this corresponds to breaking a bond. In the paper of Rijpkema *et al* (1982) bond energies corresponding to $\phi'_{ij}$ given in equations (11)–(14) are used. This is the convention employed in magnetism, where also the so-called coupling $J_{ij}$

$$J_{ij} = \beta\phi'_{ij} \tag{15}$$

is introduced ($\beta = 1/kT$).

The relation between $\phi_{ij}$ and a corresponding generalised $\phi^{ss}_{ij}$ is, according to equation (3), given by

$$\phi_{ij} = -\tfrac{1}{2}\phi^{ss}_{ij} \tag{16}$$

and between $\phi'_{ij}$ and $\phi^{ss}_{ij}$ by

$$-2\phi'_{ij} = -\tfrac{1}{2}\phi^{ss}_{ij}. \tag{17}$$

Between $\phi'_{ij}$ and a generalised bond energy $\phi_{ij}$ (see equation (1)),

$$\phi_{ij} = \phi^{sf}_{ij} - \tfrac{1}{2}(\phi^{ss}_{ij} + \phi^{ff}_{ij}) \tag{18}$$

the relation

$$-2\phi'_{ij} = \phi_{ij} \tag{19}$$

is valid. Dimensionless (critical) temperatures are expressed (after substituting equations (18) and (19) in equation (8)) as

$$\theta = \frac{kT}{|\phi'_{ij}|} = \frac{2kT}{\phi_{ij}}. \tag{20}$$

This explains the factor 2 in equations (5) and (8). In theories on crystal surfaces and crystal growth $\phi_{ij}$ given by equations (1) and (18) is used. In the theory in Kadanoff and Ceva (1971), and Rijpkema *et al* (1982), $\phi_{ij}$ corresponding to $\phi'_{ij}$ is employed. Within the framework of this theory the prime sign is dropped and energies indicated with the notation $\phi_{ij}$, $\phi_x$ or $\phi_y$ play their well defined role in the formalism of Rijpkema *et al* (1982), calculated by determining when the edge free energy vanishes. We will not treat the theory of Rijpkema *et al* (1982) here and refer to the original reference. For a discussion of the mean field approximation originally introduced by Jackson (1958) and generalised to more complex structures, we refer to a forthcoming paper (Bennema 1990).

## 3 THEORY OF HARTMAN AND PERDOK: CRYSTAL GRAPHS AND CONNECTED NETS

### 3.1 Procedure to derive morphology

So far only Ising interface models and two-dimensional nets have been discussed. No connection was made with the world of crystallography, as described for example by the crystallographic morphological theory of Hartman and Perdok (Hartman and Perdok 1955a, b, c, Hartman 1973, 1987). In this section general principles will be given and in the following sections it will be shown how the theory can be applied to organic and inorganic crystals.

In order to derive the theoretical morphology (habit) of crystals the following procedure consisting of three steps has to be carried out.

(i) *Determination of growth units and bonds from the crystal structure*
After studying carefully the supposedly known crystal structure, bond energies between growth units are calculated.

Growth units are ions, molecules, complexes, etc, occurring in the mother phase from which the crystal grows. Bonds within the growth units are not taken into account since only the overall bonds between growth units are relevant. These bonds may consist of many atom–atom pair potentials of atoms that are part of the growth units. In principle, growth units are defined after considering the information of the complexes occurring in the mother phase and by comparing these complexes with the crystal structure. Very often growth units are not known and then *ad hoc* hypotheses concerning the growth units must be made. Also, alternative growth units can be chosen which may lead to different morphologies. Comparing the alternative theoretical morphologies with the real one, conclusions concerning the structure of growth units in the mother phase may be drawn *a posteriori*.

After growth units are defined, interaction energies referenced to vacuum ($\phi_{jk}^{ss}$) between growth units need to be calculated.

Most sophisticated calculations were carried out by Berkovitch-Yellin (1985) starting with precise electron density maps. In the case of crystal structures of organic crystals, such as naphthalene, paraffin, fat-crystals, etc, overall bonds between molecules are calculated by adding up all pair potentials, such as C–C, H–H and C–H potentials, occurring between molecules, and calculating for each of these pairs energies obtained from reliable Lennard-Jones-like potentials, based on precise atomic distances which follow from structure determinations by x-rays (see next sections).

(ii) *Determination of crystal graph*
Once the set of overall bonds between molecules or growth units is known a crystal graph has to be defined. A graph is a set of elements (points) with relations between them. A crystal graph is an infinite set of points fulfilling the symmetry of one of the 230 space groups with relations (bonds) between the growth units. In order to define the crystal graph of the crystal structure under consideration the molecules or growth units are reduced to centres of gravity or points and these form the set of points of the crystal graph. Next the strongest (first-nearest-neighbour) bonds are chosen as relations between the points.

(iii) *Determination of connected nets, F faces*
From the thus defined crystal graph the so-called connected nets now have to be determined. For complex crystal structures this may be a tedious job. (Nowadays computer programs are used.) Connected nets have an overall thickness $d_{hkl}$ which is the interplanar distance of the net planes, corresponding to the crystallographic plane $(hkl)$, corrected for the extinction conditions of the space group of the crystal structure and crystal graph. A connected net is defined as a net where all points within the thickness $d_{hkl}$ are connected by all kind of arbitrary uninterrupted paths of bonds. Such connected nets show an order–disorder phase transition at a definite Ising temperature. To these nets the theory discussed in section 2.2 applies. We note that the requirement of connectedness is essential for the occurrence of an order–disorder phase transition and roughening transition because within a slice of thickness $d_{hkl}$ non-connected nets may occur and these according to equations (6) and (7) roughen up at $T = 0$ K, since in at least one direction $\gamma_{step} = 0$. In particular, this occurs if indices $(hkl)$ are large and $d_{hkl}$ small.

Within connected nets paths of uninterrupted bonds occur having a period $[uvw]$ of the lattice. Such paths of bonds are called periodic bond chains (PBCs) in the Hartman–Perdok theory. Connected nets consist of at least two different connected sets of parallel PBCs.

### 3.2 The role of the interplanar thickness $d_{hkl}$ for connected nets

In the following we will explore the role of the interplanar thickness $d_{hkl}$ in more detail. Assume that we have a crystal graph of a particular crystal structure and that we want to determine the connected nets corresponding to a certain crystallographic direction $(hkl)$. Let us assume for the time being that we can find an unambiguous cut corresponding to the elementary two-dimensional cell of the two-dimensional net having the lowest cut energy of possible alternative cuts. It may well be that the surface changes due to reconstruction or interaction with the mother phase. But for the time being we will assume that somehow a 'cheapest cut' corresponding to the lowest surface energy exists. Then, due to the crystallographic periodicity, after a repeat distance $d_{hkl}$ exactly the same cut occurs (which may, for example, be rotated due to a screw axis). If no connected net can be made no unambiguous cheapest cut occurs and a stepped (S) or kinked (K) face occurs which is rough at $T = 0$ K. (We refer to discussions on this subject in Bennema and van der Eerden (1987), Hartman and Perdok (1955a,b,c), Hartman (1973, 1987), Berkovitch-Yellin (1985)). We note that for complex crystal structures alternative connected nets parallel to one face $(hkl)$ are possible. After calculating the Ising temperature, or $E^{\text{slice}}$ (the energy per growth unit within a slice), it can be calculated which slice is the strongest and in this case it is assumed that the strongest slice will dominate the crystal growth form. Recently the theory has been applied to garnet (Rijpkema *et al* 1983, Bennema *et al* 1983, Cherepanova *et al* 1989, 1990), apatite (Terpstra *et al* 1986a,b), bismuth germanium oxide (Smet *et al* 1989), gypsum (Weijnen *et al* 1987), potassium titanyl phosphate (KTP) (Bolt and Bennema 1990), 1,2,3 high-$T_c$ crystals (van de Leemput *et al* 1989), naphthalene (studying among others the phenomenon of kinetical roughening (Human *et al* 1981, Jetten *et al* 1984, Elwenspoek and van der Eerden 1987, Elwenspoek *et al* 1987), potassium hydrogen phthalate (Hottenhuis *et al* 1988), paraffin crystals (Bennema *et al* 1990b), fat crystals (Bennema *et al* 1990a), $\beta$-lactose (Visser and Bennema 1983) and benzophenon (Docherty *et al* 1990).

### 3.3 Determination of Ising temperatures; rectangularisation of connected nets

Once the connected nets are determined, Ising temperatures have to be calculated using a computer program giving a solution of the matrix equation which follows from the theory of Rijpkema *et al* (1982).

The formalism derived in Rijpkema *et al* (1982) is a formalism developed for rectangular nets. Very often connected nets are not rectangular. It is, however, possible to transform non-rectangular connected nets into rectangular connected nets having the same partition function as the original net by introducing bonds of infinite strength and no (zero) strength. In order to 'rectangularise' nets, points of the original net may be split into two with a bond of infinite strength connecting these points. As an example we give the transformation of an hexagonal net consisting of three types of bonds into a rectangular net (see figure 3).

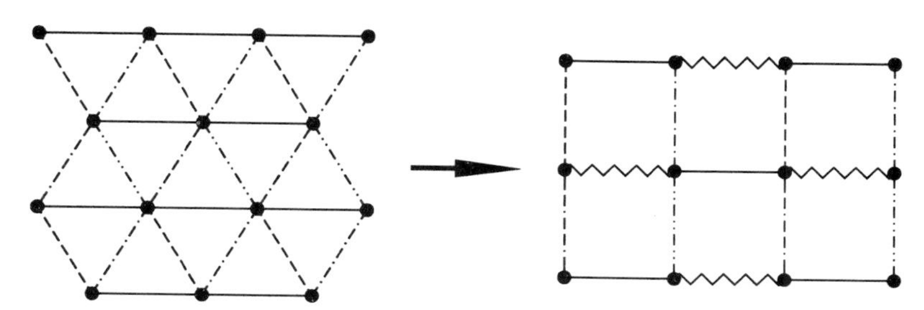

**Figure 3** Change of a hexagonal net to a rectangular net by introducing bonds of infinite strength, indicated as wiggled bonds.

Sometimes it is not possible to calculate Ising temperatures from connected nets, because the connected nets are not real planar, in the sense that crossing bonds occur or that nets consist of subnets which are within the slice thickness $d_{hkl}$ stacked upon each other and connected to each other. In this case by omitting weak bonds or adding extra (strong) bonds a weaker and a stronger rectangular net can be made, allowing an upper and a lower bound of the Ising temperature to be calculated.

### 3.4 Construction of crystal growth forms

Unlike the construction of equilibrium forms (see Kern (1987)) there are no unambiguous ways of constructing growth forms from a series of calculated Ising temperatures. It can be stated, however, that the higher the Ising temperature $\theta^c_{hkl}$ in a series $\theta^c_{h_1k_1l_1}$, $\theta^c_{h_2k_2l_2}$, ... of connected nets, obtained from a crystal graph, the stronger the connected net, the lower its growth rate $R_{hkl}$ and the higher its morphological importance (MI). MI is a statistical measure of the relative frequency of occurrence of a crystal face, parallel to a connected net, and its relative size. The larger the frequency and/or the size, the larger MI.

A crystal growth form may be constructed from a Wulff-like plot, taking the vectors $\boldsymbol{R}_{hkl}$, representing the growth rate of a face $(hkl)$, proportional to $(\theta^c_{hkl})^{-1}$, or if the actual dimensionless $\theta$ can be estimated, taking $R_{hkl} \sim (\theta^c - \theta)^{-1}_{hkl}$. For complex nets, $\theta$ and $\theta^c$ are defined as

$$\theta = \frac{2kT}{\phi_{str}} \qquad \theta^c = \left(\frac{2kT}{\phi_{str}}\right)^c \tag{21}$$

where, by convention, $\theta$ and $\theta^c$ are defined in reference to $\phi_{str}$, the strongest bond energy of the crystal graph. In order to obtain Ising temperatures of connected nets the ratios of bond energies must be known. According to the so-called proportionality hypothesis, these are assumed to be proportional to the overall $\phi^{ss}$ bond energies (see equations (1), (3), (18) and (19)).

A traditional way of constructing growth forms is to take $\boldsymbol{R}_{hkl}$ proportional to $E^{att}_{hkl}$, where according to the concepts used in the Hartman–Perdok theory $E^{att}_{hkl}$ is the energy per growth unit which is cut by the boundary of the connected net. A better way to define $E^{att}_{hkl}$ is to define it as the energy required to remove a growth unit in its crystallographic position from a flat surface $(hkl)$. $E^{slice}_{hkl}$ is defined as the energy of a growth unit within a slice. It is a kind of complement to $E^{att}$ because it corresponds to the non-broken bonds. Using these concepts from the Hartman–Perdok theory it can be seen that

$$E^{att}_{khl} + E^{slice}_{hkl} = E^{cr} \tag{22}$$

where $E^{cr}$ is the crystallisation energy, which is a bulk property; it is independent of the crystallographic face.

The justification that there is a parallel relationship between $\boldsymbol{R}_{hkl}$ and $E^{att}_{hkl}$ was demonstrated in a paper by Hartman and Bennema (1980). In practice, in order to analyse possible growth forms $\boldsymbol{R}_{hkl}$ is taken to be proportional to $E^{att}_{hkl}$. This recipe has been applied with great success by Berkovitch-Yellin (1985) from the group of Lahav and Leisserovitch from the Weissman Institute in Rehovot, Israel and by Hartman and co-workers (Hartman 1973, 1987, Strom 1980, 1981, 1985, van der Voort 1990, Woensdregt 1990).

## 3.5 Proportionality condition and equivalent wetting condition

### *3.5.1 Proportionality condition*

In order to calculate the dimensionless bond energies $\phi_i/kT$ of a crystal graph, where $i$ stands for the first, second, etc, bond, the *ad hoc* so-called proportionality hypothesis is introduced:

$$\phi_i : \phi_j : \phi_k : \ldots = \phi_i^{\text{ss}} : \phi_j^{\text{ss}} : \phi_k^{\text{ss}} : \ldots \, . \tag{23}$$

This implies that the bond energies occurring at the surface $\phi_i$ having the shape of equation (1), which may be a few per cent of the (absolute) values of the bond energies $|\phi_i^{\text{ss}}|$, are supposed to be proportional to the $\phi_i^{\text{ss}}$ bond energies.

### *3.5.2 Justification of the proportionality relation*

It is a remarkable fact that bond energies occurring at the surface $\phi_i$ have the shape of equation (1), $\phi_i = \phi_i^{\text{sf}} - \frac{1}{2}(\phi_i^{\text{ss}} + \phi_i^{\text{ff}})$, and having values of a few per cent of $\phi_i^{\text{ss}}$ seem to fulfil the proportionality condition, equation (23). This condition has to be introduced in order to make the morphology (apart from phenomena such as surface roughening and kinetical roughening) independent of the character of the mother phase from which the crystal grows. It makes the 'vacuum morphology' equal to 'mother phase morphology'. This is in most cases a good approximation.

The following justifications for the proportionality hypothesis can be given.

(i) Let us first assume that the bond $\phi_i^{\text{ss}}$ corresponds to a simple covalent bond between atoms. From a quantum mechanical point of view it can be said that the larger the overlap between the wavefunctions forming the bond, the stronger the bond. Let us assume that in the solid phase bonds as, bs, cs, etc, occur between atoms A, B, C, etc, and atoms S and that at the surface these bonds are broken and replaced by bonds af, bf, cf, etc, between the A, B, C atoms and an atom F from the fluid phase. Also assume that due to differences in overlap between A, B, C, etc, and S

$$|\phi_{\text{as}}^{\text{ss}}| > |\phi_{\text{bs}}^{\text{ss}}| > |\phi_{\text{cs}}^{\text{ss}}| > \ldots \, . \tag{24}$$

We may then suppose that the same differences occur in the overlap of the A, B, C, etc, atoms with atoms F, leading to

$$|\phi_{\text{af}}^{\text{sf}}| > |\phi_{\text{bf}}^{\text{sf}}| > |\phi_{\text{cf}}^{\text{sf}}| > \ldots \tag{25}$$

and that as an approximation the proportionality condition holds:

$$\phi_{\text{as}}^{\text{ss}} : \phi_{\text{bs}}^{\text{ss}} : \phi_{\text{cs}}^{\text{ss}} : \ldots = \phi_{\text{af}}^{\text{sf}} : \phi_{\text{bf}}^{\text{sf}} : \phi_{\text{cf}}^{\text{sf}} : \ldots \, . \tag{26}$$

The strict proportionality between the $\phi'$ equal to $\phi^{\text{sf}} - \frac{1}{2}(\phi^{\text{ss}} + \phi^{\text{ff}})$ and $\phi^{\text{ss}}$

$$\phi_{\text{as}} : \phi_{\text{bs}} : \phi_{\text{cs}} = \phi_{\text{af}}^{\text{sf}} : \phi_{\text{bf}}^{\text{sf}} : \phi_{\text{cf}}^{\text{sf}} : \ldots \tag{27}$$

implies that equation (26) also holds for the bond energies $\phi_i^{ff}$ corresponding to the average bonds between fluid cells. As mentioned above, we repeat that it is an inherent weakness of the application of Ising cell models to the interface crystal–mother phase that fluid cells and their contents are not well defined nor are the ff bonds between fluid cells. So a proportionality according to equation (26) as an approximation, which can be justified from the parallel relation between inequalities (27) and (28), cannot be justified for the ff bonds. Therefore the proportionality relation equation (26) strictly speaking may only hold if the $\phi^{ff}$ interaction energies can be neglected as compared to the $\phi^{ss}$ and $\phi^{sf}$ interaction energies.

(ii) In cases where as, bs, cs, etc, bonds correspond to positive–negative ion interactions a similar argument applies. Assuming that due to higher charges and/or higher surface charges of ions (i.e. the smaller the ionic radii the higher the surface charge) inequality (27) holds, it is then reasonable to assume that inequality (25) also holds after replacing the 'solid ion' S by the 'fluid ion' F. Then the same arguments can be applied as above.

Another way to justify the proportionality hypothesis for ionic interaction in reference to a vacuum is the following. Let us assume that all Coulomb interactions at the surface crystal–mother phase have to be divided by the relative dielectric constant of the mother phase. It then follows that, notwithstanding the interaction energies may be reduced, the proportionality is maintained within an order of magnitude. The proportionality hypothesis, equation (23), holds approximately if the density lowering of the mother phase adjacent to surfaces, to be discussed below, at surfaces for different F faces are more or less the same. As mentioned above, if a special interaction occurs between the mother phase and a particular surface, the proportionality breaks down for this particular surface.

(iii) In case of crystals of organic molecules like naphthalene and paraffin crystals the proportionality hypothesis can be better justified. Crystals of organic molecules are close-packed structures with short-range van der Waals interaction between the molecules and, due to the special shapes of the cells of the Ising lattice (each cell contains one molecule), it can be said that the larger the surface area shared by two adjacent cells, the more atom–atom pair potentials they share, and the larger the overall bond energies $\phi_i^{ss}$. If solid cells are replaced by fluid cells the same argument holds for the interactions of the solid and the fluid cells. It also holds for the fluid cells mutually which have the same shape as the solid cells. The cells are imposed on the mother phase in an artificial way, but the argument given above holds for the average contents of the fluid cells as well. So for all ss, ff and sf bonds the proportionality hypothesis (equation (23)) may be considered to be a good approximation.

### *3.5.3 Equivalent wetting condition*

Introducing equation (26) only gives the proportionality of the bond energies and not their absolute values. Traditionally the so-called equivalent wetting condition is introduced for this case (see Bennema and van der Eerden (1987)). Assuming that the sf interaction between solute and solvent molecules is the same as the sf interaction between solid molecules (cells, points) at the surface and the solution and that the solution cells at the surface are exactly the same as in the bulk, the following relation holds:

$$N \sum_{ij} \phi_{ij} = \Delta H^{\text{diss}} \tag{28}$$

where $N$ is Avogadro's number and $\Delta H^{\text{diss}}$ is the molar enthalpy of dissolution. The summation is over all bonds of a point at site $i$ (where we assume in order to simplify equation (28) that only one type of growth unit occurs). (For a more elaborate treatment we refer to Bennema and van der Eerden (1987).) Combining the proportionality condition and the equivalent wetting condition leads to absolute values of all $\phi_i$ bond energies.

*3.5.4 Failure of the equivalent wetting condition, scaling using $\theta^{\text{c}}$*

The application of surface Ising models suffer from the fact that it is difficult to obtain the bond energies of the bonds occurring on the surface. There is a great deal of evidence that the equivalent wetting condition does not work (This is true especially for growth from solvents and fluxes.) In most cases the values of $\phi_i/kT$ calculated from the proportionality and equivalent wetting conditions give values that are much too low and almost all faces would grow as rough faces without any crystallographic orientation above their roughening temperature. An explanation was given by Groot (1988) and Groot *et al* (1986) on bases of a hard sphere model for the fluid phase against a hard flat wall. It was shown that a (slight) reduction of the density of the spheres occurs against the flat wall due to ordering. Translating this into Ising language, this means that the interaction between the solid molecules at the surface with the fluid molecules is somewhat less strong than according to the equivalent wetting condition. This makes $\phi_{ij}^{\text{sf}}$ less negative and makes $\phi_{ij}$ given by

$$\phi_{ij} = \phi_{ij}^{\text{sf}} - \tfrac{1}{2}(\phi_{ij}^{\text{ss}} + \phi_{ij}^{\text{ff}})$$

larger. This is because less is subtracted from the positive number $-\frac{1}{2}(\phi_{ij}^{\text{ss}} + \phi_{ij}^{\text{ff}})$. Since $\phi_i$ is the result of small differences between large numbers, a density change at the surface of a few per cent may give a reduction of $\phi_{ij}^{\text{sf}}$ of a few per cent, and this may give an increase in $\phi_{ij}$ of 500 to 1000 per cent. An example will be given below for yttrium iron garnet growing from a lead oxide flux.

Keeping the proportionality condition and abandoning the equivalent wetting conditions leaves us with the unsolved problem of how to calculate the $\phi_i$ (or $\phi_{ij}$) values. Now there seems to be only one possibility to calculate the $\phi_i'$. Assume that from the connected nets $(h_1k_1l_1)$, $(h_2k_2l_2)$,..., $(h_ik_il_i)$,... of a crystal graph the corresponding Ising temperatures are calculated: $\theta_1, \theta_2, \ldots, \theta_i, \ldots$, respectively. If it can be observed that a face $(h_jk_jl_j)$ is close to its roughening temperature, then for this face the following relation holds:

$$\theta_{(h_jk_jl_j)} \cong \theta^{c}{}_{(h_jk_jl_j)}. \tag{29}$$

It is also possible that from a calculated decreasing series of $\theta^{c}$

$$\theta^{c}{}_{(h_1k_1l_1)} > \theta^{c}{}_{(h_2k_2l_2)} > \ldots > \theta^{c}{}_{(h_ik_il_i)} > \theta^{c}{}_{(h_jk_jl_j)} > \ldots \tag{30}$$

the faces of the form $\{h_ik_il_i\}$ are observed on a crystal growth form or a sphere of a single crystal growing from a slightly supersaturated solution, but the faces of the next weakest from $\{h_ik_il_i\}$ are not. This implies that these faces are growing above their roughening temperature. Then the following relation holds for the actual dimensionless temperature $\theta$:

$$\theta^{c}{}_{(h_jk_jl_j)} < \theta < \theta^{c}{}_{(h_ik_il_i)}. \tag{31}$$

If $\theta$ is known or can be estimated the value $\phi_{str}$ can be calculated from the temperature of the mother phase from which the crystal grows. From the value of $\phi_{str}$ all other bond energies of the crystal graph can be calculated, using the proportionality relation equation (23). These considerations only hold if the actual dimensionless temperatures ($\theta = 2kT/\phi_{str}$) are the same for all faces. This means that all bond energies $\phi_{str}$, etc, are the same for all faces. It will be shown below that for the case of a special garnet growing from a special flux this condition is not fulfilled for one type of face. The technique of calculating bond energies directly or indirectly from roughening temperatures has been applied or will be applied in Smet *et al* (1989), Weijnen *et al* (1987), Bolt and Bennema (1990), Bennema *et al* (1990a,b), in garnet (see below).

## 4 MORPHOLOGY OF ORGANIC AND SOME INORGANIC CRYSTALS

In the following a few examples of recent research will be discussed briefly.

### 4.1 Organic crystals

#### *4.1.1 Cyclohexane*

Cyclohexane, $C_6H_6$, forms plastic crystals having a cubic close-packed structure in which the $C_6H_6$ molecules rotate almost freely and switch continuously from a chair to a bed shape. They occupy on the average 24 positions but from our point of view the structure can be considered as a cubic close-packed structure, where the cyclohexane molecules occupy an average position corresponding to a sphere (Sangster *et al* 1988). Reducing the spheres to centres of gravity and taking the first-nearest-neighbour bonds, the unit cell of the crystal graph corresponding to the FCC cubic close-packed structure is presented in figure 4($a$) and the two resulting connected nets (111) and (100) in figures 4($b$) and 4($c$), respectively.

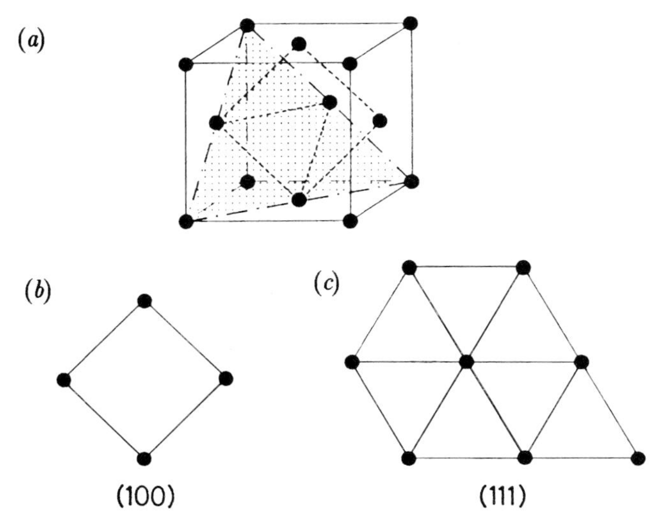


**Figure 4** (*a*) Crystal graph of FCC structure with first-nearest-neighbour bonds. (*b*) Connected net (111). (*c*) Connected net (100). (*a*)–(*c*) also represent crystal graphs and connected nets of the FCC cycloxhexane structure, after reducing the cyclohexane molecules to centres of gravity.

Since the mobility of the cyclohexane molecules within the lattice close to the melting point of 6 °C is very high, the enthalpy of melting $\Delta H^{\mathrm{m}}$ is only 2.68 kJ mol$^{-1}$. (This low value results from the small entropy difference between the molten and solid states.) Whether the equivalent wetting condition (equation (28)) is valid or not, we may assume that, as can indeed be seen from figure 5, that cyclohexane crystals grow with rough faces from their melt even at very low undercoolings. It is very interesting to see that the bond structure of the crystal graph still shows up in the anisotropy of the surface or growth rates. The resulting perturbations can be considered as the corners of a rounded-off octahedron. The three corners on top and at the bottom of the octahedron behave in different ways due to different undercoolings. It is interesting to see that dentritic-like growth forms develop induced by the anisotropy of the crystal graph. The treatment of the most fascinating subject of morphological instability and dentritic growth of unfaceted crystals growing above the roughening transition is beyond this review and we refer for this to the beautiful survey paper of Langer (1980). Anisotropy of growth rates occurring for thionaphthene crystals was studied by Elwenspoek and Boerhoff (1987) and Cherepanova *et al* (1988). It was shown that in an '$\boldsymbol{R}_{hkl}$ Wulff plot' no cusps occur as should be the case for rough faces (Bennema and van der Eerden 1987). Both cyclohexane and thionaphthene are examples of crystals growing above the roughening temperature even at the lowest driving forces for crystallisation.

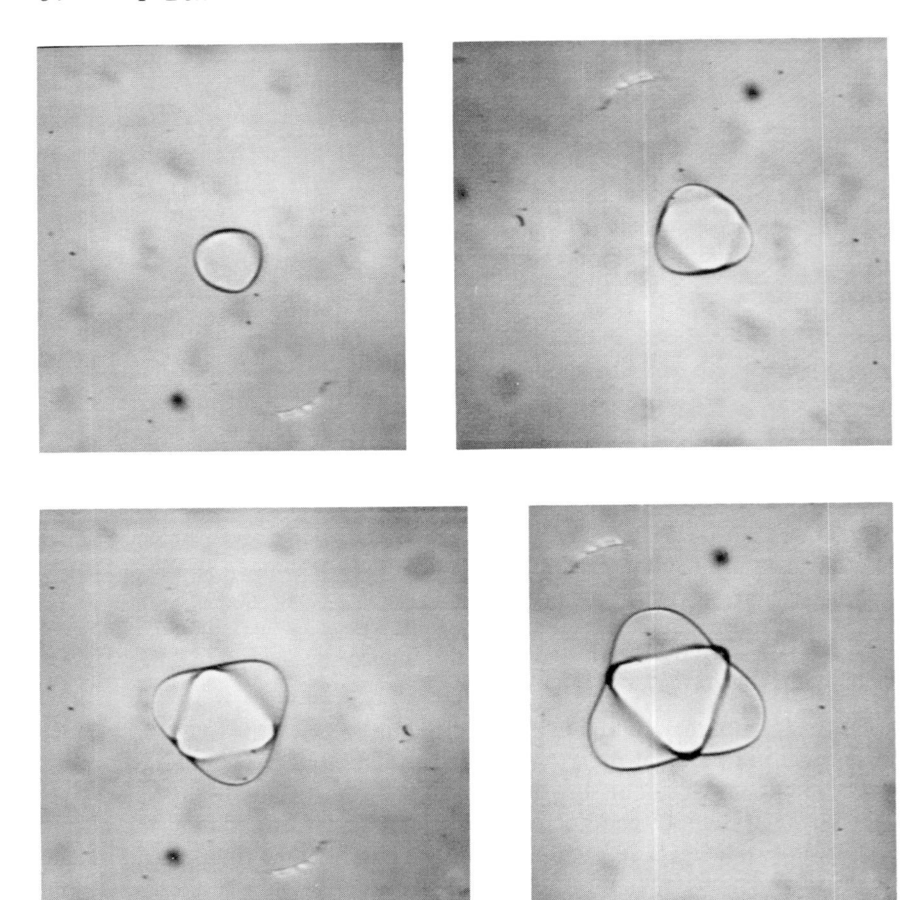

**Figure 5** Photographs of a growing cyclohexane crystal growing from a melt at very low undercooling. $\Delta T = 0.122\,^{\circ}\mathrm{C}$. Cyclohexane is growing above the roughening temperature with rounded faces. Due to the bond structure it grows in an octahedron-like shape. The six corners of the rounded-off octahedron are bulging out. Pictures are taken from a TV screen. Enlargement is about hundred times. Temperature about 6 °C. Thanks to A van der Heijden and R D Geertman for providing these pictures (work is in progress).

### *4.1.2 Naphthalene and biphenyl*

Extensive work has been done on naphthalene and biphenyl growing from different solvents, mixtures of solvents and the melt (Human *et al* 1981, Jetten *et al* 1984, Elwenspoek and van der Eerden 1987, Elwenspoek *et al* 1987). In figure 6(*a*) the crystal graph of naphthalene is presented. The bonds d and a correspond to the strong 'body–body' contacts and the bonds g and f to weak 'head–tail' contacts. From this crystal graph the five nets indicated in figures 6(*c*), (*d*), (*e*), (*f*), (*g*) are indicated. Naphthalene crystals will be dominated by strong (001) F faces of the form {001} consisting of two parallel planes (001) and (00$\bar{1}$), and also by four F faces of the form {110} forming a parallelogram and two F faces of the form {20$\bar{1}$}, which will truncate the parallelogram. The other faces of the forms {100} and the form {11$\bar{1}$} have even weaker nets than the nets of {110} and {20$\bar{1}$} and have less chance of appearing on the crystal.

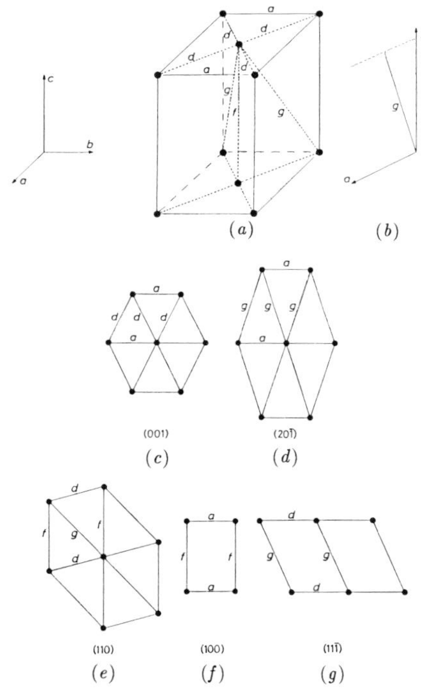


**Figure 6** (*a*) Crystal graph of naphthalene structure with two 'lateral' bonds d and a, and weak 'vertical' bonds f and g. (*b*) Crystal graph of (*a*) seen from a side in the direction of the b axis. Due to the obtuse angle g corresponds to a relativity small distance. (*c*) Hexagonal connected net of (001). (*d*) Hexagonal connected net of (110). (*e*) Hexagonal connected net of (20$\bar{1}$). (*f*) Rectangular connected net of (100). (*g*) Rectangular connected net of (11$\bar{1}$). It is interesting to compare (*a*)–(*g*) with the figures of the naphthalene structure given by Hartman (1973).

Looking at the experimental growth forms of figure 7(*a*) of a naphthalene crystal growing from the organic solvent toluene, it can be seen that a crystal is indeed limited by the strongly dominant {001} faces and the four faces of the form {110}, but also by the faces of the form {20$\bar{1}$}

(figures 7(*b*), (*c*)). The weak faces of the form {100} and {11$\bar{1}$} do not occur and it may be assumed that they grow above the roughening temperature (Human *et al* 1981, Jetten *et al* 1984, Elwenspoek and van der Eerden 1987, Elwenspoek *et al* 1987). It is most interesting that if the supersaturation increases, first the weakest face (having the lowest $\theta^c$ or $E^{slice}$ and average edge (free) energy) roughens up kinetically and then if the supersaturation increases further, the {110} faces roughen up. If the supersaturation decreases the reverse happens at exactly the same supersaturations. This holds for the solvent toluene, where obviously due to the similarity of solvent molecules and solid molecules a good match occurs, giving rise to strongly negative $\phi^{sf}$ bonds and hence low $\phi$ bonds (equation (1)). If naphthalene is growing from a solvent which is not very similar to the molecules of the crystal, like hexane, $\phi^{sf}$ bonds are much less negative, the $\phi$ bonds are larger and indeed, as has been shown by Elwenspoek *et al* (1987), naphthalene crystals remain faceted; {110} and {20$\bar{1}$} faces never roughen up.

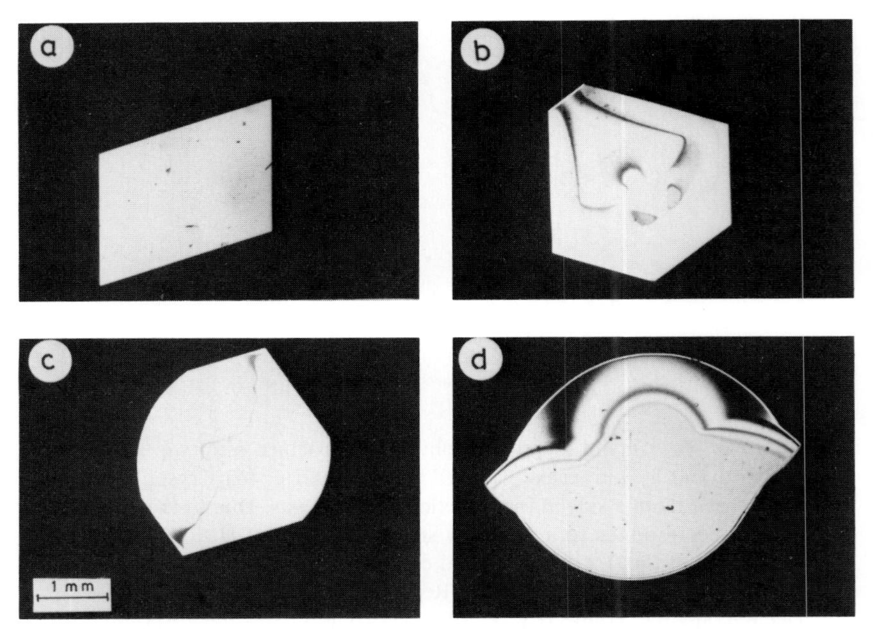


**Figure 7** (*a*) A crystal of naphthalene growing from toluene solution is presented. The crystal is lying on one (001) plane. The four {110} faces can clearly been seen. In (*a*) the relative supersaturation $\sigma = 0.88\%$, In (*b*) the faces {201} become visible because $\sigma$ is lower ($\sigma = 0.32\%$). For (*c*) $\sigma$ increases ($\sigma = 1.14\%$) and the {20$\bar{1}$} faces become rounded off, obviously due to kinetic roughening. (*d*) If the supersaturation increases further ($\sigma = 1.47\%$) the {110} faces also become rounded off.

Most interesting is the observation of Elwenspoek (1988) that if naphthalene is growing from a mixture of toluene and hexane the edge free energy of the side faces increases with an increase in the concentration of hexane. This was explained by Elwenspoek by assuming that a kind of ordered toluene layer occurring on the naphthalene surface, which may be liquid crystalline like, is perturbed by hexane molecules. This gives a less good fit and consequently a less negative $\phi^{sf}$ and larger edge (free) energy.

### *4.1.3 Paraffin*

In figure 8(*a*) the crystal graph of orthorhombic paraffin with an odd number of C atoms per molecule is presented. This orthorhombic paraffin structure can be considered as a close-packed structure of 'cigars'. Each cigar is first of all surrounded by six neighbouring cigars giving rise to six strong 'horizontal' bonds of two times a, d and d′ bonds. To this a weaker horizontal bond b is added.

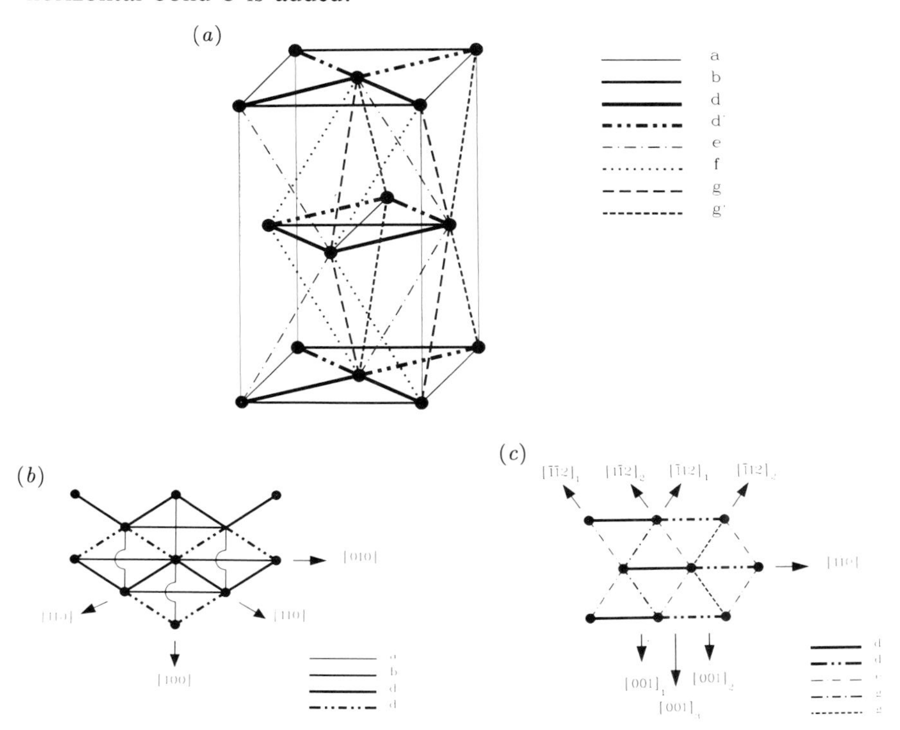


**Figure 8** (*a*) Crystal graph of orthorhombic paraffin. Bonds are indicated. The crystal graph is a pseudo FCC orthorhombic crystal graph. The bonds b, d, d′ are the strong bonds between paraffin molecules laying parallel to each other. The bonds e, g, and g′ are the most important head tail bonds. (*b*) The dominant very strong connected net (001). (*c*) One of the two alternative (110) nets. In (*b*) and (*c*) PBCs are indicated.

In addition it can be seen from figure 8(*a*) that one point (molecule) is again surrounded by three molecules in an adjacent plane, giving rise to three much weaker head–tail bonds e, g and g′. To this the even weaker bond f is added. Calculations on bond energies have been carried out by K Lewtas from Paramin (Exxon) laboratories Abingdon (England) as a part of a theoretical and experimental study of paraffin (Bennema *et al* 1990b). It follows from these calculations that the bond energies corresponding to the bonds a, d, d′ (and b) increase linearly with the chain length of the paraffin molecules. Depending on the length of the paraffin molecules, the lateral bonds a, d and d′ are about thirty to forty times larger than the 'vertical' bonds e, g and g′. Notwithstanding the weakness of these bonds, they are essential for the crystal graph and for the structure of the connected nets of the fast growing side faces.

It is easy to see how connected nets can be derived from the crystal graph (see figures 8(*a*)–(*c*)). The six other nets will be published in Bennema *et al* (1990b). A crystal growth form is made by using a computer based on the criteria $\boldsymbol{R} \sim E^{\mathrm{att}}_{hkl}$ or $(\theta^{\mathrm{c}}_{hkl})^{-1}$.

Typical figures like figures 9(*a*)(*b*) show that paraffin crystals will be limited by flat (001) faces, four faces of the form {110} and small faces of the form {010}. These are indeed the real shapes of real paraffin crystals. In a similar way as for naphthalene (figure 7) quite often the {010} faces are absent or roughened. It can be shown theoretically that if the supersaturation increases the faces of the form {110} will also roughen up and this is in principle also observed (Bennema *et al* 1990b). In figures 9(*c*), (*d*) the very recently observed roughening transition on {110} faces of paraffin with 23 C atoms per molecule, growing from a hexane solution, is demonstrated. In passing we mention that also fat crystals show a great similarity to the paraffin crystals (Bennema *et al* 1990a).

## 4.2 Inorganic crystals, application to garnets

### *4.2.1 Introduction*

We will only briefly mention the work on garnet and refer the reader to other papers and forthcoming papers (Bennema and van der Eerden 1987, Rijpkema *et al* 1982, Bennema *et al* 1983, Cherepanova *et al* 1989, 1990).

Using the PBC (periodic bond chain) theory of Hartman and Perdok, ten connected nets in the complex crystal graph of garnet have been identified (Bennema *et al* 1983). These nets have been labelled as $\{211\}_{1,2}$, $\{220\}_{1,2}$, {400}, $\{321\}_{1,2}$, {210}, $\{332\}_{1,2}$. The faces {211}, {110}, {321} and {332} are parallel to two types of connected nets, which are labelled with 1 and 2. Using a statistical mechanical so-called star–tetrahedron transformation (van der Eerden and Bennema 1982), where the real garnet structure with oxygen is replaced by a garnet structure without oxygen, the connected nets can be simplified (Bennema and van der Eerden 1987, Onsager 1944, Bennema *et al* 1983, Cherepanova *et al* 1989, 1990). The complex nets are transformed into rectangular nets by—wherever necessary—splitting nodes in two nodes, connected by a bond with infinite energy and by allowing bonds with energy zero, as demonstrated in figure 3. All predicted forms ($hkl$) were observed on spheres during sphere experiments by Tolksdorf and Bartels (1981).

It was demonstrated for the special case of YIG growing from a PbO-rich flux that the equivalent wetting condition does not hold (Tolksdorf and Bartels 1981).

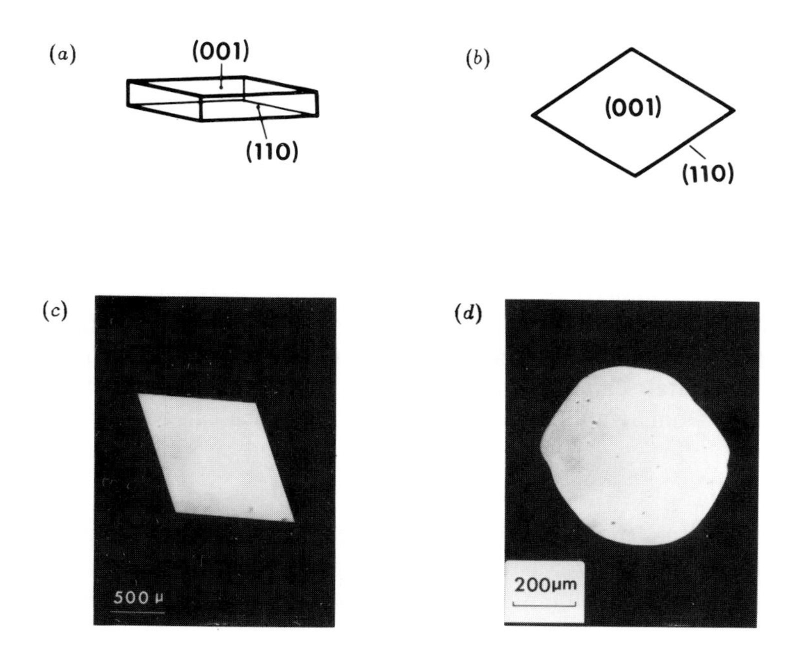


**Figure 9** (*a*) A typical paraffin crystal limited by the large faces of the form {001}, four faces of the form {110} and small faces of the form {010}. Here the criterion $R_{hkl} \sim (\theta^{c}_{hkl})^{-1}$ applies. (*b*) Same crystal seen from above. (*c*) A paraffin crystal with 23 C atoms per molecule growing with straight {110} faces below the measured roughening temperature of 11 ± 2°C of {110} faces. (*d*) Same crystal as in (*c*), but now above the roughening temperature of {110} faces. (Courtesy of Liu Xiang Yang.)

The proportionality relation, however, seems to hold except for the faces of the form {100}, due to a special interaction between PbO flux and this orientation. The example of the {100} faces of YIG growing from a PbO flux demonstrates that small subtle effects due to special interactions of the mother phase with that particular orientation may produce a drastic effect.

They may partially destroy a hierarchy of MI of faces predicted by the Hartman–Perdok theory or the hierarchy of MI predicted from a calculated sequence of $\theta^c$ values.

In order to discuss briefly the development of studies on the morphology of quasi- and modulated crystals we start by summarising classical crystallographical morphology in some more detail in the next section.

## 5 MORPHOLOGY OF MODULATED AND QUASI-CRYSTALS

The science of crystallography started with the discovery that crystallographic faces obey the law of rational indices (Burke 1966, Schneer 1977). This law implies that each crystal face is characterised by a set of integral indices $(hkl)$ which are components of a vector reciprocal $\boldsymbol{k}$ perpendicular to the face:

$$\boldsymbol{k} = h\boldsymbol{a}^* + k\boldsymbol{b}^* + l\boldsymbol{c}^* \tag{32}$$

where $\boldsymbol{a}^*, \boldsymbol{b}^*, \boldsymbol{c}^*$ are the basis vectors of the reciprocal lattice and $(hkl)$ are the Miller indices.

The set of symmetry-equivalent faces to which $(hkl)$ belongs is called a crystal form and is denoted by $\{hkl\}$. The relative morphological importance (MI) of the crystallographic forms $\{hkl\}$ on growth or equilibrium forms of crystals can be derived from the law of Bravais (1850), Friedel (1911) and Donnay and Harker (1937) (BFDH), which can be formulated as follows:

$$|k_1| > |k_2| \rightarrow \mathrm{MI}_1 < \mathrm{MI}_2 \tag{33}$$

that is, the smaller the norm $|k|$, the higher the MI. The corresponding interplanar distance $d_{hkl}$ is proportional to the inverse of $|k|$. Allowed values of $k$ obey the same space-group selection rules as the corresponding Fourier wave vectors of the crystal structure. The BFDH law has been quite successful in predicting the morphology of numerous crystals (Phillips 1963, Hartman 1973, 1987).

More recently, the law of rational indices and to a certain extent also the BFDH law has been generalised to describe crystal faces occurring on incommensurate displacively modulated crystals, such as $Rb_2ZnBr_4$ (Janner *et al* 1980), $(CH_3)_4N_2ZnCl_4$ (Dam and Janner 1985, see also Dam and Bennema 1983) and the mineral calaverite ($AuTe_2$) (Dam *et al* 1985, Janner and Dam 1989). For these incommensurately modulated crystals, the vector $\boldsymbol{k}$ perpendicular to a crystal face does not belong to a three-dimensional reciprocal lattice, but to a so-called Fourier module instead, which consists of all integral linear combinations of the three basic vectors $\boldsymbol{a}^*, \boldsymbol{b}^*, \boldsymbol{c}^*$ as above and the modulation wave vectors $\boldsymbol{q}_1, \ldots, \boldsymbol{q}_d$. Accordingly, the corresponding crystal face can be labelled by $3 + d$ integral indices

$(hklm_1 \ldots m_d)$ such that

$$k = h\boldsymbol{a}^* + k\boldsymbol{b}^* + k\boldsymbol{c}^* + m_1\boldsymbol{q}_1 + \ldots + m_d\boldsymbol{q}_d. \tag{34}$$

The crystals mentioned above are one-dimensionally modulated, so that in those cases $d = 1$ and the modulation wave vector is indicated by $\boldsymbol{q}$. Incommensurability implies that in

$$\boldsymbol{q} = \alpha\boldsymbol{a}^* + \beta\boldsymbol{b}^* + \gamma\boldsymbol{c}^* \tag{35}$$

at least one of the coefficients $\alpha, \beta, \gamma$ is an irrational number. Note that allowed values of $\boldsymbol{k}$ again obey the same selection rules imposed by the symmetry of the crystal as do the corresponding Fourier wave vectors. This can be seen as an extension of the BFDH law. The morphological importance of a crystal face, however, seems to involve more properties than the (average) interplanar distance. This is not surprising if one considers in particular the new types of face occurring on modulated crystals; these are called satellite faces and are characterised by having some non-vanishing index $m$. Indeed, one then expects that the orientation and magnitude of the modulation wave also represent relevant parameters for the MI of these faces.

In any case, morphology reflects modulation properties and it has been possible, from the orientation of satellite crystal faces, to determine the modulation wave vectors $\boldsymbol{q}$ with high precision (Dam and Janner 1983, Dam *et al* 1985, Janner and Dam 1989). (See also the review by Janssen and Janner 1987.)

The selection rules following from a four-dimensional space group (the superspace group), leaving invariant the incommensurate crystals given above, could be shown to be compatible with the morphological data available. (See previous papers and the review by Janssen and Janner (1987).) In some cases, experimental information on the MI is even sufficient to fix the superspace group itself. One can therefore claim that, at the present level of knowledge, the superspace group approach is supported not only by x-ray structure determination but also by crystal morphology.

The geometrical morphological laws considered up to now represent a first approximation only. As shown in this paper it is possible to go much further by taking into account chemical bonds and their energy as done in the crystallographic morphological theory of Hartman and Perdok (1955a,b,c) and Hartman (1973, 1987), and to combine such an approach with a statistical mechanical theory, for example the recently developed theory of roughening transition and the theory of order–disorder phase transition for complex nets.

We repeat that the geometrical BFDH law (equation (33)) can be given a qualitative physical interpretation using the following argument: the smaller $|k|$, or the larger $d_{hkl}$, the higher the energy content of a slice (denoted as $E^{\text{slice}}$ in the Hartman–Perdok theory), the higher the roughening transition temperature, the lower the surface energy and the growth rate and, finally, the higher the MI. The BFDH law is now generalised to modulated crystals and, as will be shown below, to quasi-crystals as well.

In addition to flat satellite faces observed on modulated crystals, a spiral-like growth mechanism has also been found on those faces (Dam 1985). The theory of connected nets and roughening transitions has not yet been applied to modulated crystals, but it is currently being worked on (work in progress).

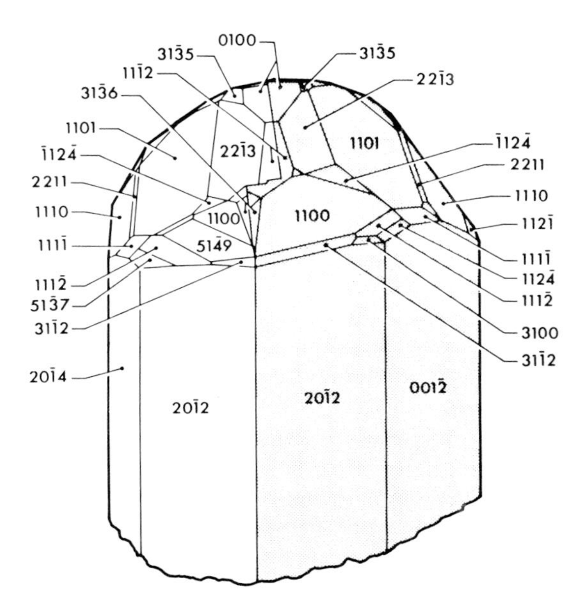


**Figure 10** Faces (*hklm*) on a twin twin of the mineral calaverite $AuT_2$ with some Ag.

In figure 10 we present a picture of a twin crystal of calavarite. The indices (*hklm*) correspond to faces. This is taken from a paper by Janner and Dam (1989). In the period from the end of the nineteenth century to 1931 mineralogical crystallographers became aware of the problem that from the incredibly rich morphology of calaverite consisting of 92 different crystal forms, only ten of the 92 forms could be described in a normal way by Miller indices {*hkl*}. (See the beautiful paper of Goldschmidt *et al* (1931).) Thanks to cooperation between three generations of scientists: Donnay about 85 years, Janner about 58 years and Dam 28 years, all the remaining crystal forms could be indexed with indices {*hklm*}. This was possible because van Tendeloo *et al* (1983, 1984) and later Schutte and de Boer (1988) showed that calaverite is modulated. The ***q*** vector determined from x-ray data (Schutte and de Boer 1988) and independently from morphological data agree quite well (Dam *et al* 1985, Janner and Dam 1989).

Because quasi-crystals belong to the family of incommensurate crystals

(Janssen 1988), it is expected that similar phenomena will be encountered. Indeed, it is then not surprising to see regular flat faces on quasi-crystals, as shown for example by Dubost *et al* (1968) and by Denoyer *et al* (1987).

In a recent paper by Janssen *et al* (1989) the following was shown.

(i) In a similar way as for modulated crystals, the law of rational indices can be generalised to quasi-crystals where forms of crystal faces labelled by six integral indices $\{hklmno\}$ can be identified.

(ii) A generalisation of the law of BFDH is in principle possible.

(iii) The concepts of PBC, crystal graph, connected net and flat face can, in principle at least, be generalised to quasi-crystals. As an illustration a two-dimensional Penrose-like net is given. We will discuss point (iii) in some more detail.

In figure 11 it is shown how a two-dimensional quasi-crystal can be partitioned into equal strips of quasi-PBCs. Contrary to the usual method of presentation we consider in this figure the edges of the two elementary cells (the thick and the thin lozenges) as bonds and the corners of these cells as 'atoms'. So the Penrose pattern is now considered as a two-connected net of a two-dimensional quasi-crystal.

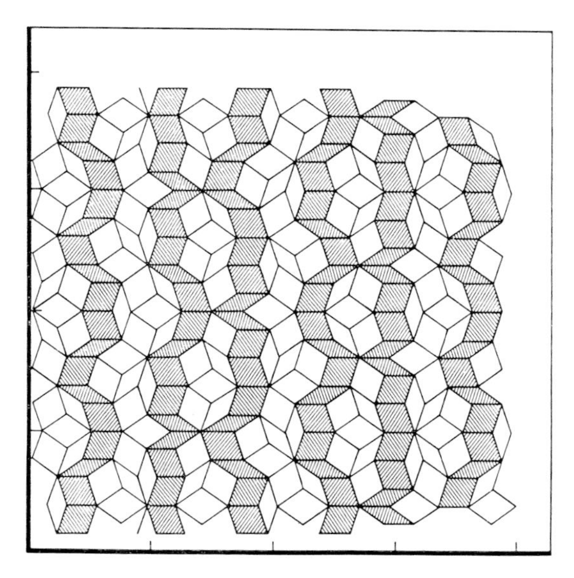

**Figure 11** Two-dimensional crystal graph of a Penrose net. The net can be unambiguously partitioned in statistically equal quasi periodic bond chains.

In addition, a three-dimensional quasi-crystal graph can be defined in an analogous way and it can be shown that such a quasi-crystal graph can be partitioned unambiguously in (statistically) equal parallel connected nets.

It is reasonable to assume that such connected nets also show a roughening or order–disorder phase transition and that quasi-crystals may grow with flat faces as indeed is observed by the authors mentioned above.

These faces have to be indexed with six indices (*hklmno*) because quasi-crystals can be considered as special three-dimensional cuts out of a six-dimensional 'Fourier structure'. As an example, we give a quasi-crystal of which the form {110000} consists of 30 faces (see figure 12). The point group is the non-classical point group 532, consisting of five fold axes, three fold axes and two fold axes. Such quasi-crystals are indeed observed (Dubost *et al* 1968, Denoyer *et al* 1987).

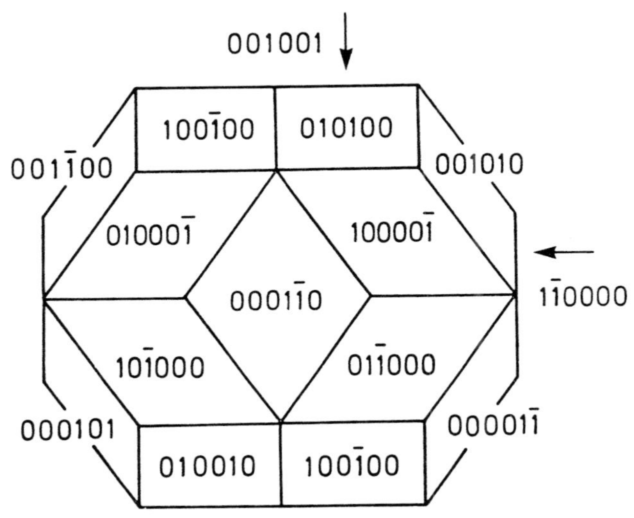


**Figure 12** A form {110000} consisting of thirty faces of a quasi-crystal with a point group symmetry 532.

## 6 CONCLUSION, SUMMARY, LIMITATIONS OF THE THEORY

In the introduction, two types of Ising interface models were discussed: (i) the SOS model giving rise to a very weak phase transition of infinite order, the so-called roughening transition, and (ii) the two-dimensional lattice gas or Ising model giving rise to a second-order order–disorder phase transition.

The results of a theory were discussed which led to the calculation of the dimensionless Ising temperature for the order–disorder phase transition in the two-dimensional lattice gas model. This theory is a kind of generalisation of the theory of Onsager. Since with the theory presented complex two-dimensional connected nets can also be tackled an integration of this statistical mechanical theory and the theory of Hartman and Perdok becomes possible. Calculated dimensionless Ising temperatures become relative measures for the morphological importance of crystal forms $\{hkl\}$.

Recent generalisations of the old law of rational indices which allows the indexing of faces with Miller indices $(hkl)$ for modulated and quasi-crystals were discussed and it was shown that forms to be indexed with four or more indices $\{hklm...\}$ have been observed recently on modulated and quasi-crystals, respectively. It was also shown that in principle for quasi-crystals the Hartman–Perdok theory could be generalised.

The integrated theory presented above does not give a description of the precise structure of the interface crystal–mother phase. It was mentioned that in the case of garnet (YIG) growing from a lead oxide flux a special interaction with the {100} orientation occurs, so that these faces grow as rough faces above the roughening temperature. Much more study has to be done on the problem of the subtle interaction between mother phase and particular $\{hkl\}$ orientations. The integrated Hartman–Perdok–statistical mechanical–thermodynamic theory, in which the proportionality condition (equation (26)) plays a key role, probably has to be used as a reference first-order theory for these studies.

The theory presented above can serve as a basis for the study of the difference between faces $\{hkl\}$ and their antipodes $\{\overline{hkl}\}$ in polar crystals such as benzophenone (Docherty *et al* 1990) and for example, alanine (Buthul 1987), to study the role of impurities and man-made (Berkovitch-Yellin 1985) and natural (Visser and Bennema 1983) tailor-made additives (van Rosmalen and Bennema 1990), and the specific role of the mother phase and surface reconstruction.

## ACKNOWLEDGMENTS

The author wishes to thank Drs G H Gilmer, H F Knops and J P van der Eerden for stimulating discussions on the background of Ising models, and Mrs D U Heeres, Mrs D D van der Wey and Mrs J Kogelman for typing the manuscript.

## REFERENCES

Bennema P 1984 *J. Crystal Growth* **69** 182–97

—— 1989 *Proc. 9th Symp. Ind. Crystallization* ed S J Jančič and E J de Jong p 267

—— 1990 *NATO Workshop, September 1989, Aqua Fredda di Marathea*, Italy to be published

Bennema P, Giess E A and Weidenborner J E 1983 *J. Crystal Growth* **62** 41

Bennema P and Gilmer G H 1973 *Crystal Growth: an Introduction* ed P Hartman (Amsterdam: North-Holland) pp 263–327

Bennema P and van der Eerden J P 1987 *Morphology of Crystals* part A, ed I Sunagawa (Dordrecht: Reidel) pp 1–75
Bennema P, Vogels L P J and de Jong S 1990a to be published
Bennema P, Liu Xiang Yang, Rijpkema J J M, Lewtas K, Tack R D and Roberts K J 1990b to be published
Berkovitch-Yellin Z 1985 *J. Am. Chem. Soc.* **107** 8239
Bolt R and Bennema P 1990 *J. Crystal Growth* **102** 329
Bravais A 1850 *J. Ecole Polytech. Paris* **19** 1
Burke J G 1966 *Origins of the Science of Crystals* (Berkeley: University of California Press)
Burton W K, Cabrera N and Frank F C 1951 *Phil. Trans. R. Soc.* A **243** 299–358
Buthul O 1987 *Master Thesis* Weizmann Institute, Rehovot, Israel
Jenn-Shing Chen, Nai-Ben Ming and Rosenberger F 1986 *J. Chem. Phys.* **84** 2365
Cheng V K, Tang E C M and Tang T B 1989 *J. Crystal Growth* **96** 293
Cherepanova T A, Bennema P, Yanson Yu A and Tsukamoto K 1990 to be published
Cherepanova T A, Didrihsons G T, Bennema P and Tsukamoto K 1989 *Oji Seminar, 1985* ed I Sungawa to be published
Cherepanova T A, Vaivods A V, Bennema P, Elwenspoek M and Boerhof J 1988 *J. Crystal Growth* **92** 671
Chernov A A and Lewis J 1967 *J. Phys. Chem. Solids* **23** 2185
Clarke S, Wilby M R, Vvedensky D D and Kawamura T 1989a *Phys. Rev.* B **40** 1369
—— 1989b *Appl. Phys. Lett.* **54** 2417
Dam B 1985 *Phys. Rev. Lett.* **55** 2806
Dam B and Bennema P 1987 *Acta Crystall.* B **43** 64
Dam B and Janner A 1983 *Z. Krist.* **165** 247
—— 1985 *Acta Crystall.* B **42** 69
Dam B, Janner A and Donnay J D H 1985 *Phys. Rev. Lett.* **55** 2301
de Haan S W H, Meeuwsen V J A, Veltman B P, Bennema P, van Leeuwen C and Gilmer G H 1974 *J. Crystal Growth* **24/25** 491
Denoyer F, Heger G, Lambert M, Lang J M and Saintfort P 1987 *J. Physique* **48** 1355
Docherty R, Roberts K J, Jetten L M A J and Bennema P 1990 to be published
Donnay J D H and Harker D 1937 *Am. Mineral.* **22** 446
Dubost B, Lang J M, Tanaka M, Saintfort P and Audier M 1968 *Nature* **234** 48
Elwenspoek M 1988 *Mol. Phys.* **64** 229
Elwenspoek M, Bennema P and van der Eerden J P 1987 *J. Crystal Growth* **83** 297
Elwenspoek M and Boerhof W 1987 *Phys. Rev.* B **36** 5326
Elwenspoek M and van der Eerden J P 1987 *J. Phys. A: Math. Gen.* **20** 669
Frank F C 1949 *Discuss. Faraday Soc.* **48** 67
Friedel G 1911 *Bull. Soc. Fr. Miner. Lecons de Cristallographice* (Paris: Hermann)
Gilmer G H 1976 *J. Crystal Growth* **36** 15
Gilmer G H and Bennema P 1972 *J. Appl. Phys.* **43** 1347
Goldschmidt V, Palache Ch and Peacock M 1931 *Neues Jahrb. Minderal. Geol. Paleont. Ref. Bul-Bd* **63** 1
Groot R D 1988 *Thesis* University of Nijmegen
Groot R D, Elwenspoek M and Bennema P 1986 *J. Crystal Growth* **79** 817

Groot R D and van der Eerden J P 1988a *J. Electroanal. Chem.* **247** 73
—— 1988b *Phys. Rev.* A **38** 296
Hartman P 1973 *Crystal Growth: an Introduction* ed P Hartman (Amsterdam: North-Holland) pp 367–402
—— 1987 *Morphology of Crystals* part A, ed I Sunagawa (Dordrecht: Reidel) pp 271–319
Hartman P and Bennema P 1980 *J. Crystal Growth* **49** 145
Hartman P and Perdok W G 1955a *Acta Crystall.* **8** 49
—— 1955b *Acta Crystall.* **8** 521
—— 1955c *Acta Crystall.* **8** 525
Hottenhuis M H J, Gardeniers J G E, Jetten L A M J and Bennema P 1988 *J. Crystal Growth* **92** 171
Human H J, van der Eerden J P, Jetten L A M J and Odekerke J G M 1981 *J. Crystal Growth* **51** 589
Jackson K A 1958 *Liquid Metals and Solidification* (Cleveland: ASM) p 174
Janner A and Dam B 1989 *Acta Crystall.* A **45** 115
Janner A, Rasing Th, Bennema P and van der Linden W H 1980 *Phys. Rev. Lett.* **45** 1700
Janssen T 1988 *Phys. Rep.* **168** 55–113
Janssen T and Janner A 1987 *Adv. Phys.* **36** 519
Janssen T, Janner A and Bennema P 1989 *Phil. Mag.* B **59** 233
Jetten L A M J, Human H J, Bennema P and van der Eerden J P 1984 *J. Crystal Growth* **68** 503
Kadanoff L P and Ceva H 1971 *Phys. Rev.* B **3** 3918
Kern R 1987 *Morphology of Crystals* part A, ed I Sunagawa (Dordrecht: Reidel) pp 78–206
Langer J L 1980 *Rev. Mod. Phys.* **52** 1
Leamy H J and Gilmer G H 1974 *J. Crystal Growth* **24/25** 766
Müller-Krumbhaar H 1978 *Current Topics in Material Science* ed E Kaldis and H J Scheel (Amsterdam: North-Holland) pp 1–46
Mutaftschief B 1958 *Adsorbtion et Croissance Cristalline* ed R Kern (Paris: CNRS) pp 231–53
Onsager L 1944 *Phys. Rev.* **65** 117
Phillips F C 1963 *An Introduction to Crystallography* (London: Longmans)
Povarenykh A S 1972 *Crystal Chemical Classification of Minerals* vol I (New York: Plenum)
Rijpkema J J M, Knops H J F, Bennema P and van der Eerden J P 1982 *J. Crystal Growth* **61** 295
Sangster J, Talley P K, Bale C W and Pelton A D 1988 *Can. J. Chem. Eng.* **66** 881
Schneer C J 1977 *Crystal Forms and Structure* (Dowden: Hutchinson and Ross)
Schutte W J and de Boer J L 1988 *Acta Crystall.* B **44** 486
Smet F M, Bennema P, van der Eerden J P and van Enckevort W J P 1989 *J. Crystal Growth* **97** 430
Strom C S 1980 *Z. Krist.* **153** 99
—— 1981 *Z. Krist.* **154** 31
—— 1985 *Z. Krist.* **172** 11
Swendsen R 1978 *Phys. Rev.* B **17** 3710
Temkin D E 1966 *Crystallization Processes* (New York: Consult. Bureau) p 15

Terpstra R A, Bennema P, Hartman P, Woensdregt C F, Perdok W G and Senechal M L 1986a *J. Crystal Growth* **78** 468
Terpstra R A, Rijpkema J J M and Bennema P 1986b *J. Crystal Growth* **76** 494
Tolksdorf W and Bartels T 1981 *J. Crystal Growth* **54** 417
van Beijeren H 1977 *Phys. Rev. Lett* **38** 993
van der Eerden J P 1979 *Thesis* University of Nijmegen
van der Eerden J P and Bennema P 1982 *J. Crystal Growth* **61** 45
van der Eerden J P, Bennema P and Cherepanova T A 1978 *Prog. Crystal Growth Charac.* **1** 219–54
van der Eerden J P and Knops H F J 1978 *Phys. Lett* **66A** 334
van der Eerden J P, van Leeuwen C, Bennema P, van der Kruk W L and Veltman B P Th 1977 *J. Appl. Phys.* **48** 2124
van de Leemput L E C, van Bentum P J M, Driessen F A J M, Gerritsen J W, van Kempen H, Schreurs L W M and Bennema P 1989 *J. Crystal Growth* **98** 551
van der Voort E 1990 *Thesis* University of Utrecht
van Dijk D J, van Leeuwen C and Bennema P 1974 *J. Crystal Growth* **23** 81
van Erk W 1979 *J. Crystal Growth* **46** 539
van Rosmalen G M and Bennema P 1990 *J. Crystal Growth* to be published
van Tendeloo G, Greoriades P and Amelinckx S 1983 *J. Sol. Chem.* **50** 321
—— 1984 *J. Sol. Chem.* **53** 281
Visser R A and Bennema P 1983 *Neth. Milk Dairy J.* **37** 109
Voronkov V V and Chernov A A 1966 *Proc. Int. Conf. on Crystal Growth (Boston) J. Phys. Chem. Solids* 593
Weeks J D and Gilmer G H 1979 *Adv. Chem. Phys.* **40** 157
Weijnen M P C, van Rosmalen G M, Bennema P and Rijpkema J J M 1987 *J. Crystal Growth* **82** 509
Woensdregt C F 1990 *Thesis* University of Utrecht
Liu Guang Zhao 1988 *J. Crystal Growth* **89** 478
Liu Guang Zhao, van der Eerden J P and Bennema P 1982 *J. Crystal Growth* **58** 152

# On Central-polar Crystals

F R N Nabarro

## 1 INTRODUCTION

The Abbé Haüy, whose analysis was based on the principle (Haüy 1784) 'ORGANIZATION. Les minéraux en sont absolument dépourvus', found problems when he came to study topaz. In the first edition (Haüy 1801) of his *Traité de Minéralogie* he already recognised that the pyroelectric properties of crystals from Brazil and from Saxony were different, and saw that this could be a structure-sensitive property:

> Mais cette propriété, déjà très-foible dans certains topazes du Brésil, pourroient bien n'être ici qu'une manifestation accidentelle qui ne tînt pas au fond de la substance.

The problem involved the centrosymmetry or otherwise of the ends of the prismatic crystals:

> L'analogie semble indiquer que dans les topazes électriques par la chaleur, les sommets, s'ils existoient tous les deux, devroient différer par leur configuration. Dans l'incertitude où je suis à cet égard, n'ayant observé jusqu'ici que des cristaux terminés d'un seul côté, j'ai supposé que tout étoit égal de part et d'autre. On pourra trouver, dans la suite, des topazes complètes, et il sera curieux d'examiner si la propriété de s'électriser par la chaleur y est jointe ... à des différences entre les formes des deux parties dans lesquelles résident les centres d'actions des électricités contraires. (vol 2, p 507, fn.)

The problem is stated clearly in the second edition (Haüy 1822):

> J'ai remarqué que ses deux extremités étaient l'une et l'autre à l'état résineux, tandis que la partie intermédiaire donnait des signes d'électricité vitrée. Ce fait a beaucoup de rapport avec celui que présentent certains aimans qui renferment une succession de pôles contraires. (vol 2, p 154).

Erman confirmed all of Haüy's observations (Erman 1832):

> Aber die Art dieser Vertheilung ist eine ganz eigenthümliche, von den bisher bekannten Analogien total abweichende. Der einer Thätigkeit, nämlich die $-E$, herrscht... mit der Axe: die andere $(+E)$ hat ihre Richtung senkrecht die Axe, und ihr Sitz ist überall an der perimetrischen Oberfläche aller Seitenflächen. ... Freilich entstände dann die Frage: Warum brasilianischer Topas so entschieden elektrisch erregbar ist durch Temperatur, während die Verfasser beim sibirischen kaum Spuren davon, und beim sächsischen durchaus keine wahrnehmen könnte.

Hankel, who initially included only one Brazilian topaz among his samples, claimed that Erman's observations were incorrect (Hankel 1840). However, after further extensive studies of pyroelectricity, he was able (Hankel 1842) to reconcile all the observations. If one takes the prismatic axis of the crystal to be the $z$ direction, the charge distribution on the surface of a complete symmetrical crystal which corresponds to the anomalous pyroelectricity of Haüy and Erman has the symmetry of $3z^2 - x^2 - y^2$. There is, however, a second distribution with the symmetry $x^2 - y^2$, and these are often superposed. Moreover, most crystals are not symmetrical, but have grown out from a solid surface; their predominant charge distribution has the symmetry $z$.

One year later, Riess and Rose proposed that pyroelectric crystals could be divided into two classes, terminal-polar, such as tourmaline, where the charge distribution has the symmetry $z$, and central-polar, such as prehnite and topaz, where the charge distribution has the symmetry $3z^2 - x^2 - y^2$ (Riess and Rose 1843). They showed that the two cleavage faces of a topaz split on its natural cleavage perpendicular to the $z$ axis showed the same sign of pyroelectricity. They also verified the existence of a charge distribution with symmetry $x^2 - y^2$. According to Beckenkamp, Hankel verified and extended the crucial cleavage experiment (Beckenkamp 1890):

> Zersprengen wir z. B. einen sächsischen Topas, bei welchem die an den beiden Enden der Hauptaxe befindlichen Krystallflächen {001} OP positive Elektricität zeigen, oder einen sibirischen Topas von Adun Tschilon bei Nertschinsk, bei welchem die Flächen des Brachydomas an den beiden Enden der Hauptaxen ebenfalls positive Polarität besitzen ... in seiner Mitte mit dem mit

{001} OP parallel Durchgange, so zeigen beide Durchgangsflächen negative Electricität. Wird dagegen bei den zuvor genannten Topasen ... in der Nähe des einen Endes der Hauptaxe nur eine dünne Platte abgesprengt, so ist der Durchgang an der dünnen Platte negativ; dagegen kann der Durchgang am grossen Stücke, falls die abgesprengte Platte nur dünn ist, noch positiv ... erscheinen; bei grösserer Dicke der abgesprengten Platte geht aber auch auf ihm die positive Spannung in die negative über.

Some crystallographers readily accepted the existence of the central-polar crystal classes. Thus Rammelsberg wrote (Rammelsberg 1852):

Zu den central-polarischen Krystallen gehören die des Prehnits und Topases. Der erstere hat zwei gegeneinander gekehrte electrische Axen in der Richtung der Krystallaxe *a*, so dass die beiden antilogen Polen längs den stumpfen Seitenkanten... liegen, die scharfen dagegen unelektrisch sind.

Hankel, who fully accepted Haüy's building-block model, would have none of this. He wrote (Hankel 1844) (italics in the original):

Was aber vor Allem gegen die Annahme der central-polarischen Krystalle spricht und sie *unmöglich* macht (sofern sie einfache Krystalle und nicht Zwillinge sind), ist Folgendes: Es möchte wohl als Grundgesetz in Betreff der elektrischen Krystalle gelten, dass *jedes Stück eines Krystalls dieselbe Anzahl und Lage der Pole zeigt, wie ein vollstandiger Krystall*; eben weil die Krystallbildung und krystallinische Structur nicht etwas bloss örtliches ist, *sondern durch die ganze Masse gleichmässig hindurchgeht.*

The apparent conflict between the basic theory of crystal structure and the experimental observations has disturbed crystallographers for a long time. Thus Groth wrote (Groth 1908):

Die nicht seltenen optischen Anomalien und die pyroëlektrischen Eigenschaften der Topaskrystalle lassen es als möglich erscheinen, dass dieselben nur pseudorhombisch sind, in Wahrheit aber eine niedere symmetrie besitzen.

Cady wrote (Cady 1946):

*Topaz* ... should possess neither piezo- nor pyroelectricity. Nevertheless, various authors have reported pyroelectric properties. ... The last statement suggests that if topaz is pyroelectric at all the effect is tertiary, dependent on the direction of the temperature gradient. ... One may also raise the question [W A Wooster] whether the irregular pyroelectric results... may not be the result of twinning.

Bragg and Claringbull avoid speculation, simply remarking (Bragg and

Claringbull 1965):

> The pyroelectricity found in some topaz is not explained by this non-polar structure.

## 2 THE CLASSES OF EXPLANATION

Since Haüy's model of the structure of a perfect crystal cannot explain the observations, the explanation must lie in deviations from this model, which may be classified as crystal defects. One defect present in all real crystals is the free surface, and it is clear that the surface can introduce central-polar effects. Point defects will also be present, and these may align along growth directions to produce central-polar effects. Finally, dislocations aligned roughly along the growth direction can produce similar effects. We discuss these possibilities in the following sections.

## 3 SURFACE ELECTRICAL DOUBLE LAYERS

Two classes of surface electrical double layers come into question. The first, as Kuhlmann-Wilsdorf has pointed out, should be present in any ionic crystal. The second is associated with the presence of charged point defects (vacant or interstitial ionic sites, impurity ions).

On the free surface of any crystal there is an inward or outward relaxation of the atomic layers from their regular periodic positions. Madelung pointed out in 1919 that in an ionic crystal the positive and negative ions would in general relax by different amounts, so that an electrical double layer would be produced. It is clear that, in a crystal which does not have cubic symmetry, these double layers will have a central-polar symmetry, and their strength will presumably change with change of temperature.

Madelung gave a formal calculation of the surface effect, which decays exponentially towards the interior of the crystal. However, the appropriate interionic potentials were not known at that time, and he did not attempt to obtain numerical results. When suitable potentials were available, the Bristol group (Lennard-Jones and Dent 1928) made quantitative estimates, but they constrained the {100} faces of crystals of the NaCl type to relax rigidly, without relative displacements of the $Na^+$ and $Cl^-$ sublattices. The first detailed calculations were those of Verwey and Asscher (1946). The relative displacements, which they assumed to be confined to the outermost layer of ions, were quite large, of the order of 0.2 Å. However, owing to an accidental cancellation, the surface drop in potential is rather small, of order 0.2 V, except in the case of LiF, where it reaches 0.95 V. These surface displacements are almost certainly confined to a few surface layers, and there is no reason to believe that the strength of the surface dipole layer will have a strong dependence on temperature. The mechanism predicts external electrical fields possibly of the observed order of magnitude, but confined to regions of atomic dimensions along the edges of the crystal.

A quite distinct mechanism is that of Frenkel (1946). Suppose the crystal contains two kinds of point defect, positive and negative ion vacancies. If these are uniformly distributed throughout the crystal, the free energies of formation of positive and negative ions will in general be different, and the crystal will have a non-zero charge density, leading to a very large electrostatic energy. Frenkel's solution was to show that the distributions of positive and negative ions are not uniform. They adjust themselves so that an electrical double layer is formed at the surface. The electrostatic potential in the interior of the crystal is no longer zero, and this potential alters the free energies of formation of positive and negative ion vacancies in opposite directions until they become equal. The equilibrium concentrations of the two kinds of vacancy in the interior are then equal, and charge neutrality is achieved except in the double layer.

This double layer is of the Debye–Hückel type, and its thickness is inversely proportional to the square root of the equilibrium concentration of either type of vacancy in the interior. This is very sensitive to changes in temperature, and, as Frenkel points out, 'can be of the same order of magnitude as the linear dimensions of the crystal.' However, central-polar pyroelectricity is normally observed in crystals such as topaz which are extremely good electrical insulators. The electrical double layer present in the crystal at a given temperature will therefore not usually be that which is in thermal equilibrium at that temperature, but that which has been frozen in at some higher temperature. The thickness of the double layer is proportional to the square root of the dielectric constant, which in an anisotropic crystal is a tensor, so that the double layer has central-polar symmetry. The thickness of the double layer will be altered if the crystal is doped with ions having a different charge from those of the matrix, so that the differing properties of specimens from different sources is readily explained. It is by no means clear that this mechanism can be excluded as a possible cause of the observed central-polar pyroelectricity.

## 4 ORDERED POINT DEFECTS

An alternative source of central-polar behaviour has been suggested by Frank. Suppose the structure of a crystal contains isolated pairs of ions of the same kind (not necessarily close pairs, but uniquely recognisable pairs), and that these pairs are aligned (closely or precisely) parallel to each other in a direction coinciding with or close to the direction of growth. For example, topaz contains (Bragg and Claringbull 1965) pairs of $Al^{3+}$ ions parallel to the *c* axis (which in this case is also the principal growth axis), separated by rather less than a third of the *c* lattice parameter. Suppose small quantities of a foreign ion of a different valency but of roughly the same ionic radius are incorporated in the crystal during its growth. In the bulk of a perfect crystal, the two ionic sites of a pair are energetically equivalent. In the growth front, this is no longer true. There will in general be a tendency for the foreign ion to be deposited preferentially in the leading (or the trailing) site of a pair. The resulting crystal is electrically central-polar. It may be chemically homogeneous, and, if it is, its external morphology should be regular.

## 5 DISLOCATIONS

It was Frank (see his review (Frank 1952)) who first showed that the growth of a pure crystal could be greatly accelerated if it contained dislocations whose lines and Burgers vectors were roughly parallel to the growth direction. Later, he considered a crystal growing in a fluid containing an impurity which could enter the crystal lattice and alter the lattice parameter. In an isolated system, the concentration of impurity in the surrounding fluid is likely to increase as the crystal grows, and the concentration of impurity in the crystal correspondingly increases, leading to a progressive alteration in the lattice parameters. This progressive change can only be accommodated by the incorporation of edge dislocations. A plausible configuration of a growth face and of the intersections of these edge dislocations with the face is shown in figure 1. The screw dislocations which facilitate the growth of this face produce displacements perpendicular to the face, which do not appear in the figure.

Experimentally, the existence of growth zones in mineral crystals has long been known. Thus Becke in 1890, studying etch pits on fluorite, found that

> Aus dem Verlauf der Sectorgrenzen ergibt sich klar, dass die 6 Theile, in welche den Aetzfiguren gemäss der Fluoritwürfel von Cumberland zerfällt, genau jenen Antheilen des Krystalls entsprechen, welche durch Stoffansatz auf den 6 Würfelflachen des Krystalls entstanden sind. ... Es ist nun so, als ob jeder Anwachskegel eine Art Erinnerung an die Richtung in der er gewachsen ist, bewahrte. ... Die Erscheinungen beim anomalen Flusspath von Cornwall sind nun von der Art, dass sie erklärt werden können durch die Annahme von linearen Störungen des Molekularbaues, welche parallel der Axe des Anwachskegels orientirt sind, und längs welcher das Aetzmittel rascher vordringen kann, als in anderen Richtungen. Diese linearen störungen, mann kann sie als Canäle vorstellen, werden von dem Aetzmittel dem Molekularbau entsprechend ausgeweitet; sie geben Anlass von Aetzfiguren.

The distinction between 'growth' and 'misfit' dislocations became clear only much later. McLaren *et al* (1971) made x-ray topographs of synthetic quartz crystals in which [0001] was the growth direction, and the (0001) faces were covered with growth hillocks called cobbles. At the summit of each cobble was a dislocation, which usually had the Burgers vector $\boldsymbol{a}+\boldsymbol{c}$ $[1\bar{2}13]$, with a substantial component in the growth direction. Their observations, in general agreement with the earlier observations of Lang and Miusov (1967), showed that the majority of the dislocations present had lines close to a cone with generators about 10° from the $c$ axis, and Burgers vectors $\boldsymbol{a}\langle 11\bar{2}0\rangle$, which correspond to the dislocations of figure 1.

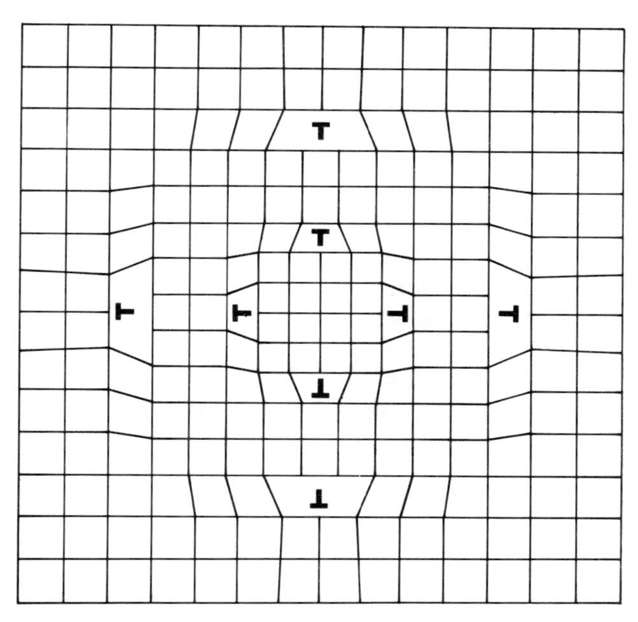

**Figure 1** Illustrating the incorporation of edge dislocations with their lines diverging slightly from the growth direction in a crystal growing in a medium in which the concentration of a large impurity ion increases with time.

The inserted half plane of an edge dislocation marks a polar direction. If the dislocations of figure 1 are not normal to the growth face, as is implied by that figure, the set of dislocations shown form a central-polar axis along the growth direction, and their presence will induce central-polar pyroelectricity. This mechanism depends on the incorporation into the crystal of a concentration of impurity ions which steadily increases as growth proceeds, as is likely to occur if the crystal grows in a closed cavity.

The mechanism depending on the ordering of matrix and impurity ion pairs operates ideally when the concentration of impurity ions incorporated is constant during growth. If the concentration is constant, the morphology of the crystal is not disturbed, but if the concentration varies the external form of the crystal will be distorted. Some evidence that the latter mechanism may be acting, at least in some cases, comes from the fact that the minerals in which central-polar pyroelectricity has most frequently been studied are topaz and prehnite. The latter has normally a distorted growth form. Riess and Rose say (Riess and Rose 1843):

> Die grössern [Krystalle] erscheinen gewöhnlich als eine Zusammenhäufung vieler Krystalle, bei welchen die Axen... divigieren, was oft so stetig geschieht, dass die stumpfen Seitenkanten eine volkommenen Bogen beschreiben... Ausserdem findet sich der Prehnit auch kugelig und nierenförmig.

As this model predicts, the central-polar pyroelectric axes point towards these obtuse edges:

> die beiden antilogen Pole längs den stumpfen Seitenkanten...oder deren Abstumpfungsflächen liegen, die scharfen dagegen unelectrisch sind (Rammelsberg 1852).

## REFERENCES

Becke F 1890 *Tschermaks mineralogische u. petrographische Mittheilungen* **11** 349
Beckenkamp J 1890 *Z. Kryst.* **17** 321
Bragg W L and Claringbull G F 1965 *Crystal Structures of Minerals* (London: Bell) pp 199–200
Cady W G 1946 *Piezoelectricity* (New York: McGraw-Hill) p 708
Erman P 1832 *Ann. Phys. u. Chem.* **25** 607
Frank F C private communication
—— 1952 *Adv. Phys.* **1** 91
Frenkel J 1946 *Kinetic Theory of Liquids* (Oxford: Clarendon)
Groth P 1908 *Chemische Kyrstallographie* vol 2 (Leipzig: Engelmann)
Hankel W 1840 *Ann. Phys. u. Chem.* **50** 237
—— 1842 *Ann. Phys. u Chem.* **56** 37
—— 1844 *Ann. Phys., Lpz.* **61** 281
Haüy, l'Abbé 1784 *Essai d'une Théorie sur la Structure des Crystaux* (Paris)
—— 1801 *Traité de Mineralogie* (Paris)
—— 1822 *Traité de Mineralogie* 2nd edn (Paris) (4 vols + atlas)
Kuhlmann-Wilsdorf D private communication
Lang A R and Miuscov V F 1967 *J. Appl. Phys.* **38** 2477
Lennard-Jones J E and Dent B M 1928 *Proc. R. Soc.* A **121** 247

Madelung E 1919 *Phys. Z.* **20** 494
McLaren A C, Osborne C F and Saunders L A 1971 *phys. status solidi* a **4** 235
Rammelsberg C F 1852 *Lehrbuch der Krystallkunde* (Berlin: Jeanrenaud)
Riess P and Rose G 1843 *Ann. Phys. u. Chem.* **59** 353
Verwey E J W with the collaboration of Asscher J E 1946, *Rec. Trav. Chim. Pays-Bas* **65** 521

# Evolving Crystal Forms: Frank's Characteristics Revisited

J W Cahn, J E Taylor and C A Handwerker

## 1 INTRODUCTION

Models of interface-controlled crystal growth or dissolution that assume that the velocity of growth $v$ is a function only of the surface normal $\boldsymbol{n}$ and system parameters, such as temperature and matrix concentration, have a long history (see for example the historical overview of dissolution kinetics of minerals by Litsakes and Ney (1985)). An early proponent of this model was Gross (1918), who explored some of its predictions. In 1958 Frank discovered a simple set of predictions for the evolution of crystal shapes under these circumstances, based on the method of characteristics. He studied step motion using an analogy between steps moving across a surface and the mathematics of traffic flow. In the limit of small steps, this method is valid, for then crystal growth can be dealt with as being continuous in time and a function of normal direction only. The theory was further explored by Chernov (1963, 1984) in the USSR. In 1972, Frank returned to this problem and treated it without using steps, assuming only that some $v(\boldsymbol{n})$ was given. He derived all his results from scratch, without tying them into a more general mathematical framework. We show here what this framework is.

The advantage of putting this problem into a larger context is (i) that the types of extension of the problem become clear, and (ii) the 'reason' that the results hold is seen not to be something special to this problem but to be part of a greater whole.

We will also address the issue of growth versus dissolution, both in formulation and in prediction of limiting end forms. It has been clear for some time that there is a well defined limiting form to a growing crystal independent of the form of the initial finite seed, but that the dissolution forms depend on starting shapes (Moore 1986). Well established rules, such as that orientations growing fast enough disappear from a finite form by growing out, appear to have no counterpart in dissolution. We will demonstrate that the shape of a dissolving finite form, when the velocity depends only on the normal direction, depends strongly on the initial form right up to the instant of complete dissolution. No matter how slowly a given orientation dissolves relative to other orientations, it is possible to find an initial shape such that that orientation will not only not disappear but will predominate in the last moments prior to dissolution.

Even though our understanding of the physical microscopic basis for deriving $v(\boldsymbol{n})$ is also considerable, especially for crystal growth from the vapour phase, we will in this paper primarily assume that $v(\boldsymbol{n})$ is given, and explore the techniques for calculation of the evolving shapes, concentrating on the method of characteristics. As we shall show, the method of characteristics is applicable because assuming that the velocity depends only on $\boldsymbol{n}$ leads to a particularly simple, although non-linear, first-order partial differential equation. If the velocity depends on time and position of the interface as well, as it might if temperature has an imposed time and space dependence, the formulation still leads to a first-order PDE and the method of characteristics. We will include examples where the velocity is an explicit function of $\boldsymbol{n}$, position and time.

While the most important applications of the theory are to crystal growth and dissolution, there are many examples of other physical processes in which a surface moves with a velocity that depends on its orientation. In 1973 Frank collaborated in applying these methods to ion-bombarded surface topographies (Barber *et al* 1973). This application has been extensively studied by the group in the Thin Film and Surface Research Centre in Salford, UK (Nobes *et al* 1987). We will give examples from our own work, of diffusion-induced grain boundary migration (DIGM), migration of thin liquid film (LFM) (Handwerker *et al* 1985) and discontinuous coarsening (Livingston and Cahn 1974).

There is a connection between crystal growth as a function of normal direction and the surface energy minimisation problem. As Chernov (1984) observed, the construction for the bounded steady state growth shape, using $v(\boldsymbol{n})$, is identical to the Wulff construction for the shape having the least surface energy for the volume it encloses, using the surface tension function $\gamma(\boldsymbol{n})$. There are connections of this growth problem to other physical problems as well. For example, sometimes the PDE for normal growth can be interpreted as an eikonal equation for the 'wave front' of some hyperbolic PDE. In particular, elastic and optical waves in crystals

(Kelvin 1904) as well as flame propagation (Sethian 1985) share a common mathematical base. These and other connections are explored in the last section of this paper.

### 1.1 Growth versus dissolution

There is a long and confused history of treating the mathematics of dissolution differently from growth, in each case given a $v(\boldsymbol{n})$. Our LFM work caused us to re-examine this subject, because here both the growing and dissolving crystals share an interface. If there was a difference in the theory for growing and dissolving crystals, which theory was applicable?

Frank dealt with the dissolution/growth problem in his 1972 paper without finding a satisfactory generalisation. In the end, the normal vector $\boldsymbol{n}$ was defined to be the outward normal from the crystal and $v(\boldsymbol{n})$ was defined to be positive for growth and negative for dissolution. But this sign convection is artificial: the mathematics is the same for the two processes. One decides (mathematically arbitrarily) which of the two regions into which the surface divides space is to be officially the 'crystal', and takes $\boldsymbol{n}$ to be the exterior unit normal to that region (i.e. to point into the other region). The velocity $v$ is then positive or negative according to whether the crystal grows or shrinks. (Of course for a given crystal, the dissolution velocity function may be different from the growth velocity function, not only in sign and magnitude but also in orientation dependence).

It should be noted that the problem of the dissolution of any particular crystal is identical to the problem of the growth of its complement. (There is a mathematical problem about whether to include the points of the bounding surface in the crystal or its complement that has no counterpart in physics. We will choose always to have the growing body contain its bounding surface. Then the only difference between growth of the crystal or its complement is whether the bounding surface of the 'crystal' is in the 'crystal' or its complement.) The attempt to distinguish between growth and dissolution when one is dealing with a surface between two crystals and where neither is convex (as often occurs for grain boundaries) is particularly unsuited to the conventions previously advocated. There is a basic symmetry in the problem. One could use the opposite orientation for the surface and $v'(\boldsymbol{n}) = -v(-\boldsymbol{n})$ as the velocity of growth; this does not change the problem, but it converts, for example, a dissolution problem with a negative $v$ into a problem with a positive $v$. A dissolution problem described in terms of a negative velocity $v$ is fully equivalent to a growth problem of another 'crystal' whose growth velocity is $v'$, and which occupies the space of the complement of the original crystal.

The geometric constructions of the shape evolution described below are accomplished more easily with positive $v$. Thus if it is possible to orient the surface so that $v$ is always positive, we will do so.

Finally, trying to do dissolution by using negative time is particularly hazardous. Whenever there are shocks (see below) present in growth with positive time, there is non-uniqueness for the crystal in negative time. Letting $v'$ be $-v$ and running time forward is not the same as recovering the previous growth history of a form containing shocks.

There is also the question of limiting shapes. Chernov (1984) observed that 'the steady state growth shape is described by the envelope of the family of planes given by the equation $\boldsymbol{n}\cdot\boldsymbol{x} = v(\boldsymbol{n})t$.' This is indeed the only steady state growth shape that is bounded at each time; this is a consequence of a more general theorem (whose proof we intend to provide in a later paper) that any bounded initial seed crystal which grows with positive normal velocity $v$ and which at time 0 contains $s_0 W_\infty$ and is contained in $s_1 W_\infty$ will, for all positive $t$, contain $(s_0 + t)W_\infty$ and be contained in $(s_1 + t)W_\infty$, where

$$W_\infty = \{\boldsymbol{x} : \boldsymbol{x}\cdot\boldsymbol{n} \leqslant v(\boldsymbol{n}) \qquad \text{for every unit vector } \boldsymbol{n}\}.$$

Note that $W_\infty$ is (by definition!) convex. There are, however, a number of infinite-extent surfaces (many of which are neither completely convex nor completely concave) that also retain their shape on growth.

Chernov (1984) asserts that 'from the geometric procedure of constructing Wulff's envelope ($W_\infty$ here) it follows that the higher the growth rate of a face (or of a rounded area), the smaller its size on the convex steady state growth shape must be.' This is not quite accurate; the areas depend on the values of $v$ on neighbouring directions. In fact, if one defines a new growth velocity function $v_b$ by $v_b(\boldsymbol{n}) = v(\boldsymbol{n}) + \boldsymbol{n}\cdot\boldsymbol{b}$, then the $W_\infty$ for the two growth velocities differ only by translation, and the limiting end shapes for growth of finite seeds are the same. Chernov goes on to say 'Analogously, faces with maximal growth rates should form the concave growth shape' (i.e. with the cavity decreasing in size). But there is no analogy! The concave growth shapes depend strongly on the shape of the initial seed (and the details of the dependence of $v$ on $\boldsymbol{n}$); there are exceptions to most general rules one might try to formulate, as we will demonstrate. In particular, there are a number of concave bounded-hole steady state growth shapes.

## 2 FORMULATION AND SOLUTION WITHIN THE FRAMEWORK OF PDE

Assume that $v$ is a function defined on unit vectors $\boldsymbol{n}$ giving the normal velocity of growth of a crystal. That is, if the crystal at a certain time $t$ and at a point $\boldsymbol{x} = (x_1, x_2, x_3)$ on its surface has exterior unit normal $\boldsymbol{n}$, then the instantaneous velocity of growth in direction $\boldsymbol{n}$ at $\boldsymbol{x}$ is $v(\boldsymbol{n})$ (or more generally, $v(\boldsymbol{x}, t, \boldsymbol{n})$). In order to convert this statement into a differential equation let $\tau(\boldsymbol{x})$ be the time that the surface of the crystal arrives at $\boldsymbol{x}$, so that $\tau(\boldsymbol{x}) = t_1$ is an implicit equation for the shape (the set of $\boldsymbol{x}$ in the surface) at time equal to $t_1$, $\tau(\boldsymbol{x}) = t_2$ is the equation of the shape at time equal to $t_2$, etc. (For the moment, let us restrict our attention to a region and time interval where $v$ is always positive or always negative so that $\tau$ is a uniquely defined function of $x$ there.) By regarding $\tau$ as a function of $\boldsymbol{x}$, a time sequence of crystal forms, such as is shown in figure 1, can be considered as a contour map of the function $\tau$. If $v$ is positive, then the gradient of $\tau$ at any point $\boldsymbol{x}$ where that gradient is defined is a vector whose direction is that of the oriented surface normal $\boldsymbol{n}$ (since the surfaces are level surfaces in $\tau$) and whose magnitude is inversely proportional to $v(\boldsymbol{n})$. Thus $\nabla\tau / |\nabla\tau| = \boldsymbol{n}$ and $v(\boldsymbol{n}) = 1/|\nabla\tau|$, i.e.

$$|\nabla\tau|\, v(\nabla\tau / |\nabla\tau|) = 1.$$

It is convenient, especially for the PDE which follows, to extend $v$ to a function on all vectors by defining $v(r\boldsymbol{n}) = rv(\boldsymbol{n})$ for all $r > 0$. When we do this, the above statement becomes

$$v(\nabla\tau) = 1.$$

(Note that $v$ has become dimensionless by this extension.) If $v$ is negative, then $\nabla\tau$ is in the direction of $-\boldsymbol{n}$ and its magnitude is inversely proportional to $-v(\boldsymbol{n})$, so that

$$-v(-\nabla\tau) = -|\nabla\tau|\, v(\boldsymbol{n}) = 1.$$

If we let $v'(\boldsymbol{n}) = -v(-\boldsymbol{n})$ when $v$ is negative, then the above statement is just

$$v'(\nabla\tau) = 1.$$

The mathematical equivalence of growth and dissolution problems should now be completely clear; in order to treat the negative-velocity case, one in fact switches to the equivalent positive-velocity problem.

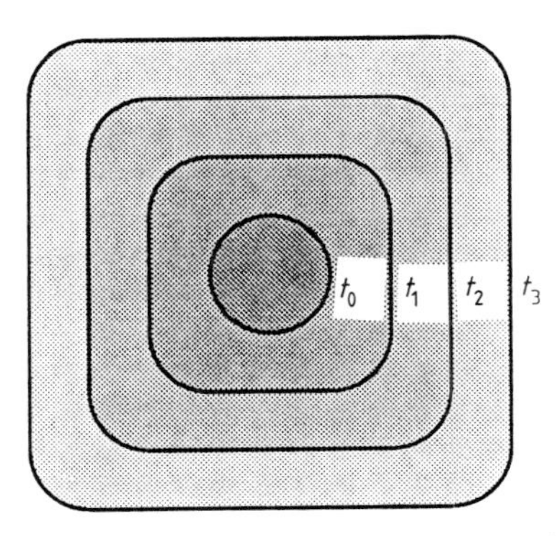


**Figure 1** A sequence of crystal forms can be considered a contour map in which each contour is the surface at a constant time. The gradient of the time when the surface reaches a point $\boldsymbol{x}$ is $\boldsymbol{n}/v(\boldsymbol{n})$.

Thus the statement that the normal velocity is given by $v(\boldsymbol{n})$ becomes equivalent when $v > 0$ to the statement

$$v(\nabla\tau) - 1 = 0$$

and when $v < 0$ to

$$v'(\nabla\tau) - 1 = 0.$$

This equation may look much more cryptic than the equivalent statement in words, but it clearly presents the problem as a first-order partial differential equation of the form

$$F(\boldsymbol{x}, \tau, \nabla\tau) = 0.$$

In fact, one sees that if $v$ depends on the position $\boldsymbol{x}$ of the point on the surface and the arrival time $\tau$ as well as on the normal direction $\boldsymbol{n}$ at that point, the resulting equation is

$$F(\boldsymbol{x}, \tau, \nabla\tau) = v(\boldsymbol{x}, \tau, \nabla\tau) - 1 = 0$$

(or the same with $v'$, if $v$ is negative), which is again just a first-order PDE.

If the normal velocity is not always of one sign, one has to let $\tau(\boldsymbol{x})$ be the set of arrival times at which the crystal surface passes through $\boldsymbol{x}$. Sometimes (e.g. if this set of arrival times at $\boldsymbol{x}$ contains no intervals or accumulation points) $\tau$ can still be regarded locally as a function of $\boldsymbol{x}$ if time is restricted to an appropriate interval. In any case, one always has to be careful about and treat separately the places where $v = 0$ and $\nabla\tau$ is undefined.

The theory of first-order PDE (John 1975) asserts (under conditions of smoothness of the given function $F(\boldsymbol{x}, \tau, \boldsymbol{p})$ in all its variables and of the initial data to be specified shortly) that the initial value problem can be solved by the use of curves called *characteristics*. These curves, parametrised by some variable $t$ and thus written as $x(t)$ (and with $\tau = \tau(t)$ and $\nabla\tau = \boldsymbol{p}(t)$ at $\boldsymbol{x}(t)$), emanate from each point in the initial surface and are determined by the following set of seven ordinary differential equations:

(1) $\mathrm{d}x_i/\mathrm{d}t = \partial F/\partial p_i$ for $i = 1, 2, 3$ (i.e. $\mathrm{d}\boldsymbol{x}/\mathrm{d}t = \nabla_p F(\boldsymbol{x}, \tau, \boldsymbol{p})$)
(2) $\mathrm{d}\tau/\mathrm{d}t = \Sigma_i\, p_i \partial F/\partial p_i$ (i.e. $\mathrm{d}\tau/\mathrm{d}t = \boldsymbol{p} \cdot \nabla_p F(\boldsymbol{x}, \tau, \boldsymbol{p})$)
(3) $\mathrm{d}p_i/\mathrm{d}t = -\partial F/\partial x_i - p_i \partial F/\partial\tau$ for each $i$
(i.e. $\mathrm{d}\boldsymbol{p}/\mathrm{d}t = -\nabla_x F - \boldsymbol{p}\partial F/\partial\tau$).

Because $\boldsymbol{p} \cdot \nabla_p F = \boldsymbol{p} \cdot \nabla_p v = v(\boldsymbol{p}) = 1$ (as is discussed further in the appendix), (2) above will always reduce to $\mathrm{d}\tau/\mathrm{d}t = 1$ for us, and so the parameter $t$ along the characteristics can be always (for this crystal growth problem) be taken to be the arrival time $\tau$.

Note that since only derivatives of $F$ appear in these equations, we can replace $F$ by $v$ (or $v'$, if $v$ is negative) in the above equations.

If $v$ does not depend explicitly on $\boldsymbol{x}$ then (3) becomes

(3′) $\mathrm{d}\ln(p_i)/\mathrm{d}t = \partial v/\partial\tau$ (unless $p_i$ is initially 0, in which case $p_i$ stays 0 for all time on that characteristic) for each $i$.

Since ln $p_i$ has the same derivative for each $i$ (or $p_i = 0$), we conclude that the ratios of the $p_i$ are constant along characteristics, and thus that the normal $\boldsymbol{n} = \boldsymbol{p}/|\boldsymbol{p}|$ to the crystal stays constant along characteristics (though $\boldsymbol{p}$ may change in magnitude). Thus in this no-spatial-dependence case, a characteristic is the trajectory of a point with a given normal as the crystal grows. (Such is not the case when $v$ depends on $\boldsymbol{x}$, as the temperature-gradient example below will demonstrate.)

If $v$ does not explicitly depend on either the arrival time $\tau$ or spatial position $\boldsymbol{x}$, then (3) becomes

(3″) $\mathrm{d}\boldsymbol{p}/\mathrm{d}t = -\nabla_x v = 0$, which says that $\boldsymbol{p}$ as well as $\boldsymbol{n} = \boldsymbol{p}/|\boldsymbol{p}|$ is constant along characteristics, and (1) becomes

(1″) $\mathrm{d}\boldsymbol{x}/\mathrm{d}t = \nabla_p v$, which says that characteristics are straight lines of the form

$$\boldsymbol{x} = \boldsymbol{x}_0 + t\nabla v(\boldsymbol{n}_0)$$

where $\boldsymbol{x}(t)$ is on the surface of the crystal at time $t$. (Here $\nabla v(\boldsymbol{n}_0)$ means that we are evaluating the gradient at $\boldsymbol{n}_0$, not that we are taking any kind of surface gradient. As discussed in the appendix, $\nabla v$ is constant in radial directions.)

The crystal shape at time $t$ is the locus of all the points $\boldsymbol{x}(t)$ on all the characteristics from initial points $\boldsymbol{x}_0$. When $v$ depends only on $\boldsymbol{n}$ and not on $\boldsymbol{x}$ or $t$, if we follow an element of surface of a given orientation $\boldsymbol{n}$, it will travel with constant velocity, here given by $\nabla v(\boldsymbol{n})$. Since the plot of $\boldsymbol{n}/v(\boldsymbol{n})$ is the level set $v(\boldsymbol{p}) = 1$ for the function $v(\boldsymbol{p})$, and since $\boldsymbol{p}\cdot\nabla v(\boldsymbol{p}) = v(\boldsymbol{p})$, this is equivalent to Frank's result (1958), which said that the element moves in the direction $\boldsymbol{n}'$ of the normal to Frank's polar plot of the slowness vectors $\boldsymbol{n}/v(\boldsymbol{n})$ and with a speed given by $v(\boldsymbol{n})/(\boldsymbol{n}\cdot\boldsymbol{n}')$.

## 2.1 A simple two-dimensional example

We illustrate the use of characteristics in figure 2 by taking the two-dimensional example of an initial circle and growth velocities $v$ which are of the form

$$v_{a,b}(\boldsymbol{n}) = a + b(\boldsymbol{n}\cdot(1,0))^2 = a + bn_1^2.$$

For $a > 0$ and $b > -a$, $v$ will be positive; $b = 0$ is the isotropic case, and the case $b = -a$ will be dealt with below under discontinuous coarsening. Multiplying by $|\boldsymbol{p}|$ to create $v(\boldsymbol{p})$, we obtain

$$v(\boldsymbol{p}) = a|\boldsymbol{p}| + (b/|\boldsymbol{p}|)(\boldsymbol{p}\cdot(1,0))^2 = a|\boldsymbol{p}| + (b/|\boldsymbol{p}|)(p_1)^2.$$

Setting $a = 1$, polar plots of $v$ are circles for $b = 0$ which pinch towards the centre when $b$ is negative as shown in figure 2. Taking the gradient, we obtain

$$\nabla v = (an_1 + bn_1(2 - n_1^2),\ an_2 - bn_2 n_1^2).$$

The plot of $\nabla v$ is a circle for $b = 0$, but becomes distorted when $b$ is non-zero. For $b < -\frac{1}{2}a$ or $b > a$ it becomes self-intersecting and develops two swallowtails, as shown in figure 2. For $b > a$, the second component of $\nabla v$ changes sign at $n_1 = \pm (a/b)^{1/2}$ and $\pm 1$; there are six crossings of the $x_1$ axis. Similarly, for $-a < b < -\frac{1}{2}a$, there are six crossings of the $x_2$ axis. The tips of the swallowtails are spinodes; the curves turn back on themselves without a change of slope.

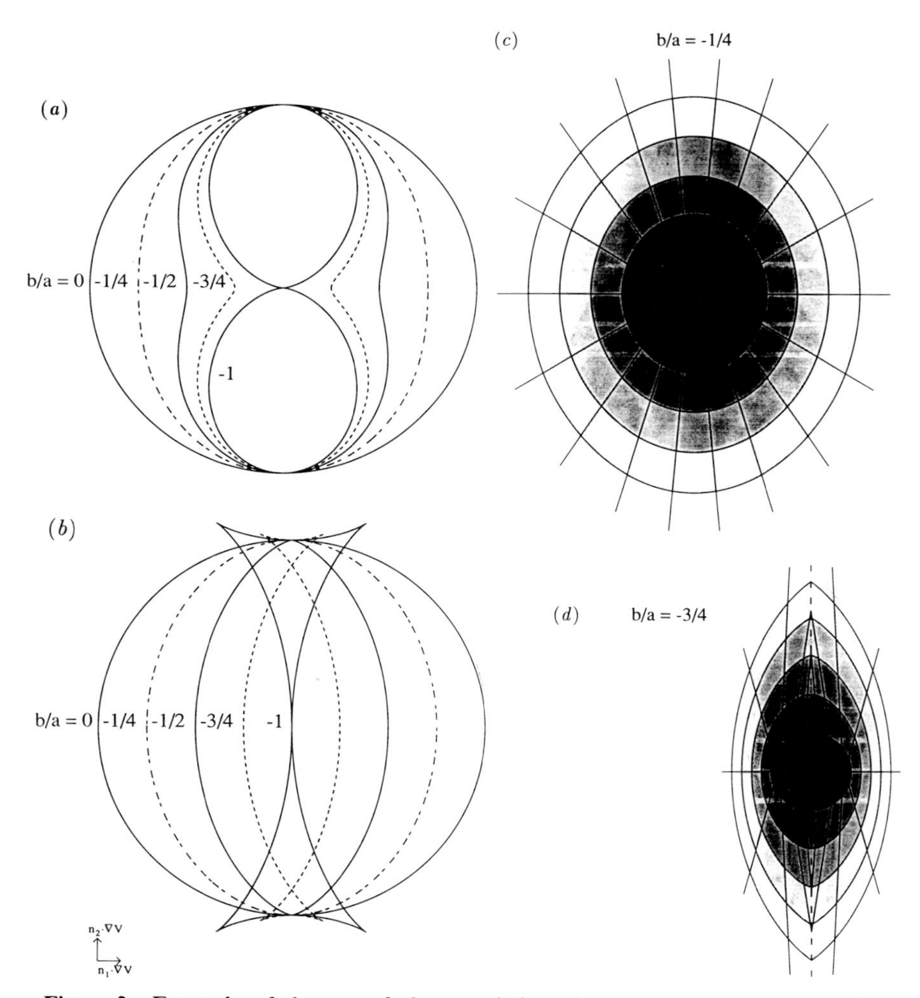


**Figure 2** Example of the use of characteristics when the growth rate $v$ is of the form $v_{a,b}(\boldsymbol{n}) = a + bn_1^2 (= a + b\cos^2\theta)$ for various ratios of $b/a$. ($a$) Polar diagrams of $v(\boldsymbol{n})$. ($b$) Diagrams of $\nabla v(\boldsymbol{n})$. ($c$) and ($d$) Evolving crystal forms from an initial circle for $b/a = -\frac{1}{4}$ and $-\frac{3}{4}$.

For $-\frac{1}{2}a < b < a$, $\nabla v$ is smooth and does not intersect itself.

The solutions are shown for times $t > 0$ by connecting the equal-time points on the characteristics. For the curve with $b/a = -\frac{1}{4}$, the outward characteristics diverge and the solution is valid for all times. However, for the curve for $b/a = -\frac{3}{4}$, the characteristics converge and cross. Continuing along the characteristics leads to a swallowtail shape as a perfectly valid mathematical solution to the PDE. However, the physics of crystal growth does not allow this; a point $\boldsymbol{x}$ enters the crystal only once, unless $v$ changes sign and the crystal melts back through $\boldsymbol{x}$. This suggests that characteristics that cross must be terminated, in a way that will be discussed next under shocks.

Note that if we assume that the $v$ of this example is for inward growth from the circle, characteristics that diverged for outward growth converge for inward growth, and vice versa. The orientations on the circle that eventually disappear on outward growth by having their characteristics terminate in a shock do not disappear in a similar shock on inward growth (though they may disappear due to their characteristics intersecting with an entirely different set of characteristics).

### 2.2 The question of regularity: shocks and fans

An important theorem that holds with minor modification in any dimension is clearly and simply presented in two dimensions by John (1975): If $F$ has continuous second derivatives with respect to its variables $\boldsymbol{x}, \tau, \boldsymbol{p}$, and if along an initial curve $\boldsymbol{x}(s) = \boldsymbol{x}_0(s)$, $0 \leqslant s \leqslant 1$, the value $\tau(s) = t_0$ is assigned, with $\boldsymbol{x}_0$ having continuous second derivatives, and if one also has $\boldsymbol{p}_0(s)$ which is continuously differentiable and satisfies $F(\boldsymbol{x}_0(s), t_0, \boldsymbol{p}_0(s)) = 0$ and $\boldsymbol{p}(s) \cdot \mathrm{d}\boldsymbol{x}(s)/\mathrm{d}s = 0$ for each $s$, and if $\mathrm{d}\boldsymbol{x}/\mathrm{d}s$ and $\nabla_p F$ are never zero and never parallel, then in some neighbourhood of the initial curve there will exist one and only one solution $\tau(\boldsymbol{x})$ of $F(\boldsymbol{x}, \tau, \boldsymbol{p}) = 0$ with $\tau(\boldsymbol{x}_0) = t_0$ and $\nabla\tau(\boldsymbol{x}_0(s)) = \boldsymbol{p}_0(s)$ for all $0 \leqslant s \leqslant 1$.

This solution is the one constructed by the method of characteristics; it extends for precisely the open time interval such that characteristics do not cross. An examination of the proof shows that the condition $\nabla_p F \neq 0$ is used only in the proof that the characteristics together form a surface with no gaps or overlaps for short times; one can easily see that the motion of the surface remains solvable by characteristics even if there are points or intervals along which $\nabla_p F = 0$, if we leave such points stationary. The requirement that $\mathrm{d}\boldsymbol{x}_0/\mathrm{d}s \neq 0$ together with the requirement that $\boldsymbol{x}_0(s)$ be twice continuously differentiable is the requirement that the curve itself be smooth. The condition that $\mathrm{d}\boldsymbol{x}/\mathrm{d}s$ and $\nabla_p F$ are never parallel is automatic from the fact that the normal component of $\nabla_p F$ has magnitude $1/|\boldsymbol{p}| = v(\boldsymbol{n})$. Remembering that $\tau(\boldsymbol{x})$ is the arrival time of the crystal at the point $\boldsymbol{x}$ (i.e the time at which the point $\boldsymbol{x}$ is in the interface), from this one can prove in our case that if $v$ has continuous second derivatives and if the entire initial crystal shape has a normal vector field which is continuous and piecewise-continuously differentiable, then the evolving crystal shape is uniquely defined for some open interval of time containing the initial time and each shape in this continuum will also have a continuous and piecewise-continuously differentiable normal vector field. Entirely analogous results hold in three dimensions, both in terms of the theorem (John 1975) and its application to crystal growth.

We thus see that there are several ways in which this theorem can fail to cover cases we would like to consider. The first is that we want to consider growth velocities $v$ that may be continuous with respect to normal direction but not have a continuous gradient. (As we will see, such discontinuities in the gradient of $v$ do occur at facet orientations of $W_\infty$.) The second is that we want crystal growth to be determined for all positive time (including when characteristics cross) and for shapes with corners and edges. These issues are resolved by the use of shocks and fans, provided we assume that both our initial surface and $v$ are continuous everywhere and piecewise $C^2$ (i.e. first and second derivatives are piecewise continuous).

Shocks must be used whenever characteristics cross. A shock is a discontinuity along the surface, in this case in $\nabla v$ and in $\boldsymbol{n}$ as well, and thus gives rise to a corner or an edge on the crystal surface. The jump condition at a shock is determined from the physics of the problem, not the PDE; here the physics requires that the crystal surface (as determined by the points $\boldsymbol{x}(t)$ the characteristics reach at time $t$) be continuous, and (provided $v$ is positive) that once the crystal has grown past a point, that point should remain part of the crystal. Thus a shock occurs when two characteristics arrive at the same point at the same time. Differentially, our shock condition at a point $\boldsymbol{x}$ along a shock between normals $\boldsymbol{n}_\alpha$ and $\boldsymbol{n}_\beta$ is, in two dimensions, that

$$\mathrm{d}\boldsymbol{x}/\mathrm{d}t = s_\alpha t_\alpha + \nabla v(\boldsymbol{p}_\alpha) = s_\beta t_\beta + \nabla v(\boldsymbol{p}_\beta)$$

where $t_\alpha$ is a unit tangent vector to the $\alpha$ surface and is thus perpendicular to $\boldsymbol{p}_\alpha$, $t_\beta$ is defined similarly for the $\beta$ surface, and $s_\alpha$ and $s_\beta$ are numbers determined by solving the vector equation (which is a system of two linear equations in those two variables). In the special case where each gradient lies in the other plane, this becomes

$$\mathrm{d}\boldsymbol{x}/\mathrm{d}t = \nabla v(\boldsymbol{p}_\beta) + \nabla v(\boldsymbol{p}_\alpha).$$

In three dimensions, at a point in an edge of a growing crystal we have three naturally defined tangent vectors, one along that edge and the other two perpendicular to it and into the two surfaces meeting along that edge. The equation for $\mathrm{d}\boldsymbol{x}/\mathrm{d}t$ is similar to the above, except there are two $s$ for the $\alpha$ side and two for the $\beta$ side. The three-dimensional vector equation thus yields three linear equations in four unknowns, but the extra degree of freedom corresponds to the fact that one has a shock surface rather than a shock line emanating from an edge, and the degree of freedom is in the direction of the tangent line to the edge. At a corner where three surfaces meet, there are six $s$ and two three-dimensional vector equations, yielding six equations in six unknowns, and thus a differential equation for the propagation of the corner. If a corner with more than three surfaces is to propagate, then special relations must exist among the $v(\boldsymbol{n})$ in order to enable a solution to the equations to exist; this can happen in crystals because of their symmetry. (In order for such a shock to form with more than three surfaces related by symmetry, there must be appropriate symmetry in the initial crystal shape as well.)

Shocks in particular appear when a non-convex crystal grows so that two different portions of surfaces come into contact. At that instant, the shock starts from the contact point(s) and spreads out all around them.

If contact occurs on a whole piece of surface at a given time, then all the characteristics going into that piece of surface from both sides terminate. (It is convenient to put the crystal surface mathematically into the region which grows with positive velocity when it is called the 'crystal', so that if contact occurs on a whole piece of surface, that portion automatically disappears into the interior of the crystal.) Similarly, all remaining characteristics collide and terminate at the instant that a hole in a crystal (or a dissolving crystal) disappears.

Fans of characteristics are used at points where characteristics are not uniquely defined. One way that this can happen is that there is a point on the growing crystal with normal $\boldsymbol{n}_0$ and $v$ has a discontinuous gradient at $\boldsymbol{n}_0$. Then in place of $\nabla v$ one uses all the convex combinations of the limits of $\nabla v(\boldsymbol{n})$ as the normals $\boldsymbol{n}$ approach $\boldsymbol{n}_0$. An example where this happens is $v_{\text{cube}}(\boldsymbol{n}) = |n_1| + |n_2| + |n_3|$; see the appendix. This results in a facet developing there (or propagating, if that point is already in a facet): note that if one has a facet with a direction $\boldsymbol{n}_0$ where $\nabla v$ is discontinuous, then there are fans of characteristics emitted from each point, and fans from neighbouring points in the facet cross each other.

It is as if there is a whole continuum of shocks. However, they all in fact give the same two items of information: the facet is moving forward at velocity $v(\boldsymbol{n}_0)$, and is spreading out no further than the rate allowed by the outer characteristics of the fans (and perhaps less, if there are shocks at the edges of the facet).

Another way that characteristics can be non-uniquely defined is at a corner or edge of the crystal surface. Here one puts in characteristics corresponding to all normals in the convex hull of the normals around the edge or corner (using the extended definition above if $\nabla v$ is undefined for some of those normals). Note that it is quite possible that $\nabla v$ is the same on all those normals (for example, this can happen with $v = v_{\text{cube}}$), so that the fan from all the limiting normals around the edge or corner reduces to one uniquely defined characteristic after all. It is also possible that none of those characteristics overlap; in that case, the corner becomes rounded instantly as growth proceeds. Finally, some of the members of the fan of characteristics may overlap other members; this can become quite complicated, and the above shock condition of continuity of the surface is often not adequate to determine the way the corner grows because there may be more than one solution which produces a continuous crystal surface. The appropriate condition seems to be that the crystal should grow as fast as possible, subject to the condition that the crystal surface, as determined by the points the characteristics reach at time $t$, be continuous. We believe that this is the correct shock condition because we believe that this is also the condition for such a solution to be stable under perturbations (in the sense that a perturbation to such a solution will not grow without bound).

That is, look at the tangent cone to the initial surface $C_0$ at a point $\boldsymbol{x}_0$ which is on an edge or in a corner of $C_0$; this cone consists of rays through the origin and is

$$\{y: \text{there exist } s_k \downarrow 0 \text{ and } \boldsymbol{y}_k \to \boldsymbol{y} \text{ with } \boldsymbol{x}_0 + s_k\boldsymbol{y}_k \text{ in } C_0\}.$$

If $C_0$ is piecewise $C^2$, then this tangent cone is composed of planar segments. For each $\boldsymbol{y}$ in the tangent cone such that there is a normal $\boldsymbol{n}_y$ to the tangent cone at $\boldsymbol{y}$, plot $\boldsymbol{y} + t\nabla v(\boldsymbol{n}_y)$ for $t = 1$. At each point $\boldsymbol{y}$ in an edge or corner of the tangent surface, plot $\boldsymbol{y} + t\nabla v(\boldsymbol{n})$ for $t = 1$ and all $\boldsymbol{n}$ omitted at that edge or corner. (Note the similarity of Huygens' wavefront construction (Morse and Feshbach 1953); see below for an analogy of this crystal-growth problem to wavefront propagation.) The resulting surface will be continuous, because of the convention regarding discontinuities in $\nabla v(\boldsymbol{n})$, but will possibly be self-intersecting. If along the edges or at the corner it has portions which stick out beyond the rest of the surface, in the sense that the boundary of the solid formed by using $0 \leqslant t \leqslant 1$ rather than $t = 1$ has portions that are not in the surface obtained with $t = 1$ or the initial surface, then the corresponding $\nabla v(\boldsymbol{n})$ must not be used to construct a characteristic at $\boldsymbol{x}_0$ in the original problem. This continuity condition means that any 'ears' which stick out beyond the resulting planar surfaces will not be able to be used for convex corners, but portions of them will be able to be used for concave corners; for saddle-shaped corners with both concave and convex edges, it is quite possible that some parts of the 'ears' will be used but others will not. This construction determines which values of $\nabla v$ are to be used in constructing characteristics on the original surface. It will coincide with choosing among possible growth paths the one that is stable, because it inherently includes all perturbations of the corner by including all values of $\nabla v$ at that corner (values of $\nabla v$ where $\boldsymbol{n}$ is not a direction omitted at the corner or edge cannot contribute to growth—they are automatically behind the surfaces obtained from the plane segments of the tangent cone).

Solutions that are not piecewise $C^2$ but have curvatures approaching infinity in some of their pieces, e.g. with portions of circular cones, do arise, either as part of $C_0$ or as growth features. One example occurs in figure 2(*d*), if we assume that these solutions are for a problem of cylindrical symmetry; the tangent cone at the shock for $b/a = -\frac{3}{4}$ is then a circular cone that is not piecewise $C^2$. Another example occurs at the termini of a shock surface, places on the surface where the sharp crease of an edge becomes rounded. Such singular points are part of swallowtail development in $R^3$. If $C_0$ is not piecewise $C^2$, then the decision of which characteristics to use is more complicated, but is still based on fastest growth subject to continuity.

We do not know of any other physical model in which a shock condition is imposed that necessarily (at time $t = 0$) involves both continuity and shortest time, even though characteristics have long been used. But the right way mathematically to describe the solution is probably as the viscosity solution of the PDE (Chen *et al* 1989), rather than through specifying the details of the shocks and fans.

## 3 EXAMPLES

### 3.1 Some examples of growth of finite seed crystals

The isotropic case, $v(\boldsymbol{n}) =$ constant, arbitrarily set to 1, is intuitively simple and is useful for introducing shocks and fans. It leads to a $v(\boldsymbol{p}) = |\boldsymbol{p}|$, and thus to a $\nabla v(\boldsymbol{n}) = \boldsymbol{n}$, which is constant in magnitude and always parallel to $\boldsymbol{n}$. The limiting outward growth form $W_\infty$ is a unit sphere; any starting finite shape growing outward will eventually tend to a sphere. For example, the faces of a starting cube will grow by translation without an area increase, the fans on edges and corners will round the edges into growing cylinders, and the eight corners will grow octants of spheres. After a large enough elapsed time the corner spherical pieces will dominate, so that it will look like a sphere, but the eight spherical pieces will have different centres (the corners of the original cube), and there will be cylindrical strips separating the eight octants and six little flat facets (of the same size as the faces of the original cube) at the ends of the cylindrical strips.

An initial crystal consisting of two balls of different radii $r_1$ and $r_2$ joined by a thin cylinder of radius $r_0$ about the axis joining their centres will grow as balls of radius $r_1 + t$ and $r_2 + t$ joined by a cylinder of radius $r_0 + t$ until the spheres touch; the spheres will meet the cylinder along circles in a sharp edge. The shock surfaces will early on consist of the union of all these circles, forming surfaces which are surfaces of revolution of parabolas. When the balls touch, the cylinder is entirely eliminated and the crystal continues to grow as intersecting balls, the shock surfaces having collided and formed a new shock surface which happens to be a rotated hyperbola.

The shock will get weaker and weaker (the angle between the normals at the discontinuity will approach zero) but will never entirely disappear.

For $v = v_{\text{cube}} = |n_1| + |n_2| + |n_3|$, starting from any bounded seed the end form will be a cube (since $W_\infty$ is the cube with corners $(\pm 1, \pm 1, \pm 1)$), but it is instructive to examine how an initial sphere will evolve in time. Because $\nabla v_{\text{cube}}$ is constant (it is $(\pm 1, \pm 1, \pm 1)$) in each of the eight open octants, the eight spherical segments in each of the open octants translate without distortion or expansion in these $\langle 111 \rangle$ directions to become rounded corners of the cube; the fans emanating from the six points along the $\langle 100 \rangle$ directions create square facets expanding in size, and fans from the circular arcs at the octant boundaries generate pieces of circular cylinders, not growing in arc length but lengthening along their axis to become curved edges of the cube (see figure 3). Ultimately the facets dominate and the form is a growing cube with rounded edges and corners whose radii are those of the starting sphere. If $v$ were any other normal velocity function with the same $W_\infty$, but containing ears in the plot of $\nabla v$, then the crystal might grow faster than the cylindrical and spherical portions, possibly forming corners or edges there due to shocks, but it would never grow beyond the growing cube initially surrounding it. An example of the effect of 'ears', which are also called 'swallowtails', in growth was shown in figure 2 for $v_{a,b}$.

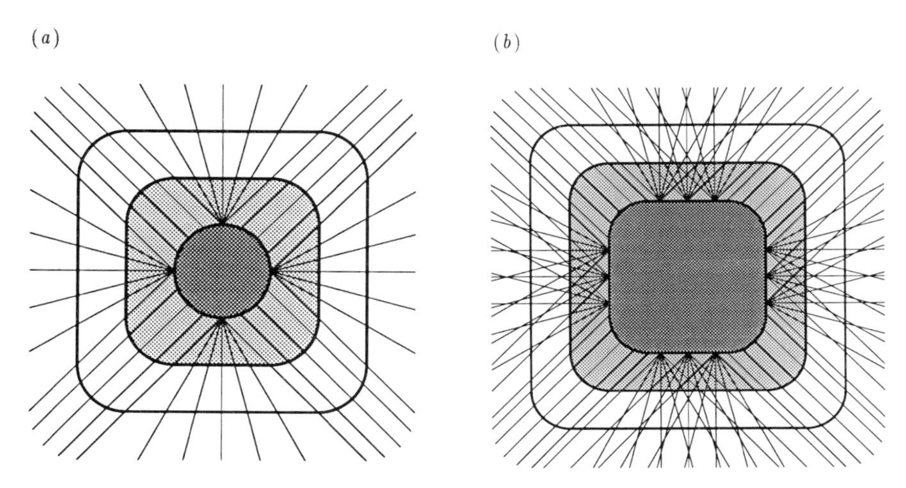


**Figure 3** Evolving crystal forms for $v = v_{\text{square}} = |n_1| + |n_2|$, starting from a circle and from a rounded square. In each of the four open quadrants, $\nabla v_{\text{square}}$ is constant and the characteristics are parallel. The four circular segments in each of the open quadrants translate without distortion upon expansion to become rounded corners of the square. (*a*) If $C_0$ is a circle, fans emanating from the $\langle 10 \rangle$ directions create facets. (*b*) If $C_0$ is a rounded square, fans emanating from the facets overlap and lead to translation and expansion of the facets.

### 3.2 Examples of shapes of inward growth from a finite form

Simply because crystal growth in this model remains well defined until the surface entirely disappears, there is a limit to the shape of any particular hole in a crystal as the hole volume goes to zero, or to the shape of any crystal as it dissolves, given $v$ (we orient the surface inward in either case so that we can use a positive velocity $v$). This shape is just the set of limit points of the surface as time approaches the instant of disappearance—the points on the characteristics at the final moment of collision. One can also often characterise the shape of a piece just before it disappears. But these shapes are not independent of the initial shape. A non-concave starting shape will often become disconnected on shrinking, with separate pieces having different limiting shapes as they disappear, usually at different times. Also, dissolving forms need not be more rounded than growth forms, as simple examples will show. It is furthermore the case that the actual value of $v$ on all directions matters a great deal, rather than just the values on the directions that appear in $W_\infty$. We illustrate some of what can happen by example.

Again let us consider the case where $v(\boldsymbol{n}) = 1$, but we are now letting the 'crystal' be on the outside of some finite region, so that the exterior normals point into the finite region. If the initial shape $C_0$ is exactly spherical, the shape of the hole will stay spherical. But a shape close to that of a sphere deviates increasingly from spherical, developing shocks. A starting parallelepiped will remain a parallelepiped with the same angles; the shocks from the intersecting characteristics from the faces will maintain the corners and edges, and, because they grow too slowly, the inward-directed fans from the corners and edges will be eliminated. When the shortest altitude has shrunk to zero, the limiting form will be a two-dimensional parallelogram. An initial shape $C_0$ that is a convex polyhedron remains a convex polyhedron, with surviving faces parallel to the faces of $C_0$ as it decreases in size. Shocks due to intersecting characteristics from the faces will maintain the corners and edges; nothing will be contributed by the inward-directed fans from the corners and edges. Intersecting shocks will disappear into a new shock, causing corners and edges to combine into new corners and edges; some faces and edges may disappear. A cube or a rhomboid remains a cube or a rhomboid, but a rectangular solid that is not a cube becomes increasingly elongated; the end form is a rectangular plate unless the two shorter sides were initially of the same length, in which case the end form is a line segment. These simple counterexamples disprove the general statement that dissolving forms are more rounded than growth forms.

In general, when one starts with an initial shape $C_0$ exactly similar to the central inversion of $W_\infty$, and when $v(\boldsymbol{n})$ has the smallest values it can to have that $W_\infty$ (which is equivalent to saying that $v(\boldsymbol{p})$ is a convex function, or that the plot of the function $\boldsymbol{n}/v(\boldsymbol{n})$ is the boundary of a convex region), then the shape does not change as it shrinks, and the end form is the same as that of the initial shape. Track the points inward along the $\nabla v$ vector(s) corresponding to the normal at the point and the shape stays the same; the omitted directions move exactly at the speed so as to stay omitted at the corners. But it is not just the initial shape that matters. If $v$ is a little bigger on the omitted directions and the starting shape is the central inversion of $W_\infty$, then some of those directions will appear, both immediately and in the limiting shape, thereby blunting the edges and/or corners and possibly even entirely consuming all the directions occurring in the original hole. But even when the starting shape is the central inversion of $W_\infty$, it is not always the case that 'faces with maximal growth rates should form the concave growth shape' (since the growth rates on omitted normals of $W_\infty$ are usually larger than on the directions which occur). Rather new directions can show up here if their growth rates are greater than that minimally allowed by $W_\infty$.

Of course, things can be quite different if one has different starting shapes. If the velocity function $v$ is $v_{\text{cube}}$ as defined above, any parallelepiped oriented so its normals point inward (whether its faces have special orientations—those where $\nabla v$ is discontinuous—or not) will have each of its faces translate inward with constant speed with no new orientations appearing at the corners and edges. The fans of characteristics introduced at some of the edges and corners may not give rise to shocks at all (when $\nabla v$ is the same on all adjoining faces of the parallelepiped), whereas the fans at the other edges and corners predict a motion slower than the shock between the adjoining faces and are eliminated. The limiting form for a starting parallelepiped is thus a line segment or a 2D rhombus, unless the initial shape is a rhomboid (in which case it is a point approached via the same rhomboid). An initial sphere dissolving with this same velocity function $v(\boldsymbol{n})$ will be composed of eight segments of spheres of constant radius but decreasing extent, translating inward along the $\langle 111 \rangle$ directions at constant speed and getting chopped off by shocks which form along the octant boundaries. The limiting form as this originally spherical particle disappears into a point is an octahedron.

The only rule is that if the characteristics corresponding to a given direction are excluded by crossing characteristics at some time, then they disappear, whether one has 'growth' or 'dissolution'. Characteristics with directions corresponding to 'ears' are not necessarily excluded at concave corners (as they are at convex corners) and so regions with these normals can grow in a concave shape if the corners are not too sharp. Only in this sense can corners be said to round on dissolution. These new orientations, if they form and grow, can crowd out slower growth directions if the characteristics from another corner cross them. But a simple counter-example of an initial shape that is very flattened shows that there is no general theorem here—the end-form can consist primarily of slow-growth directions. These simple examples illustrate some of the simpler anisotropic growth results. They demonstrate that while the limiting forms from outward growth are independent of initial form, the forms from inward growth always reflect the initial form. They also illustrate how shocks and fans operate in easily visualised cases.

### 3.3 Additional growth rules

An edge or a corner in a growing crystal is the result of intersecting or overlapping characteristics. If the corner or edge is convex and identical to one in $W_\infty$, then it will stay the same shape, whether the plot of $\nabla v$ has 'ears' or not. If the corner or edge is convex and sharper than the corresponding region of $W_\infty$ (so that it omits normals that occur in $W_\infty$) then a fan of characteristics will emerge from the corner, and the corner will immediately round out or, if $W_\infty$ also omits some of the normals omitted by the corner, it will partially round out and thereby become less sharp. If the corner or edge is convex and $W_\infty$ omits all of the normals that the corners omits plus more, then the corner will stay the same shape only if the surfaces meeting at the corner are planes or if the plot of $\nabla v$ has no 'ears' at that corner; if the surfaces meeting at the corner have any positive curvature and if $v(\boldsymbol{n})$ is larger than required by $W_\infty$ for the directions present in the corner of the crystal but in $W_\infty$, then these directions will 'grow out', their characteristics will be cut off by shocks, and the corner will sharpen.

Concave corners and edges will stay the same shape unless the plot of $\boldsymbol{x}_0 + \nabla v(\boldsymbol{n})$ for the directions $\boldsymbol{n}$ omitted at the edge or corner at $\boldsymbol{x}_0$ extends past the plot of $\boldsymbol{x} + \nabla v(\boldsymbol{n}(\boldsymbol{x}))$ for points $\boldsymbol{x}$ near $\boldsymbol{x}_0$. In particular, sharp enough concave corners and edges always remain the same shape. Less sharp concave corners and edges can become more rounded if the omitted directions grow fast enough, or can become more sharp if the surfaces which meet at the corner or edge are not planar but also concave.

Saddle-shaped regions can easily become more sharply curved in some tangent directions, even developing edges and corners where none were previously, and can simultaneously become less sharply curved in others.

It should be noted that if in an initial convex polyhedron there is a planar facet with a direction in a concave part of an 'ear' of the plot of $\nabla v$, then not only will that direction disappear by growing out, but while it is growing out, the adjacent facets will usually partially round out at their edges with this facet.

### 3.4 Example of extension to case of constant temperature gradient

Suppose that the temperature is $T_m - bx_1$, where $T_m$ is the melting temperature and $b > 0$, so that growth occurs for a seed crystal in the right half-space. Suppose $v(\boldsymbol{x}, \tau, \boldsymbol{n}) = bx_1 u(\boldsymbol{n})$; as before, extend $u$ and $v$ to be defined on all vectors $\boldsymbol{p}$ rather than just unit vectors $n$. We will give solutions in the two-dimensional isotropic case of $u(\boldsymbol{n}) = 1$ for all $\boldsymbol{n}$.

As before, we define $v'(\boldsymbol{x}, \tau, \boldsymbol{p}) = v(\boldsymbol{x}, \tau, \boldsymbol{p})$ where $v > 0$ and $v'(\boldsymbol{x}, \tau, \boldsymbol{p}) = -v(\boldsymbol{x}, \tau, -\boldsymbol{p})$ where $v < 0$, and then the PDE is $v' - 1 = 0$. In our example, $v'(\boldsymbol{x}, \tau, \boldsymbol{p}) = b \mid x_1 \mid \mid \boldsymbol{p} \mid$ and so the PDE is

$$F(\boldsymbol{x}, \tau, \boldsymbol{p}) = b \mid x_1 \mid \mid \boldsymbol{p} \mid - 1 = 0.$$

If we let $s = \text{sign}(v) = \text{sign}(x_1)$ and $b' = sb$, then the PDE is equivalent to

$$b' x_1 \mid \boldsymbol{p} \mid = 1.$$

Since $\boldsymbol{p}$ is $\nabla\tau$ in the PDE, and $\nabla\tau = \text{sign}\ (v) \mid \nabla\tau \mid \boldsymbol{n}$ for $\boldsymbol{n}$ the normal to the surface, we have $\boldsymbol{p} = s \mid \boldsymbol{p} \mid \boldsymbol{n} = \boldsymbol{n}/(bx_1)$. If an initial point $\boldsymbol{x}_0 = (x_{10}, x_{20})$ has normal $\boldsymbol{n}_0$, then $\boldsymbol{p}_0 = (p_{10}, p_{20}) = (n_{10}/bx_{10}, n_{20}/bx_{10})$.

Since $\partial \mid \boldsymbol{p} \mid / \partial p_i = p_i / \mid \boldsymbol{p} \mid = sn_i$, the equations for the characteristics become

$$\mathrm{d}x_1/\mathrm{d}t = b' x_1 p_1 / \mid \boldsymbol{p} \mid (= bx_1 n_1)$$

$$\mathrm{d}x_2/\mathrm{d}t = b' x_1 p_2 / \mid \boldsymbol{p} \mid (= bx_1 n_2)$$

$$\mathrm{d}p_1/\mathrm{d}t = -sb \mid \boldsymbol{p} \mid = -b'(p_1^2 + p_2^2)^{1/2}$$

$$\mathrm{d}p_2/\mathrm{d}t = 0$$

so along characteristics, $p_2$ is constant. If $p_{20}$ is non-zero, the equation for $p_1$ integrates to

$$p_1 = -\mid p_{20} \mid \sinh(b't - a)$$

where $a$ depends on $\boldsymbol{n}_0$ and $\boldsymbol{x}_0$ and satisfies $\sinh a = p_{10} / \mid p_{20} \mid = sn_{10} / \mid n_{20} \mid$ (and thus $\cosh a = 1/ \mid n_{20} \mid$). If $p_{20} = 0$, then

$$p_1 = p_{10} \exp(-\text{sign}(p_{10}) b't).$$

Using these formulae for $p_1$, the equation for $x_1$ becomes separable, but it is easier to use the original PDE and the solution for $\boldsymbol{p}$ to get, for $p_{20} \neq 0$,

$$x_1(\mathrm{t}) = 1/(b' \mid \boldsymbol{p} \mid) = 1/[b' \mid p_{20} \mid \cosh(b't - a)]$$
$$= x_{10} \mid \boldsymbol{p}_0 \mid / [\mid p_{20} \mid \cosh\ (b't - a)] = x_{10} / [\mid n_{20} \mid \cosh(b't - a)]$$

and, for $p_{20} = 0$,

$$x_1(t) = x_{10} \exp(\text{sign}(p_{10}) b't) = x_{10} \exp(\text{sign}(n_{10}) bt).$$

The differential equation for $x_2$ is thus

$$\mathrm{d}x_2/\mathrm{d}t = b' x_1 p_2 / \mid \boldsymbol{p} \mid = \mid x_{10} \mid b' / [n_{20} \cosh^2(bt - a)]$$

so

$$x_2(t) = (1/b' p_{20}) \tanh(b't - a) + x_{20} + x_{10} p_{10} / p_{20}$$
$$= (\mid x_{10} \mid / n_{20}) \tanh(b't - a) + x_{20} + x_{10} n_{10} / n_{20}.$$

Using the hyperbolic function identities (or integrating $\mathrm{d}x_2/\mathrm{d}x_1 = p_2/p_1$) one obtains

$$x_1^2(t) + (x_2(t) - C)^2 = (x_{10}/n_{20})^2$$

the equation of a circle with centre $(0, C)$ and radius $|x_{10}/n_{20}|$, where $C$ is $x_{20} + x_{10}n_{10}/n_{20}$. Thus all characteristics are arcs of circles.

If the initial crystal is entirely in the region $x_1 > 0$, then the size of the crystal grows exponentially, and it approaches the $x_2$ axis very slowly. If the crystal is entirely in the region $x_1 < 0$, it would disappear in finite time. If it straddles the $x_2$ axis, then part will grow and part will shrink, but it will always straddle the $x_2$ axis. For example, if the initial crystal is the unit circle centred at the origin, then the characteristics are portions of non-intersecting circles which fill the plane; trajectories in the left half of the plane are inward along the circles towards the portion of the $x_2$ axis within the initial crystal (corresponding to that part of the crystal shrinking), whereas in the right half of the plane they are outward along the circles and back to the portions of the $x_2$ axis which lie outside the original crystal. The points $(0, 1)$ and $(0, -1)$, where $v = 0$, are stationary. Furthermore, one can show that for each point $(x_1(t), x_2(t))$ on a characteristic from any initial point $(x_{10}, x_{20})$ on that unit circle,

$$(x_1(t) - \sinh(bt))^2 + x_2^2(t) = \cosh^2(bt)$$

so that, at time $t$, the surface of the growing crystal is itself a circle, one with centre $(\sinh(bt), 0)$ and radius $\cosh(bt)$.

Since crystals which start out as circles remain as circles, one might wonder whether this spatially dependent velocity were equivalent to a velocity $v'$ of the form $v'(t, \boldsymbol{n}) = a(t) + \boldsymbol{b}(t) \cdot \boldsymbol{n}$. Indeed, if one uses

$$v'(t, \boldsymbol{n}) = b \sinh(bt) + \boldsymbol{n} \cdot (b \cosh(bt), 0)$$

then an initial circle of radius 1 centred at the origin does grow as circles with centres $(\sinh(bt), 0)$ and radii $\cosh(bt)$, and one checks that indeed at any point $(x_1(t), x_2(t)), v'(t, \boldsymbol{n}) = v(\boldsymbol{n}) = bx_1(t)$. Note that the crystal shapes—the level sets of the arrival time function—are the same for the two velocities, but that the characteristics for the two different PDE are not. Also, $v'$ reproduces the same growth as $v$ only for initial circles centred at the origin; for an initial circle centred at $(a, 0)$ of radius $r$, one would have to use the different velocity

$$v'' = ab \cosh(bt) + rb \sinh(bt) + \boldsymbol{n} \cdot (ab \sinh(bt) + rb \cosh(bt), 0)$$

as the velocity to reproduce the growth given by $v$ (circles with centres $(a\cosh(bt) + r \sinh(bt), 0)$ and radii $a \sinh(bt) + r \cosh(bt)$), and for initial forms that are not circles, in general no growth law depending only on time will reproduce the same growth as $v(\boldsymbol{x}, \boldsymbol{n}) = bx_1$.

The idea of 'limiting shape' is also much looser here. If one uses barrier arguments (as will be proven in our subsequent paper), putting a circle inside and outside an arbitrary shape and growing each, the circles do not get closer together when translated back to the origin and rescaled, but rather their radii remain in the same constant ratio as they had initially, if they initially had the same centre. Thus one cannot say that the limiting form is a circle, for an arbitrary initial finite seed with at least part of it in the right half-plane.

### 3.5 Example of $v(\boldsymbol{n})$ determined by anisotropic elasticity

When stresses develop in multicomponent crystals because of diffusion from the surrounding medium into the crystals of some component that changes the lattice parameters, the local solubility of the crystals in the surrounding medium changes in a way that depends primarily on the local stress at the surface. As a result, the crystals can grow or dissolve in response to these stresses. Because the elastic stresses depend on anisotropic elastic moduli, and the details of the stress patterns depend on the normal to the surface $\boldsymbol{n}$ but not on the details of the diffusion, elasticity theory can be used to predict a growth velocity $v$ depending only on the normal direction $\boldsymbol{n}$.

This situation has been observed in a variety of settings, usually involving two crystals, one dissolving, the other growing (Handwerker *et al* 1985, Yoon *et al* 1986). It is called diffusion-induced grain boundary migration (DIGM) when the surfaces are pre-existing grain boundaries, and diffusion-induced recrystallisation (DIR) when new crystals nucleate, usually at the interface between an existing crystal and a fluid, and the new crystals grow into the old as in DIGM. Migration resulting from self-stresses is known as liquid film migration (LFM) when there is a liquid film between the two crystals; the pair of crystal–liquid surfaces can be treated as a single surface.

For cubic crystals, the coherency strain energy is a function only of the strain $\delta$ and an orientation-dependent elastic parameter, $Y(\boldsymbol{n})$,

$$f_{\text{coherent}} = Y(\boldsymbol{n})\delta^2.$$

When the change in lattice parameter $a$ with composition is constant (Vegard's law)

$$f_{\text{coherent}} = Y(\boldsymbol{n})\eta^2(C^{\text{s}} - C_0)^2$$

where $C^{\text{s}}$ is the composition of the solid at the interface, $C_0$ is the initial composition and $\eta = (\mathrm{d}a/\mathrm{d}C)/a(C_0)$.

Hilliard (1970) has given an expression for $Y$ for cubic crystals. Cubic elasticity is determined by three elastic constants $c_{11}$, $c_{12}$ and $c_{44}$; it is useful to rearrange his expression to be in terms of the modulus of compressibility

$K = c_{11} + 2c_{12}$, which is isotropic, and two ratios that best describe the orientation dependence of the elastic properties for cubic crystals: the 'anisotropy', $A = (c_{11} - c_{12})/(2c_{44})$, and $B = c_{12}/c_{44}$, which is 1 for any solid composed of particles held by central pairwise forces (Cauchy's condition):

$$Y(\boldsymbol{n}) = 2K \left( \frac{3(A-1)^2\theta + A(A-1)\phi + A}{4(2A+3B+1)(A-1)^2\theta + 4(A-1)(A+B)\phi + 2A + B} \right)$$

where the two cubic invariants of $\boldsymbol{n}$, $\theta$ and $\phi$ are given by $\theta = n_1^2 n_2^2 n_3^2$, $\phi = n_1^2 n_2^2 + n_2^2 n_3^2 + n_3^2 n_1^2$, where the coordinates of $\boldsymbol{n}$ are defined relative to the cube axes.

In liquid film migration, DIR and DIGM, there are two solid grains, 1 and 2, of the same phase and the same initial composition but differing by rotation $R$. Since there is a single surface (or two parallel surfaces) between the grains, the outward normals of grain 1 and grain 2 are in opposite directions. If we let the common coordinate system be relative to the cube axes of 1, then the normal relative to the cube axes of 2 is $-R\boldsymbol{n}$. The migration velocity depends on the difference in $f_{\text{coherent}}$ between 1 and 2:

$$v(\boldsymbol{n}, R) = M[Y(\boldsymbol{n})\eta^2(C^s - C_{0,1})^2 - Y(-R\boldsymbol{n})\eta^2(C^s - C_{0,2})^2]$$

where $M$ is a combination of parameters that can be assumed to be isotropic for a given experiment. From these relationships between migration rate and interface orientation, interface shapes can be predicted as a function of the elastic anisotropy for various conditions of stress in adjacent grains.

Often the stress in the growing grain becomes completely relaxed; the grain tends to become uniform in composition during LFM and in equilibrium with the liquid. Effectively $C_0$ for that grain approaches $C^s$, and the strain energy tends to zero in that grain (e.g. grain 2). The migration velocity becomes:

$$v(\boldsymbol{n}) = M' Y(\boldsymbol{n})$$

where $M' = M\eta^2(C^s - C_0)^2$ is also assumed to be isotropic. The symmetry of the limiting shape of a stress-free growing grain (2) embedded in a self-stressed dissolving matrix is that of the dissolving grain (1). Observations on DIR (Handwerker 1988) show that the new growing grains are almost dislocation-free while the dissolving matrix grains contain a high density of defects, including dislocations. The shape of the growing grains is cuboidal but oriented with the axes of the dissolving grains.

For an elastically isotropic material, that is, for $A = 1$ and for all $B$, $Y(\boldsymbol{n}) =$ constant and $W_\infty$ is a sphere. Computing $\nabla v$ leads to a prediction of how $W_\infty$ varies with $A$ and $B$. The limiting grain shapes $W_\infty$ obtained from $\nabla Y(\boldsymbol{n})$ for KCl, Mo and Fe (figure 4), represent two forms for $W_\infty$ for cubic materials. Sections of $v(\boldsymbol{n})$ and $\nabla v$ for both types of shapes are shown in figure 5. For Fe, where $A < 1$, the shape is cuboidal with corners at

⟨111⟩ with curved edges and faces. For KCl and Mo, where $A > 1$, the shape is an octahedron with corners at ⟨100⟩ and curved edges and faces. The creases in the edges join the edges for the KCl, but for the Mo the middle of the edges around the ⟨110⟩ are rounded. There are critical points along the edges where the creases terminate that are part of the description of a swallowtail in catastrophe theory (Thom 1975). Such morphologies have been observed (Handwerker 1988; see figures 17 and 19).

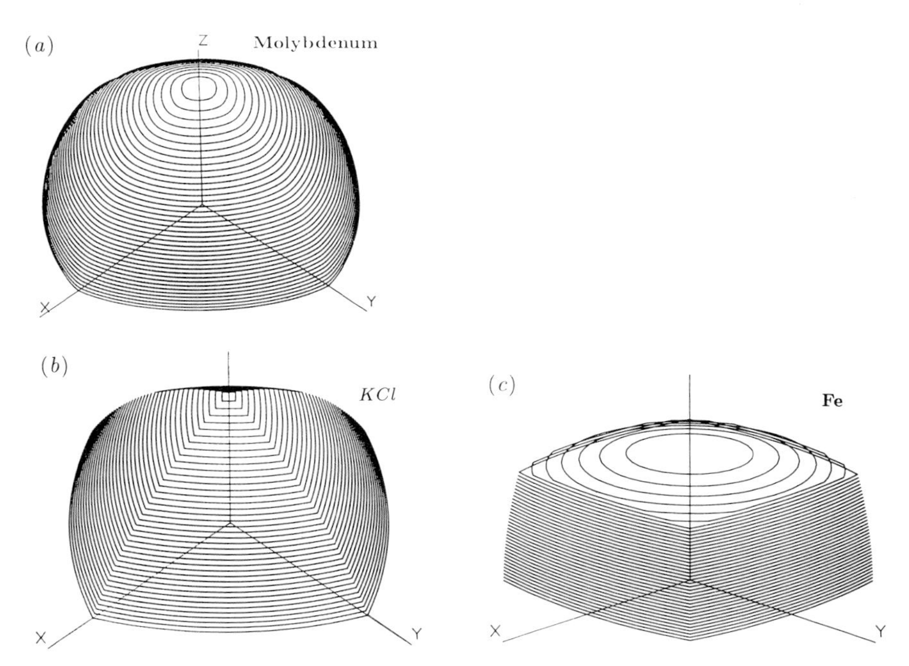


**Figure 4** Limiting grain shapes $W_\infty$ determined by the elastic anisotropy of $Y(\boldsymbol{n})$ for (*a*) Mo, (*b*) KCl and (*c*) Fe for $n_3 \geqslant 0$.

### 3.6 Example of $v(n)$ for discontinuous coarsening models

Discontinuous coarsening is a solid state process in which a region with a fine lamellar structure dissolves and is replaced with a coarser one. The process is driven by capillary forces (reduction of surface energy between the lamellae). In a linear macroscopic theory where we consider only the capillary forces due to the fine lamellae, averaged over distances large compared to the lamellar spacing, the velocity of the process should be proportional to the product of the density of the lamellae ending along the interface and to the resolved 'surface tension' forces of each lamella along the smoothed interface normal. The density is $|\sin\phi|/\lambda$, where $\lambda$ is the lamellar spacing and $\phi$ is the angle between the normal to the interface $\boldsymbol{n}$

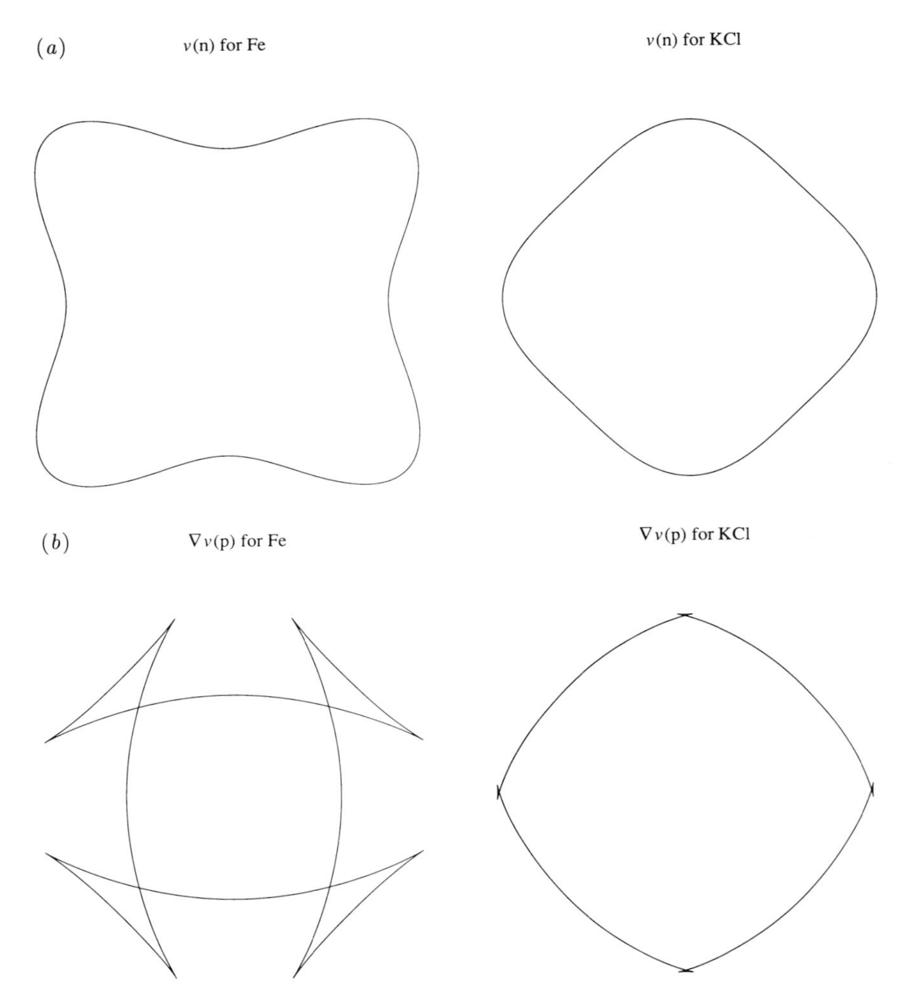


**Figure 5** Sections of ($a$) polar plots of $v(\boldsymbol{n})(\boldsymbol{n}v(\boldsymbol{n}))$ and ($b$) $\nabla v$ for Fe and KCl for $n_3 = 0$, showing ears at $\langle 110\rangle$ and $\langle 100\rangle$, respectively.

and the normal to the lamellae, which we may take to be $(0, 0, 1)$, so that $\cos(\phi) = n_3$. The second factor depends on whether or not the lamellar ends can rotate, so that the trijunctions, where lamellar surfaces meet the grain boundary, can attain equilibrium angles. If rotation occurs until equilibrium is reached, the second factor is simply twice the energy of the surface of the lamellae, $\gamma$. If not, and we assume isotropic surface energies, then the surface tension force resolved in the direction of motion is reduced by another factor of $\sin\phi$. Since lamellar rotations were not observed, the $\sin\phi$ in the second factor is a possibility. However, the steady state motion required to rotate enough of the tips to eliminate this factor was well below the microscope resolution. Furthermore, equilibrium may be attained without rotation, if these lamellar surfaces are so highly anisotropic that they resist rotating over a range of $\boldsymbol{n}$, the sum of shrinkage and 'torque' terms add to a force equal to the surface energy $\gamma$, independent of $\sin\phi$ over that range. We conclude that the velocity of growth $v$ of the coarser region into the lamellar region is proportional to $(2\gamma/\lambda)|\sin^k\phi|$ where $k$ is probably 1, but could be as large as 2. Therefore we should contrast two cases, $|\sin\phi|$ and $\sin^2\phi$. A section through a radial plot of these two functions is shown in figure 6. These plots are like doughnuts without a complete hole. In either case $v$ is zero when the surface is parallel to the lamella. This obviously suggests that growth is confined to the caliper diameter of the initial form $C_0$, the range between the highest and lowest point of the initial form transverse to the lamellae. The orientation dependence of $v$ has not been measured, and $C_0$ was not controlled. However, evolving growth forms suggested that $v$ was indeed zero perpendicular to the lamella, and an average $v$ measurement confirmed the dependence on $\lambda$. Let us examine the predictions for these two cases.

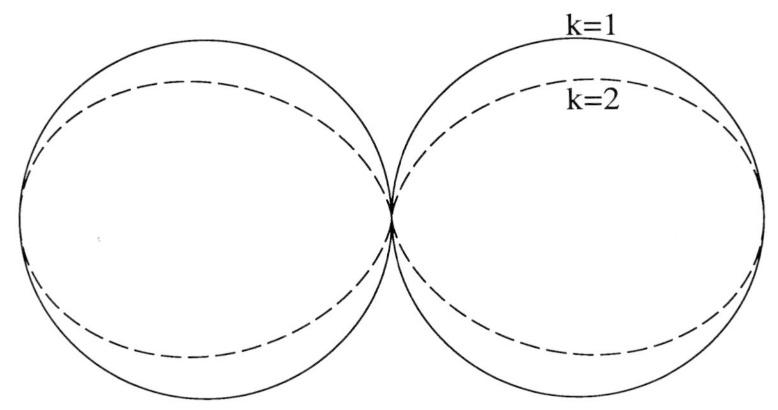


**Figure 6** Sections through radial plots of $|\sin\phi|$ and $\sin^2\phi$; two possibilities for $v(\boldsymbol{n})$ for discontinuous coarsening. $\nabla v$ for these two cases are respectively vectors of constant magnitude in the plane $n_3 = 0$, and the double swallowtail in figure 2(*b*) with $b/a = -1$.

Given $\boldsymbol{p}$ and $\boldsymbol{n} = \boldsymbol{p}/|\boldsymbol{p}|$, we set

$$p_r = (p_1^2 + p_2^2)^{1/2}$$

and

$$\boldsymbol{n}_r = (p_1/p_r, p_2/p_r).$$

Note that $\boldsymbol{n}_r$ is a unit vector in the direction of the projection of $\boldsymbol{n}$ into the plane of the lamellae.

For the first case of $k = 1$,

$$v(\boldsymbol{n}) = a[1 - ((0, 0, 1) \cdot \boldsymbol{n})^2]^{1/2} = a(1 - n_3^2)^{1/2}$$

$$v(\boldsymbol{p}) = a[\,|\boldsymbol{p}|^2 - (p_3)^2]^{1/2} = ap_r$$

$$\nabla v(\boldsymbol{p}) = 2a(p_1/p_r, p_2/p_r, 0) = 2a\boldsymbol{n}_r.$$

Note that $\nabla v$ is a radial vector of constant magnitude in this plane. For the case of $k = 2$,

$$v(\boldsymbol{n}) = a - a(\boldsymbol{n} \cdot (0, 0, 1))^2 = a(1 - n_3^2)$$

$$v(\boldsymbol{p}) = (a/|\boldsymbol{p}|)(|\boldsymbol{p}|^2 - p_3^2) = (a/|\boldsymbol{p}|)p_r^2$$

$$\nabla v(\boldsymbol{p}) = a[2(1 - n_3^2)^{1/2}\boldsymbol{n}_r - (1 - n_3^2)^{3/2}\boldsymbol{n}].$$

Note that $\nabla v$ is composed of two terms, an outward radial term along the lamellae, that is dependent on the inclination, plus a smaller inward term that is along $\boldsymbol{n}$. In component form the significance of the second term is more apparent:

$$\nabla v(\boldsymbol{p}) = a(1 - n_3^2)^{1/2}((2 - (1 - n_3^2)n_1), (2 - (1 - n_3^2)n_2), -(1 - n_3^2)n_3).$$

The first case, $k = 1$, is a good physical example of constant radial characteristics. The evolving forms tend to circular discs with thickness equal to initial caliper diameter; all the elements of $C_0$ translate radially in the direction of the projection of their $\boldsymbol{n}$ with constant velocity. A vertical cross section of the evolving growth forms would look blocky; this was similar to the shapes observed in the Co–Si alloys (Livingston and Cahn 1974, see their figure 1c). In the second case, the characteristics have a vertical component that is opposite in sign to $n_3$. With $k = 2$ the stronger orientation dependence of $v$ is such that, in a convex front, the fastest growth direction is grown out and the slower growth directions take over. A convex front (as viewed from the coarser region) sticks out further and develops shocks and in vertical cross section will look pointed; a concave front flattens. Such features were seen in the Ni–In and Cu–In alloys (Livingston and Cahn 1974, see their figures 2 and 3). Thus the two cases give quite different predictions for the evolving forms; hints of both predictions were seen.

## 4 CONNECTIONS AND EXTENSIONS OF THIS THEORY

### 4.1 Analogy with surface energy

The gradient of (the extension of) $v$ turns out to be of central importance in this model. It is similarly of central importance (and called $\xi$ there) in determining equilibrium shapes of curved surfaces with anisotropic surface free energy $\gamma(\boldsymbol{n})$ (Hoffman and Cahn 1972, Cahn and Hoffman 1974). We can apply some of the known relations between $\gamma$, its gradient, and its

Wulff shape

$$W = \{x : x \cdot n \leqslant \gamma(n) \qquad \text{for every unit vector } n\}$$

to the case of $v$, its gradient, and the shape

$$W_\infty = \{x : x \cdot n \leqslant v(n) \qquad \text{for every unit vector } n\}$$

since they are simply results about certain functions defined on unit normal vectors; this is done in the appendix. If $v$ depends on position and time as well as the normal vectors, then those variables must be held fixed for the analogy to hold, and the dependence of $v$ on those variables will be temporarily suppressed. Similarly, $v$ ought to be non-negative.

If $\gamma(n)$ is replaced by $\gamma_b(n) = \gamma(n) + n \cdot b$, then the Wulff shapes for $\gamma$ and for $\gamma_b$ differ only by a translation by $b$, since $\nabla\gamma_b(n) = \nabla\gamma(n) + b$; this was first observed by Lee *et al* (1975) for isotropic $\gamma$, and is a counter-example to Chernov's conjecture that the area of a face decreases as $\gamma$ increases. Also, by Stokes' theorem, the integral of $\gamma_b$ over any piece of surface differs from the integral of $\gamma$ over that piece of surface only by the integral of $b \cdot m$ over the two-dimensional boundary of that piece of surface, where $m$ is the exterior normal of the surface, so that $\gamma_b$ and $\gamma$ have the same surface-energy-minimising surfaces. This was first noted by Arbel and Cahn (1975), and theorems showing that two surface energy functions have the same minimal surfaces if and only if they differ by the addition of $n \cdot b(x)$ for $b(x)$ a divergence-free vector field were proved by Taylor (1982). For crystal growth, the parallel results are that the growth shapes using $v(t, n)$ and $v(t, n) + n \cdot b(t)$ are indistinguishable by shape alone, as at each time one shape is just a translation of the other if they start with the same initial shape.

An important feature about the $W$ corresponding to a given surface free energy function $\gamma$ is that it has less surface energy than any other shape having the same volume. Since the surface energy is the integral of $\gamma$ over the surface, and the integral of $v$ over a surface is the rate at which volume is swept out during growth, we conclude that volume is consumed at a slower rate by the growth of $W_\infty$ than by the growth of any other shape of the same volume, assuming that the growth rate is given by $v$. Similarly, results about the structure of surface-energy-minimising surfaces under other constraints can be interpreted as results about the structure of surfaces which instantaneously consume volume at the slowest rate under those constraints; such surface-energy-minimising shapes (for example, with the constraint of having a fixed boundary curve) have been extensively studied theoretically (Taylor 1983, Taylor and Cahn 1986a,b, Zia *et al* 1988). The direct utility of these other results is not clear, however, since such surfaces rarely remain of the same shape as they grow.

However, many results for $\gamma$ do not carry over to growth with $v$ because they explicitly involve minimisation in ways that growth does not. For example, if a planar surface has direction $n$ such that $\nabla\gamma(n)$ is part of an 'ear' of the plot of $\nabla\gamma$, then that plane is not surface-energy minimising; a corrugated surface would have less surface energy (Herring 1951, Frank 1963). Furthermore, if the plot of $\nabla\gamma$ is negatively curved in any direction at $\boldsymbol{n}$, then the plane is also unstable to perturbations by sine waves of arbitrarily small height, and if the plot is negatively curved in all directions, then the plane is unstable to perturbations by localised bumps (Mullins 1963). Thus these normal directions do not occur in surfaces governed by surface-energy minimisation (except possibly as infinitesimally corrugated varifold surfaces). None of these results holds in any way for a plane with normal $\boldsymbol{n}$ growing with velocity $v(\boldsymbol{n})$—planar growth is always stable, in the sense that any perturbation may spread sideways but always has a limited growth in the normal direction. Planes with such directions do not occur in the limiting growth shape of a bounded seed crystal, but they can certainly occur in other situations.

### 4.2 Linkage with eikonal equations

Associated to a hyperbolic equation such as the wave equation

$$\partial^2 u(\boldsymbol{x}, t)/\partial t^2 = c^2 \sum_i \partial^2 u(\boldsymbol{x}, t)/\partial x_i^2$$

is an eikonal equation, which is an equation for the propagating wave front; for the (isotropic) wave equation it is

$$c^2 = 1/(\nabla\tau \cdot \nabla\tau).$$

The square root of this equation is of the form of our equation for normal growth. That is, $v = c$ can be regarded as the speed of light ($v$ can also depend on position and time), and the arrival surface at time $t$, which is $\{\boldsymbol{x} : \tau(\boldsymbol{x}) = t\}$, can be thought of as the wave front—the points where light has reached at time $t$ if the initial surface suddenly started radiating light from all its points at time $t = 0$.

Anisotropic wave propagation is discussed in many books, one being Whitham (1974) (the use of shocks and fans of characteristics is also presented elsewhere in that book). Musgrave (1954, 1970) gives plots, for a wide variety of elastically anisotropic crystals, of zonal sections of longitudinal and transverse wave velocities $\boldsymbol{n}v(\boldsymbol{n})$, the corresponding slowness surfaces $\boldsymbol{n}/v(\boldsymbol{n})$, and the wave surfaces by a method equivalent to forming $\nabla v(\boldsymbol{n})$. These wave surfaces show all the features expected, including ears. However, only certain types of dependence of $v$ on orientation can arise in this way; for the anisotropic propagation of light, the dependence is at most quadratic in the components of $\boldsymbol{n}$, and other hyperbolic PDE yield at best algebraic dependence of $v$ on $\boldsymbol{n}$. Additionally, in this analogy there are no shocks, and thus no analogue to crystal edges and corners, since there is nothing forbidding crossing of characteristics ('light rays'). Light rays simply cross (focus), giving rise to 'caustics'. Thus many of the more interesting features of crystal growth with a prescribed normal velocity do not appear in the solution of the anisotropic wave equation.

### 4.3 Alternative numerical techniques for solution

The method of characteristics to construct the family of growing crystal surfaces is nice for existence theorems and for constructing the solution 'by hand', but is relatively difficult to implement numerically when shocks are present. Sethian (1985, 1990) has investigated alternative numerical techniques for solving the isotropic case $v(\boldsymbol{n}) = 1$ (which he interprets as a simple model for the case of flames burning at a constant speed normal to themselves). The condition at shocks is the same as the one we use, but he formulates it as 'once a particle is burnt, it remains burnt'. It is described as an entropy condition because of the irreversibility of the motion at shocks. By replacing $v$ by $v_\varepsilon = 1 - \varepsilon K$, where $K$ is the curvature of the interface, he 'regularises' the equation to a second-order partial differential equation. Now the method of characteristics no longer applies, because the PDE is no longer first order. But the advantage of doing so is that curvature plays a smoothing role, preventing the occurrence of shocks. He proves (at least in the two-dimensional case (Sethian 1985)) that, as $\varepsilon$ approaches zero, the limit of these smooth solutions is the solution to the problem with $\varepsilon = 0$ (which may of course no longer be smooth but have shocks). Viewing, as we do, the $(N-1)$-dimensional moving surface as a level set of a time-dependent function of $N$ dimensions, he observes that 'the equation of motion for this function resembles an initial value Hamilton–Jacobi equation with parabolic right-hand-side and is closely related to a viscous hyperbolic conservation law' and 'numerical schemes designed to approximate hyperbolic conservation laws may be used to approximate the motion of the propagating surface'. (The model of having $v$ be $1 - \varepsilon K$ is certainly also of interest in itself for crystal growth.)

### 4.4 Limitations

Our formulation, which assumes that the normal velocity is a function only of $\boldsymbol{n}, \boldsymbol{x}$ and $t$, is an idealisation. It excludes all factors that alter the type of equation, such as a dependence on surface curvature which would make the equation second order and render the method of characteristics inapplicable. In particular, the formulation is inapplicable where two surfaces with the same normal $\boldsymbol{n}$ have different velocities $v$. This may arise, for example, from the presence or absence of dislocations with screw components normal to the surface, or from differences in large-scale geometry; a facet on a convex surface encounters a different physical problem to nucleate and/or grow a new layer than a facet on a concave surface.

A different formulation is also needed to consider cases that arise from diffusion control, in which the velocity does change with time, but only implicitly because the crystal significantly alters its environment during growth. An attempt by Chernov (1963) to treat this case as part of a unified formulation in terms of orientation and supersaturation seems of limited usefulness because it requires explicit knowledge of how the supersaturation or temperature varies with orientation.

## ACKNOWLEDGMENTS

We are most grateful to Craig Carter for generous and tireless help with the symbolic calculations and with the graphics, and for critical readings of the manuscript, and to Robert Parker for help with some of the early crystal growth literature. We gratefully acknowledge support of the ONR, DARPA, Air Force and NSF.

## APPENDIX. SUMMARY OF RESULTS RELATING $v(\boldsymbol{n}), v(\boldsymbol{p})$, $\nabla v$ and $W_\infty$

First, we extend $v$ to a function on all vectors $\boldsymbol{p}$, rather than just unit vectors $\boldsymbol{n}$, by defining $v(\boldsymbol{p}) = |\boldsymbol{p}|\, v(\boldsymbol{n})$ for all $\boldsymbol{p}$ (when we use $\boldsymbol{p}$ and $\boldsymbol{n}$ in the same formula or description in this paper, we use implicitly assuming that $\boldsymbol{p} = |\boldsymbol{p}|\,\boldsymbol{n}$). One can easily verify that $\nabla v$ is thereby constant in the radial direction, i.e. $\nabla v(\boldsymbol{p}) = \nabla v(\boldsymbol{n})$; whenever we write $\nabla v$, we mean $\nabla_p v = (\partial v/\partial p_1, \partial v/\partial p_2, \partial v/\partial p_3)$ (whether or not $v$ depends additionally on $\boldsymbol{x}$ and $t$) and $\nabla v(\boldsymbol{n})$ means that gradient evaluated at the unit normal $\boldsymbol{n}$, not any type of surface gradient. Note that $\boldsymbol{p} \cdot \nabla_p v = v(\boldsymbol{p}) = 1$, the first equation holding because $v(\boldsymbol{p}) = |\boldsymbol{p}|\, v(\boldsymbol{n})$ and the last equation holding because of the PDE.

Secondly, for any $\boldsymbol{n}_0$ such that $\nabla v(\boldsymbol{n}_0)$ is undefined (which happens where $\nabla v$ is discontinuous), we use in place of $\nabla v(\boldsymbol{n}_0)$ the set of all convex combinations of the limits of $\nabla v(\boldsymbol{n})$ as $\boldsymbol{n}$ approaches $\boldsymbol{n}_0$. For example, if we define $v_{\text{cube}}(\boldsymbol{n}) = |n_1| + |n_2| + |n_3|$, then for every $\boldsymbol{n}$ in the open first octant, $\nabla v_{\text{cube}} = (1, 1, 1)$. Because of the absolute values in the orientation dependence of $v_{\text{cube}}$, there are discontinuities in the gradient at the octant boundaries. If $\boldsymbol{p} = c(1, 0, 0)$ for some $c > 0$, then wherever $\nabla v_{\text{cube}}(\boldsymbol{p})$ is called for, one uses instead all values $(1, s, t)$, where $|s|, |t| \leqslant 1$, and if $\boldsymbol{p} = c(1, 1, s)$, one uses all values $(t, (1-t^2)^{1/2}, 0)$, where $-1 \leqslant t \leqslant 1$, for $\nabla v_{\text{cube}}(\boldsymbol{p})$ (figure 3). The plot of $\nabla v_{\text{cube}}$, when understood as above, is a cube; each corner is the value of $\nabla v_{\text{cube}}(\boldsymbol{n})$ for all $\boldsymbol{n}$ in its octant. This is the same as using the 'subgradient' of $v$ (Rockafellar 1981).

The plot of $\boldsymbol{n}/v(\boldsymbol{n})$ is a level set of the function $v(\boldsymbol{p})$, namely, the set of $\boldsymbol{p}$ such that $v(\boldsymbol{p}) = 1$. Therefore $\nabla v$ is normal to this plot. $1/v(\boldsymbol{n})$ has been called the 'slowness' not only by Frank in crystal growth but also in crystal optics and elastic waves.

The relationship between $v$ and $W_\infty$ has been well studied, especially in the case where $v$ is a convex function (which is equivalent to that plot being convex) (see, for example, Rockafellar (1970, 1981); Cahn and Hoffman (1974) rediscovered many of these results for the case when $v$ is interpreted as a surface-energy function $\gamma$). When $v$ is a convex function, $v$ is the support function of its $W_\infty$, and $W_\infty$ is the plot of $\nabla v$. Note that if $\boldsymbol{x}$ is a boundary point of $W_\infty$ and the plane with normal $\boldsymbol{n}$ through $\boldsymbol{x}$ is a support plane of $W_\infty$, then $\boldsymbol{x} \cdot \boldsymbol{n} = v(\boldsymbol{n})$ when $v$ is convex. One can regard $W_\infty$ itself as the plot of $\boldsymbol{m}/w(\boldsymbol{m})$ for some function $w$ defined on unit vectors and then extend $w$ by $w(r\boldsymbol{m}) = rw(\boldsymbol{m})$. This can be written as

$$w(\boldsymbol{x}) = v^*(\boldsymbol{x}) = \sup_{\boldsymbol{p}} \{\boldsymbol{p} \cdot \boldsymbol{x} - v(\boldsymbol{p})\}.$$

The mapping $v \to w = v^*$ is called the *Fenchel transform* (or *Legendre transform* when $v$ is differentiable); the Fenchel transform of $w = v^*$ takes one back to $v$. Thus the 'Wulff shape' for the function $w$ is the plot of $\boldsymbol{n}/v(\boldsymbol{n})$; Frank (1963) refers to this plot as the pedal of $W_\infty$.

When $v$ is not a convex function, the plot of $\nabla v$ (interpreted as above) coincides with the surface of its $W_\infty$ except for additional 'ears' at corners and/or edges of $W_\infty$ (as is the case for a surface energy function $\gamma$ and its Wulff shape $W$).

Corners and sharp edges occur in $W_\infty$ where the plot of $\nabla v$ is self-intersecting, i.e. where $\nabla v(\boldsymbol{n}_1) = \nabla v(\boldsymbol{n}_2)$ for some $\boldsymbol{n}_1 \neq \boldsymbol{n}_2$; there are no 'ears' if and only if $\nabla v(\boldsymbol{n})$ is constant on all directions $\boldsymbol{n}$ omitted at that corner or edge. Facets and straight edges on $W_\infty$ correspond to discontinuities in $\nabla v$.

As a simple example, the $W_\infty$ for $v_{\text{cube}}(\boldsymbol{n}) = |n_1| + |n_2| + |n_3|$ is a cube, and so is the plot of $\nabla v_{\text{cube}}$. This $v$ is the smallest it can be and still have its $W_\infty$ be that cube; there are no 'ears' on its $\nabla v$ plot (with $\nabla v$ being interpreted as described above). In some situations, theories predict that the surface energy $\gamma$ should be a constant times this $v$; in those theories, the plot of $\boldsymbol{n}\gamma(\boldsymbol{n})$ is sometimes called a 'raspberry' figure. Small changes in this $v$ can either round the edges or corners of its $W_\infty$ or add swallowtails there.

The most important property of $W_\infty$ for our purposes is that it is the limiting growth shape for any finite seed.

## REFERENCES


Arbel E and Cahn J W 1975 *Surf. Sci.* **51** 305–9

Barber D J, Frank F C, Moss M, Steeds J W and Tsong I S T 1973 *J. Mater. Sci.* **8** 1030–40

Burton W K, Cabrera N and Frank F C 1951 *Phil. Trans. R. Soc.* A **243** 299–358

Cahn J W and Hoffman D W 1974 *Acta Metall.* **22** 1205–14

Chen Y-G, Giga Y and Goto S 1991 *J. Diff. Geom.* to be published

Chernov A A 1963 *Sov. Phys.–Cryst.* **7** 728–30

—— 1984 *Modern Crystallography III, Crystal Growth* (Berlin: Springer) p 219ff

Frank F C 1958 *Growth and Perfection of Crystals* ed R H Doremus, B W Roberts and D Turnbull (New York: Wiley) pp 411–19

—— 1963 *Metal Surfaces* (Metals Park, OH: ASM) pp 1–15

—— 1972 *Z. Phys. Chemie N.F.* **77** 84–92

Gross R 1918 *Abhandl. d. K. Sachsische Gesellsch. d. Wissensch., math.-phys. Kl.* XXXV iv. 12 p. 137–202

Handwerker C A 1988 *Diffusion Phenomena in Thin Films and Microelectronic Materials* ed D Gupta and P S Ho (Park Ridge, NJ: Noyes) pp 245–322

Handwerker C A, Cahn J W, Yoon D N and Blendell J E 1985 *Diffusion in Solids: Recent Developments* ed M A Dayananda and G E Murch (Warrendale, PA: TMS) pp 275–92

Herring C 1951 *Phys. Rev.* **82** 87–93

Hilliard J E 1970 *Phase Transformations* (Metals Park, OH: ASM) pp 497–560

Hoffman D W and Cahn J W 1972 *Surf. Sci.* **31** 368–88

John F 1975 *Partial Differential Equations* 2nd edn (Berlin: Springer) pp 38–40

Kelvin, Lord 1904 *Baltimore Lectures on Molecular Dynamics and the Wave Theory of Light* (London: Cambridge University Press) pp 131–9

Lee J K, Aaronson H I and Russell K C 1975 *Surf. Sci.* **51** 302–4

Litsakes C N and Ney P 1985 *Fortshr. Miner.* **63** 135–54

Livingston J D and Cahn J W 1974 *Acta Metall.* **22** 495–503

Moore M 1986 *Miner. Mag.* **50** 331–2

Morse P M and Feshbach H 1953 *Methods of Theoretical Physics* (New York: McGraw-Hill) p 847

Mullins W W 1963 *Metal Surfaces* (Metals Park, OH: ASM) pp 44–5

Musgrave M J P 1954 *Proc. R. Soc.* A **226** 239–355

—— 1970 *Crystal Acoustics* (London: Holden-Day) pp 83–137

Nobes M J, Katardjiev I V, Carter G and Smith R 1987 *J. Phys. D: Appl. Phys.* **20** 870–9

Rockafellar R T 1970 *Convex Analysis* (Princeton: Princeton University Press)

—— 1981 *The Theory of Subgradients and its Applications to Problems of Optimization. Convex and Nonconvex Functions* (Berlin: Heldermann Verlag)

Sethian J A 1985 *Commun. Math. Phys.* **101** 487–99

—— 1990 *J. Diff. Geom.* to be published

Taylor J E 1982 *Indiana Univ. Math. J.* **31** 789–99

—— 1983 *Seminar on Minimal Submanifolds* (Princeton: Princeton University Press) pp 275–92

Taylor J E and Cahn J W 1986a *Acta Metall.* **34** 1–12

—— 1986b *Science* **233** 548–51

Thom R 1975 *Structural Stability and Morphogenesis* (Reading, PA: Benjamin) p 66

Whitham G B 1974 *Linear and Nonlinear Waves* (New York: John Wiley)

Yoon D N, Cahn J W, Handwerker C A, Blendell J E and Baik Y J 1986 *Int. Symp. on Interface Migration and Control of Microstructure* (Metals Park, OH: ASM) pp 1–13

Zia R K P, Avron J E and Taylor J E 1988 *J. Stat. Phys.* **50** 727–36

# Interfacial Dislocations in Epitaxy

J H van der Merwe

## 1 INTRODUCTION

Interfacial dislocations occur in a variety of systems, for example, in epitaxial interfaces, in grain boundaries and in precipitate–matrix interfaces. This paper shall be mainly concerned with epitaxial interfaces where the dislocations have commonly become known as *misfit dislocations*.

In 1949 Frank and van der Merwe (F&M) published two papers, based on what has become known as the Frenkel–Kontorowa (FK) model (1938). N F Mott suggested that van der Merwe, a PhD student at the time, exploit this model with the view of using it as a first step in developing a theoretical description of intercrystalline boundaries. With the ingenious leadership, insight and collaboration of his supervisor F C Frank this culminated in the abovementioned papers addressing the role of interfacial dislocations in epitaxial interfaces.

The objectives of this paper are to show: (i) that the abovementioned work has laid the foundations for the description of the role of interfacial dislocations in the accommodation of misfit at submonolayer to multilayer epitaxial film interfaces; foundations which are still valid after 40 years—Hibma writes for example in 1988 ' ... the most successful theory is the original theory of Frank and van der Merwe ...', (ii) that the description lends itself ideally for extensions to more complicated systems and (iii) that there are further important aspects which the authors did not deal with at the time.

In relation to item (i) above it is of interest that several subsequent investigators have introduced alternative names to 'interfacial dislocations'. An appropriate one, coined by J W Matthews and to be used regularly in this paper, is 'misfit dislocations' (MDs). Others are 'solitons, discommensurations, walls and static distortion waves'.

Significant stimuli for the investigations of F&M in epitaxy came from the works of (i) Royer (1928), who concluded from x-ray studies of epitaxial systems that epitaxy only occurs when the misfit between the partners is less than about 15% and (ii) Finch and Quarrell (1934) who concluded that ZnO grows pseudomorphically on Zn, i.e. that the ZnO is strained homogeneously to match the Zn substrate.

There exists a vast and ever growing literature on MDs in epitaxy. The reason for this is that epitaxy is one of the most successful mechanisms for growing crystals of high perfection needed in the fabrication of devices for the industries of electronics, optoelectronics, etc.

The work of F&M is a very typical case in which the theory was well ahead of observation, the main reason being that experimental techniques were not sophisticated enough. Interfacial dislocations were observed for the first time in 1961 by Delavignette *et al*, more than a decade after they were predicted. The phenomenon of interfacial dislocations was taken seriously by Matthews (1979) who contributed more than any other to the development of the field, the main motivation being the fabrication of devices. He addressed the question of minimising the density of dislocations as they are detrimental to the devices in one way or another.

## 2 THEORY: F&M

In their approach to MDs in monolayers (MLs) Frank and van der Merwe (1949a,b) implicitly assumed that the first ML grows as an extensive ML and that subsequent growth proceeds in a ML-by-ML fashion, a growth mode which subsequently earned the name 'Frank–van der Merwe' growth (Bauer 1958).

The F&M theory is an equilibrium theory (assumption I), i.e. assumes that the substrate temperature $T_A$ is high enough for the relevant kinetic processes to equilibrate the MD configuration within the duration of the experiment. It also assumes (assumption II) that, in the equilibrium free energy criterion, the free energy may be approximated by the internal energy. This is valid when either $T_A$ is small enough and/or the entropy, essentially the configurational entropy of the MDs, is negligible. This will be more or less the case if MDs are straight and regularly spaced.

For the accomplishment of our objectives it is necessary to briefly describe the Frenkel–Kontorowa (1938) model and the results obtained by Frank and van der Merwe (1949a,b). In the FK model a linear chain of particles (atoms) connected by elastic (the harmonic approximation) springs (force constant $\bar{\mu}$ and natural length $b$) simulates a monolayer (ML), and a periodic sinusoidal potential (overall amplitude $W$ and wavelength $a$) the interaction between atoms of the ML and substrate (atomic spacing $a$). Frank has proposed a mechanical model, displayed in figure 1, to simulate the FK model. The total energy of the F&M system is of the form

$$E = \tfrac{1}{2}\bar{\mu}a^2 \sum_n [(\xi_{n+1} - \xi_n - f)^2 + \tfrac{1}{2}W(1 - \cos 2\pi\xi_n)] \tag{1}$$

where $n$ enumerates both the substrate potential troughs and the ML atoms and $a\xi_n$ is the displacement of atom $n$ from trough $n$. The ratio

$$f = (b - a)/a = 1/P_0 \tag{2}$$

defines the misfit; $P_0$ is the natural vernier period, i.e. the number of atoms between consecutive positions of registry, when both crystals have their equilibrium spacings.

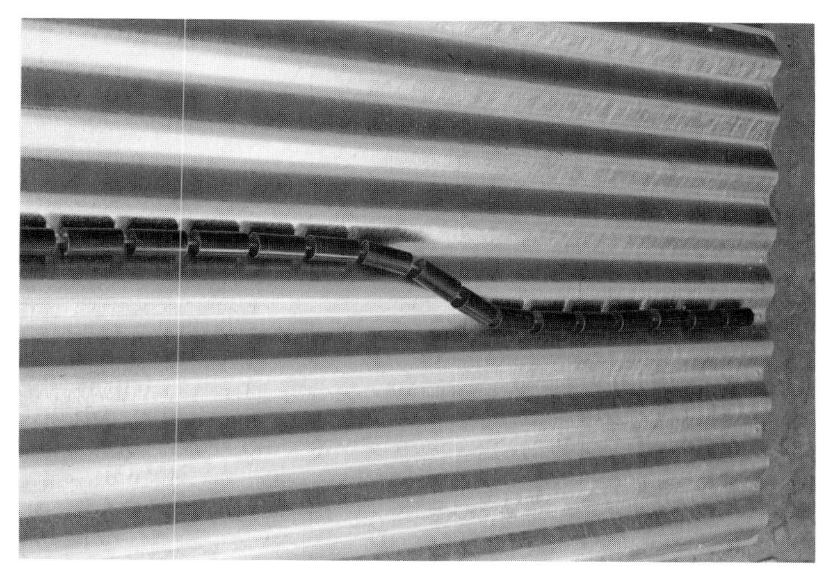

**Figure 1** Photograph of Frank's mechanical design of the Frenkel–Kontorowa model displaying a screw dislocation. The particles (atoms), periodic potential and harmonic forces are simulated by the metal cylinders, the corrugated metal plate and the stiffness of the piano wire threading the cylinders, respectively.

With assumptions I and II the governing equations for the positions of interior atoms are second-order non-linear difference equations. These equations approximate to the well-known sine-Gordon equation

$$\frac{d^2\xi}{dn^2} = \frac{\pi}{2l_0^2} \sin 2\pi\xi \tag{3a}$$

$$l_0^2 = \bar{\mu}a^2/2W \tag{3b}$$

when $l_0$ is large enough, or equivalently $x_n$ varies slowly with $n$. This is known as the continuum approximation (assumption III). The parameter $l_0$ is seen to combine important mechanical properties ($\bar{\mu}$ and $W$) of the overlayer–substrate system.

Frank and van der Merwe (1949a,b) found a value of about 7 for $l_0$ if they used the force constant $\bar{\mu}$ for a close-packed (CP) row of atoms and an overall amplitude $W$ calculated for a single atom moving along a CP edge of a CP plane of atoms, when $\bar{\mu}$ and $W$ are calculated on the assumption that the atoms interact according to 6–12 Lennard-Jones (LJ) pair potentials. In a somewhat different approach van der Merwe (1988) obtained a value of about $2\pi$ for the equivalent of $l_0$ applying the harmonic approximation only. These values can be regarded as rough approximations when overlayer and substrate atoms are of the same kind. It is seen from equation (3$b$) that $l_0$ will be larger when atomic bonding within the overlayer ($\bar{\mu}$) is large compared to overlayer–substrate bonding ($W$), and vice versa.

Equation (3$a$) may be integrated to give

$$\frac{\mathrm{d}\xi}{\mathrm{d}n} = + \frac{1}{l_0 k}(1 - k^2 \cos^2 \pi\xi)^{1/2} \qquad 0 \leqslant k \leqslant 1 \tag{4a}$$

$$\frac{\pi n}{l_0 k} = + \pi \int_{1/2}^{\xi} (1 - k^2 \cos^2 \pi\xi)^{-1/2}\, \mathrm{d}\xi = + F(k, \pi\xi - \tfrac{1}{2}\pi) \tag{4b}$$

where $k$ is an integration constant, $F$ is an incomplete elliptic integral of the first kind and $n = 0$ when $\xi = \frac{1}{2}$. The plus signs on the right-hand side apply to 'positive' ($b > a$) misfit; when $b < a$ the plus sign must be replaced by a minus sign. For simplicity we restrict the considerations to one type, choosing arbitrarily the positive case.

The dependence of $\xi$ on $n$ in (4$b$) is illustrated in figure 2. All integer values of $\xi$ correspond to positions of exact 'fit'—particles in potential minima. It is seen that there are sizable intervals of good fit and relatively narrow intervals of bad fit around half-integer values of $\xi$. These constitute the centres of what have been rightly named misfit (interfacial) dislocations (MDs). It follows from equations (4) and figure 2, that the width (interval of bad fit) $L_{\mathrm{D}}$ and the spacings $P$ (number of atomic spacings) of the sequence of MDs are given by

$$L_{\mathrm{D}} = l_0 k \qquad P(k) = 2 l_0 k K(k)/\pi \tag{5a}$$

in terms of $k$ as parameter. $K$ is a complete elliptic integral of the first kind. When $k = 1$, $L_{\mathrm{D}} = l_0$ and $P = \infty$. This is equivalent to having a single MD (width $l_0$). Since the chain is infinite this also defines the limiting situation in which there is registry 'everywhere'. This is referred to as the *pseudomorphic* or *coherent* configuration. In a pseudomorphic chain all the misfit $f$ is taken up ($\bar{b} = a$) by homogeneous or misfit strain (MS) $e$, $\bar{b}$

being the average particle spacing. When MDs are introduced a fraction

$$\bar{f} = (\bar{b} - a)/a = 1/P \tag{5b}$$

is accommodated by MDs $P$ atoms apart and it follows from (2) and (5$b$) that

$$\begin{aligned} f &= \bar{f} + |e| b/a \\ &\approx \bar{f} + |e| \qquad b/a \approx 1. \end{aligned} \tag{5c}$$

Thus in the F&M approach, the introduction of a sequence of dislocations (MDs) to accommodate the misfit at the interface between two different crystals is a natural consequence of the analysis. The single dislocation is characteristic of the initial stage of the introduction process.

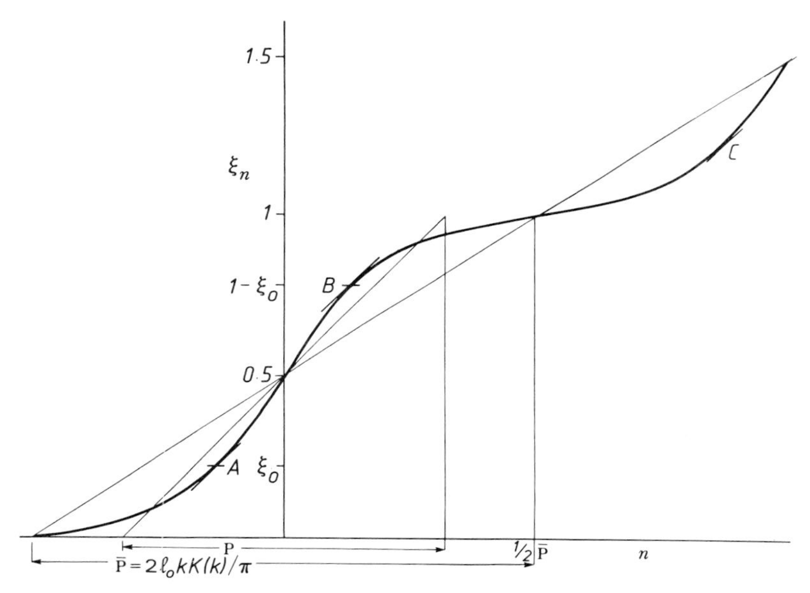


**Figure 2** Curve illustrating the variation of particle displacement $\xi(n)$ in equation (4$b$) with position $n$ along the chain. The particles are in perfect registry at $\xi = 0, 1, \ldots$ and out of registry at $\xi = \frac{1}{2}, \frac{3}{2}, \ldots$ representing the centres of the MDs. (From Frank and van der Merwe (1949a) by permission of the Royal Society.)

Of physical significance is the stability of a ML (chain). F&M have considered the stability of a chain with free boundaries, to the introduction of MDs. Two situations are of particular interest: (i) the stable (minimum-energy) configuration and (ii) the configuration at which MDs will enter spontaneously, needing no activation energy.

F&M have considered the introduction of MDs by glide from the free boundaries (edges). The relevant questions are most easily answered in terms of the work which is needed to introduce one more MD. The force needed to do this is the tension in an additional spring (number $n+1$), tied to the particle at the free end, and used to pull it over the relevant potential hill. For a given displacement $\xi_n$ of the atom this tension is given by

$$t_n = \bar{\mu}a(\xi_{n+1} - \xi_n - f) \approx \bar{\mu}a(\partial\xi/\partial n - f) \tag{6}$$

using the continuum approximation.

When $b > a$ the atom at the free end of the chain will ride up the substrate potential hill to a displacement $a\xi_0$ from the trough, where

$$t(\xi_0) = \frac{2Wl_0^2}{a}\left(\frac{1}{kl_0}(1 - k^2\cos^2\pi\xi_0)^{1/2} - f\right) = 0. \tag{7}$$

If the end atom is displaced forward reversibly from here positive work $W_A$ will be needed to take it to the point $1 - \xi_0$ where again $t(1 - \xi_0)$ vanishes but $1 - \xi_0$ is now a point of unstable equilibrium. Energy is gained from point $1 - \xi_0$ onwards up to $1 + \xi_0$, the neighbouring stable position; the total work from $\xi_0$ to $1 + \xi_0$ is $W_D$. The quantities $W_A$ and $W_D$ are the activation energy and energy of formation of one more MD, respectively. They are easily calculated by integrating the infinitesimal work $t(\xi)ad\xi$ to give

$$W_D = 2Wl_0^2\ \{2E(k)/\pi kl_0 - f\} \tag{8a}$$

$$W_A = 4Wl_0^2\{E(k, \tfrac{1}{2}\pi - \pi\xi_0)/\pi kl_0 - (\tfrac{1}{2} - \xi_0)f\} \tag{8b}$$

where $E(k)$ and $E(k, \phi)$ are complete and incomplete elliptic integrals of the second kind, and $\xi_0$ is given by equation (7). For the first MD that is formed in a pseudomorphic ($k = 1$) chain,

$$W_D^\infty = 2Wl_0^2(2/\pi l_0 - f) \tag{9a}$$

$$W_A = 4Wl_0^2\{\cos\ \pi\xi_0/\pi l_0 - (\tfrac{1}{2} - \xi_0)f\} \quad \sin\ \pi\xi_0 = l_0 f. \tag{9b}$$

The stability of the system clearly depends strongly on the misfit. It follows from (8) and (9) that the stable (minimum energy) will be realised when $W_D = 0$, i.e.

$$f = f^e \equiv \frac{2E(k)}{\pi kl_0} \underset{1}{\overset{k}{\rightarrow}} \frac{2}{\pi l_0} \equiv f_c. \tag{10a}$$

It also follows from (8$b$) that $W_A$ vanishes when $\xi_0 = \frac{1}{2}$ and hence from (7) that MDs will enter spontaneously when

$$f = 1/kl_0 \underset{1}{\overset{k}{\rightarrow}} 1/l_0 \equiv f_s. \tag{10b}$$

It follows from (10$a$) that the pseudomorphic configuration is the stable one when $f < f_c$ but that the activation energy $W_A$ to the introduction of

MDs only vanishes when $f = f_s > f_c$. At $f = f_c$

$$W_A = \frac{4Wl_0}{\pi}\left((1 - 4/\pi^2)^{1/2} - \frac{1}{2} + \frac{1}{\pi}\sin^{-1}\frac{2}{\pi}\right) \tag{11}$$

which is about $4W$.

The conditions ($10a$) for stability could also be obtained, more laboriously though, by minimising the average total energy

$$\bar{V}_D = \frac{V_D}{P(k)} = \frac{Wl_0^2}{P(k)}\left(\frac{4E(k)}{\pi k l_0} - \frac{2(1-k^2)K(k)}{\pi k l_0} - 2f + P(k)f^2\right) \tag{12}$$

per atom, which we calculate from equation (1), using the continuum approximation and equations ($4a$) and ($5a$). The dependence of the equilibrium value of $\bar{V}_D$ on $f$, using ($5a$), ($10a$) and (12) and $k$ as parameter, is shown in figure 3.

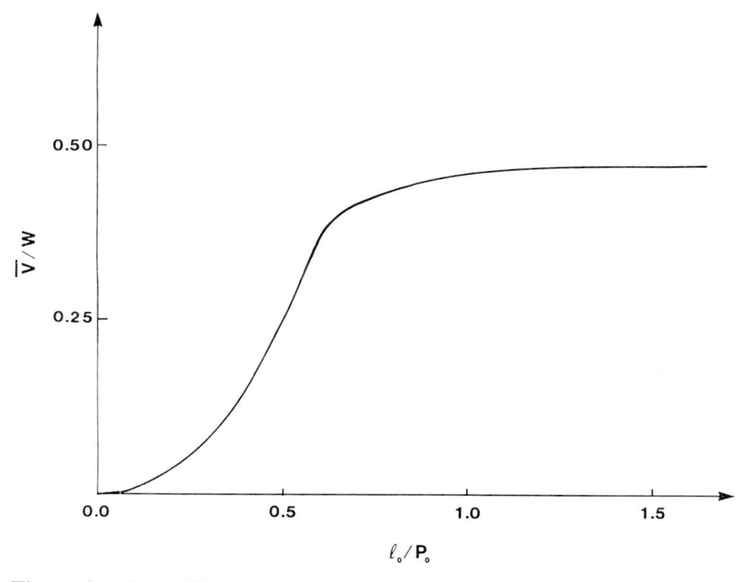


**Figure 3** Curve illustrating the dependence of the equilibrium average energy per atom $\bar{V}_D$ on relative misfit $l_0 f = l_0/P_0$, as determined by equations ($10a$) and (12). Note the abrupt change of curvature at $l_0 f = l_0 f_c$, where coherency breaks down.

We briefly summarise the essence of the foregoing on MDs, analysed in detail by Frank and van der Merwe (1949a,b), before proceeding to extensions of the theory.

F&M have analysed the equilibrium distribution of misfit between MDs and MS in the 1D FK model with free boundaries. In this model the (film)

ML atom–atom interaction is expressed in terms of the harmonic approximation ('strength' $\bar{\mu}$) and the atom–substrate interaction by a sinusoidal potential; a periodic potential ('strength' $W$) with the periodicity of the substrate. The authors have demonstrated the relevance of the analysis to a 2D system. The analysis has dealt with the introduction of MDs by glide under quasi-equilibrium conditions and predicted (i) that below a critical misfit $f_c$ ($\sim 9\%$ average) a ML is pseudomorphic, (ii) that above $f_c$ MDs enter so that $f$ is now accommodated by MDs and MS jointly, (iii) that MDs enter spontaneously above $f = f_s$ ($\sim 14\%$ average), (iv) that the introduction of MDs in the interval $f_c < f < f_s$ requires an activation energy $W_A$ that is normally provided thermally; at $f = f_c$, $W_A \approx 4W$ per atom length of MD, (v) that a film that grows ML-by-ML-wise, starting with a subcritical ($f < f_c$) ML, will become unstable to the introduction of MDs at a critical thickness $t_c$, which the authors crudely estimated by assuming that the strain gradient in the multilayer normal to the interface vanishes everywhere, and (vi) that the calculated critical quantities will be different for different combinations ($\bar{\mu}$ and $W$) of crystals.

Frank and van der Merwe (1949a,b) have also (vii) proposed a theory of oriented (epitaxial) growth in which subcritical misfit ($f < f_c$) for a ML is a prerequisite to epitaxy, the contention being that the initial orientation will be preserved in a ML-by-ML growth process and considered (viii) the introduction of MDs between a surface ML and the underlying crystal as a consequence of 'surface misfit' (van der Merwe 1988) and concluded that the surface misfit that obtains from 6–12 Lennard-Jones forces is well below $f_c$. Empirical data on multilayer relaxation (Jayanthi *et al* 1985) suggest that in some crystals surface misfit may attain critical values.

## 3 EXTENSIONS: F&M THEORY

The F&M theory of interfacial dislocations is incomplete in three categories. In category I the authors were aware of the inadequacies and the needs for extensions or refinements and have even gone some way in satisfying the needs. In category II the authors had touched on certain aspects but did not identify them or their significance fully. In category III are those aspects not considered. Although completeness will not be aimed at an effort will be made to cover essentials.

### 3.1 Category I: extensions and refinements

An important item in this category is the model, particularly its dimensionality which is important in both (a) the Fourier modelling of the atomic interaction across the interface, and (b) the atomic interaction in the crystals modelled by the harmonic approximation. The authors have dealt with two features. Firstly (Frank and van der Merwe 1949b) they have demonstrated, using the 1D model, that a truncation at the second harmonic has quantitative effects on the predicted energy of the MDs, for example, and that second harmonic terms with relatively large amplitudes may introduce additional features such as stacking faults, for example. It is fortunate though that the magnitude of the Fourier coefficients fall off rapidly with harmonic order (Stoop and Snyman 1988, Stoop *et al* 1990a,b). Secondly they have demonstrated how the analysis could be extended to 2D by using a first-order 2D Fourier truncation in a system with quadratic symmetry. Van der Merwe (1970) has elaborated on this by modelling a ML as an arrangement of atoms embedded in an isotropic elastic sheet that is elastically and crystallographically a truncation of the bulk. The main features of this work were (i) an extension from quadratic to rectangular interfacial symmetry, (ii) the approximation of ML elastic and lattice parameters by those of the bulk in estimating critical quantities and (iii) the establishment of the effect of Poisson's ratio $\nu$, for example, that the critical misfit $f_c$ in (10$a$) generalises to

$$f_c^{(2)} = 2/\pi l_0(1+\nu). \tag{13}$$

In the foregoing, stability with respect to the introduction of MDs by glide along the interface from the free edge of the ML was summarised. The formation of MDs by climb of adatoms from or to the free surface of the ML was not considered by F&M and will be dealt with in section 3.3 below.

An important 2D extension (van der Merwe 1980, 1982, Stoop and van der Merwe 1982, Braun 1987) was that to rhombic interfacial symmetry to describe MDs at $\{111\}$FCC–$\{110\}$BCC epitaxial interfaces (Bruce and Jaeger 1978). A useful Fourier truncation at second-order harmonics for the $\{110\}$BCC surface (van der Merwe 1982) is

$$V = W\left\{1 + A_1\left[\cos 2\pi\left(\frac{x}{a_x}+\frac{y}{a_y}\right) + \cos 2\pi\left(\frac{x}{a_x}-\frac{y}{a_y}\right)\right] + A_3\cos\frac{4\pi x}{a_x} + A_4\cos\frac{4\pi y}{a_y}\right\}. \tag{14}$$

The Fourier coefficients $WA_i$ have recently been calculated by Stoop *et al* (1990a,b) for a $W$ adatom and estimated by Stoop (1990) for an atom within a $W\{110\}$ ML, both on a $W\{110\}$ surface. Whereas the relative values of the $A_i$ are largely determined by the symmetry the scale factor $W$ was significantly different. The values obtained by Stoop (1990) for a ML atom are

$$W = 0.5869 \text{ eV} \quad A_1 = -0.5079 \quad A_3 = 0.1263 \quad A_4 = -0.0080. \tag{15}$$

The calculations (van der Merwe 1982) predicted two important properties of the Fourier description. Firstly, when close matching of parallel rows exists on either side of the interface—equivalently the close matching of reciprocal lattice vectors (Braun 1987)—energy is gained from the interfacial interaction if the crystals align so that the two sequences are parallel and MDs form by elastic relaxation to accommodate the misfit. The identification of the Burgers vectors of the MDs are facilitated by the reciprocal lattice formalism (Braun 1987). Secondly, the amount of energy gain is proportional to the Fourier coefficient with the matching wavelength. Although the MD energy constitutes a positive contribution there is a net gain in energy when the misfit is small enough. This result emphasises the importance of the quantification of Fourier coefficients. Also, different sets of close matching row sequences constitute different relative orientations and different MD structures.

Two important orientations are the Nishiyama–Wassermann (NW) and Kurdjumov–Sachs (KS) orientations (Bruce and Jaeger 1978). In the former the diagonals of the rhombi are parallel and there are two (near) row matching possibilities, whereas in the latter the rhombi sides are parallel and closely matching, and there are two equivalent orientations differing by about 5°. The MD Burgers vectors can be determined by reciprocal lattice vector construction (Braun 1987). An important feature of the systems is that the deviation from row matching, i.e. the corresponding misfit, is uniquely determined by the nearest-neighbour distance ratio $r$ of the participating crystals (van der Merwe 1982). For perfect matching $r_{KS} \approx 1.0887$ and $r_{NW} \approx 0.9428$, 1.1547.

The mathematical expressions involved in the analysis for systems with rhombic symmetry, and their final results, for critical misfit for example, are different from the simple ones for quadratic and rectangular symmetry and more so when anisotropy is taken into account (Braun 1987). Thus the critical misfit for an isotropic $\{111\}_{FCC}$ ML in NW orientation on a $\{110\}_{BCC}$ substrate for matching in the $\langle\bar{2}11\rangle_{FCC}$ $\langle\bar{1}10\rangle_{BCC}$ directions (Stoop and van der Merwe 1982) would depend on many factors, for example, on whether the ML–substrate interaction is strong enough to maintain coherency in the perpendicular direction or is so weak that there are many MDs, approximating to a vernier in the perpendicular direction.

The other important extension of dimensionality is in a direction normal to the interface. This is necessary when the film grows into a multilayer. F&M have foreseen this need and indicated for example in the case where a ML is subcritical, that a critical thickness $t_c$ in the case of small multiplicity $n$ can be estimated by assigning a value $n\mu$ to the force constant of the chain. This approximation is useful only when $n < 3$. For larger $n$ strain gradients normal to the interface become important in both the film and substrate (Ball 1970).

The extreme case of this problem, i.e. the one where both film and substrate are 'infinitely' thick, has been analysed by van der Merwe (1950, 1963) using an extension of the Peierls–Nabarro (PN) model (Nabarro 1947), which itself may be seen as a generalisation of the FK model. In this model the authors expressed the interaction between two crystal halves in terms of a periodic shear stress. In the van der Merwe (1950) generalisation

$$p_{zx}(U) = (\mu c/2\pi d)\sin(2\pi U/c) \qquad U = U(x) \tag{16}$$

where $\mu$ is an interfacial shear modulus, $d$ is the separation of the atomic planes on either side of the interface, and $U$ is the relative displacement of opposing atoms. The relation (16) models the shear stress along $x$ for the interface of two SC crystals $A$ and $B$ respectively with lattice parameters $a$ and $b$ and shear moduli $\mu_a$ and $\mu_b$; the $x$ and $z$ axes are parallel to a cube axis in the interface and normal to the interface, respectively. The quantity

$$c = \frac{ab}{\frac{1}{2}(a+b)} \tag{17a}$$

is the lattice parameter of a reference lattice for $A$ and $B$ and

$$f = \frac{c}{p} = \frac{(b-a)}{\frac{1}{2}(b+a)} \tag{17b}$$

the misfit. In this case the misfit is entirely accommodated by a cross-grid of MDs. It is assumed that the crossing energy is negligible so that the two sequences can be treated independently. For either sequence the problem is one of plane strain and reduces to a boundary value problem in linear isotropic elasticity with boundary stress (16) and hence can be analysed in terms of an Airy stress function. The average energy per MD then takes the form

$$E_{\mathrm{D}} = (\mu c^2/4\pi^2 d)\{1 + \beta - (1+\beta^2)^{1/2} - \beta \ln[2\beta(1+\beta^2)^{1/2} - 2\beta^2]\} \tag{18a}$$

$$\beta = \frac{2\pi d\lambda}{\mu p} \qquad \frac{1}{\lambda} = \frac{1-\nu_a}{\mu_a} + \frac{1-\nu_b}{\mu_b} \qquad p = \frac{ab}{b-a} \tag{18b}$$

where $p$ is the natural spacing of MDs and $\lambda$ is an effective shear modulus. $\beta$ is the equivalent of $l_0/P_0$ and combines the elastic and misfit parameters of the system.

Two additional results are of particular interest: (i) when MDs are far apart ($p$ large) $E_{\mathrm{D}}$ takes on the asymptotic form

$$E_{\mathrm{D}}^{\infty} = \frac{\lambda c^2}{2\pi} \ln\left(\frac{p}{d}\frac{e\mu}{4\pi\lambda}\right) \tag{19}$$

and (ii) less than 2% of the energy is stored at distances from the interface exceeding $\frac{1}{2}p$, i.e. half the MD spacing. The implication of (ii) is that the expression in (18$a$) can be used for all crystal halves whose thickness exceeds $\frac{1}{2}p$.

Several attempts have been made to solve the problems for films of finite thickness $t$ that cannot be approximated in terms of the two extreme cases (i) a ML on a thick substrate and (ii) a bicrystal of infinite ($t > \frac{1}{2}p$) halves.

The first attempt was made, by employing instead of (16) the parabolic model (van der Merwe 1963)

$$p_{zx}\{U(x)\} = \mu U(x)/d \qquad |U| \leqslant \tfrac{1}{2}c \tag{20}$$

which is a straight line segment with the same slope as (16) at $U = 0$. The segment is repeated in every period of (16) in a saw-tooth-like manner. The parabolic model is evidently a more crude approximation of the interfacial interaction than the sinusoidal approximation (16) but has the merit that the governing equations can be solved exactly. When integrated for the energy (20) takes a parabolic instead of sinusoidal form as in equation (1). The author approximated the energy minimisation procedure by an equality of energies, the assumption being that a transition from coherency to incoherency will occur when the misfit strain energy becomes equal to the energy of the MDs when there is no MS. It was subsequently shown (Jesser and van der Merwe 1988) that this approximation may introduce errors up to 25%. Other approximations undertaken by Ball (1970) are based on the sinusoidal form (16). The results were meaningful but complicated and not easy to apply.

The most successful attempt to analyse the stability of a thickening overlayer to the introduction of MDs is the approach developed by Matthews (1979). This approach is analytically relatively simple and can deal with cases that are far too difficult for the approach that invoke specific (periodic) forms of interfacial interaction. It describes the MDs in terms of the Volterra dislocation model using two concepts: (i) the cancellation of overlapping dislocation stress fields proposed by Nabarro (Kuhlmann-Wilsdorf 1986) and (ii) the core energy of a dislocation that can be manipulated to account, amongst others, for the strength of interfacial interaction (Hirth and Feng 1990). Matthews (1979) defines the misfit for a thin overlayer as

$$f = (b - a)/b \tag{21a}$$

instead of (2), so that exactly

$$f = \bar{f} + |e| \tag{21b}$$

for MDs of edge type. In this approach the average energies per atom of interface due to MS and MDs respectively take the forms

$$\varepsilon_{\mathrm{e}} = 2\mu_b(1 + \nu_b)b^2te^2/(1 - \nu_b) \tag{22a}$$

$$\varepsilon_{\mathrm{D}} = 2\bar{f}\lambda b^2[\ln(R/r_0) + 1]/2\pi \tag{22b}$$

for an overlayer $B$ of thickness $t$ and a thick substrate crystal $A$. $\lambda$ is defined in (18$b$). $R$ and $r_0$ are the outer and inner cut-off radii. A value of $r_0 \approx 4.7c$ is obtained from (19) by comparison with (22$b$). A value $r_0 = b$, where $b$ is the Burgers vector, seems to have acquired general acceptance. The term 1 in square brackets together with an appropriate value of $r_0$

determine the core energy of the dislocation so that $r_0$ may be manipulated to account for different strengths of interfacial interaction. On the basis of the cancellation of overlapping stress fields Matthews (1979) proposed that

$$R = \begin{cases} \frac{1}{2}\bar{p} = c/2\bar{f} & \text{when } t > \frac{1}{2}\bar{p} \\ t & \text{when } t < \frac{1}{2}\bar{p} \end{cases} \tag{23}$$

where $\bar{p}$ is the spacing of MDs of density $\bar{f}$. By minimising the total energy per atom with respect to $\bar{f}$, eliminating $e$ with the help of (21$b$) the equilibrium MD density may be obtained. It follows from (23) that there are two possibilities for the equilibrium MD density $\bar{f}_{eq}$:

$$\bar{f}_{eq} = f - A[\ln(t/r_0) + 1] \qquad t < \tfrac{1}{2}\bar{p} \tag{24a}$$

$$\bar{f}_{eq} - A \ln \bar{f}_{eq} = f - A \ln(2r_0/b) \qquad t > \tfrac{1}{2}\bar{p} \tag{24b}$$

$$A = \lambda(1 - \nu_b)/4\pi\mu_b(1 + \nu_b)t. \tag{24c}$$

Matthews (1979) has continued to simplify the analysis by replacing the energy minimisation process by a balance of force criterion. This was particularly helpful in the case where the MDs (inefficient MDs) were generated from threading dislocations (TDs) whose Burgers vectors were not in the interface. A variety of forces are relevant, for example, the line tensions of the MDs and steps and Peach–Koehler MS forces. We also note that the mechanism of MD formation from an existing (threading) dislocation in a multilayer differs fundamentally from the mechanism of nucleation at the edge of a ML, analysed by F&M.

Frank and van der Merwe (1949a,b) estimated the parameters (Fourier coefficients ($W$), and lattice ($a$,$b$) and elastic ($\bar{\mu}$) constants) for the quantification of their predictions based on the phenomenological FK model, using Lennard-Jones (LJ) 6–12 pair potentials. LJ pair potentials were also used by Stoop and Snyman (1988) in a more complete analysis to calculate Fourier coefficients for an isolated Ar atom on an Ar crystal substrate. However, in generalisations of the F&M model, and the approach of Matthews, elastic and lattice constants of a thin film were crudely identified with those of the bulk crystal. The interfacial interaction as characterised by the Fourier coefficients was more difficult to quantify. They were either assigned values on the basis of hand-waving arguments (van der Merwe 1982) or were defined (Volterra) by the condition that the atomic planes terminate in 'registry' at the interface, the crystals being modelled as elastic continua. While the inadequacies of the latter approximations are evident and were realised all along, it was also established that the pair potentials had serious shortcomings in modelling interatomic potentials in metals (Johnson 1973). The possibility of a fundamental approach using quantum mechanics has been ruled out by the magnitude of the problem. More recently semi-empirical $N$-body potentials (Finnis and Sinclair 1984, Johnson and Oh 1989, Gollisch 1986), the embedded-atom methods (EAM), with a more fundamental basis have been developed that offer a meaningful intermediate between the pair potentials and the purely fundamental approach.

Embedded-atom methods were used to calculate (Stoop *et al* 1990a,b) self-absorption parameters, including Fourier coefficients, for a $W\{110\}$ surface, yielding values in good agreement with empirical data, where they exist. Like the calculations of F&M they were for an isolated adatom, whereas in the present considerations the atom forms part of a ML or multilayer, where the interaction of an adatom with the substrate is influenced by neighbouring atoms of the film. Preliminary results (Stoop 1990) indicate that the relative values of the Fourier coefficients for an isolated atom and an atom in a ML are determined by the symmetry of the surfaces and that the strength of bonding as contained in a scale factor can be significantly different.

The calculations (Stoop and Snyman 1988, Stoop *et al* 1990a,b) show that the magnitudes of the Fourier coefficients fall off rapidly with harmonic order and accordingly justify the simplifying practice of truncating Fourier representations at low harmonic order.

### 3.2 Category II: aspects touched on only

In this section we briefly consider aspects that F&M were clearly aware of but did not enter into and often even did not explicitly name. Important items in this category are equilibrium aspects of the calculations. Firstly, while the authors clearly have been aware of the thermodynamic nature of the considerations they did not discuss the conditions under which the free-energy equilibrium criterion can be replaced by the mechanical minimum-energy criterion. Some of these conditions were briefly mentioned in section 2.

Secondly, the authors have also tacitly assumed that MDs will be available where and whenever they are needed to effect an equilibrium distribution. The existence of barriers to the motion (Peierls) and formation of nucleation of MDs constitute an important obstacle that can only be overcome by thermal agitation or/and the appropriate driving forces; as equilibrium is approached the driving forces fall off to zero. In many cases the duration of an experiment is by far inadequate to equilibrate the MD distribution. Accordingly, in many observations (People and Bean 1985) the apparent critical misfit has been well above the equilibrium prediction, more than could be explained by the crudeness of the models. Also this fact was fully appreciated, and was addressed briefly by Matthews (1979). The problem was also more recently considered by Tsao *et al* (1987) and by Dodson and Tsao (1987). Tsao *et al* have identified the equilibration driving force with the amount by which the Peach–Koehler MS force on a threading dislocation exceeds the MD line tension in metastable $Si_xGe_{1-x}$ strained multilayers grown on Ge $\{001\}$ substrates and expressed the force in terms of an excess stress

$$\tau_{\text{exc}} = \frac{2e\mu(1+\nu)}{1-\nu} - \frac{\mu(1-\nu\cos^2\alpha)}{2\pi(1-\nu)}\frac{\ln(4t/b)}{t/b} \qquad e = f$$

for 60° ($\alpha = 60°$) dislocations. The authors could correlate $\tau_{\text{exc}}$ with the substrate temperature at which the equilibration is fast enough to become observable. Dodson and Tsao have analysed the dynamics of MS relief in the same system by postulating that a local stress which can be expressed in terms of residual mismatch—misfit not taken up by MDs—acts as a driving force for the dislocations, whose motion is opposed by the existing energy barriers present.

An interesting phenomenon briefly dealt with by F&M is that of 'surface misfit' $f_s$, the difference in the equilibrium spacing of a free planar ML and the bulk surface (van der Merwe 1988). The authors concluded that in their 6–12 LJ crystal the surface misfit of a close-packed surface is only about 2% and thus well below the critical misfit above which MDs are needed. The pair potentials have serious shortcomings though (Johnson 1973). Indeed, one may conclude from empirical data on multilayer relaxation that cases exist where $f_s$ is appreciably greater than 2%. Though perhaps still too small to stabilise the presence of MDs it may effect surface reconstruction and sufficient elastic relaxation at steps (van der Merwe 1988) on the crystal surface to induce long-range stress fields with significant consequences for surface phenomena, for example phase transitions (Bauer 1990).

F&M proposed a theory of oriented overgrowth in which pseudomorphic ML-by-ML growth is a prerequisite, as consecutive MLs will stay put and preserve the orientation, whereas MDs introduce mobility that would be detrimental to the formation of a unique orientation. Mobility, both in translation and rotation, has been confirmed for island overgrowths, both experimentally and theoretically. A rotation implies the introduction of screw-type MDs or MDs of mixed character. The energetics of island overgrowths in rotation and the behaviour of MDs were investigated by Jesser and Kuhlmann-Wilsdorf (1967). These considerations suggest that rotational thermal fluctuations do occur so that a continuous film will display a distribution of orientations with a MD structure depending on the average island size at coalescence. The energetics was also investigated by Reiss (1968) and van der Merwe (1982) for a model with weak interfacial bonding. All these considerations show that off orientations are generally of higher energy and that islands with MDs are significantly more mobile.

### 3.3 Category III: aspects not dealt with

The F&M calculations implicitly assumed a ML-by-ML (FM) growth. Equilibrium criteria for the different growth modes were proposed by Bauer

(1958) in terms of the surface free energies for the growing film, the interface and the substrate crystal, respectively $\sigma_b$, $\sigma_i$ and $\sigma_a$, as

$$\Delta\sigma \equiv \sigma_b + \sigma_i - \sigma_a \leqslant 0 \qquad \text{FM}$$
$$> 0 \qquad \text{VW}$$

for FM and Volmer–Weber (VW: island) growth. Stranski–Krastanov (SK) growth obtains when a transformation occurs from FM to VW growth after completion of a few MLs and may be accompanied by the introduction of MDs in the areas covered by islands. A film can be coerced into growing FM-like (Bauer and van der Merwe 1986) by a high enough supersaturation and the realisation of a high island nucleation density in which the critical nucleus is one atom. While the concepts of MDs and MS also apply to isolated islands the 'critical thickness' will be size dependent. One can expect that the MD structures for continuous overlayers will also depend on the relevant growth mode.

An important mechanism for the formation of MDs, not dealt with by F&M, is the climb mechanism. A pseudomorphic ML may acquire MDs by the climb of atoms from the interior of the ML 'up' to its free surface or conversely by the climb of adatoms from the free surface 'down' to the ML interior. The former (latter) will evidently be energetically favourable when the misfit is positive (negative) provided the misfit exceeds the critical value. An important consequence of this mechanism is that MDs of 'finite' length occur, i.e. MDs that terminate at least at one end within the ML in a Taylor-type dislocation in the atomic structure of the ML. The other end may likewise terminate in a Taylor dislocation or at the edge of the ML.

The energy of formation of MDs in MLs by climb has recently been studied by Tönsing (1990) within a model that is a combination of a quadratic 2D isotropic FK model for the ML-substrate system and a Peierls–Nabarro model for the Taylor dislocation within the ML. By linearisation with a parabolic approximation the complicated non-linear governing equations became soluble. The calculations predicted: (i) a critical MD nucleation length $L_c \sim 1$ atomic spacing, (ii) a nucleation energy of about $0.15\,\mu\Omega$ ($\approx 4.3$ eV for Cu), $\mu$ being the shear modulus of the ML and $\Omega$ the volume per atom, (iii) that a supercritical MD will grow (lengthen) when $f > f_c$ and (iv) that the interaction energy for dislocation unbinding (receding) is relatively small and will fall off exponentially. An important difference with the nucleation of MDs by glide from the ML edge is that the critical nucleation loop length by glide must be at least a few atomic diameters and that the nucleation energy is therefore significantly larger. The climb process is therefore energetically favourable. Finite length MDs, when captured by the growth of additional MLs, would grow into dislocation half loops and form sources for the multiplication of dislocations within the growing film.

The nucleation and growth of MDs by climb is an important mechanism in phase transitions of adlayers on crystal surfaces, for example, in commensurate–incommensurate transitions and dislocation-mediated melting. The latter has been the subject of extensive theoretical study in weakly incommensurate MLs (Coppersmith *et al* 1982, Pokrovsky and Talapov 1983), i.e. in MLs where the distance between MDs is much larger than the MD width $l_0$ as defined for example in equation (3*b*). At low temperatures there exists a more or less regular lattice of parallel finite length MD lines (dislocation dipoles), that unbind when the temperature exceeds some critical value. The theories concerned themselves mainly with the interaction between the dislocations of a dipole. The dislocation unbinding creates large numbers of free dislocations in the ML leading to an exponential decay of translational order. The condition for dislocation unbinding is that the interaction between the poles is repulsive, or zero. It has been shown that unbinding is always favourable when the MD length exceeds the critical length.

In conclusion we wish to mention three more matters of interest. These concern the harmonic approximation for the atomic interaction within the overlayer: anisotropy, unharmonicity and structural metastability. The preliminary analysis of anisotropy by Braun (1987) predicted that anisotropy significantly influences the MD structure and could also affect the epitaxial orientation, for example in orientational epitaxy. The consequences of unharmonicity were investigated by Milchev and Markov (1984) for a 1D chain as in the FK model. The essential finding is that the predicted critical quantities are asymmetric with respect to the sign of the misfit. The critical misfit increases with increasing unharmonicity when the misfit is negative whereas it decreases for positive misfit. Furthermore, the magnitude of the former was the greater of the two and could deviate as much as 25% from the standard value predicted by Frank and van der Merwe (1949a,b). The case of overlayer materials that can adopt a bulk metastable crystal structure with a lattice constant significantly different from that of the stable ground state structure was analysed by Bruinsma and Zangwill (1986). In this case there will be an energy barrier between the ground state and the metastable state. The substrate interaction could stabilise the overlayer in the metastable phase with an apparent large MS which cannot be understood on the basis of the harmonic approximation.

## 4 CONCLUDING REMARKS

A feature of great interest in systems with competing crystallographic periodicities is the accommodation of misfit by misfit (interfacial) dislocations (MDs) and misfit strain (MS). In epitaxy great fundamental and technological importance is attached to the registry–disregistry transition of accommodation by MS alone to one with MDs and MS jointly. The acquisition of MDs has accordingly become a topical subject. The work of Frank and van der Merwe has laid the foundations for the understanding of the phenomena involved and the quantification of the important parameters, e.g. critical misfit and thickness for coherency breakdown. Most important contributions to finding analytical expressions of practical use in device fabrication have been made by Matthews. Important aspects not dealt with by F&M are the formation of MDs by climb and non-equilibrium effects related to barriers of motion and formation of MDs. Attempts at analysing these have briefly been considered. The dynamics of MS relief is a subject that still needs considerable attention.

Whereas empirical data provide qualitative and semi-quantitative confirmation of the predictions of the phenomenological theories referred to above, particularly for thick overlayers, the modelling of ultrathin overlayers is still rather crude, for example, the assumption that a ML can be identified with a truncation of the bulk. Work is in progress using embedded-atom methods with the view of developing descriptions suitable for application to ultrathin overlayers. Much insight into the dynamics of the processes involved have already been acquired using Monte Carlo and molecular dynamic techniques, and more can be expected.

## REFERENCES


Ball C A B 1970 *phys. status solidi* **38** 335–44, **42** 357–68

Bauer E 1958 *Z. Kristall.* **110** 423

—— 1990 Private communication

Bauer E and van der Merwe J H 1986 *Phys. Rev.* B **33** 3657–71

Braun M W H 1987 *PhD Thesis* University of Pretoria

Bruce L A and Jaeger H 1978 *Phil. Mag.* A **38** 223–40

Bruinsma R and Zangwill A 1986 *J. Physique* **47** 2055–73

Coppersmith S N, Fisher D S, Halperin B I, Lee P A and Brinkman W S 1982 *Phys. Rev.* B **25** 349–63

Delavignette P, Tournier J and Amelinckx S 1961 *Phil. Mag.* **6** 1419–20

Dodson B W and Tsao J Y 1987 *Appl. Phys. Lett.* **51** 1325–7

Finch A I and Quarrell A G 1934 *Proc. Phys. Soc.* **48** 148–62

Finnis M W and Sinclair J E 1984 *Phil. Mag.* A **50** 45–55

Frank F C and van der Merwe J H 1949a *Proc R. Soc.* A **198** 205–25

—— 1949b *Proc. R. Soc.* A **200** 125–34

Frenkel J and Kontorowa T 1938 *Phys. Z. Sowjet.* **131** 1

Gollisch H 1986 *Surf. Sci.* **166** 97–100, **175** 249–62

Hibma T 1988 *The Structure of Surfaces II* eds J F van der Veen and M A van Hove (Berlin: Springer)

Hirth J P and Feng X 1990 *J. Appl. Phys.* **67** 3343–9

Jayanthi C S, Tosatti E, Fasolino A and Pietronero L 1985 *Surf. Sci.* **152/153** 155–61

Jesser W A and Kuhlmann-Wilsdorf D 1967 *phys. status solidi* **19** 95–105

Jesser W A and van der Merwe J H 1988 *J. Appl. Phys.* **63** 1928–35
Johnson R A 1973 *J. Phys. F: Met. Phys.* **3** 295–321
Johnson R A and Oh D J 1989 *J. Mater. Res.* **4** 1195
Kuhlmann-Wilsdorf D 1986 *S. Afr. J. Phys.* **9** 46–54
Matthews J W 1979 *Dislocations in Solids* ed F R N Nabarro (Amsterdam: North-Holland) pp 462–545
Milchev A and Markov I 1984 *Surf. Sci.* **136** 503–31, **145** 313–28
Nabarro F R N 1947 *Proc. Phys. Soc.* **59** 256–72
People R and Bean J C 1985 *Appl. Phys. Lett.* **47** 322
Pokrovsky V L and Talapov A L 1983 *Theory of Incommensurate Crystals, Soviet Scientific Reviews* vol 1 (New York: Harwood) pp 90–162
Reiss H 1968 *J. Appl. Phys.* **39** 5045–61
Royer L 1928 *Bull. Soc. Fr. Minéral Crystallogr.* **51** 7
Stoop L C A and van der Merwe J H 1982 *Thin Solid Films* **91** 257–75
Stoop P M 1990 *PhD Thesis* University of Pretoria
Stoop P M and Snyman J A 1988 *Thin Solid Films* **158** 151–66
Stoop P M, van der Merwe J H and Braun M W H 1990a *Phil. Mag.* A to be published
—— 1990b *Vacuum* **41** 195–8
Tönsing D L 1990 *Phys. Rev.* B to be published
Tsao J Y, Dodson B W, Picraux S T and Cornelison D M 1987 *Phys. Rev. Lett.* **59** 2455–8
van der Merwe J H 1950 *Proc. Phys. Soc.* A **63** 616–37
—— 1963 *J. Appl. Phys.* **34** 117–22
—— 1970 *J. Appl. Phys.* **41** 4725–31
—— 1980 *Thin Solid Films* **74** 129–51
—— 1982 *Phil. Mag.* A **45** 127–70
—— 1988 *Phys. Rev.* B **37** 2892–901

# Seminal Ideas on the Structure of Interfaces

R C Pond

## 1 INTRODUCTION

According to my pocket dictionary, seminal means 'pregnant with consequences'. The contribution of F C Frank's ideas to the study of interfacial structures in crystalline materials certainly merits the use of this adjective. Revealing the fine details of interfacial structure by experimental and theoretical means is actually proving to be a challenging task, and hence the gestation period for this knowledge is not yet complete. However, very significant progress has been made, and Frank's work in the early development of this research field remains an essential part of our present understanding. Moreover, his recent contributions, such as the treatment of orientation mapping, continue to influence the direction of interfacial research.

The objective of this paper is to celebrate Professor Frank's eightieth birthday by describing his contribution to our current understanding of the structure of interfaces. This will not be approached from a biographical point of view, but rather by discussing recent investigations, and showing how Frank's work forms the cornerstone in the interpretation of the observations. In order to focus the paper, we shall develop two particular themes, both of which touch on other research fields where Frank has made distinctive contributions. Firstly, we shall consider some aspects of the relationship between a free surface on a crystal and the resulting interface following deposition of a second crystal on the initial substrate. This issue has, of course, gained immense technological significance due to the applications of epitaxial films, particularly in microelectronics. Secondly, we shall concentrate on interfaces where at least one of the adjacent crystals exhibits hexagonal or rhombohedral symmetry. The motivation for this is that the crystallography of such materials is still regarded with trepidation by many, and treatments in the literature are not always clear. In 1965 Frank published a paper in which he explained the mathematical basis for the Miller–Bravais system of indexing, and thereby introduced an approach which simplifies crystallographic calculations in direct and reciprocal space.

The next section of this paper is a summary of Frank's scheme for crystallographic calculations in hexagonal and trigonal crystals. We include this account at some length in order to promote more widespread use of Frank's elegant method, which the present author has found to be extremely useful. In the subsequent section, this approach is used in the analysis of reflection-high-energy-electron-diffraction (RHEED) patterns obtained from an $(01\bar{1}2)$ surface of sapphire ($\alpha$-$Al_2O_3$). It will be seen that surface and bulk diffraction effects can be analysed effectively using this comprehensive method. The nature of surface steps on unrelaxed surfaces is discussed in the following section, and is illustrated by examples from hexagonal-close-packed (HCP) metals and sapphire. Next we turn our attention from surfaces to interfaces, and consider the description of orientation relationships between adjacent crystals in the cases of grain boundaries in HCP metals and the interphase boundary separating silicon deposited epitaxially on sapphire. The final section is a discussion of interfacial defects. In particular, the role of arrays of defects in interfaces will be considered, and Frank's ideas concerning the accommodation of misfit and small orientation differences at low-angle boundaries will be applied to the interface between silicon and sapphire.

## 2 HEXAGONAL AND TRIGONAL CRYSTALLOGRAPHY

Frank pointed out in 1965 that 'although all who have to do with hexagonal crystals employ the Miller–Bravais notation, few exploit its capabilities to the full'. The principal advantage of the Miller–Bravais scheme is, of course, that the indices of planes and directions reflect the true symmetry of the crystal in question, and hence crystallographically equivalent planes or directions have similar indices. On the other hand, this advantage may appear to have been gained at the expense of increased complexity in crystallographic calculations because the coordinate frame used is not orthogonal. However, by relating the Miller–Bravais scheme to the more general system of mathematics, Frank (1965) showed that these apparent complexities are removed, and that the commonly performed calculations can be carried out readily. The device exploited is a four-dimensional orthogonal frame such that the Miller–Bravais symbols can be interpreted as vectors confined to a three-dimensional subspace of this 4-space.

### 2.1 Projection from a higher dimension

It is possibly the invocation of a four-dimensional space which has discouraged the widespread use of Frank's method by crystallographers. For this reason, we present here, in a slightly different manner to that in the original paper, Frank's introduction to the use of a higher dimensional frame for the representation of figures in a lower dimension. Imagine first that it is necessary to devise an indexing scheme for the hexagon AEBFCD illustrated in figure 1. One choice would be to index the direction **OA** as [1, 0], and **OB** as [0, 1]. Clearly, this procedure would not reflect the symmetry of the figure because, for example, the index for **OE** would be [1, 1]. A second possibility is to regard the hexagon as the two-dimensional projection of a cube viewed along its body diagonal OO′, where O′ is vertically above O in the figure. Now, making use of a three-dimensional cartesian frame, we can index **O′A′** as [100], **O′B′** as [010] and **O′C′** as [001] where A′, B′ and C′ represent vertices of the cube located vertically above A, B and C, respectively. Since the projection direction O′O is [111], the components of these primitive vectors resolved parallel to the plane of the projection are $\mathbf{OA} = \frac{1}{3}[2\bar{1}\bar{1}]$, $\mathbf{OB} = \frac{1}{3}[\bar{1}2\bar{1}]$ and $\mathbf{OC} = \frac{1}{3}[\bar{1}\bar{1}2]$, respectively. We note that the scalar product formed between any vectors in the two-dimensional space and the projection direction, [111], must be zero or, in other words, the three indices of the projected vector must sum to zero. Thus, by exploiting the auxiliary dimension, a much more symmetrical algebraic representation is obtained, while still retaining the advantages of a cartesian frame. We note that the three-fold symmetry exhibited by the hexagon is also a 'true' symmetry element of the cube, but that the six-fold symmetry arises only in the two-dimensional space.

Now consider the analogous procedure of projecting a four-dimensional hypercube, with edges indexed as [1000], [0100], [0010] and [0001], along the direction [1110]. The first three unit vectors projected into three dimensions along this direction are $\frac{1}{3}[2\bar{1}\bar{1}0]$, $\frac{1}{3}[\bar{1}2\bar{1}0]$ and $\frac{1}{3}[\bar{1}\bar{1}20]$, whereas the fourth, being perpendicular to [1110], remains invariant, i.e. [0001]. Thus, we have obtained a three-dimensional subspace comprising three vectors in a basal plane, and a fourth vector, [0001], which is perpendicular to this plane. As was the case for the previous example, the scalar product formed between vectors in the three-dimensional subspace and the projection direction must be zero, and hence the first three indices of vectors in the subspace must sum to zero.

The relationship between the subspace cell defined by the vectors $\frac{1}{3}[2\bar{1}\bar{1}0]$, $\frac{1}{3}[\bar{1}2\bar{1}0]$, $\frac{1}{3}[\bar{1}\bar{1}20]$ and [0001] and the conventional hexagonal unit cell is now apparent. However, it is necessary to scale this cell in order that its dimensions correspond to the lattice parameters, $a$ and $c$, of a real hexagonal or trigonal crystal. Let the unit 4-space vectors have magnitude $e$. It follows that the magnitude of each of the vectors $\frac{1}{3}\langle 2\bar{1}\bar{1}0\rangle$ is equal to $(2/3)^{1/2}e$, and therefore that the ratio of the magnitude of [0001] to $\frac{1}{3}\langle 2\bar{1}\bar{1}0\rangle$ is $(3/2)^{1/2}$. In other words, the cell obtained by projection of a hypercube has a particular $c/a$ ratio. Now, by increasing the length of 4-vector components parallel to [0001] by the appropriate factor, the three-dimensional cell will exhibit the $c/a$ ratio required for a particular problem. The required factor is simply equal to $c/e = (2/3)^{1/2}c/a$, and this was designated $\Lambda$ by Frank.

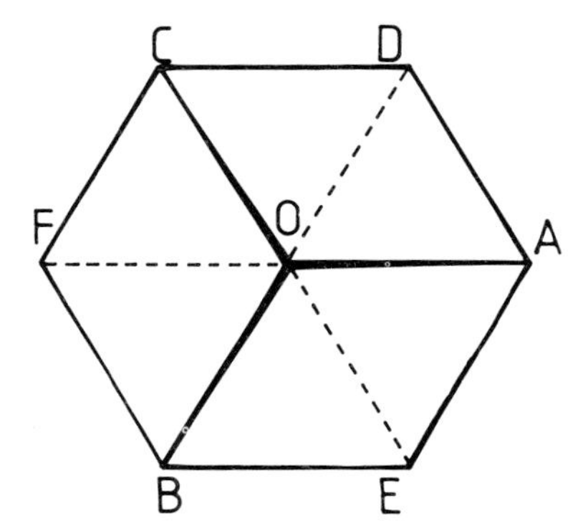


**Figure 1** Schematic view of a cube viewed along a body diagonal illustrating the hexagonal form in projection.

## 2.2 Crystallographic calculations

The mathematical scheme is now complete; it can be seen that a direction symbolised by $[uvtw]$ according to the Miller–Bravais system corresponds to the cartsian 4-vector $[uvt\Lambda w]_4$, where the subscript is used here to distinguish 4-vectors. Thus, for crystallographic directions the transformation between the Miller–Bravais symbolism and the 4-vector notation is very simple. Crystallographic calculations can be carried out straightforwardly using the latter formulation:

(i) the magnitude of the vector $[uvt\Lambda w]_4$ is equal to $(u^2 + v^2 + t^2 + \Lambda^2 w^2)^{1/2}$ in units of $e$ (or alternatively $(3/2)^{1/2}\,a$);

(ii) the scalar product of $[u_1v_1t_1\Lambda w_1]_4$ and $[u_2v_2t_2\Lambda w_2]_4$ is equal to $u_1u_2 + v_1v_2 + t_1t_2 + \Lambda^2 w_1w_2$ in units of $e^2$;

(iii) the angle between these two vectors is given by

$$\cos\theta = \frac{u_1u_2 + v_1v_2 + t_1t_2 + \Lambda^2 w_1w_2}{(u_1^2 + v_1^2 + t_1^2 + \Lambda^2 w_1^2)^{1/2}(u_2^2 + v_2^2 + t_2^2 + \Lambda^2 w_2^2)^{1/2}}$$

(iv) the vector product of these two vectors can be obtained by expanding the following matrix (Pond *et* al 1987b):

$$\begin{vmatrix} \mathbf{e}_1 & \mathbf{e}_2 & \mathbf{e}_3 & \mathbf{e}_4 \\ u_1 & v_1 & t_1 & \Lambda w_1 \\ u_2 & v_2 & t_2 & \Lambda w_2 \\ \bar{3}^{-1/2} & \bar{3}^{-1/2} & \bar{3}^{-1/2} & 0 \end{vmatrix}$$

where $\mathbf{e}_1, \mathbf{e}_2$, etc, are the cartesian unit vectors in 4-space. We note that the resulting 4-vector can be regarded as that which is simultaneously perpendicular to $[u_1v_1t_1\Lambda w_1]_4$ and $[u_2v_2t_2\Lambda w_2]_4$, and is also perpendicular to $3^{-1/2}[\bar{1}\bar{1}\bar{1}0]_4$ and hence exists in the projected subspace. The resultant 4-vector can also, of course, be transformed back into Miller–Bravais symbolism, if required, by simply dividing the final index by $\Lambda$.

Calculations involving planes can also be performed readily. Frank (1965) showed that the 4-vector perpendicular to the plane with Miller–Bravais index $(hkil)$ is $[hkil/\Lambda]_4$. (This can be proved, for example, by identifying two 4-vectors in the plane, and forming their vector product as described above.) It is then straightforward to determine the angle between this plane normal and any other crystal direction by using expression (iii) above. It can also be seen that the factor $\Lambda$ will not appear in the scalar product of these two directions. Moreover, this formulation shows why the 'zone law', i.e. $hu + kv + it + lw = 0$, is valid for directions $[uvtw]$ lying in the plane $(hkil)$. It is also straightforward to show that the interplanar spacing, $d$, for the planes $(hkil)$ is given by

$$d = (h^2 + k^2 + i^2 + l^2/\Lambda^2)^{-1/2}$$

in units of $e$. (This can be shown, for example, by determining $\phi$, the angle between [0001] and the plane normal, using expression (iii) above, and then using $\cos\phi = dl/c$, where $c/l$ is the intercept of the plane on the $c$ axis.)

### 2.3 Reciprocal space

Conventional treatments do not appear to relate direct and reciprocal lattices for hexagonal crystals in the customary way. On the other hand, Frank's method makes this connection in a straightforward and rigorous manner. The unit vectors of the direct, $\mathbf{e}_i$, and reciprocal, $\mathbf{e}_j^*$, 4-spaces are simply related by

$$\mathbf{e}_i \cdot \mathbf{e}_j^* = \delta_{ij} \qquad i, j = 1, 2, 3, 4$$

where $\delta_{ij}$ is the Kronecker delta. Thus, the reciprocal unit vectors are parallel to the direct ones and have magnitude $e^{-1}$. It follows that the usual crystallographic notation for the reciprocal vector, $\mathbf{g} = hkil$, associated with the planes $(hkil)$, corresponds to the reciprocal 4-space vector $\mathbf{g}_4 = hkil/\lambda$. In other words, the Miller–Bravais symbol of a direction in direct space is transformed to the 4-vector by multiplying the last index by $\Lambda$, and in reciprocal space the corresponding procedure is to divide the last index by $\Lambda$. Clearly, the magnitude of this latter vector is equal to $(h^2 + k^2 + i^2 + l^2/\lambda^2)^{1/2}$ in units of $e^{-1}$, and this is equal to $d^{-1}$. Moreover, the vector $\boldsymbol{g}_4$ can be seen to be parallel to the normal to the plane, i.e. parallel to $[hkil/\lambda]_4$, as is required by definition. Students are frequently confused by the apparent rotation of 30° between the diffraction pattern,

**Table 1** Symmetry matrices for group 6/m mm.

| | | | | | |
|---|---|---|---|---|---|
| $1$ | $\begin{bmatrix} 1&0&0&0\\0&1&0&0\\0&0&1&0\\0&0&0&1 \end{bmatrix}$ | $\bar{3}^-$ | $\begin{bmatrix} 0&\bar{1}&0&0\\0&0&1&0\\1&0&0&0\\0&0&0&\bar{1} \end{bmatrix}$ | $6^+$ | $\frac{1}{3}\begin{bmatrix} 2&\bar{1}&2&0\\2&2&\bar{1}&0\\\bar{1}&2&2&0\\0&0&0&3 \end{bmatrix}$ |
| $3^+$ | $\begin{bmatrix} 0&0&1&0\\1&0&0&0\\0&1&0&0\\0&0&0&1 \end{bmatrix}$ | $\mathrm{m}_{(2\bar{1}\bar{1}0)}$ | $\begin{bmatrix} \bar{1}&0&0&0\\0&0&\bar{1}&0\\0&\bar{1}&0&0\\0&0&0&1 \end{bmatrix}$ | $6^-$ | $\frac{1}{3}\begin{bmatrix} 2&2&\bar{1}&0\\\bar{1}&2&2&0\\2&\bar{1}&2&0\\0&0&0&3 \end{bmatrix}$ |
| $3^-$ | $\begin{bmatrix} 0&1&0&0\\0&0&1&0\\1&0&0&0\\0&0&0&1 \end{bmatrix}$ | $\mathrm{m}_{(\bar{1}2\bar{1}0)}$ | $\begin{bmatrix} 0&0&\bar{1}&0\\0&\bar{1}&0&0\\\bar{1}&0&0&0\\0&0&0&1 \end{bmatrix}$ | $\mathrm{m}_{(1\bar{1}00)}$ | $\frac{1}{3}\begin{bmatrix} \bar{2}&1&\bar{2}&0\\1&\bar{2}&\bar{2}&0\\\bar{2}&\bar{2}&1&0\\0&0&0&3 \end{bmatrix}$ |
| $2_{[2\bar{1}\bar{1}0]}$ | $\begin{bmatrix} 1&0&0&0\\0&0&1&0\\0&1&0&0\\0&0&0&\bar{1} \end{bmatrix}$ | $\mathrm{m}_{(\bar{1}\bar{1}20)}$ | $\begin{bmatrix} 0&\bar{1}&0&0\\\bar{1}&0&0&0\\0&0&\bar{1}&0\\0&0&0&1 \end{bmatrix}$ | $\mathrm{m}_{(\bar{1}010)}$ | $\frac{1}{3}\begin{bmatrix} \bar{2}&\bar{2}&1&0\\\bar{2}&1&\bar{2}&0\\1&\bar{2}&\bar{2}&0\\0&0&0&3 \end{bmatrix}$ |
| $2_{[\bar{1}2\bar{1}0]}$ | $\begin{bmatrix} 0&0&1&0\\0&1&0&0\\1&0&0&0\\0&0&0&\bar{1} \end{bmatrix}$ | $2_{[1\bar{1}00]}$ | $\frac{1}{3}\begin{bmatrix} 2&\bar{1}&2&0\\\bar{1}&2&2&0\\2&2&\bar{1}&0\\0&0&0&\bar{3} \end{bmatrix}$ | $\mathrm{m}_{(0\bar{1}\bar{1}0)}$ | $\frac{1}{3}\begin{bmatrix} 1&\bar{2}&\bar{2}&0\\\bar{2}&\bar{2}&1&0\\\bar{2}&1&\bar{2}&0\\0&0&0&3 \end{bmatrix}$ |
| $2_{[\bar{1}\bar{1}20]}$ | $\begin{bmatrix} 0&1&0&0\\1&0&0&0\\0&0&1&0\\0&0&0&\bar{1} \end{bmatrix}$ | $2_{[\bar{1}010]}$ | $\frac{1}{3}\begin{bmatrix} 2&2&\bar{1}&0\\2&\bar{1}&2&0\\\bar{1}&2&2&0\\0&0&0&\bar{3} \end{bmatrix}$ | $\mathrm{m}_{(0001)}$ | $\frac{1}{3}\begin{bmatrix} 1&\bar{2}&\bar{2}&0\\\bar{2}&1&\bar{2}&0\\\bar{2}&\bar{2}&1&0\\0&0&0&\bar{3} \end{bmatrix}$ |
| $\bar{1}$ | $\begin{bmatrix} \bar{1}&0&0&0\\0&\bar{1}&0&0\\0&0&\bar{1}&0\\0&0&0&\bar{1} \end{bmatrix}$ | $2_{[0\bar{1}10]}$ | $\frac{1}{3}\begin{bmatrix} \bar{1}&2&2&0\\2&2&\bar{1}&0\\2&\bar{1}&2&0\\0&0&0&\bar{3} \end{bmatrix}$ | $\bar{6}^+$ | $\frac{1}{3}\begin{bmatrix} \bar{2}&1&\bar{2}&0\\\bar{2}&\bar{2}&1&0\\1&\bar{2}&\bar{2}&0\\0&0&0&\bar{3} \end{bmatrix}$ |
| $\bar{3}^+$ | $\begin{bmatrix} 0&0&\bar{1}&0\\\bar{1}&0&0&0\\0&\bar{1}&0&0\\0&0&0&\bar{1} \end{bmatrix}$ | $2_{[0001]}$ | $\frac{1}{3}\begin{bmatrix} \bar{1}&2&2&0\\2&\bar{1}&2&0\\2&2&\bar{1}&0\\0&0&0&3 \end{bmatrix}$ | $\bar{6}^-$ | $\frac{1}{3}\begin{bmatrix} \bar{2}&\bar{2}&1&0\\1&\bar{2}&\bar{2}&0\\\bar{2}&1&\bar{2}&0\\0&0&0&\bar{3} \end{bmatrix}$ |

obtained when a beam is incident perpendicular to the basal plane of a hexagonal crystal, and the atomic arrangement in that plane. The shortest reciprocal vectors have the form $\boldsymbol{g} = 1\bar{1}00$, whereas the shortest translation vectors in the crystal have the form $\frac{1}{3}\langle 2\bar{1}\bar{1}0\rangle$. Frank clarified this point by explaining that the direct unit cell can be visualised as a *projection* from a higher dimensional space, and hence fractional indices may arise, in contrast to the situation in reciprocal space where the diffraction pattern corresponds to a *section* of the reciprocal lattice, and hence non-integral indices do not arise using the crystallographic notation.

In the foregoing we have recapitulated the elementary aspects of direct and reciprocal space hexagonal crystallography using Frank's method. The present author has found this approach very helpful in both teaching and research. Further applications will be discussed later, in particular the use of matrices of 4-vectors for describing the orientation relationship between two crystals. Since these later considerations will involve the use of symmetry operators, we complete the present section by presenting the matrix formulation of such operators. The matrices corresponding to the twenty four operations in the group 6/m mm are shown in table 1, and the method of their derivation has been described elsewhere (Pond *et al* 1987b). It can be seen that the matrices have a particularly simple form and all elements are independent of $\Lambda$. Twelve of the operations comprise column vectors of the type $\langle 1000\rangle$ exclusively, and the remaining twelve matrices involve columns of the form $\frac{1}{3}\langle 22\bar{1}0\rangle$, which are 4-vectors of unit magnitude and include a component $\frac{2}{3}[1110]$ parallel to the projection direction from four- into three-dimensional space. Symmetry matrices can be used to operate on a given 4-vector, expressed as a column vector, in order to find the symmetry-related 4-vector.

## 3 RHEED STUDY OF THE (01$\bar{1}$2) SURFACE OF SAPPHIRE

The usefulness of Frank's appreciation of hexagonal indexing is now illustrated by application to a study of the (01$\bar{1}$2) surface of sapphire using RHEED. Now, diffraction patterns obtained by scattering from surfaces are often interpreted using a two-index scheme. However, in practice such patterns often exhibit diffraction effects arising from scattering by the underlying bulk of the crystal, particularly if the surface is not atomically smooth. Thus, it is preferable to use an indexing scheme whereby surface and bulk diffraction effects can be readily distinguished. As explained in the previous section, beams which have been Bragg reflected by a bulk crystal can be indexed using a reciprocal-lattice vector, $\boldsymbol{g}$. On the other hand, beams diffracted by a surface periodicity may not coincide with bulk crystal beams. Using Frank's scheme we shall see that such surface beams can be indexed using reciprocal-space vectors, $\boldsymbol{h}$, but that these are generally distinct from reciprocal-lattice vectors $\boldsymbol{g}$.

### 3.1 Crystallography of the unrelaxed surface

First, we consider the crystallography of the unrelaxed $(01\bar{1}2)$ surface of sapphire ($c/a = 2.729$, and hence $\Lambda = 2.229$). A schematic illustration of the two-dimensional mesh of translation vectors on the $(01\bar{1}2)$ surface is shown in figure 2($a$). The Miller–Bravais indices of the primitive translation vectors are $\boldsymbol{r}_1 = \frac{1}{3}[\bar{2}110]$ and $\boldsymbol{r}_2 = \frac{1}{3}[0\bar{1}11]$, and the corresponding 4-vector forms of these are $\boldsymbol{r}_1 = \frac{1}{3}[\bar{2}110]_4$ and $\boldsymbol{r}_2 = \frac{1}{3}[0\bar{1}1\Lambda]_4$, respectively. The reader can readily verify that the magnitudes of these vectors are equal to $a$ and $[(2+\Lambda^2)/6]^{1/2}a$, respectively, that the scalar product $\boldsymbol{r}_1 \cdot \boldsymbol{r}_2$ equals zero, and that the vector product $\boldsymbol{r}_1 \times \boldsymbol{r}_2$ is parallel to the $(01\bar{1}2)$ plane normal, $[01\bar{1}2/\Lambda]_4$. The reciprocal-space mesh for this surface is illustrated in figure 2($b$); the crystallographic notation for the primitive reciprocal-space vectors is $\boldsymbol{h}_1 = \bar{1}, \frac{1}{2}, \frac{1}{2}, 0$ and $\boldsymbol{h}_2 = 0, \alpha, \alpha, \Lambda^2\alpha$, and the corresponding reciprocal 4-vector forms are $\bar{1}, \frac{1}{2}, \frac{1}{2}, 0$ and $0, \alpha, \alpha, \Lambda\alpha$, respectively where $\alpha = 3(2+\Lambda^2)^{-1}$. It can be seen that $|\boldsymbol{h}_1| = |\boldsymbol{r}_1|^{-1}$ and $|\boldsymbol{h}_2| = |\boldsymbol{r}_2|^{-1}$. It is also clear that, in the present case, $\boldsymbol{h}_1$ and $\boldsymbol{h}_2$ are not reciprocal-lattice vectors, and hence surface diffracted beams will not coincide with bulk crystal beams. In the case of $\boldsymbol{h}_1$, although its indices are integers and rational fractions (and hence $\boldsymbol{h}_1$ is perpendicular to rational crystal planes), its magnitude is equal to $(2d_{(2\bar{1}\bar{1}0)})^{-1}$, whereas $\boldsymbol{h}_2$ has irrational indices and is therefore not perpendicular to rational crystal planes.

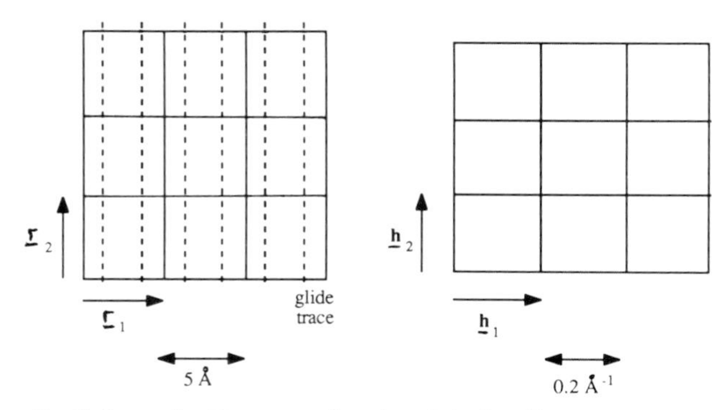


**Figure 2** Schematic diagrams showing ($a$) the direct space net of the $(01\bar{1}2)$ surface of sapphire, and ($b$) the corresponding reciprocal net.

The space group of $\alpha$-$Al_2O_3$ is R$\bar{3}$c, where the c-mirror-glide planes are parallel to $\{2\bar{1}\bar{1}0\}$. The matrices representing the operations in the point group $\bar{3}$m are included in table 1, and the intrinsic glide translation is $\frac{1}{2}[0001]$. We note that this latter vector can be re-expressed equivalently as one lying parallel to the $(01\bar{1}2)$ surface by adding a rhombohedral translation vector, i.e. $\frac{1}{2}[0001] + \frac{1}{3}[\bar{1}01\bar{1}] = \frac{1}{6}[\bar{2}021]$ (the 4-vector $\frac{1}{6}[\bar{2}02\Lambda]_4$ is perpendicular to the surface normal $[01\bar{1}2/\Lambda]_4$). The traces of these mirror-glide planes are indicated in figure 2($a$). We also note that if the atomic structure of $\alpha$-$Al_2O_3$ is projected onto $(01\bar{1}2)$, these mirror-glide planes are transformed into mirror-glide lines, and the two-dimensional space group of this configuration is pg. The presence of such symmetry would be expected to lead to the systematic absence of the odd orders of $\boldsymbol{h}_2$ in a RHEED pattern.

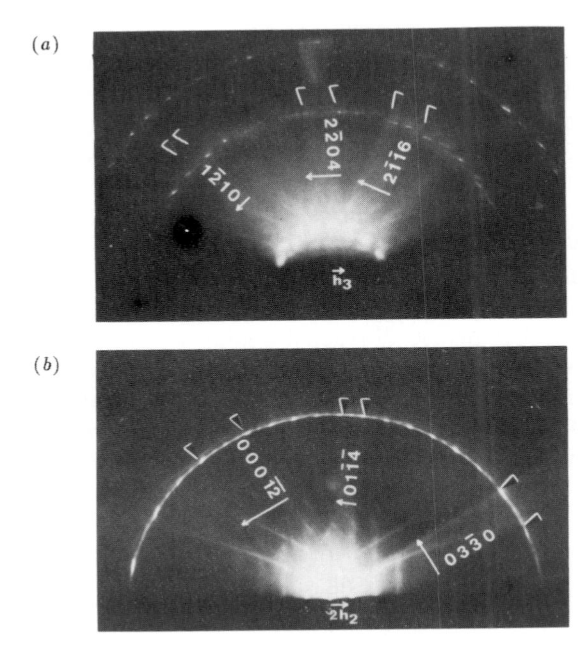


**Figure 3** Convergent beam RHEED patterns obtained from the $(01\bar{1}2)$ surface of $\alpha$-$Al_2O_3$, (*a*) $[20\bar{2}\bar{1}]$ azimuth ($\boldsymbol{h}_3 = \boldsymbol{h}_1 - \boldsymbol{h}_2$), (*b*) $[\bar{2}110]$ azimuth.

### 3.2 RHEED patterns

RHEED patterns obtained by Aindow *et al* (1987) using high purity polished specimens of sapphire are shown in figures 3(*a*) and (*b*). These patterns exhibit both strong surface reflections and somewhat diffuse Kikuchi bands arising from bulk crystal diffraction. These latter can be used to determine the azimuth of the incident beam; by measuring the width of the Kikuchi bands, and the angles between neighbouring bands, reciprocal-lattice vectors $\boldsymbol{g}$ can be assigned to each band unambiguously. The zone axis can then be found by obtaining the vector product of any two such vectors. Thus, in figure 3(*a*) the azimuth is parallel to $1\bar{2}10 \times 2\bar{2}04/\Lambda = 20\bar{2}\bar{\Lambda}$, i.e. the Miller–Bravais index for this direction in direct space is $[20\bar{2}\bar{1}]$ which is parallel to $(\boldsymbol{r}_1 + \boldsymbol{r}_2)$ in figure 2(*a*). Similarly, the azimuth in the case of figure 3(*b*) is [$\bar{2}$110] which is parallel to $\boldsymbol{r}_1$. The surface reflections, $\boldsymbol{h}$, in figures 3(*a*) and (*b*) have been indexed according to the reciprocal mesh of figure 2(*b*). Measurements of the rod spacing parallel to the shadow edge of the specimen confirm that no surface reconstruction leading to changes of the surface periodicity has occurred. Also, it is observed that in figure 3(*b*) the odd orders of $\boldsymbol{h}_2$ are missing in the zero layer, but are present in the first Laue zone, which is consistent with the space group pg mentioned above. Finally, we note that Kikuchi lines with index $\boldsymbol{g}$ present in the patterns shown here were not associated with strongly excited surface reflections, $\boldsymbol{h}$, since no reflections $\boldsymbol{g}$ and $\boldsymbol{h}$ were coincident.

## 4 SURFACE DISCONTINUITIES: FACET JUNCTIONS AND STEPS

The objective of the present section is to show how elementary symmetry theory can be used to reveal important geometrical details of surface features such as facet junctions and steps. The conclusions reached may be significant in the field of surface structure and processes, but will also serve as a foundation for the discussion of interfacial defects in a later section. Our particular concern here is to consider the nature of discontinuities separating energetically degenerate surfaces of a crystal, taking into account the full space-group symmetry of the crystal in question. Two distinct types of discontinuities can arise, namely facet junctions and steps, but we will show that the presence of screw-rotation axes or mirror-glide planes in a crystal's space group can lead to the subdivision of each of these two categories into two subclasses. The notation used for symmetry operators is that of the *International Tables for Crystallography* (Hahn 1983). An operator is designated by the augmented symbol $\mathcal{W}$, which corresponds to $(\boldsymbol{W}, \boldsymbol{w})$ using the Seitz (1936) formalism, where $\boldsymbol{W}$ stands for a matrix representing rotation, mirror reflection, or inversion, and $\boldsymbol{w}$ is any translation associated with the operator. Thus, for example, pure translation operators have the form $\mathcal{W} = (\boldsymbol{I}, \boldsymbol{t})$, where $\boldsymbol{I}$ represents the identity matrix and $\boldsymbol{t}$ is a translation vector. In the case of symmorphic crystals, i.e. those which do not exhibit any screw axes or mirror-glide planes, the form of point symmetry operators is particularly simple. If the origin associated with these operations, i.e. the point through which they operate, is chosen to be a point of maximum symmetry in the crystal, then $\boldsymbol{w} = 0$ and all point symmetry operators have the form $\mathcal{W} = (\boldsymbol{W}, \boldsymbol{0})$. The situation is slightly more complicated in the case of non-symmorphic crystals since $\boldsymbol{w} \neq 0$ for some of the operators. The vector $\boldsymbol{w}$ for a given operator depends on the choice of origin, and comprises a component equal to the intrinsic glide (if any), $\boldsymbol{w}_{(i)}$, associated with the operation in question, and also a component depending on the location of the operator, $\boldsymbol{w}_{(l)}$, which will only be zero if the operator acts through the chosen origin.

### 4.1 Facet junctions and steps

Consider a surface with orientation specified by the unit vector $\boldsymbol{n}(\boldsymbol{n} = \boldsymbol{r}/|\boldsymbol{r}|$ where $\boldsymbol{r}$ is the direction perpendicular to the surface, and hence $\boldsymbol{n}$ has the character of a unitless director). The set of degenerate surfaces related to that specified by $\boldsymbol{n}$ is given by $\mathcal{W}_j\boldsymbol{n}$, where $\mathcal{W}_j$ represents the $j$th operator in the crystal's space group and $j$ goes from one to infinity. We are interested here in the geometrical nature of the discontinuity which separates the initial and $j$th surfaces which have normals $\boldsymbol{n}$ and $\boldsymbol{W}_j\boldsymbol{n}$. Let us consider symmorphic crystals first. When $\mathcal{W}_j$ is a point symmetry operator, $\mathcal{W}_j\boldsymbol{n}$ is either (i) identical to $\boldsymbol{n}$, or (ii) given by $\boldsymbol{W}_j\boldsymbol{n}$, as illustrated in figure 4($a$). In the former case, the operator in question leaves the surface invariant, and hence a semi-infinite crystal terminated with this surface orientation would be an object exhibiting this symmetry. In the latter case, the two distinct but crystallographically equivalent surfaces would be separated by a facet junction if they were to coexist. When $\mathcal{W}_j$ represents a translation operation, $(\boldsymbol{I},\boldsymbol{t}_j)$, the two surfaces have the same orientation, but may have different locations, i.e. the initial surface passes through the chosen origin and the new surface passes through the point located by $\boldsymbol{t}_j$ with respect to the origin. The two surfaces are therefore separated by a step of height $h = \boldsymbol{n} \cdot \boldsymbol{t}_j$, as illustrated in figure 4($b$), and we note that $h$ is a topological quantity which is independent of the line direction of the step.

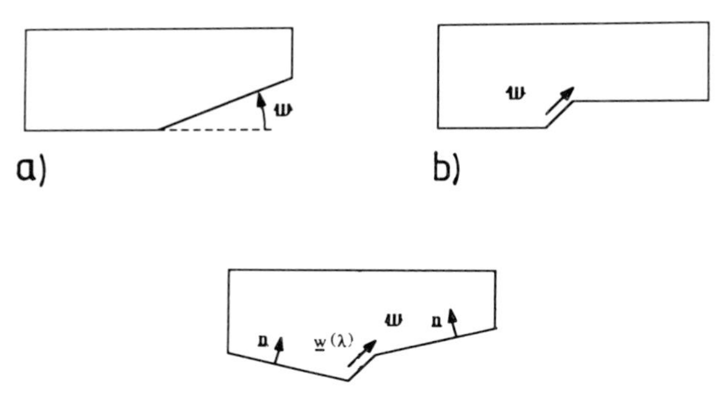


**Figure 4** Schematic illustration of surface features, ($a$) facet junction, ($b$) step, ($c$) demistepped facet junction.

Now consider the case of non-symmorphic crystals. (We note first that, in the present context, it is always feasible to choose an origin so that $\boldsymbol{w}_{(l)} = 0$ for a particular symmetry operator of interest, and we shall assume this to be the case in the remainder of this section.) Facet junctions and steps can arise on the surface of these crystals in the same way as described above, the initial and new surfaces being related by a simple point symmetry or translation operator, respectively. However, when two surfaces are related by a compound operator (screw rotation or mirror-glide plane) a supplementary step contribution of height $h = \boldsymbol{n} \cdot \boldsymbol{w}_j$ (where $\boldsymbol{w}_j$ represents the intrinsic translation component of the $j$th symmetry operator) arises. Since $\boldsymbol{w}_j$ can never be equal to a translation vector, $\boldsymbol{t}_k$, such steps are distinct from the previously considered cases, and hence warrant a distinguishing name which we chose to be 'demisteps'. Thus, for example, demistepped facet junctions can arise as illustrated in figure 4($c$). Surfaces with orientations which are left invariant by $\mathcal{W}_j$, i.e. where $\boldsymbol{W}_j\boldsymbol{n} = \boldsymbol{n}$, are of special interest. Two possibilities can occur; firstly, if $\boldsymbol{n} \cdot \boldsymbol{w}_j = 0$ then both the orientation and location of the surface in question are left invariant by the operator $\mathcal{W}_j$. The semi-infinite crystal terminated at this surface would therefore exhibit the symmetry $\mathcal{W}_j$, and the c-mirror-glide plane parallel to $(2\bar{1}\bar{1}0)$ which left the $(01\bar{1}2)$ surface of sapphire invariant, as described in the previous section, is an example (in this case $\boldsymbol{n} = 1/\beta[01\bar{1}2/\Lambda]_4$, where $\beta = (3 + 6/\Lambda^2)^{1/2} a$, $\boldsymbol{w}_j = \frac{1}{6}[\bar{2}02\Lambda]_4$, and $\boldsymbol{W}_j$ is the tenth matrix listed in table 1). The second possibility is where $\boldsymbol{n} \cdot \boldsymbol{w}_j \neq 0$, and demisteps of height $h = \boldsymbol{n} \cdot \boldsymbol{w}_j$ can then arise on these surfaces.

To summarise the foregoing, we have shown that there is a direct relationship between the individual symmetry operations in a crystal's space group and the nature of discontinuities that can arise on the unrelaxed surfaces of that crystal. Symmetry operators may leave a particular surface invariant, in which case no surface discontinuity is associated with that operator for the particular surface in question. However, if the orientation of the initial surface does not remain invariant, the initial and new surfaces would be separated by a facet junction if they were to coexist. We can say that the facet junction is characterised by this symmetry operator, and the facet angle is equal to $\widehat{\boldsymbol{n}\boldsymbol{W}_j\boldsymbol{n}}$. Steps can arise on any surface, $\boldsymbol{n}$, and are characterised by a translation operator $\mathcal{W}_j = (\boldsymbol{I}, \boldsymbol{t}_j)$, and exhibit height $h = \boldsymbol{n} \cdot \boldsymbol{t}_j$. Further possibilities can arise on the surfaces of non-symmorphic crystals; notably, a subset of specially oriented surfaces can exhibit demisteps (i.e. those with orientations for which $\boldsymbol{n} = \boldsymbol{W}_j\boldsymbol{n}$ and $\boldsymbol{n} \cdot \boldsymbol{w}_j \neq 0$) with height $h = \boldsymbol{n} \cdot \boldsymbol{w}_j$. Steps of this type separate degenerate surfaces with the same orientation, but where the structures are related by a rotation or mirror operation. Again, we may say that the discontinuity is characterised by the compound operator $\mathcal{W}_j = (\boldsymbol{W}_j, \boldsymbol{w}_j)$. The distinction between steps and demisteps is illustrated in figure 5 by examples on the $(11\bar{2}1)$ surface of an HCP metal with $c/a$ close to the ideal value ($\Lambda = \frac{4}{3}$). A schematic view along $[1\bar{1}00]$ of a step characterised by the translation vector $\boldsymbol{t}_j = \langle 11\bar{2}0 \rangle$ is depicted in figure 5($a$) and the step height is equal to $h = \boldsymbol{n} \cdot \boldsymbol{t}_j = d_{(11\bar{2}1)}$. A demistep is illustrated in figure 5($b$), characterised by the c-mirror-glide plane which is parallel to the plane of the figure (the space group of HCP metals is $P6_3/m\,mc$). The symmetry operator in the case, $\mathcal{W}_j$, has the form $(\boldsymbol{W}_j, \boldsymbol{w}_j)$, where $\boldsymbol{W}_j$ is the matrix representing reflection across $(1\bar{1}00)$, and $\boldsymbol{w}_j = \frac{1}{2}[000\bar{1}]$ (modulo a translation vector). Thus, the step height in the present case is given by $h = \boldsymbol{n} \cdot \boldsymbol{w}_j = \frac{1}{2}d_{(11\bar{2}1)}$, which is equal to a $(22\bar{4}2)$ atomic plane spacing. It can also be seen that the atomic arrangement of the two surfaces separated by the demistep are related by a mirror reflection parallel to the plane of the figure.

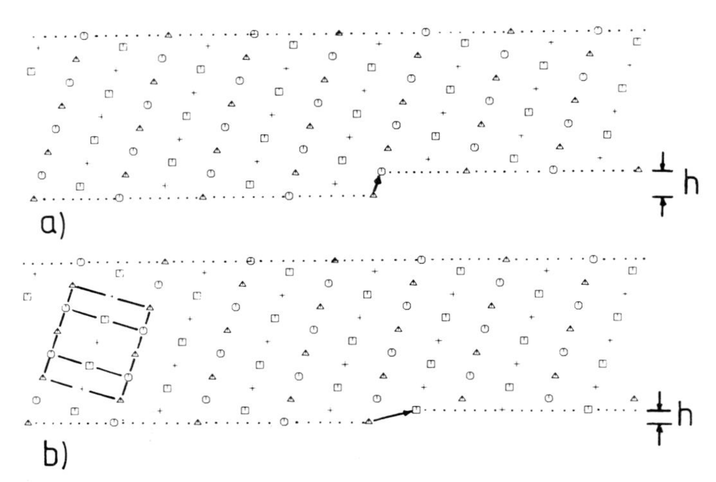


**Figure 5** Schematic illustration of unrelaxed surface features on the $(11\bar{2}1)$ surface of a HCP metal. The projection direction is $[1\bar{1}00]$, and the symbols represent atoms at heights $\frac{2}{6}$, $\frac{1}{6}$, $-\frac{1}{6}$ and $-\frac{2}{6}$ with respect to centres of symmetry in the crystal, and a unit cell is shown in outline. A step characterised by the translation operation $\mathcal{W} = (\boldsymbol{I}, \frac{1}{3}[11\bar{2}0])$ is shown in (*a*), and a demistep, characterised by $\mathcal{W} = (\boldsymbol{W}, \boldsymbol{w} + \boldsymbol{t})$ is shown in (*b*), where $\boldsymbol{W}$ represents the mirror operation parallel to $(1\bar{1}00)$, $\boldsymbol{w} = \frac{1}{2}[0001]$ and $\boldsymbol{t} = \frac{1}{3}[11\bar{2}0]$. The step heights are equal to $d_{(11\bar{2}1)}$ and $d_{(22\bar{4}2)}$ in cases (*a*) and (*b*), respectively.

### 4.2 Arrays of steps

Before leaving the topic of surface discontinuities, we wish to mention briefly the subject of arrays of surface discontinuities. Frank (1963) has pointed out that arrangements of equi-spaced steps lead to a systematic deviation of a surface orientation away from a low-energy configuration. As illustrated schematically in figure 6, the deviation, $\theta_v$, is given by $\tan \theta_v = h/m$, where $m$ is the length of the terraces separating adjacent steps (or demisteps) on the vicinal surface. More complicated arrangements of steps can be regarded as leading to surface roughness.

### 4.3 Relationship between surface and bulk discontinuities

Finally, we wish to make brief remarks about the relationship between surface discontinuities and line defects in the interior of a crystal. It will be recalled from the preceding discussion that surface features can be characterised conveniently by symmetry operators. This, of course, is also true for line defects, and Frank (1951) carried out pioneering work in this context, introducing the Burgers circuit, and also outlining some ‘topological oddities’ which we now refer to as disclinations and dispirations (Harris 1970). To put the matter very briefly, dislocations are characterised by pure translation operations, disclinations by proper rotations, and dispirations by screw-rotation operations. There are some minor difficulties in including these concepts into the general topological theory of ordered media, or homotopy theory (e.g. see Mermin 1979), but these can be ignored for present purposes. Thus, we see that the same symmetry operation can characterise both a line defect and a surface discontinuity. In other words, a dislocation with Burgers vector $\boldsymbol{b} = \boldsymbol{t}_j$ will be manifested as a step with height $h = \boldsymbol{n} \cdot \boldsymbol{t}_j$ if it emerges from the bulk of a crystal onto the surface $\boldsymbol{n}$. This possibility was identified by Frank (1949) as an inexhaustible source of steps which is known to be a crucial factor in crystal growth. For completeness, we mention that emerging disclinations would lead to facet junctions (we note that Frank referred to disclinated crystals as Moebius crystals by analogy with Moebius strips), and dispirations would lead to demistepped facet junctions or demisteps on appropriate surfaces.

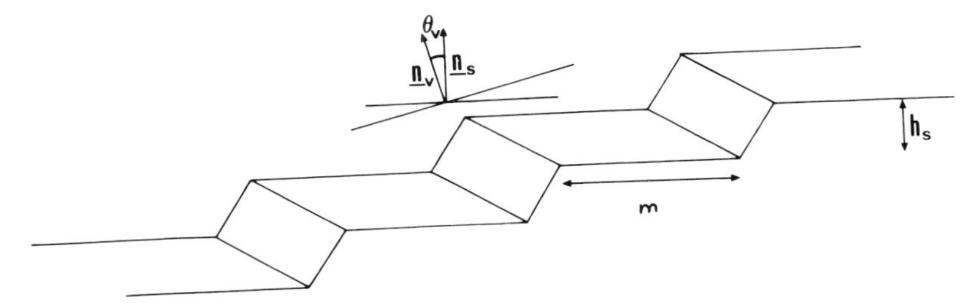


**Figure 6** Schematic illustration of the terraced structure of a vicinal surface. The vicinal angle, $\theta_V$, is defined as the angle between the normal to the substate terraces, $\boldsymbol{n}_S$, and average surface orientation $\boldsymbol{n}_V$.

## 5 ORIENTATION RELATIONSHIPS BETWEEN ADJACENT CRYSTALS, THE TOPOLOGICAL CHARACTERISATION OF INTERFACES AND ORIENTATION MAPPING

In the foregoing sections we have been concerned primarily with crystal surfaces. In the following section we shall approach some aspects of the structure of interfaces by imagining these to be created by bringing together appropriate surfaces. However, before this can be discussed, it is necessary to give an account of the geometrical relationship between two crystals. We distinguish the two crystals by designating one as being white, $\lambda$, and the other black, $\mu$. Thus the $i$th symmetry operator in the space group of the white crystal is written $\mathcal{W}(\lambda)_i = (W(\lambda)_i, w(\lambda)_i)$, and the $j$th black one is $\mathcal{W}(\mu)_j = (W(\mu)_j, w(\mu)_j)$. The two crystals may be different in the sense that their lattices and/or bases may be distinct, in which case the interface between them is referred to as an interphase (or heterophase) boundary. Alternatively, the two crystals may exhibit the same lattice and basis, and the interface is then known as a grain boundary. Special types of grain boundaries, such as twin and domain boundaries, can arise, where the relationship between the two crystals has a particularly simple form, and we shall approach the present topic from the most general and proceed to the most special form.

## 5.1 Orientation relationships

A correspondence can always be defined even between two distinct crystals; we designate this $\mathcal{P} = (P, p)$, where $P$ signifies a matrix representing deformation and/or rotation and $p$ is a displacement of the black crystal away from a reference position where the chosen origins in the two crystals coincide. $\mathcal{P}$ is taken here to be a vector transformation when acting on a vector specified in the white coordinate frame, and is therefore a coordinate transformation when acting on a vector specified in the black frame. As an illustration we consider the matrix $P$ relating silicon and sapphire, which can be grown epitaxially with the nominal orientation relationship $(01\bar{1}2)//(001)$ and $[2\bar{1}\bar{1}0]//[100]$. Taking sapphire to be white and silicon black, the matrix $P$ can be formulated readily using Frank's 4-vector scheme, and has the form

$$P = \frac{a_c}{a_h} \begin{vmatrix} \frac{2}{3} & 0 & 0 \\ -\frac{1}{3} & 2^{1/2}\alpha & \Lambda\alpha \\ -\frac{1}{3} & -2^{1/2}\alpha & -\Lambda\alpha \\ 0 & -2^{1/2}\Lambda\alpha & 2\alpha \end{vmatrix}$$

where $\alpha = (6 + 3\Lambda^2)^{-1/2}$, and $a_c$ and $a_h$ are lattice parameters of the cubic and trigonal crystals, respectively. Thus, for example, if $P$ is operated on a black (cubic) vector, the coordinates of this vector are transformed into those of a 4-vector in the hexagonal frame. We note that $\mathcal{P}$ can be regarded as a mapping between the crystals, and does not include any notion relating to the interface between the crystals; in fact the crystals can be imagined to be interpenetrating if desired. Topologically speaking therefore, we can use $\mathcal{P}$ to characterise the relationship between adjacent crystals irrespective of the boundary location and orientation. This is analogous to the previously described characterisation of a line defect or surface feature by means of a symmetry operator, which is independent of the direction of the discontinuity.

When the adjacent crystals are separated by a grain boundary, $\boldsymbol{P}$ takes the form of a proper or improper unimodular transformation. As an illustration, we consider a simple rotation of 27.79° between hexagonal crystals about [0001] which leads to the three-dimensional $\Sigma = 13$ coincidence site lattice. The formulation of this matrix using the 4-vector scheme has been described in detail elsewhere (Pond *et al* 1987b) and has the form

$$\boldsymbol{R} = \frac{1}{13}\begin{vmatrix} 12 & \bar{3} & 4 & 0 \\ 4 & 12 & \bar{3} & 0 \\ \bar{3} & 4 & 12 & 0 \\ 0 & 0 & 0 & 13 \end{vmatrix}.$$

We note the simplicity of this matrix using the 4-vector approach, and remark that the elements of $\boldsymbol{R}$ are independent of $\Lambda$ for rotations about [0001].

The relationship between the adjacent crystals separated by a grain boundary is closer than that for interphase boundaries in the sense that, whereas the latter is a correspondence, the former is an equivalence. In particular, the equivalence between grains can take the form of a symmetry operation, such as a mirror reflection or diad axis, and such crystals are said to be twinned. We designate these operations antisymmetry, i.e. operations which inter-relate black and white entities, to distinguish them from crystal symmetry operators which relate white entities to white or black to black. As an illustration, we consider two hexagonal crystals related by a reflection across $(11\bar{2}1)$, i.e. twinned crystals related by a $(11\bar{2}1)$ antimirror. Using the 4-vector scheme the transformation matrix $\boldsymbol{P} = \boldsymbol{M}'$ (the prime indicates antisymmetry) has the following form using the 4-vector scheme:

$$\boldsymbol{M}' = \frac{1}{(6\Lambda^2+1)}\begin{vmatrix} (4\Lambda^2+1) & -2\Lambda^2 & 4\Lambda^2 & -2\Lambda \\ -2\Lambda^2 & (4\Lambda^2+1) & 4\Lambda^2 & -2\Lambda \\ 4\Lambda^2 & 4\Lambda^2 & (-2\Lambda^2+1) & 4\Lambda \\ -2\Lambda & -2\Lambda & 4\Lambda & (6\Lambda^2-1) \end{vmatrix}.$$

The reader can readily verify that $\boldsymbol{M}'$ leaves 4-vectors in the $(11\bar{2}1)$ plane, such as $[1\bar{1}00]_4$ and $[\bar{1},\bar{1},2,6\Lambda]_4$, invariant and inverts the normal to the plane, $[11\bar{2}1/\Lambda]_4$.

In the case of non-holosymmetric crystals, i.e. those which exhibit lower symmetry than that of the lattice on which they are based, a further class of equivalence relationship can arise. Two grains can now be related by a symmetry operator which does not belong to the crystals but does belong to their lattice, and such crystals are said to be separated by a domain boundary. For example, inversion domains of sphalerite-structure materials are related by the inversion operator, and merohedral twins by a mirror plane of the lattice, and we note that, provided $\boldsymbol{p} = 0$, the lattice of the adjacent crystals is continuous across the interface irrespective of the location or orientation of the latter. A class of equivalence relationships closely related to those described above is the instance of domains produced by the transformation or ordering of a parent phase into a multiplicity of lower-symmetry crystals. The matrix $\mathcal{P}$ will take the form of a symmetry operator of the parent phase, but the lattice may not necessarily be continuous across interfaces separating variant products in this case. Crystallographic shear faults can also be incorporated into the present scheme; $\mathcal{P}$ takes the form $(\boldsymbol{I}, \boldsymbol{p})$ in this case, where $\boldsymbol{p}$ is the characteristic displacement and $\boldsymbol{I}$ is the identity matrix.

## 5.2 Topological characterisation of interfaces

We have introduced $\mathcal{P}$ as an operator which defines the relationship between two crystals. However, this operator can also be used to characterise topologically the interface which separates the crystals, even though the operator does not describe any aspects of the interface's orientation, location or atomic structure. This is analogous to the situation regarding line defects or surface features which are characterised topologically by symmetry operators, but these operators do not convey any information about the line direction or core structures of the discontinuities. We can now explore further the connections between the topological properties of interfaces, line defects and surface features. As discussed earlier, a line defect is characterised by a proper symmetry operation, and the same operation characterises the surface feature which arises when the line defect emerges onto a surface. On the other hand, surface features may be characterised by a proper or improper symmetry operator, but only the former can be associated with emerging line defects. In the case of grain boundaries, the characterising operator $\mathcal{P}$ has the quality of a symmetry operator insofar as it relates equivalent objects (as opposed to the situation with correspondence operators for interphase boundaries), although it is not a symmetry operation of either crystal. Moreover, it may be a proper or improper operator. In the former case, the interface in question can be terminated in the bulk by a line defect; this defect will be a dislocation if $\mathcal{P}$ has the form $(\boldsymbol{I}, \boldsymbol{p})$, a disclination if $(\boldsymbol{R}, \boldsymbol{0})$, or a dispiration if $(\boldsymbol{R}, \boldsymbol{p})$. In addition, the intersection of the interface with a free surface can be delineated by a surface feature separating crystallographically equivalent surfaces with normals $\boldsymbol{n}$ and $\mathcal{P}\boldsymbol{n}$. These features will be facet junctions or steps depending on the form of $\boldsymbol{P}$, and can be demisteps in special cases. Interfaces characterised by an improper operation cannot be terminated in the bulk by a line defect, although they can, of course, lead to surface features on emergence.

The final topic in this section is concerned with alternative formulations of the matrix $\mathcal{P}$ relating two crystals (or characterising an interface). Because the adjacent crystals exhibit symmetry, there is no unique way of describing their correspondence or equivalence. Consider the transformation of a vector expressed in the white frame, $\boldsymbol{v}^{\lambda}$, to the corresponding vector in the black crystal, $\boldsymbol{v}^{\mu}$, but expressed in the white coordinate frame; we have $\boldsymbol{v}^{\mu} = \boldsymbol{P}\boldsymbol{v}^{\lambda}$, where $\mathcal{P}$ is the initially chosen description of the correspondence (equivalence). By virtue of the white crystal's symmetry, there is an infinity of vectors equivalent to $\boldsymbol{v}^{\lambda}$, given by $\mathcal{W}(\lambda)_i\boldsymbol{v}^{\lambda}$. Similarly, because of the black crystal's symmetry, there is an infinity of vectors equivalent to $\boldsymbol{v}^{\mu}$, given by $\mathcal{P}\mathcal{W}(\mu)_j\mathcal{P}^{-1}\boldsymbol{v}^{\mu}$, where $\mathcal{P}\mathcal{W}(\mu)_j\mathcal{P}^{-1}$ is the $j$th black symmetry operator expressed in the white coordinate frame. Therefore, we can reformulate the correspondence as $\boldsymbol{v}^{\mu} = \mathcal{P}\mathcal{W}(\mu)_j^{-1}\mathcal{W}(\lambda)_i\boldsymbol{v}^{\lambda}$, showing that the set of transformations $\mathcal{P}\mathcal{W}(\mu)_j^{-1}\mathcal{W}(\lambda)_i$ comprises operations which are alternative descriptions to $\mathcal{P}$ leading to the same physical situation. Clearly, the number of alternative descriptions depends upon the number of symmetry operations $\mathcal{W}(\mu)_j$ and $\mathcal{W}(\lambda)_i$ in the black and white crystals' space groups. In the case of grain boundaries, where the two space groups are identical, i.e. $\mathcal{W}(\lambda)_k = \mathcal{W}(\mu)_k$, the alternative descriptions are given by $\mathcal{P}\mathcal{W}(\lambda)_l$, since the product of two symmetry operators in the same group must be identical to another operator in that group. Thus, if we consider the initial choice of $\mathcal{P}$ to be a rotation $(\boldsymbol{R}, \boldsymbol{0})$ in the case of two HCP crystals for example, there will be twenty three alternative descriptions, making twenty four in total, of which twelve will be proper and twelve improper. The twenty four symmetry operations in the point group 6/m mm are given in table 1, and an example of the set of alternative descriptions obtained using these operations has been presented by Pond *et al* (1987b).

In the light of this discussion of alternative descriptions, we must modify slightly the earlier statement regarding line defects delineating the termination of an interface inside a crystal. To be precise, it is necessary to state that such a configuration can arise if any of the operators in the set $\mathcal{P}\mathcal{W}(\lambda)_k$ is proper. It is possible that several descriptions in this subset are proper, and hence the interface may be terminated by a variety of line defects characterised by the individual operations in the subset. However, these proper operations will be inter-related by proper symmetry operators, and therefore the multiplicity of line defects corresponds to combinations of the initially identified defect with other crystal line defects. Similarly, the nature of the surface feature delineating an emergent interface will depend on the particular choice of description. However, it can be seen that a choice $\mathcal{P}\mathcal{W}(\lambda)_k$, rather than $\mathcal{P}$, corresponds to combining the surface feature characterised by $\mathcal{W}(\lambda)_k$ with the feature characterised by $\mathcal{P}$.

### 5.3 Orientation mapping

The multiplicity of alternative descriptions of a correspondence (equivalence) matrix $\mathcal{P}$ is also a complicating factor in the treatment of populations of crystal orientations, as occurs for example in texture studies. Frank

(1987) has recently presented an elegant solution to the problems involved in orientation mapping by describing the fascinating topological properties of what has been called the Frank–Rodrigues (FR) space. This framework is appropriate for treating the case of grain boundaries, and takes into account the subset of alternative descriptions, $\mathcal{PW}(\lambda_k)$,which are proper. (In fact, it is possible that none of the alternative descriptions are proper, and Frank (1987) has indicated a method using colour to deal with this situation, but this has not yet been developed. Nevertheless, the use of FR space will be most helpful in the analysis of the texture of many high symmetry materials currently used in engineering.) The basis of the method is to represent the matrix $\mathcal{P} = (\boldsymbol{R}, \boldsymbol{0})$ by the Rodrigues vector which is equal to $\boldsymbol{u} \tan \frac{1}{2}\theta$, where $\boldsymbol{u}$ is the unit vector parallel to the rotation axis and $\theta$ is the angle of rotation. Frank (1987) has shown that FR space is divided up into zones, and the central or fundamental zone is analogous to a Brillouin zone. The planes bounding the fundamental zone are orientationally equidistant between the origin and neighbouring points corresponding to Rodrigues vectors representing proper symmetry operations. The smallest angle description amongst a set of alternative descriptions, also known as the disorientation, will be represented by a point within the fundamental zone (or, exceptionally, by points on the edges, faces or vertices). Frank (1987) has illustrated the form of some fundamental zones for high symmetry crystals in various crystal classes; that for hexagonal crystals is a thin dodecagonal prism bounded by two dodecagons normal to the hexagonal axis and twelve square prism faces (with edge lengths equal to 2 tan $\pi/12$) normal to diad axes at unit distance from the origin. For trigonal crystals, the fundamental zone is a hexagonal prism with hexagonal faces normal to the triad axes, and square prism faces (with edge length 2 tan $\pi/6$) normal to diad axes at unit distance from the origin. We finish this section by noting that Frank's 4-vector scheme is very helpful for facilitating calculations in FR space. For example, the largest possible disorientation corresponds to the Rodrigues vector terminating at one of the vertices most remote from the origin, and we shall identify these points for hexagonal and trigonal crystals which exhibit the point groups 6/m mm, 622, $\bar{3}$m, and 32, i.e. those crystals for which the fundamental zones are as described above. In the hexagonal case, this vector can be found by combining a unit 4-vector perpendicular to a prism face, $(3a)^{-1}[2\bar{1}\bar{1}0]_4$, with a 4-vector perpendicular to an orthogonal prism face and with magnitude tan $\pi/12$, i.e. $(\tan \pi/12)/(3^{1/2}a)^{-1}[01\bar{1}0]_4$ and a 4-vector equal to half the height of the prism, i.e. $(\tan \pi/12)/c[000\Lambda]_4$. The resultant 4-vector is $(3a)^{-1}[2, (3^{1/2}\tan \pi/12 - 1), (1 - 3^{1/2}\tan \pi/12), 6^{1/2} \tan \pi/12]_4$ which is a dimensionless vector (it will be recalled that the implied units of a 4-vector are $|\boldsymbol{e}| = (3/2)^{1/2}a$, and therefore the lattice parameter cancels with that in the factor outside the bracket). The Miller–Bravais index for this direction is obtained by dividing the last index of the 4-vector by $\Lambda$ (dimensionless), and hence it is seen that the orientation of the actual rotation axis in a crystal depends on $c/a$, but is independent of this in FR space. The magnitude of the vector is equal to $(1 + 2\tan^2 \pi/12)^{1/2}$, and hence the angle associated with the largest disorientation, given by $2\tan^{-1} (1 + \tan^2 \pi/12)^{1/2}$, is equal to 93.84°. The largest disorientation for trigonal crystals can be obtained in a similar manner, the 4-vector has the same form as that for hexagonal crystals except that the angular term $\pi/12$ must be replaced by $\pi/6$, and its magnitude is $(1 + 2\tan^2 \pi/6)^{1/2}$ which corresponds to a maximum disorientation angle equal to 104.47°.

## 6 INTERFACIAL LINE DEFECTS

A recurrent theme in the foregoing sections has been that crystals are distinguished from other forms of matter by the order, or symmetry, that they exhibit. Moreover, it is by virtue of this order that they can accommodate admissible discontinuities such as line defects and surface steps, which, furthermore, are themselves characterised topologically by symmetry operations exhibited by the undistorted infinite crystal. When two crystals are joined to make a bicrystal, the nature of admissible line defects, which can exist in the interface thereby separating crystallographically equivalent (energetically degenerate) regions of interface, is also constrained by the symmetry of the component crystals. We shall show that such discontinuities are characterised by combinations of symmetry operators, one from each of the adjacent crystals. Thus, from the topological point of view, surface discontinuities, bulk and interfacial line defects are closely inter-related, and are characterised by crystal symmetry operators. The range of interfacial defect types predicted on the basis of this symmetry theory is rather broader than had been appreciated previously. However, we shall show that the present treatment is a generalisation of the seminal work by Frank and van der Merwe (1949) regarding the role of defects in the mutual accommodation of 'misfitting' crystals. In addition, we shall illustrate the fact that Frank's appreciation of the geometrical properties of arrays of line defects, first devised in 1950, is still an essential tool in the analysis of interfacial structure. This will be illustrated by reference to recent work on the interface between silicon and sapphire, where arrays of interfacial defects were found to be responsible for both the accommodation of misfit and a small angular deviation from the nominal epitaxial orientation.

### 6.1 Topological characterisation of interfacial defects

We now consider the characterisation of interfacial defects by imagining the process by which they can be introduced into a bicrystal. Initially, we consider two semi-infinite crystals, as illustrated schematically in figure 7($a$); the white and black crystals have the required structures and relative orientation and position, and are brought together so that the previously prepared surfaces are juxtaposed. Now as explained earlier, the black and white crystal surfaces can exhibit a range of discontinuities; the white discontinuities are characterised by $\mathcal{W}(\lambda)_j$ and the black ones by $\mathcal{W}(\mu)_i$, and some examples are illustrated in figure 7. When the initial black and white surfaces have been bonded together, the new surfaces may also be forced together and bonded to create an interface which is equivalent to the initial one. The surface discontinuities then become an interfacial defect, and this can be characterised by the operation involved in bringing the new surfaces together. Thus, using the white crystal's frame for the expression of the operation required to bring the new black onto the new white surface, we have the inverse of the black operation $\mathcal{P}\mathcal{W}(\mu)_i^{-1}\mathcal{P}^{-1}$, where $\mathcal{P}$ defines the correlation or equivalence of the adjacent crystals, followed by the white operation, $\mathcal{W}(\lambda)_j$. In other words, the operation characterising the interfacial defect, designated $\mathcal{Q}_{ij} = (\boldsymbol{Q}_{ij}, \boldsymbol{q}_{ij})$, is equal to $\mathcal{W}(\lambda)_j\,\mathcal{P}\mathcal{W}(\mu)_i^{-1}\,\mathcal{P}^{-1}$. This operation characterises an interfacial dislocation if $\mathcal{Q}_{ij} = (\boldsymbol{I}, \boldsymbol{q}_{ij})$, and $\boldsymbol{q}_{ij}$ is equal to the Burgers vector of the defect, $\boldsymbol{b}_{ij}$. When $\boldsymbol{Q}_{ij}$ is a proper rotation, the corresponding defect is an interfacial disclination when $\boldsymbol{q}_{ij} = 0$, and a dispiration otherwise. We note that although $\mathcal{Q}_{ij}$ must be proper in order to characterise an interfacial line defect, the component crystal symmetry operators can be either both proper, or both improper. In addition, we remark that the set of defects defined by $\mathcal{Q}_{ij}$ for a given interface is independent of the choice of alternative descriptions of $\mathcal{P}$, and also that this topological characterisation of defects is independent of their line direction.

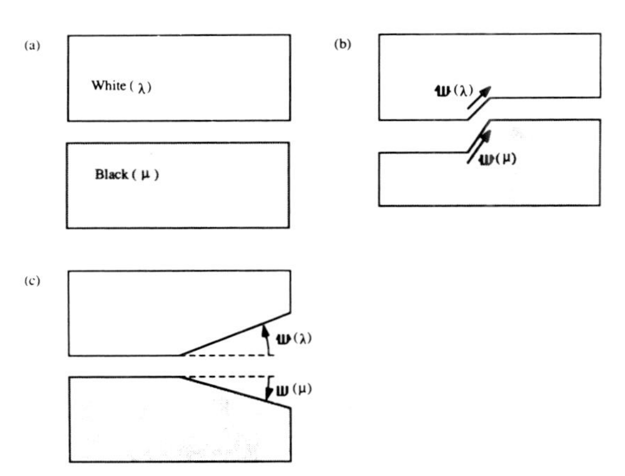


**Figure 7** Schematic illustration of the formation of defects separating equivalent regions of interface. The initial white and black surfaces to be bonded are shown in (*a*), and stepped and facetted surfaces are shown in (*b*) and (*c*), respectively. When the initial surfaces are bonded, the operation to bring the new black onto the new white surface topologically characterises the resulting defect.

The expression for $\mathcal{Q}_{ij}$ obtained above shows that the key factor governing the number of distinctive defects that can arise in a given interface is the extent to which the adjacent crystals exhibit coincident symmetry (or the extent of bicrystal connection). At one extreme we have the single crystal, i.e. where $\boldsymbol{P} = (\boldsymbol{I}, \boldsymbol{0})$ and $\mathcal{W}(\lambda)_k = \mathcal{W}(\mu)_k$, and it can be seen that $\mathcal{Q}_{ij}$ then becomes equal to a proper symmetry operation, $\mathcal{W}(\lambda)_l$. In other words, the sets of line defects which can inhabit the black and white half-spaces, and the interface, are identical in this case. At the other extreme, where $\mathcal{P}$ represents a correspondence (or equivalence) such that none of the symmetry of the two crystals is coincident, the sets of admissible line defects in the bulk of the black and white crystals are completely distinct, and a wide variety of interfacial defects, confined to exist in the interface only, can arise. Bicrystal connection can be regarded as the generalisation of the concept of misfit used by Frank and van der Merwe (1949) in their treatment of 'one-dimensional' dislocations at the interface between an overlayer and substrate which exhibit different natural lattice spacings. In this pioneering work, attention was focused on the misfit parallel to the interface, but the expression for $\mathcal{Q}_{ij}$ shows that misfit in any direction between the lattice parameters of two adjacent crystals can lead to interfacial defects. Moreover, any difference of nature, orientation or position between the symmetries of the two crystals leads to the possibility of defects. This topic has been explored elsewhere (Pond 1989) in some depth, and here we shall illustrate one example of an interfacial dislocation in an HCP metal which arises because $(1\bar{1}00)$ c-mirror-glide planes in the adjacent crystals are disconnected by virtue of being orientationally aligned, but where the intrinsic glide components are not parallel. We shall consider twinned HCP crystals related by a rotation of 180° about $[\bar{1}\bar{1}26]$; the matrix $\boldsymbol{P}$ describing this operation is the same as that given in section 2 for reflection across $(11\bar{2}1)$, but with the first two columns interchanged. A $[1\bar{1}00]$ projection of the twin boundary, with interfacial plane parallel to $(11\bar{2}1)$, is shown in figure 8, and a twinning dislocation is present in the interface. This defect can be regarded as having been created by bringing together black and white surfaces exhibiting demisteps (see figure 5(*b*)) characterised by the mirror-glide operations $(\boldsymbol{W}(\mu)_i, \boldsymbol{w}(\mu)_i)$ and $(\boldsymbol{W}(\lambda)_j, \boldsymbol{w}(\lambda)_j + \boldsymbol{t}(\lambda)_k)$, respectively, where $\boldsymbol{W}(\mu)_i$ and $\boldsymbol{W}(\lambda)_j$ represent reflection parallel to $(1\bar{1}00)$, and $\boldsymbol{w}(\mu)_i$ and $\boldsymbol{w}(\lambda)_j + \boldsymbol{t}(\lambda)_k$ are indicated in the figure. The resultant Burgers vector of the defect, $\boldsymbol{b}_{ij}$, can be found by substituting these operations into the expression for $Q_{ij}$, giving $\boldsymbol{b}_{ij} = \boldsymbol{w}(\lambda)_j + \boldsymbol{t}(\lambda)_k - \boldsymbol{P}\boldsymbol{w}(\lambda)_i + (\boldsymbol{W}(\lambda)_j - \boldsymbol{I})\boldsymbol{p}$. Using the value of $\boldsymbol{p}$ found in a recent computer simulation study of this interface (Serra *et al* 1988), we find $\boldsymbol{b}_{ij} = \frac{1}{6}(6\Lambda^2 + 1)^{-1}\ [11\bar{2}\bar{6}]$. In addition the step associated with the core of the dislocation has height equal to $\boldsymbol{d}_{(22\bar{4}2)}$, and we note that the motion of this dislocation along the interface would cause twinning without the necessity of atomic shuffles.

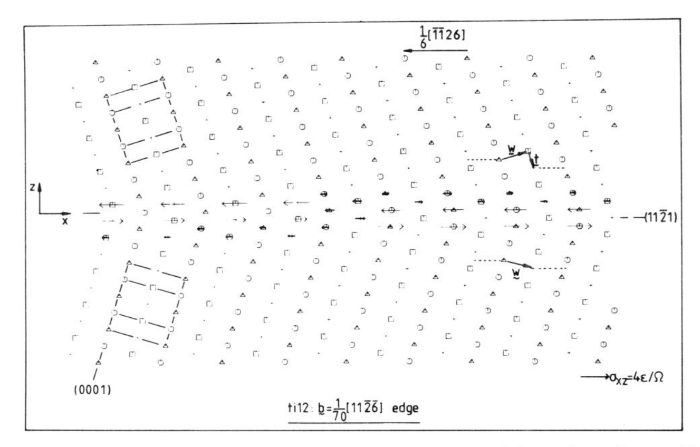


**Figure 8** Computer simulation of twinning dislocation in a (11$\bar{2}$1) twin boundary viewed along [1$\bar{1}$00]. Symbols are as for figure 5, the arrows represent the shear stress $\sigma_{xz}$, and $c/a$ is very close to the ideal value.

## 6.2 Arrays of interfacial defects

Dislocation models of grain boundaries were first proposed by Bragg (1940) and Burgers (1940), but Frank (1950) showed how the dislocation structure of general grain boundaries can be established. Frank is eponymous for the formula that relates the total Burgers vector, $\Sigma \boldsymbol{b}$, of the individual dislocations which cross a probe vector $\boldsymbol{v}$ lying in the interface to the misorientation of the crystals; $\Sigma \boldsymbol{b} = (\boldsymbol{I} - \boldsymbol{P})\boldsymbol{v}$. Dislocation configurations in a wide variety of interfaces have been explored using this expression (see Read 1953, Bilby *et al* 1955, Christian 1975, Hirth and Lothe 1968, Matthews 1979 for reviews). As is now well known, the simple tilt boundaries envisaged by Bragg (1940) and Burgers (1940) consist of a wall of discrete edge dislocation lines separated by a distance given by $|\boldsymbol{b}|/\theta$ using the low-angle approximation, where $\theta$ is the angle of rotation. As an illustration of the application of Frank's equation, we consider the interface between silicon and sapphire, mentioned previously. The (01$\bar{1}$2) surface periodicity of sapphire is illustrated in figure 2($a$), and is the substrate on which (001) silicon is deposited; the [100] and [010] directions of the overlayer are oriented parallel to $\boldsymbol{r}_1$ and $\boldsymbol{r}_2$ in figure 2($a$), leading to a misfit equal to 12.3% parallel to $\boldsymbol{r}_1$ and 5.5% parallel to $\boldsymbol{r}_2$. We first assume the silicon to be deposited pseudomorphically and then use Frank's formula to determine the network of dislocations necessary to accommodate this anisotropic strain, thereby relieving the long-range strain in the epilayer. This geometric analysis does not identify a unique network; energetic and kinetic factors need to be taken into account to find the most favourable configuration. In the present case, the network deduced, and which was also consistent with experimental observations, is shown in figure 9 (Aindow *et al* 1989). The dislocations are silicon crystal dislocations with Burgers vectors $\boldsymbol{b}_1 = \frac{1}{2}[1\bar{1}0]$ and $\boldsymbol{b}_2 = \frac{1}{2}[110]$, line directions $\zeta_1 = [0.41, 0.91, 0]$ and $\zeta_2 = [\overline{0.41}, 0.91, 0]$, and spacings $d_1 = 2.21$ nm and $d_2 = 4.94$ nm. (Note that crystal dislocations are always admissible defects in any interface since $\mathcal{Q}_{ij}$ will be equal to a white crystal translation operation, for example, if the black operation substituted into the expression for $\mathcal{Q}_{ij}$ is the identity.) It is interesting to note that this network is not orthogonal, because the Burgers vectors are not parallel to the principal strains in this case. This result, which can be readily appreciated by application of Frank's formula, resolves a controversy regarding experimental observations of this interface. Using cross-sectional high-resolution electron microscopy, the structure of the interface when observed along $[1\bar{1}0]$ or $[110]$ is apparently incoherent. However, since this direction of view is inclined to the misfit dislocation lines described in the above semi-coherent model of the interface, complicated images of the interfacial structure are to be expected.

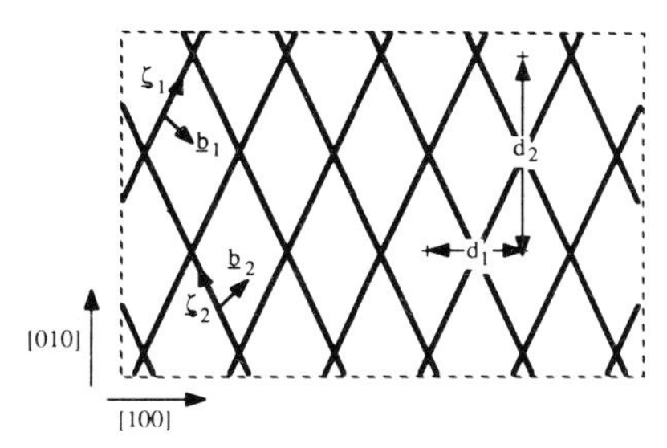


**Figure 9** Schematic diagram of an array of dislocations which would accommodate the misfit between (001) silicon and $(01\bar{1}2)$ sapphire.

## 6.3 Epitaxial growth on vicinal surfaces

Accurate measurements of the orientation of the silicon overlayer with respect to the sapphire substrates showed that a small angular deviation from the nominal orientation relationship was present (Pond *et al* 1987a). Moreover, it was shown that the magnitude and sense of this deviation is correlated with the precise orientation of the surface orientations of the substrates, these latter being vicinal surfaces oriented within a few degrees of $(01\bar{1}2)$. The origin of the angular deviation was deduced by consideration of the terraced structure of vicinal surfaces (see figure 6), as follows. When an overlayer is deposited on a vicinal surface, it must accommodate the surface steps present, which are individually characterised by the operation $\mathcal{W}(\mu)_i$. The deposited crystal will presumably accommodate these features by means of white surface features which are as nearly complementary as possible. Let these latter be characterised by the operation $\mathcal{W}(\lambda)_j$; it follows that the resulting interfacial defects, characterised by

$\mathcal{Q}_{ij} = \mathcal{W}(\lambda)_j \mathcal{P} \mathcal{W}(\mu)_i^{-1} \mathcal{P}^{-1}$, will exhibit defect nature unless $\mathcal{W}(\lambda)_j$ and $\mathcal{W}(\mu)_i$ are coincident symmetry operators. In the case of silicon-on-sapphire, the range of possible surface steps and demisteps that can be present on the two crystal surfaces would lead to the formation of interfacial dislocations with components of their Burgers vectors, $\boldsymbol{b}'$, perpendicular to the nominal interface, given by $|\boldsymbol{b}'| = h_s - h_e$, where $h_s$ and $h_e$ are the heights of the substrate and epilayer steps. Consequently, the initial array of surface steps is transformed, after deposition of the epilayer, into an array of interfacial dislocations which resemble a simple low-angle tilt boundary. Moreover, we can use Frank's formula to find the misorientation angle, $\theta_m$, expected on the basis of the spacing of the initial steps, $m$, and their effective Burgers vector, $\boldsymbol{b}'$. Furthermore, since we have $\theta_v = h_s/m$, where $\theta_v$ is the vicinal angle, we can write $\theta_m = (1 - h_e/h_s)\theta_v$. In other words, the magnitude and sign of $\theta_m$ is expected to be linearly related to $\theta_v$. Taking $h_s = d_{(01\bar{1}2)} = 0.35$ nm, and $h_e = d_{(002)} = 0.27$ nm, which are the heights of the smallest and most nearly matching surface steps, the factor $(1 - h_e/h_s)$ is equal to 0.23. A plot of $\theta_v$ against $\theta_m$ for a series of specimens taken from the work of Pond *et al* (1987a) is shown in figure 10, and it can be seen that the observed correlation is in excellent agreement with the discussion above (the straight line in the figure has gradient equal to 0.23). We note that the misorientation angles, $\theta_m$, are only of the order of 10 minutes of arc. However, despite their small magnitudes, it is believed that these deviations, which break bicrystal symmetry, can exert disproportionate influence on the evolution of defect miscrostructure in thin films. For example, there is a very pronounced anisotropy in the populations of the four twinning systems in the (001) silicon films, and this is also correlated to $\theta_m$.

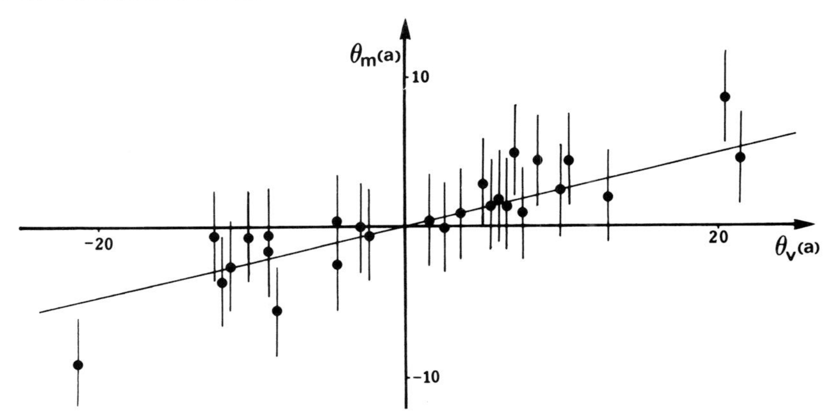


**Figure 10** Plot of vicinal angle, $\theta_v$, against misorientation angle (about $[0\bar{1}11]$), $\theta_m$, for (001) silicon grown on $(01\bar{1}2)$ sapphire. The axes are graduated in units of minutes of arc; vertical error bars are $\pm 3$ minutes of arc, and horizontal ones (not shown) are about $\pm 1$ minute of arc.

## ACKNOWLEDGMENT

I was a postgraduate student in the Department of Physics at the University of Bristol from 1969 to 1972, and this was an outstandingly stimulating and enjoyable period for me and my student colleagues. We all value enormously the experience of those years, particularly our contact with Professor Frank in the lecture theatre, coffee room and on social occasions. Professor Frank's zest for science and his encouragement to try to make progress from a fundamental basis were invaluable to us. His continued interest in my own research since that time has been immensely helpful to me, and it is with great pleasure and gratitude that I am able to make this acknowledgment on the occasion of his eightieth birthday.

## REFERENCES

Aindow M, Batstone J L, Pfeiffer L, Phillips J M and Pond R C 1989 *Mat. Res. Soc. Symp. Proc.* **138** 373–8

Aindow M, Eaglesham D J and Pond R C 1987 *Inst. Phys. Conf. Ser.* no 90 123–6

Bilby B A, Bullough R and Smith E 1955 *Proc. R. Soc.* A **231** 263–73

Bragg W L 1940 *Proc. Phys. Soc.* **52** 54–5

Burgers J M 1940 *Proc. Phys. Soc.* **52** 23–33

Christian J W 1975 *Transformations in Metals and Alloys* Part I, 2nd edn (Oxford: Pergamon) ch 7,8

Frank F C 1949 *Proc. Phys. Soc.* A **62** 131–4

—— 1950 *Carnegie Inst. of Tech. Symp. on the Plastic Deformation of Crystalline Solids* (Pittsburg: Office of Naval Research) pp 150–1

—— 1951 *Phil. Mag.* **42** 809–19

—— 1963 *Metal Surfaces* ed W W Mullins (Ohio: ASM) pp 1–15

—— 1965 *Acta Crystall.* **18** 862–6

—— 1987 *Met. Trans.* A **19** 403–8

Frank F C and van der Merwe J J 1949 *Proc. R. Soc.* A **198** 205–25

Hahn T (ed) 1983 *International Tables for Crystallography* (Dordrecht: Reidel)

Harris W F 1970 *Phil. Mag.* **22** 949

Hirth J P and Lothe J 1968 *Theory of Dislocations* (New York: McGraw-Hill) ch 19

Matthews J W 1979 *Dislocations in Solids* vol 2, ed F R N Nabarro (Amsterdam: North-Holland) p 461

Mermin N D 1979 *Rev. Mod. Phys.* **51** 591–648

Pond R C 1989 *Dislocations in Solids* vol 8, ed F R N Nabarro (Amsterdam: North-Holland) pp 1–66

Pond R C, Aindow M, Dineen C and Peters T 1987a *Inst. Phys. Conf. Ser.* no 87 181–6

Pond R C, McAuley N A, Serra A and Clark W A T 1987b *Scripta Met.* **21** 197–202

Read W T 1953 *Dislocations in Crystals* (New York: McGraw-Hill) ch 11, 12

Seitz F 1936 *Ann. Math. Stat.* **37** 17

Serra A, Bacon D J and Pond R C 1988 *Acta Metall.* **36** 3183–203

# Selective Dissolution and De-alloying of Silver–Gold and Similar Alloys

A J Forty

## 1 FOREWORD

The H H Wills Physics Laboratory of the University of Bristol in the late 1940s and early 1950s was an exciting place for a young research student. Under the inspiring leadership of Frank, Mott and Powell new discoveries were being made in the fields of cosmic rays, solid state chemistry and the micro-deformation of metals. It was in this period particularly that Bristol gained the reputation for 'intelligent observation'. The discovery of the meson by observing the tracks of cosmic rays in photographic emulsions by Powell and his team is a good example and, of course, Frank's intuitive discovery of the spiral-step process of crystal growth is another.

I had the privilege of being Charles Frank's first research student. It was from him that I discovered the art of observation and the science of intelligent deduction. It is a further privilege now to be able to contribute to his Eightieth Birthday *Festschrift*. To my knowledge he has never worked on the subject of my paper, which is how alloys dissolve selectively and decompose or de-alloy. But the approach my colleagues and I have adopted in our work in this field is the one which we would have followed under his tutorship.

## 2 INTRODUCTION

The refinement of gold from its alloys by selective dissolution of the less noble metals in *aqua regia* is a well-known industrial process (Shreir 1976). Selective dissolution is also thought to be the basis of the art of depletion gilding (*mise-en-coulour*) which was very skilfully practiced by pre-Columbian Indians in South America (Forty 1979). More recent scientific interest in the phenomenon has been focused on the corrosion of metals and the possible role of selective dissolution in embrittlement and stress corrosion of alloys (Forty and Edeleanu 1960, Pickering 1967), an area of very broad scientific and practical importance. This paper is concerned only with the basic processes of selective dissolution and, for simplicity, deals only with the de-alloying of gold alloys, and silver–gold alloys in particular.

When an alloy is exposed to a corrosive environment the components respond differentially according to their electrochemical potential. Under certain circumstances the less noble metal is oxidised or dissolved selectively leaving a residue which is enriched with the more noble element. In the case of gold alloys, depending on the initial composition of the alloy, the process of selective dissolution can lead to complete de-alloying, or parting of the components. It is found in practice that parting only occurs wholly if the initial concentration of the less noble element in the alloy is sufficiently high. Indeed, in the practice of extracting gold from base alloys it is often necessary to melt the alloy down with additional less noble metal before it can be parted chemically. There is said to be a critical alloy composition, or 'parting limit', which determines whether complete parting can occur (Shreir 1976).

The nature of the parting limit was first investigated systematically by Tammann and Brauns (1931) who found that for silver–gold in hot sulphuric acid there is no attack unless the alloy contains more than 50% silver; the parting limit is 50% gold. For copper–gold in ferric chloride solution the parting limit is said to be 30% gold (Shreir 1976).

Complete parting of the alloy must involve not only a surface reaction between the metal and the corrosive environment but also processes that require atomic transport. The importance of this was highlighted by Bakish and Robertson (1956) who showed that complete de-alloying of copper–gold in ferric chloride took place even when the specimen was a single crystal, with no possibility therefore of penetration of the solvent along grain boundaries. In order to illustrate the importance of atomic transport it is useful to consider the processes that might be expected to occur on the surface of a binary alloy, components A and B, during dissolution. For simplicity we ignore molecular adsorption from the solution, oxidation or complexing effects associated with the solvent, and assume that dissolution occurs only by ionisation and solvation of the metal atoms. Referring to the familiar terrace, ledge, kink model of the crystal surface depicted in figure 1 (Frank 1949, for example), dissolution is expected to occur most readily from kink sites (K) on surface steps where the atoms are less firmly bound. If there is a sufficiently large difference of electronegativity the less noble atoms, labelled A, will be dissolved preferentially. However, the successive dissolution of A will mean that the kink sites will become progressively occupied by the noble B atoms; thereafter, preferential dissolution can proceed only by the removal of A atoms from non-kink sites (N) or from terrace sites (T), both of which require a greater activation energy or overpotential. Eventually, if the dissolution progresses sufficiently all the surface sites should be occupied by B atoms only and the surface will become passivated against further attack unless the alloy has a high enough potential for the dissolution of B.

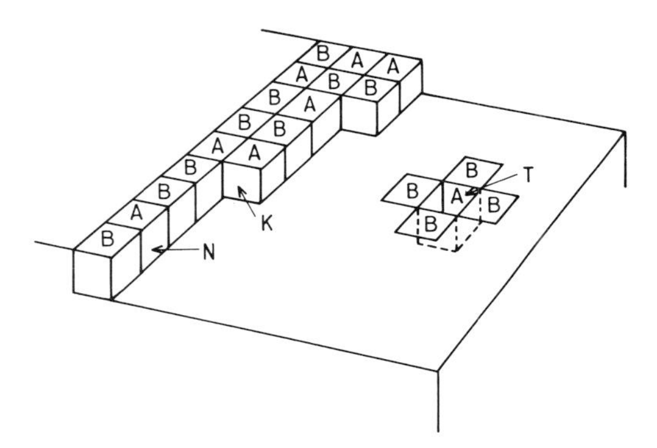


**Figure 1** Schematic representation of the surface of an alloy composed of dissolvable A atoms and noble B atoms. K is a kink site in a surface step, N is a non-kink site in a step, T is a terrace site.

It follows from the foregoing argument that the continuing dissolution of A from the alloy must involve forms of atomic transport so that there will be a replenishment of the less noble species at the metal/solvent interface. One mechanism for this would be the electrochemical transfer of B atoms from more anodic sites to more cathodic sites on the surface. However, it is doubtful whether such a process could apply to gold alloys where such a highly noble species is involved. A more realistic mechanism, involving the diffusion of A atoms from below the surface so that the supply of less noble component is continuously replenished, has been proposed by Pickering (1967). He has supported this by x-ray diffraction studies which have shown, in the case of copper–gold alloys after varying amounts of exposure to ferric chloride, a series of sub-surface structural phase changes which can be related to the loss of copper from the alloy.

He has also shown that the dissolution currents measured at anodic potentials where only copper dissolution is possible are compatible with diffusion of copper to the solid surface, but only if the diffusion coefficient is considerably enhanced above the normal room temperature value for copper–gold crystals. Pickering argued that such an enhancement can be expected because the high concentration of vacancies created in the vicinity of the alloy surface by dissolution should lead to divacancy and trivacancy clusters which are more mobile than monovacancies.

It was pointed out by Vermilyea (1967) that, as an alternative to bulk diffusion as a means of replenishing the supply of dissolvable atoms at the surface, surface diffusion and the consequent rearrangement of the alloy surface could be important. It is, of course, possible to envisage more complex atomic processes occurring at the metal/electrolyte interface, involving dissolution of A, oxidation and solvation of B, enhanced transport in the boundary layer at the interface as well as disordering of the metal surface. In order to throw further light on this the author and his co-workers have undertaken a comprehensive series of studies of the electrochemistry and the structural and micromorphological changes occurring when silver–gold alloys are subjected to selective dissolution in nitric acid. The results of this work are brought together here for the first time. The conclusion we reach is that the selective dissolution of silver by the nitric acid solution is accompanied by oxidation of the residual gold. The gold oxide is in a metastable state and its decomposition gives rise to a restructuring of the alloy surface into gold-rich islands and fresh areas of uncorroded alloy. Under certain circumstances, notably an initial alloy composition which is silver-rich and the use of strong nitric acid, the continuous process of dissolution, oxidation and restructuring gives rise to a quasi-stable surface morphology of corrosion tunnels which enable the dissolution of silver to be maintained until complete parting occurs.

## 3 ELECTROCHEMICAL STUDIES

Metal corrosion is in general a complex process involving the interaction between the metal surface and its environment. It often entails the formation of oxide or other forms of passivating layer formed on the surface of the metal. The type of corrosion depends on the electrochemical processes involved. These might, for example, give rise to anodic and cathodic sites on the surface itself, as in pitting corrosion, or electrolytic coupling between the metal and electrolyte which leads to uniform oxide film formation, as in oxidation or tarnishing. A great deal of our present understanding of corrosion has been developed on the basis of electrochemical analysis: measurements of current flow across the metal–electrolyte interface at various metal potentials which can be used to identify the changed state of the surface and the chemical processes which occur during corrosion.

Such a study has ben made for silver–gold in nitric acid (Carter 1985, Carter and Forty 1990a). In order to relate these electrochemical measurements with micromorphological observations of the effect of corrosion on the same alloys, Carter and Forty developed special techniques for preparing alloys in the form of thin films suitable for transmission electron microscopy (TEM) which could also be used for electrochemical analysis. The method adopted was similar to that developed earlier by Forty and Durkin (1980) for (TEM) studies of chemical corrosion. This entailed simultaneous evaporation of gold and silver onto a single-crystal gold substrate which itself was prepared by vapour deposition of gold onto freshly cleaved mica. The composition of the alloy was pre-determined by the relative rates of deposition of silver and gold checked afterwards by x-ray microanalysis in a scanning electron microscope. The electrochemical measurements were carried out in a standard cell consisting of the alloy film as the working electrode, a counter-electrode of pure gold sheet and a saturated calomel electrode (SCE) used as a reference electrode for measuring the potential of the gold alloy film in the electrolyte. The electrolyte was a non-aggressive 0.1 M aqueous nitric acid solution, chosen so that the electrochemical measurements could be related to the micromorphological observations made earlier by Forty and Durkin (1980) on the de-alloying of the same alloys during free corrosion in stronger nitric acid.

The thin-film geometry of the specimens limited the amount of silver that could be dissolved and the form of electrochemical analysis used by Carter and Forty had therefore to be chosen carefully. Thus, whilst the measurement of dissolution currents for various fixed anodic potentials might have been more useful for kinetic studies of selective dissolution this approach was not possible because the flow of large dissolution currents over a prolonged period would have had a significant effect on the residual alloy composition, making it difficult to relate dissolution rates unambiguously with composition. The method adopted was the alternative, possibly more useful, one of recording cell current as the working electrode (specimen) potential was ramped linearly from an initial value below the rest potential (i.e. the open-circuit potential acquired by the specimen immediately after being placed in the nitric acid—for gold this is about 50 mV on the saturated calomel electrode scale) to higher potentials. Ideally, the specimen potential should have been increased infinitesimally slowly so that the cell currents recorded would be the steady state values. However, for practical reasons, due to the thinness of the films, a finite ramp rate of 50 $mV\,s^{-1}$ had to be used. The anodic voltammograms so recorded were usually limited to terminate below an anodic potential of 1500 mV (SCE) which is the anodic breakdown potential of the aqueous electrolyte after which oxygen evolution occurs. The anodic voltammograms were found to give much useful information about the effect of alloy composition on silver dissolution. It was found also that useful information about the state of the alloy after silver dissolution could be obtained by reversing the potential ramp to record the corresponding cathodic voltammograms.

Figure 2 is the cyclic voltammogram recorded in this way for pure gold. The shallow plateau observed in the anodic sweep, commencing at about 1200 mV, is thought to be the early stage of oxidation of gold; other workers (for example, Schmid and O'Brian (1964) and Ogura *et al* (1971)) have produced evidence that an oxidation reaction takes place close to 120 mV (SCE). In this particular voltammogram the anodic potential ramp was continued to 1600 mV and the rapid rise of cell current at the higher potential corresponds to oxygen evolution in the electrolyte. The small peak commencing at about 900 mV in the reverse cathodic potential sweep is thought to correspond to the reduction of the oxides or hydroxides produced during the anodic sweep. Though not shown in the figure, hydrogen evolution in the electrolyte occurs at approximately −250 mV (SCE).

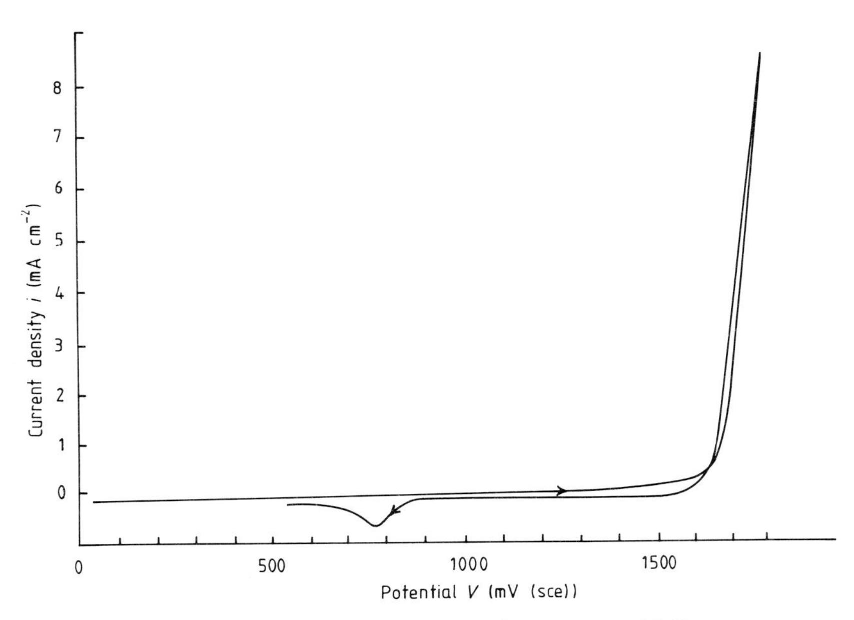


**Figure 2** Cyclic voltammogram for a pure gold film.

Figure 3 shows the cyclic voltammogram for pure silver. The rapid rise in anodic current in the forward sweep corresponds to the dissolution of silver, commencing at about 350 mV. The reverse peak in the cathodic current when the potential ramp is reversed in this case corresponds to silver deposition.

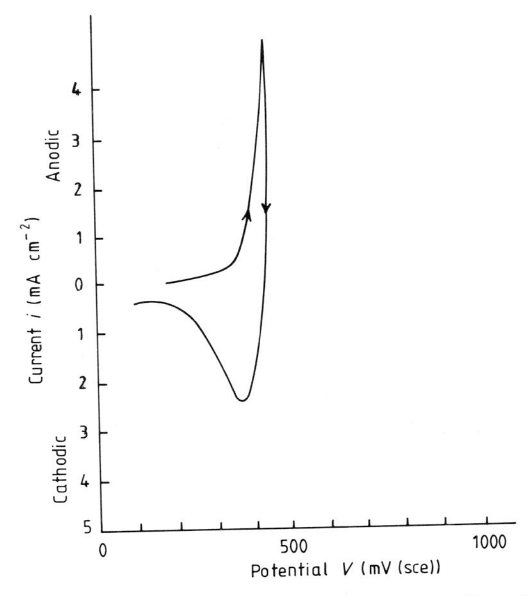


**Figure 3** Cyclic voltammogram for a pure silver film.

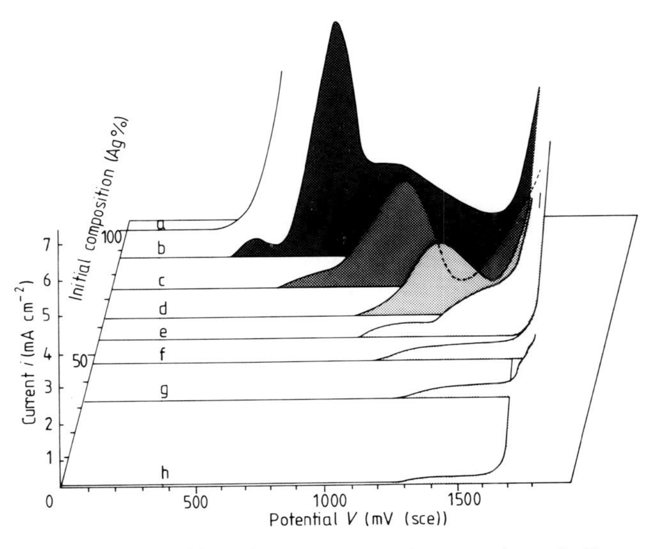


**Figure 4** Set of anodic voltammograms for a series of silver–gold alloys.

The voltammetric behaviour for silver–gold is shown in figure 4 where a series of anodic voltammograms are plotted for a number of different alloys with increasing silver concentration. For alloys containing less than about 55% silver the behaviour is very similar to that for pure gold. X-ray microanalysis of these same films after voltammetric measurement shows that there is no change of alloy composition. Parting of these alloys does not therefore occur.

At compositions above 55% silver there are strong anodic currents with complex features starting at low potentials depending on the percentage of silver. The most complex behaviour is found for alloys with more than 85% silver. Here there are three merging peaks commencing at about 380, 500 and 750 mV, respectively. These are thought to be associated with the dissolution of three different types of silver characterised by three distinct types of surface site. The first, small, peak occurs at a potential close to that for the dissolution of pure silver and is therefore thought to be associated with the dissolution of a thin surface-segregated layer. Such a layer is expected for silver-rich alloys as shown by Tokutaka *et al* (1981) on theoretical grounds and confirmed experimentally using Auger electron analysis. The much larger anodic features at higher potentials consisting of two merged peaks correspond to dissolution of silver from more tightly bound surface sites. Segregation at grain boundaries or other defects could be responsible for the dissolution peak commencing at 500 mV and near-surface crystalline surface sites could be the source of silver for the higher potential dissolution peak commencing at 750 mV. The association of these anodic peaks with selective dissolution of silver has been confirmed by x-ray microanalysis of the films after the anodic treatment.

For alloys containing less than 85% silver the initial small peak is no longer evident, indicating that the surface segregation of silver is insignificant at these compositions. The two larger peaks are now merged and commence at progressively higher potentials as the initial silver concentration of the alloy decreases.

Whilst there is no silver dissolution for the gold-rich alloys, the potential at which oxidation commences decreases as the percentage of silver in the alloy increases, indicating that the silver content of the alloy might have a significant influence on the process by which the gold is oxidised. It is not evident from the voltammograms in figure 4 that gold oxidation occurs for the highly silver-rich alloys because of the dominance of the silver dissolution peaks. The results do suggest, however, that there is a cross-over of the variation of silver dissolution potential and the variation of gold oxidation potential with alloy composition at about 65% silver. This raises the important question of whether the dissolution of silver and the oxidation of gold in the alloy might be coupled in some way.

Further light is shed on this by measuring the decay of 'open circuit' potential (i.e. the potential of the alloy under open-circuit conditions in the electrochemical cell) with time for a series of alloys following an anodising treatment at a potential of 1200 mV. The anodising treatment was controlled in each case at a charge density of 7 mC $cm^{-2}$. This small anodising charge was chosen so that the anodising treatment should be completed in a sufficiently short time to avoid ageing effects in the alloys and also to ensure that the anodising current could be supported by silver dissolution without running out of available silver in the thin alloy films. The anodising potential of 1200 mV was chosen to be sufficiently below the electrolyte breakdown (1500 mV) to avoid oxidation by oxygen evolution, yet high enough to ensure that any gold oxidation accompanying silver dissolution can occur.

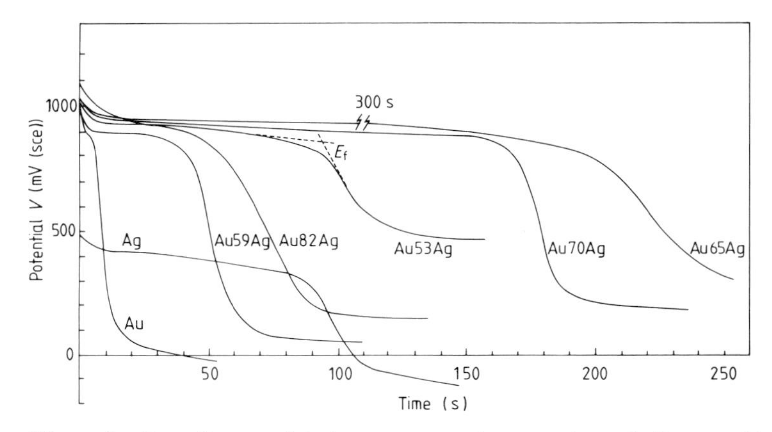


**Figure 5** Set of potential decay curves for a series of silver–gold alloys following anodisation at 1200 mV (SCE) and charge density 7 mC $m^{-2}$.

The results so obtained are shown in figure 5. The general behaviour is similar for all alloys. There is an initial period when the open-circuit potential is steady at about 900 mV. This is then followed by a fall to a new steady potential whose value is dependent on the initial composition of the alloy. The potential at which the extrapolations of the plateau region and the final rapid decay of potential intersect is known as the Flade potential. This kind of behaviour is generally associated with the dissolution or decomposition of a chemically unstable surface species, such as an oxide formed by the previous anodic treatment. The value of the Flade potential for all the alloys examined in figure 5 is close to that for pure gold (about 860 mV), which implies that the anodic species formed during the anodic dissolution of silver from silver–gold is similar to that formed during the anodisation of gold. Carter and Forty also found that silver–gold alloys which are allowed to corrode freely in nitric acid solutions reach a stable potential (of about 820 mV) which is close to the composition-independent Flade potential. This suggests that the surfaces of the alloys reach the same state after open-circuit chemical corrosion as that arrived at as a result of anodically driven electrochemical corrosion.

Figure 5 shows that the length of time spent on the potential plateau, i.e. at the Flade potential, clearly varies with alloy composition. There is a pronounced maximum value of the elapsed time at a composition of about 65% silver. This suggests that the amount of anodic species formed on the silver–gold alloys is greatest at this composition. This same conclusion can be reached by an analysis of the cyclic voltammograms. On the assumption that the peak in the reverse cathodic sweep represents the reduction of oxide previously formed during silver dissolution in the anodic sweep, the area contained by the cathodic peak represents the amount of anodic oxide. Figure 6 shows how this quantity varies with composition of the alloy. There is a very marked maximum when the alloy contains about 60% silver. We see therefore that the area enclosed by the cathodic reduction peak and also the time spent at the Flade potential during the decay of the anodised surface have a maximum value when the alloy has a silver content of 60–65%. It is interesting that, as was noted in connection with figure 4, the silver dissolution potential and the gold oxidation potential for the alloys cross over at this same composition. We may conclude therefore that the dissolution of silver from a silver–gold alloy is accompanied by oxidation of the residual gold on the surface of the alloy and that the coupling of these processes is greatest when the alloy contains about 65% silver.

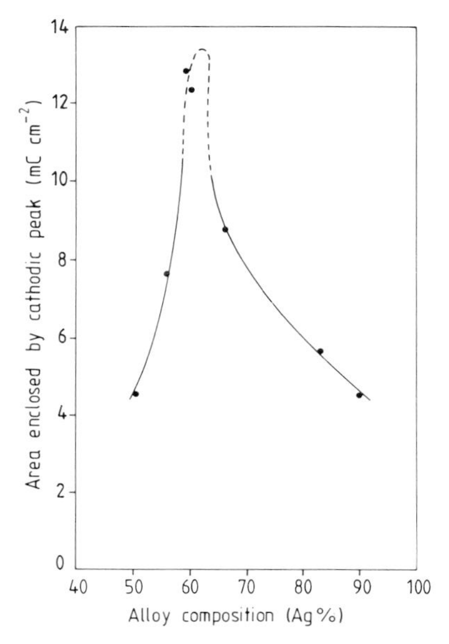


**Figure 6** Variation of area enclosed by cathodic peak in cyclic voltammograms with alloy composition.

The areas under the peaks in the voltammograms are representative of the amounts of charge passed. The ratio of the areas under the peak in the anodic sweep (largely silver dissolution) and the peak in the cathodic reverse sweep (reduction of gold oxide) should therefore correspond to the ratio of silver and gold in the surface of the alloy. Carter and Forty have compared the ratios estimated from the voltammograms with those measured by x-ray analysis for a wide range of alloys. The result is consistent with the conclusion that the dissolution of silver is accompanied by the formation of a gold (III) oxide.

The conclusion to be drawn from these electrochemical studies can therefore be summarised as follows: alloys containing more than 50% silver undergo selective dissolution and the residual gold is left in an oxidised state, probably in the form of gold (III) oxide. In the case of anodic oxidation, at least, the oxidation occurs most efficiently when the alloy contains about two-thirds silver.

The important question of how de-alloying continues beyond the surface layers of the metal is still not answered. It will be seen in the next sections that extensive micromorphological changes accompany the selective dissolution of silver. Under certain conditions a tunnelling morphology develops and this promotes in-depth de-alloying. In order to establish a mechanism for de-alloying we must therefore examine the relationship between selective dissolution, oxidation and micromorphological change.

## 4 MICROMORPHOLOGICAL STUDIES OF ELECTROCHEMICALLY CORRODED SILVER–GOLD

Carter and Forty (1990b) have extended their electrochemical analysis by a transmission electron microscope (TEM) study of the morphological changes accompanying de-alloying. For this purpose the silver–gold films were formed on very thin (20 nm) films of pure gold pre-mounted on gold electron microscope grids. The films were corroded electrochemically under potentiostatic control in 0.1 M nitric acid at an anodic potential of 1200 mV (SCE). The earlier work indicated that this potential is high enough to ensure that silver dissolution should occur for all the alloys to be investigated and is sufficiently below the electrolyte breakdown potential to avoid oxygen evolution. Morphological and structural changes should therefore be related to silver dissolution and any accompanying anodic oxidation.

Films of pure silver were found to simply dissolve under these conditions. The gold substrate did not appear to be affected. Likewise, films of pure gold showed no structural change. However, if the anodic potential was raised to a value well beyond the oxygen evolution potential the film became deeply corroded as shown by the micromorphology in the TEM image (figure 7). This reveals a complex structure of overlapping tunnels.

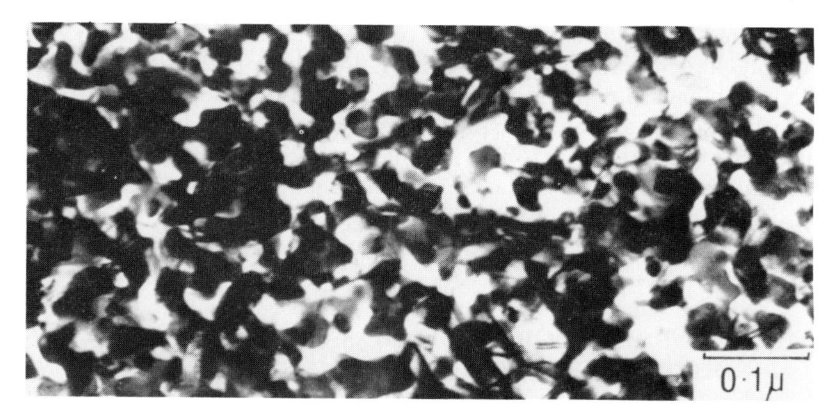


**Figure 7** Transmission electron micrograph of a gold film after anodic corrosion at a potential of 2.6 V.

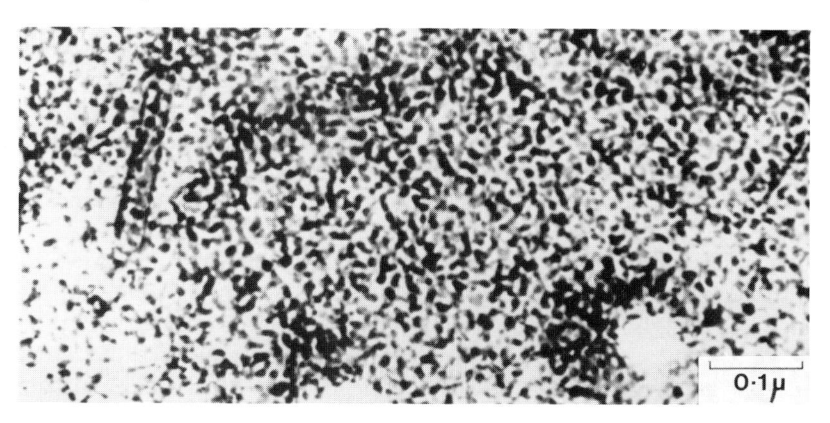


**Figure 8** Transmission electron micrograph of a 65% silver/35% gold alloy after anodic corrosion at 1200 mV and charge density 24 mC $cm^{-2}$.

The electron diffraction patterns for these heavily corroded specimens show that the gold films, initially good quality single crystals, become polycrystalline and there is evidence of a new phase which could be a form of gold oxide. Similar results were found for the gold-rich (i.e. > 50% gold) alloys.

In the case of silver-rich (i.e. > 50% silver) alloys the electrochemical corrosion at 1200 mV took the form of pitting. The density of pits increased as the period of corrosion, and therefore the amount of charge passed in the cell, was increased. After the passage of sufficient charge (12 mC $cm^{-2}$ in the case of a 65% silver alloy) the pitted structure developed into a network of interconnected channels (width about 5 nm) and then, after a charge of 24 mC $cm^{-2}$, the tunnel morphology, illustrated by the TEM micrograph in figure 8, was established. The electron diffraction patterns for the electrochemically corroded alloys showed no evidence of structural change at the normal anodising potential of 1200 mV. However, at slightly higher potentials, 1350 mV in the case of an alloy with 55% silver, additional diffraction spots were found which corresponded in position to the extra rings found for the heavily corroded gold specimen. This is evidence that a new phase, probably a gold oxide, is formed on the alloy under these conditions. It is interesting to note that 1350 mV is the potential at which silver dissolution commences for this particular alloy.

## 5 MICROMORPHOLOGICAL STUDIES OF CHEMICALLY CORRODED SILVER–GOLD

In order to use these electrochemically produced results to gain a better understanding of de-alloying under normal corrosion conditions it is necessary to compare the micromorphological changes described above with those observed in the earlier study by Forty and Durkin (1980) of the micromorphology found after chemical corrosion. In that work single-crystal films of the various alloy compositions were prepared by vapour deposition onto previously prepared thin-film substrates of pure gold. The composite films were immersed in nitric acid solution of various strengths and for various periods of a few seconds to a few minutes, and then examined by TEM.

The observations are summarised by the following micrographs. Figure 9(*a*) shows a characteristic morphology of islands, thought to be gold-rich, formed on a slightly silver-rich alloy (55% silver) after exposure to 35% nitric acid solution for 30 s. Figure 9(*b*) shows the same area of the alloy after a further immersion in nitric acid for another 30 s. A comparison of the two micrographs shows that islands have grown extensively as a result of the more extended corrosion, leaving a surface morphology in the form of interconnected channels. With more extensive exposure to the acid the islands grow further whilst the channels shrink to form isolated pits. In the case of more silver-rich alloys pits are found to develop earlier. If concentrated nitric acid is used an extensive tunnelling morphology is established, as illustrated by figure 10. It appears therefore that pitting and tunnelling are characteristic of both chemical and electrochemical corrosion of silver–gold in nitric acid.

One of the reasons for choosing the silver–gold/nitric acid system as a model for microcorrosion studies was the expected absence of insoluble corrosion products. The reaction products formed when pure silver is dissolved in nitric acid are primarily gaseous nitric oxide and silver nitrate, which is very soluble in an aqueous environment. On the assumption that a similar chemical reaction occurs in the selective dissolution of silver from silver–gold alloys, the corrosion reaction should not therefore be restricted

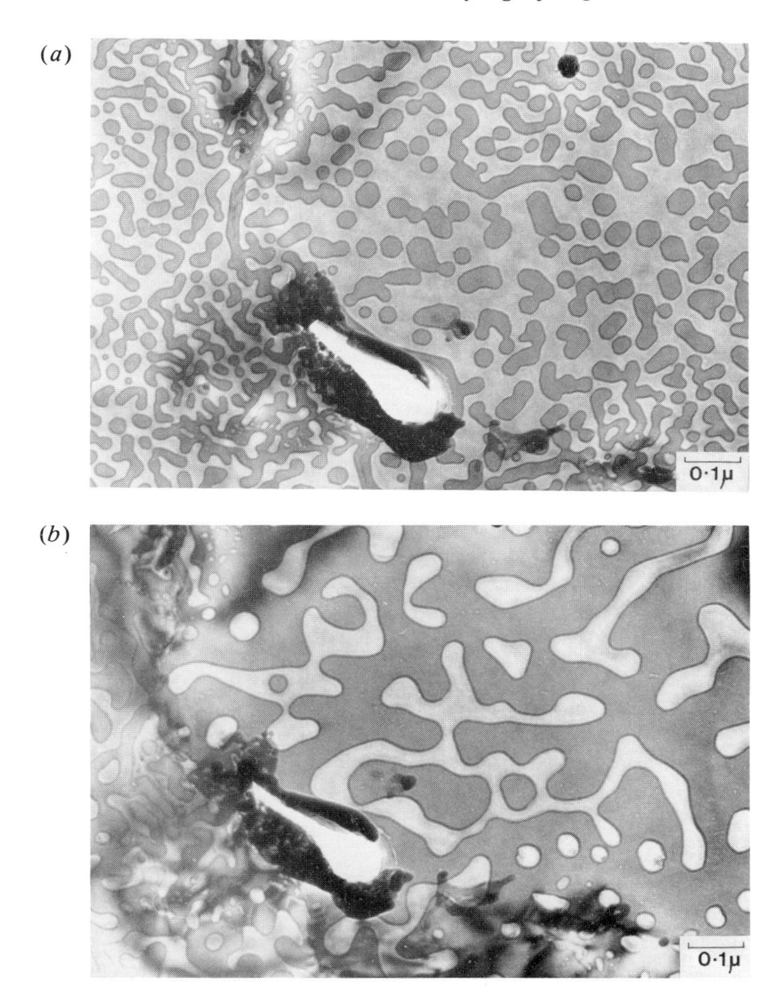


**Figure 9** Transmission electron micrographs illustrating the formation and growth of gold-rich islands during the chemical corrosion of a 55% silver/45% gold alloy film in 35 per cent nitric acid solution: (*a*) is after exposure to acid for 30 s and (*b*) is the same area of surface after immersion in acid for a further 30 s.

by the formation of a passivating film of corrosion product. There are several other useful features of the silver–gold alloy system. For example, the change of lattice parameter across the complete range of alloy composition from pure silver to pure gold is less than 0.2%. Furthermore, the alloy remains a single phase (FCC) at all compositions. There should not therefore be any significant microstructural change accompanying selective dissolution and the interpretation of the TEM images should be a fairly straightforward matter. This is certainly not the case for other binary alloys, such as copper–gold and copper–zinc, which have been used for microstructural and micromorphological studies of corrosion in the past (Pickering and Swann 1963, Pickering 1967).

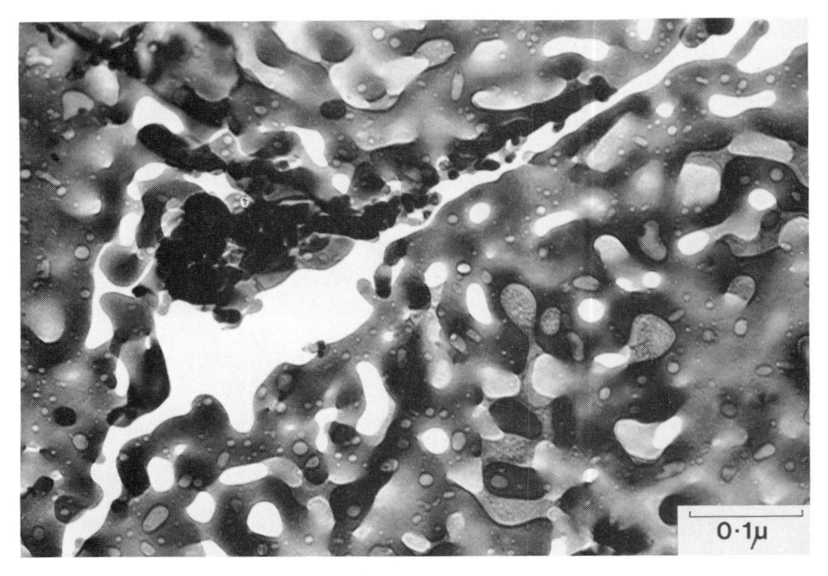


**Figure 10** Transmission electron micrograph of 70% silver/30% gold alloy film after exposure to concentrated nitric acid.

In view of these considerations it is both surprising and interesting that, under certain circumstances, the electron diffraction patterns for the silver–gold films are found to change significantly as a result of corrosion in nitric acid. A most striking example is shown in figure 11 which is the diffraction pattern for a 75% silver alloy after corrosion in 35% nitric acid solution. In a further investigation Durkin and Forty (1982) have analysed this diffraction pattern and have shown that it corresponds to an overlayer, about 2 nm in thickness, formed on the alloy after corrosion. The nature of the overlayer is of considerable interest. Silver nitrate can be ruled out for reasons given earlier; so too can silver oxide which is known to be soluble in nitric acid. Tests were also carried out to confirm this by treating the corroded film with other acids; the diffraction pattern disappeared after the film had been immersed in hydrochloric acid but was quite stable in concentrated nitric acid. This behaviour is unexpected for silver oxide but is consistent with the reported properties of monovalent gold (I) oxide (Brauer 1951). It might be concluded therefore that the overlayer is a form of gold oxide formed during corrosion. The complex diffraction pattern can, in fact, be analysed in terms of a simple cubic crystal structure with lattice parameter $a = 4.91$ Å in epitaxial relationship with the FCC silver–gold alloy such that (110) $[1\bar{1}0]$ in the corrosion phase is parallel to (111) $[1\bar{1}0]$ in the alloy. There are three distinct ways in which this relationship can be achieved on a (111) alloy surface and the complex pattern arises from the superposition of three distinct patterns, each corresponding to one of the three different types of oxide domain (Durkin and Forty 1982). Forty and Durkin (1980) showed that the gold-rich islands formed on silver–gold after corrosion in nitric acid could be caused to extend across the surface by heating in a good vacuum to about 400 °C. It is also found that the diffraction pattern due to the gold oxide overlayer disappears under this treatment. This implies that the gold oxide is unstable and the separation of gold into the island structure is a result of its decomposition.

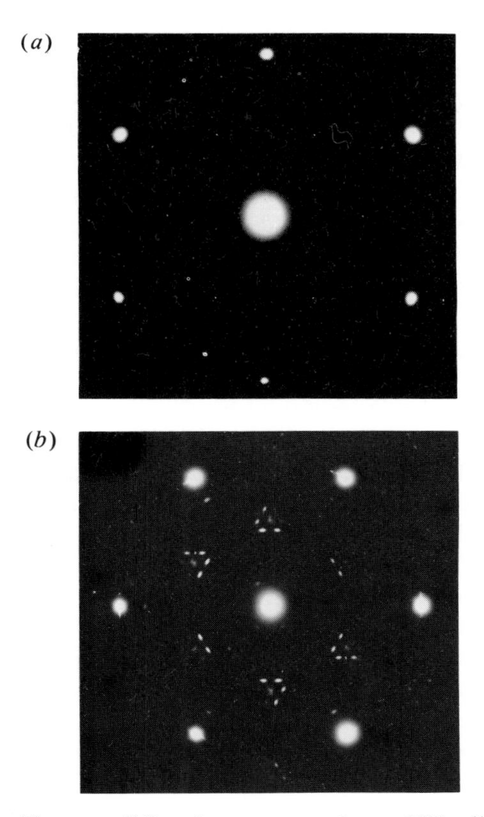


**Figure 11** Electron diffraction patterns for a 75% silver/25% gold alloy film: (*a*) is before corrosion and (*b*) is the same film after corrosion in dilute nitric acid. The additional diffraction spots in (*b*) are due to the formation of an oxide film.

## 6 THE CHEMISTRY OF SELECTIVE DISSOLUTION

The results of the electrochemical studies and these microstructural studies show that selective dissolution of silver is accompanied by oxidation of the residual gold. At sufficiently high anodic potentials or for open-circuit corrosion in sufficiently strong nitric acid, gold-rich islands form and spread over the alloy surface and, for certain alloy compositions, a pitting and tunnelling morphology develops. Tunnelling appears to be a necessary micromorphological development for in-depth de-alloying. We now attempt to develop a self-consistent chemical model for silver dissolution and gold oxide/gold island growth. In the next and final section we shall use this to develop a micromorphological model for tunnelling and de-alloying.

Pure gold is stable in nitric acid unless it is raised to a sufficiently high potential for electrolyte breakdown and oxygen evolution to occur. Pure silver is readily soluble in nitric acid. Dissolution of silver follows a set of autocatalytic chemical reactions in which $NO_2$ and $NO_2^-$ are formed (Evans 1960). The dissolution of gold in aqueous solutions requires a strong oxidising agent (e.g. nitric acid) and a complexing agent ($Cl^-$ is an efficient complexing agent for $Au^{3+}$) to stabilise the ion so formed (Finkelstein and Hancock 1974) and hence gold is soluble in *aqua regia* ($HNO_3$ and HCl). In the case of a silver-rich silver–gold alloy in nitric acid the selective dissolution of silver can be expected to produce an aqueous environment enriched in $NO_2$ and $NO_2^-$ which can at least partially oxidise the residual gold atoms. Thus, free corrosion of silver–gold in nitric acid creates a surface state similar to that of anodically controlled corrosion. It is interesting to note that the potential reached by all silver–gold alloys during free corrosion has been found by Carter and Forty (1990a) to be close to that reached by anodically corroded alloys immediately after the electrochemical cell is made open circuit (the Flade potential, about 860 mV (SCE)). Under conditions of high silver concentration in the alloy or, in the case of electrochemical corrosion, at high anodic potentials an oxide layer can be formed. In other circumstances (i.e. low silver concentration or lower anodic potential) the oxidised gold atoms might just be taken into the first few 'layers' of solution. In either case the 'gold oxide' is metastable and can be expected to continuously form, decompose (i.e. be reduced to gold) and reform as further silver dissolution occurs. The observed growth of gold-rich islands might therefore be the product of the reduction of gold oxide formed as a result of selective dissolution of silver.

The dependence of alloy corrosion on alloy composition can now be understood in terms of this model. For low silver concentration the oxidising conditions necessary for gold oxidation are not reached in open circuit corrosion and the dissolution of silver is inhibited by the rapid accumulation of residual gold at the surface of the alloy. For high concentration of silver there is a high activity of silver dissolution which creates a strongly oxidising aqueous environment in which the gold component of the alloy is readily oxidised and general corrosion occurs. At intermediate alloy compositions silver dissolution creates a sufficiently oxidising environment for gold oxidation and reduction to occur locally on the surface of the alloy giving rise to island growth.

A detailed kinetic treatment of the silver dissolution, gold oxidation and gold reduction reactions should lead to an explanation of the various alloy composition dependencies observed in the electrochemical measurements. For example, the pronounced maximum for the area enclosed by the cathodic peak (gold oxide reduction) in the cyclic voltammograms at an alloy composition of 60% silver should correspond to a maximum rate of formation of gold oxide in the previous anodic treatment. The existence of a critical alloy composition for complete de-alloying, the parting limit, should also be evident from the kinetics of these reactions. Further advance in these more quantitative directions must await the availability of better data for the individual reactions involved.

## 7 A MICROMORPHOLOGICAL MODEL FOR DE-ALLOYING

The micromorphological studies made by Forty and Durkin (1980) have been used by Forty and Rowlands (1981) as a basis for an empirical description of corrosion pitting and tunnelling. They envisaged that silver is selectively dissolved from the surface of the alloy leaving the gold-rich near-surface layer in a highly disordered state. In the case of gold-rich alloys the surface disorder may be regarded as an abnormally high concentration of vacancies. For silver-rich alloys it is more appropriately described as a high concentration of gold adatoms. The later work of Carter and Forty indicates that for sufficiently silver-rich alloys the residual gold atoms are left in an oxidised state and the disordered layer might therefore be more realistically described as a metastable oxide covering the surface. In any case the residual gold atoms can be expected to be in an unstable state, and a reordering of the structure should occur at a significant rate even at the ambient temperature at which corrosion occurs. In the case of silver-rich alloys having a high concentration of gold 'adatoms' after corrosion, possibly in an oxidised state, nucleation of gold-rich islands can occur. These island nuclei will continue to grow by the accretion of migrating 'adatoms'.

The surface migration of the residual gold atoms exposes fresh alloy to the corrosive environment. This fresh surface becomes disordered by further selective dissolution of silver and further growth of the islands occurs. We can envisage a continuous process of surface disordering and reordering by which the alloy becomes progressively covered by gold-rich islands. As the islands grow they thicken, whilst the channels between them shrink and deepen. Finally, the islands coalesce and the channels degenerate into deep pits, or tunnels. The complete sequence of morphological change is illustrated schematically in figure 12.

In order to simplify the analysis of this process Forty and Rowlands treated the surface disordering and reordering as a sequence of discrete steps. This is valid if the surface reordering occurs sufficiently rapidly. They ignored any local electrolytic action involved in the deposition of the gold 'adatoms' and assumed that only those less-noble atoms situated on the surface of the alloy are dissolved, that is diffusion of silver from the underlying layers is sufficiently slow to be ignored. The rate-controlling process is therefore surface diffusion. Oxidation or partial oxidation, if it occurs, can be expected to enhance this, as will be discussed later.

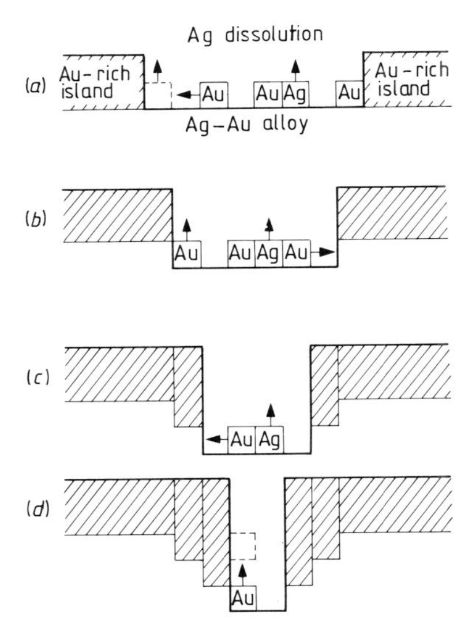


**Figure 12** Schematic representation of the corrosion disordering surface reordering model of island–channel–pit formation during corrosion by selective dissolution.

According to the model, the surface of the alloy undergoing selective dissolution becomes progressively covered with noble metal. This restricts further corrosion to a smaller area so that the alloy gradually becomes passivated. According to the TEM images the initial island nuclei appear to have a thickness estimated at about 25 atomic layers. There is a striking similarity between the island structure observed by Forty and Durkin and that formed in the initial stages of electro-deposition (Dickson *et al* 1965) and in vapour deposition of gold (Pashley 1965) where nuclei of similar thickness are observed. This seems to confirm the surface disordering/

reordering model since surface diffusion of gold atoms is an important feature of these other processes.

After nucleation the islands can be expected to spread by accretion of gold. As a result of both the reduction in the corroding surface area between islands and the thickening of the island walls as the channels between deepen, the coverage of the surface by each island occurs in successively smaller steps (see figure 13, where $x_1$, $x_2$, $x_3$, ... are the successive fractional coverages of the surface, $c$ is the concentration of gold in the alloy and $m$ is the thickness, in monolayers, of the island nucleus—about 25 monolayers). By a process of summation it can be shown that the total fraction of surface covered after $n$ layers of alloy have been corroded is

$$X(n) \simeq 1 - \left(\frac{m}{m+n}\right)^c$$

which shows that if the surface disordering and reordering continues the surface should eventually be completely covered with gold. The alloy will, therefore, become passivated unless the protective film is ruptured mechanically or microscopic defects persist to allow the continued access of the acid to the underlying alloy.

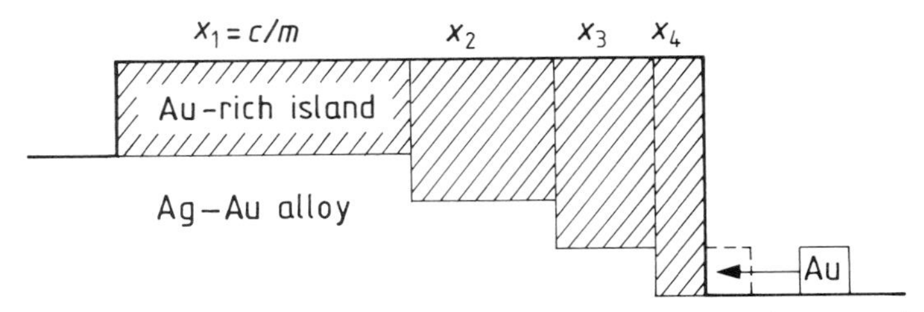


**Figure 13** Schematic representation of island growth in successive steps $x_1$, $x_2$, $x_3$, ... by the corrosion disordering surface reordering model.

The channel between a pair of islands can be shown to shrink by a similar progression so that after $n$ layers of alloy have been corroded the width becomes

$$W(n) \simeq 2L\left(\frac{m}{m+n}\right)^c$$

where $2L$ is the original width before corrosion, i.e. the distance between island centres.

It can readily be shown that pits, formed by the incomplete coalescence of islands in the later stages of channel closure, have a radius $R(n)$ given by

$$R^2(n) \simeq R_0^2\left(\frac{m}{m+n}\right)^c$$

where $R_0$ is the starting size of the surface before corrosion.

Since this function approaches zero radius infinitely slowly it might be inferred that the pits are never completely filled in under practical conditions with specimens of finite thickness, even for very gold-rich alloys. However, a significant difference in the behaviour of gold-rich and silver-rich alloys becomes apparent if, instead of simply considering the spatial progression of the islands, the time taken to reach each stage of development in the island morphology is considered. As already indicated, the rate-controlling process in corrosion disordering and surface reordering is surface diffusion of the disordered gold over the surface of the channel towards the islands. This occurs with a characteristic time $\tau(n)$, given by $W(n)/2D$, where $D$ is the surface diffusivity of gold adatoms and $W(n)$, as previously defined, is the width of the channel between islands after $n$ atomic layers of alloy have been corroded. It is assumed that $D$ is a constant, for all $n$, so that $\tau(n)$ decreases simply because the distance over which the gold atoms must diffuse to reach the islands decreases. Following the process, layer-by-layer, it can be shown that the time required for the channel between a pair of islands to shrink to a width $W(n)$ is

$$t(n) \simeq \frac{L^2}{D} \sum_{p=0}^{n-1} \left(\frac{m}{m+p}\right)^{2c}.$$

The behaviour predicted by this is illustrated graphically in figure 14. This is a set of shrinkage curves for various alloy compositions. The significant conclusion is that there is a change of kinetic behaviour at $c = 0.5$, giving the result that the alloy surface can become completely passivated by a covering of gold only if the concentration of gold exceeds 50%. This is consistent with the micromorphological observation of Forty and Durkin (1980) and Carter and Forty (1990b) that the island–channel–pit morphology is not found for silver–gold alloys with greater than 50% gold. Forty (1981) has developed the analysis further to show that, for silver-rich (i.e. $c < 0.5$) alloys, corrosion by selective dissolution of silver gives rise to a stable pit morphology which permits in-depth de-alloying to continue indefinitely. This clearly establishes a criterion for the existence of a parting limit close to 50% silver.

It is implicit in the model that the key factor in determining whether de-alloying can occur is the ability of the residual gold to become reordered. This is possible of course if the residual gold is mobile at the ambient temperature during corrosion. The oxidising environment created by silver dissolution is therefore an important factor since transport of oxidised gold species can be expected to occur more readily across the metal/electrolyte interface. If, as the electrochemical studies have demonstrated, the local environment is sufficiently strongly oxidising (or if the anodic potential is sufficiently high, in electrochemical terms) a layer of gold oxide might be formed during corrosion. This could, of course, passivate the alloy surface against further dissolution. However, the oxide is likely to be metastable particularly at the potential attained by the alloy under open circuit conditions (free corrosion) which, at 820 mV, is marginally below the anodic potential at which anodically formed gold oxide decomposes (the Flade potential observed by Carter and Forty (1990a)). We can therefore envisage a situation for the more silver-rich alloys where there is a continuous process of formation and decomposition of oxide. The decomposition product, gold, is then deposited in island growth. Clearly, under such circumstances the corrosion disordering and surface reordering processes are more complex but the basic model for corrosion tunnelling is still applicable.

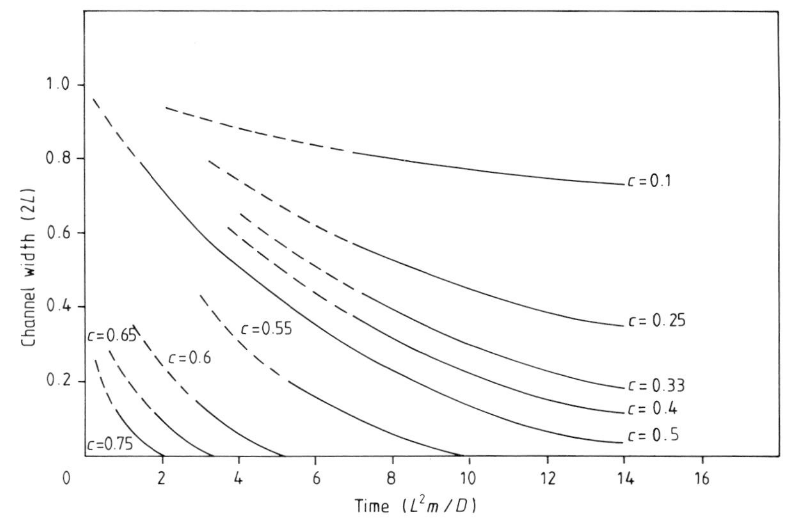


**Figure 14** The reduction of corrosion channel width (as a fraction of the initial distance between islands $2L$) with time (in units of $L^2m/D$) for a series of silver–gold alloys according to the corrosion disordering surface reordering model.

It is interesting to compare the Forty and Rowlands model with that proposed earlier by Swann (1969) for corrosion tunnelling. The latter envisages the formation of small pits at structural irregularities, or at sites of electrochemical instability, and their preferential growth into the metal due to the higher electrochemical potential at the base of a pit. Following anodic dissolution, the noble-metal atoms are redeposited by a cathodic reaction or by surface diffusion onto the side walls of the pits, which are therefore partially passivated. Stable corrosion tunnels are formed when the number of noble-metal atoms dissolved from the core and then redeposited is just sufficient to fill all the available sites on the freshly exposed walls. This simple geometrical criterion leads to a critical value for the radius

$$R_c = 2d\left(\frac{1-c}{c}\right)$$

below which a pit will tend to widen because the amount of redeposited noble metal is insufficient to passivate the walls, and above which it will tend to narrow because a thicker deposit can be formed. The parameter, $d$, is the thickness of an atomic layer in the metal and $c$ is, as before, the concentration of gold in the alloy. $R_c$ is effectively the radius for a stable tunnel morphology.

Table 1 compares the values of $R_c$ calculated using this criterion with those actually observed for a range of gold alloys. It can be seen that, except for very low gold concentration, the Swann model greatly under-estimates the tunnel radii.

**Table 1** Observed and calculated values of the radii of stable corrosion tunnels in various gold alloys.

| Alloy atomic composition | Observed $R_c$ (nm) | Calculated $R_c$ (Forty and Rowlands) (nm) | Calculated $R_c$ (Swann) (nm) |
|---|---|---|---|
| 50Ag/50Au | 28 | 30 | 0.3 |
| 67Ag/33Au | 26 | 25 | 0.6 |
| 95Ag/ 5Au | 4 | 5 | 6 |
| 75Cu/25Au | 10 | 15 | 1 |
| 60Cu/40Au | 30 | 40 | 0.5 |
| 84Nc/16Au | 4.5 | 5 | 1.8 |

According to the Forty and Rowlands model the critical radius of pit for the development of a stable tunnelling morphology occurs when the rate of penetration of pit into the alloy begins to exceed the rate of shrinkage of its diameter. This corresponds to the stage at which the amount of gold released by selective dissolution from a monolayer at the bottom of the pit is only just sufficient to give monolayer coverage of the side walls. This leads to the geometrical condition:

$$R_c = \frac{2N_c d}{c}$$

where $N_c d$ is the critical pit depth, measured in number of alloy monolayers of thickness $d$, and $c$ is the concentration of gold in the alloy. The values of $R_c$ calculated on this basis, also shown in table 1, are much closer to those observed. Thus, the island–channel–pit model is not only more consistent with the micromorphological studies, but is also more successful in accounting quantitatively for the measured radii of corrosion tunnels.

We can conclude therefore that for silver-rich alloys a stable corrosion tunnel morphology is established if the alloy contains less than 50% gold. Tunnels can be expected to penetrate the alloy indefinitely. Branching into secondary tunnels might occur at structural inhomogeneities and consequently the final morphological state can be expected to consist of a labyrinth or interconnected porous structure of gold-rich alloy. This corresponds to the so-called gold 'sponge' commonly found after parting has been completed (Shreir 1976).

## REFERENCES

Bakish R and Robertson W D 1956 *Stress Corrosion Cracking and Embrittlement* ed W D Robertson (New York: Wiley) pp 32–47

Brauer G 1951 *Handbuch der Preparation Anorganischen Chemie* (Stuttgart: Enke)

Carter D 1985 *PhD thesis* University of Warwick

Carter D and Forty A J 1990a to be published

—— 1990b to be published

Dickson E W, Jacobs M H and Pashley D W 1965 *Phil. Mag.* **11** 575

Durkin P and Forty A J 1982 *Phil. Mag.* A **45** 95–105

Evans U R 1960 *The Corrosion and Oxidation of Metals* (London: Edward Arnold) p 326

Finkelstein N P and Hancock R D 1974 *Gold Bull.* **7** 72

Forty A J 1979 *Nature* **282** 597–8

—— 1981 *Gold Bull.* **14** 25–35

Forty A J and Durkin P 1980 *Phil. Mag.* A **42** 295–318

Forty A J and Edeleanu C 1960 *Phil. Mag.* **5** 1029–40

Forty A J and Rowlands G 1981 *Phil. Mag.* A **43** 171–88

Frank F C 1949 *Discuss. Faraday Soc.* **5** 48

Ogura K, Haruyama S and Nagasaki K 1971 *J. Electrochem Soc.* **118** 531

Pashley D P 1965 *Adv. Phys.* **14** 327

Pickering H W 1967 *Fundamental Aspects of Stress Corrosion Cracking* ed R W Staehle, A J Forty and D van Rooyen (Houston: NACE) pp 159–74

Pickering H W and Swann P R 1963 *Corrosion* **19** 369

Schmid G M and O'Brian R N 1964 *J. Electrochem. Soc.* **108** 726

Shreir L L 1976 *Corrosion* vol I (London: Newnes-Butterworths)

Swann P R 1969 *Corrosion* **25** 147–50

Tammann Z and Brauns E 1931 *Z. Anorg. Chem.* **200** 209

Tokutaka H, Nishimori K, Tanako K, Takashima K, Hency J and Langeron J P 1981 *J. Appl. Phys.* **52** 6109

Vermilyea D A 1967 *Fundamental Aspects of Stress Corrosion Cracking* ed R W Staehle, A J Forty and D van Rooyen (Houston: NACE) p 176

# Speculation concerning the Causes of Twinning during Czochralski Growth of Crystals having Diamond Cubic or Zinc-blende Structure

D T J Hurle

## 1 INTRODUCTION

At some stage during the evolution of techniques for the commercial production of single crystals of the diamond cubic and III–V zinc-blende semiconductors, yield problems caused by the occurrence of twinning during growth have been encountered. As the technology has matured, this has ceased to be a problem with some materials but stubbornly persists with several of the III–V compounds, notably the indium-containing ones.

This is a problem which has been studied since the early days of semiconductor technology but remains one for which there is no complete and convincing explanation. First attempts at an explanation drew analogy with deformation twinning in metals but Billig (1954/5), in pioneering work, showed this not to be the cause. He grew crystals of germanium, copper and aluminium and noted that twins occurred only with germanium and that this was related to the appearance of {111} facets on the growing crystal surface. He speculated that the crystal exhibited 'a tendency to twin in such a way that, after twinning, it will contain one set of {111} planes more closely coinciding with the main temperature gradient than before twinning'. From this he deduced that crystals grown on a [100] axis were the most prone to twin. Deformation twinning has been reconsidered subsequently (Churchman *et al* 1956, Brown *et al* 1980) but it is now universally accepted that the single growth twin (figure 1) occurs during growth on a {111} facet, existing on the crystal–melt interface.

**Figure 1** Single growth twin on a {111} edge facet of an LEC InP crystal (courtesy of W Bonner).

Charles Frank, in his classic chapter on the 'The Geometrical Thermodynamics of Surfaces' (Frank 1963) described the relationship between facets on the equilibrium form of the crystal and the polar plot of the reciprocal of its surface energy ($\gamma$). On raising the crystal temperature, specific surfaces pass through a roughening transition (Frank described it as 'a rapid rise in the texture depth'—a phrase borrowed from highway engineering) corresponding to the rounding of vertices in the $\gamma^{-1}$ plot. However, some vertices can persist right up to the melting point, even when the solid is in contact with its own melt. He postulated that such singular surfaces control the growth under isothermal conditions but in the presence of a temperature gradient, as is experienced in the case of growth from the melt, the singular faces will be limited in extent because, at some distance behind the most advanced part of the growth front, the supercooling will have become large enough to nucleate new layers.

In the case of the melt growth of diamond cubic and III–V zinc-blende semiconductors, these singular faces are the {111} planes. In Czochralski growth they can be seen at the periphery of the growth surface by rapidly 'snatching' the crystal from the melt (figure 2). Twinning on these planes consists of a 60° rotation of the lattice about the [111] plane normal. It can apparently be caused by a variety of factors which make controlled experimentation and the acquisition of statistically significant data difficult. It is perhaps not surprising, therefore, that the subject is not well documented.

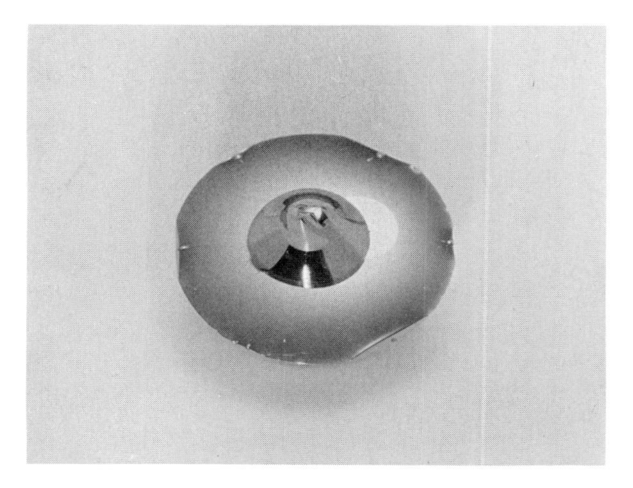

**Figure 2** Edge facets on a crystal of indium antimonide snatched from the melt (Hulme and Mullin 1962).

In the case of the III–V compounds, the incidence of growth twinning can depend markedly on the seed crystal orientation. The situation with indium antimonide has been well reviewed by Hulme and Mullin (1962) who have shown a definite correlation of twinning incidence with the size and inclination of growth facets at the edge of the crystal. The twinning probability depends markedly on whether the exposed {111} facets terminate in In or Sb atoms to the point that it is virtually impossible to grow single crystals in the $[\bar{1}\bar{1}\bar{1}]$ indium-terminating direction because of twinning on the $\{\bar{1}11\}$ antimony-terminating facets formed at the periphery of the interface. Such growth twins are particularly prone to occur with the Czochralski technique during growth of the shoulder region of the crystal.

Hulme and Mullin (1962) overcame this problem in indium antimonide by the use of large diameter seed crystals. Experience with indium phosphide (Bonner 1981) shows a similar proneness to twinning in the shoulder region which can be minimised by growing with either a very small cone angle or by using a very large angle (rapid grow-out). A significant increase in twinning probability occurs for intermediate cone angles.

Specific residual impurities can also markedly affect twinning incidence in ways which are not understood. It is, however, known that foreign impurities can influence facet development (Mullin 1962). Some years ago a particular grade of commercially available silicon would consistently yield crystals twinned towards the end of growth unless pre-purified by fractional crystallisation or zone melting (Green 1970). Growth twinning has been reviewed recently by Buzynin *et al* (1988). The role of dislocations was considered by Simon and Kern (1967).

In this paper the concepts of surface thermodynamics are used to predict conditions under which the formation of a nucleus of twin orientation can occur on a facet at the crystal–melt interface. It is hypothesised that a necessary, but not sufficient, condition for twinning is the presence of a facet on the crystal–melt interface which extends to the point of intersection with the three-phase boundary. In the next section we use a model expounded by Voronkov (1975), together with a description of the equilibrium state of the three-phase boundary due to Frank, the author and colleagues (Bardsley *et al* 1974) to derive criteria for the intersection of a facet with the three-phase boundary. In section 3 we apply this theory to the problem of twinning and the results are summarised in section 4.

## 2 FORMATION OF A FACET AT THE THREE-PHASE BOUNDARY

The attachment of the meniscus to a crystal growing by the Czochralski technique is characterised by two angles, $\theta_L$ and $\theta_I$, as shown in figure 3. At first sight one might suppose that, to grow a cylindrical crystal, the meniscus should contact the crystal vertically (i.e. $\theta_L = 0$) since, one might argue, any other angle would necessitate surface diffusion if the crystal was to continue to grow at the same diameter. This is indeed the case for metal systems, which are completely wetted by their own melt (i.e. which exhibit surface melting (Kristensen and Cotterill 1977)) but is very definitely not the case for the group IV and group III–V semiconductors. This was first shown by Antonov (1968), who photographed the profile of the meniscus during the Czochralski growth of germanium crystals and measured $\theta_L^0$ (the value of $\theta_L$ with respect to the vertical when growing a cylindrical crystal) from the photographs, obtaining values in the range of 10–20°. More recent and precise measurements using another technique by Surek (1976) give a value of 13 ± 1°. Measured values for this and some other semiconductors are given in table 1.

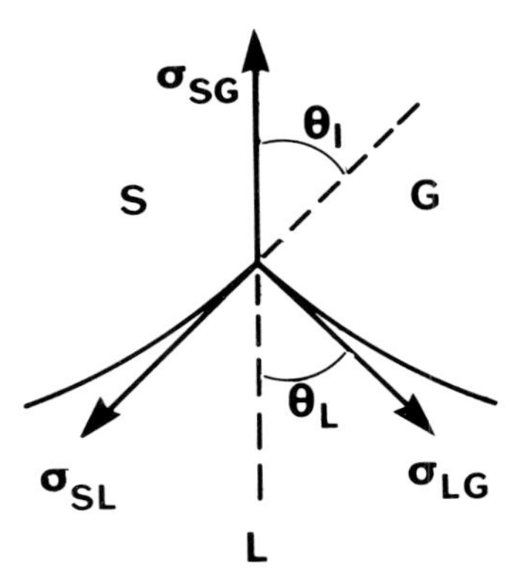


**Figure 3** Angles defining the three-phase boundary.

Applying criteria derived by Herring (1951) in connection with the theory of sintering, one can write the conditions for obtaining equilibrium at the three-phase boundary as

$$\sum_{i=1}^{3}\left(\sigma_i t_i + \frac{\partial \sigma_i}{\partial t_i}\right) = 0 \quad (1)$$

where $\sigma_i$ are the interfacial energies and $t_i$ is the unit vector normal to the line of intersection lying in the plane of the $i$th interface (Bardsley *et al* 1974). Expressing the components of equation (1) separately, we obtain expressions for the two angles $\theta_L^0$ and $\theta_I^0$ in terms of the three interfacial energies, $\sigma_{SG}$, $\sigma_{LG}$ and $\sigma_{SL}$, where the subscripts S, L and G stand for solid, liquid and gas, respectively;

$$\sigma_{SG} - \sigma_{LG}\cos\theta_L^0 + (\partial\sigma_{SL}/\partial\theta)_{\pi-\theta_I^0}\sin\theta_I^0 = 0 \quad (2)$$

$$\sigma_{LG} - \sigma_{SG}\cos\theta_L^0 + \sigma_{SL}\cos(\theta_I^0+\theta_L^0) + (\partial\sigma_{SG}/\partial\theta)_0\sin\theta_L^0 - (\partial\sigma_{SL}/\partial\theta)_{\pi-\theta_I^0}\sin(\theta_I^0+\theta_L^0) = 0 \quad (3)$$

where angles are measured in a clockwise direction from the vertical. In writing these relations we have ignored the curvature of the crystal $1/R$ (where $R$ is the crystal radius), out of the plane of figure 3.

Because of the dependence of $\sigma_{SL}$ and $\sigma_{SG}$ upon crystallographic orientation, we may expect the angles $\theta_L^0$ and $\theta_I^0$ to vary around the periphery of the crystal. Indeed it is known that there are small variations in the height of the meniscus around the periphery of a growing germanium

**Table 1** Thermodynamic properties.

| | Ge | Si | GaAs | InP | Units |
|---|---|---|---|---|---|
| $\Phi_0$ | 45 (Voronkov 1974) | 33 (Voronkov 1974) | 50 (Schultheiss and Rowland 1973) | 53[c] | Degrees |
| $\theta_L^0$ | 13 (Surek 1976) | 11 (Surek 1976) | 15 (Kashkooli *et al* 1974) | 17[e] | Degrees |
| $\sigma_{LG}$ | 616 (Bardsley *et al* 1977) | 874 (Hardy 1984) | 442 (Karataev *et al* 1966) | 396 (Popov and Demberel 1977) | erg $cm^{-2}$ |
| $\sigma_{SL}$ | 216 (Bardsley *et al* 1977) | 273 (Zadumkin 1960) | 155[b] | 139[b] | erg $cm^{-2}$ |
| $\sigma_{SG}$ | 766[e] | 1074[e] | 532[e] | 458 (Zadumkin 1960) | erg $cm^{-2}$ |
| $\sigma_{SG}^{0(c)}$ | 652 | 1006 | 439[c] | 376[d] | erg $cm^{-2}$ |
| $W$ | 2.5 (Voronkov 1973) | 3.85 (Voronkov 1973) | 3.28[a] | 3.08[a] | $10^{-13}$ erg |
| $T_m$ | 1209 | 1688 | 1511 | 1343 | K |
| $a_0$ | 5.66 | 5.43 | 5.65 | 5.87 | $10^{-8}$ cm |
| $q$ | 4.9 | 4.4 | 4.4 | 4.0 | $10^{22}$ atm $cm^{-3}$ |
| $\gamma$ | 36 | 81 | 97 | 62 | erg $cm^{-2}$ |

[a] Estimated from latent heat (see text).
[b] Estimated to be $0.35\sigma_{LG}$ (see Zadumkin 1961).
[c] Calculated from equation (9).
[d] Noting that $\sigma_{SG}^0/\sigma_{SG}$ is 0.85 and 0.82 for Ge and GaAs, respectively, we take $\sigma_{SG}^0 = 0.82\ \sigma_{SG}$ for InP.
[e] Calculated from equation (4) with neglect of torque terms (see Bardsley *et al* 1974).

crystal (Antonov 1968). However, this orientation dependence is relatively small and we will take an average value. Denoting the average values of the surface energy by a superscripted bar, we obtain from equations (2) and (3) an expression for $\theta_L^0$:

$$\cos\theta_L^0 = \frac{\bar{\sigma}_{SG}^2 + \sigma_{LG}^2 - \bar{\sigma}_{SL}^2}{2\bar{\sigma}_{SG}\sigma_{LG}}$$

$$+ \frac{\bar{\sigma}_{SL}}{2\sigma_{LG}}\left(\frac{\sigma_{LG}}{\bar{\sigma}_{SL}}\sin\theta_L^0 + \sin\theta_I^0\right)\frac{1}{\bar{\sigma}_{SG}}\left(\frac{\partial\sigma_{SG}}{\partial\theta}\right)_0$$

$$- \left(\frac{\sigma_{LG}}{\bar{\sigma}_{SG}}\sin(\theta_L^0 + \theta_I^0) - \sin\theta_I^0\right)\frac{1}{\bar{\sigma}_{SL}}\left(\frac{\partial\sigma_{SL}}{\partial\theta}\right)_{\pi - \theta_I^0}. \tag{4}$$

A similar expression can be obtained for $\theta_I^0$ (Bardsley *et al* 1974).

Far from the three-phase boundary, the shape of the crystal–melt interface is governed by the shape of the freezing point isotherm. Over a characteristic capillary length adjacent to the three-phase boundary, however, the Gibbs/Thomson effect will act to produce a curvature of the interface which is matched to the local supercooling. This capillary length

$$a = (2\sigma_{SL}T_m/LG_L)^{1/2} \tag{5}$$

is of the order of a few microns. $T_m$ is the melting point of the material, $L$ its latent heat and $G_L$ the temperature gradient normal to the interface. If this curved portion of the interface is tangent to a {111} plane then a facet will form on it.

The conditions where such a facet shall intersect the three-phase boundary (see figure 4(*a*)) have been considered by Voronkov (1975). When this occurs the possibility exists that nucleation will occur at the three-phase boundary, a position of great susceptibility to defect introduction into the crystal as a result of such extraneous factors as residual scum and particulate matter on the melt surface, disturbances due to gravity waves on the melt surface, etc.

Voronkov argued that a growth facet will be present at the three-phase boundary if, and only if, the change of free energy associated with the absorption of a laterally growing step at the boundary is negative, i.e.—with reference to figure 4—if

$$(\gamma_e - \gamma)h < 0 \tag{6}$$

where $\gamma h$ is the free energy of unit length of the step, having height $h$ and

$$\gamma_e h = h\,(\sigma_{SG} - \sigma_{SL}^0\cos\nu - \sigma_{LG}\cos\theta_L^0)/\sin\nu \tag{7}$$

is the free energy change associated with the change in the geometrical configuration upon absorption of the step. (The angle $\nu$ is depicted in figure 4. In general $\nu \neq \theta_I$.) This change in free energy is proportional to the surface energy associated with the length AB minus that associated with BC and BD; see figure 4(*b*). The superscript 0 applied to a surface energy is taken to indicate its value on the facet. Un-superscripted values relate to the rough non-facetted surfaces. Voronkov (1975) has shown that $\sigma_{SL} - \sigma^0_{SL} < \gamma$ and in the absence of other information we take $\sigma_{SL} = \sigma^0_{SL}$. (Values of $\sigma^0_{SG}$ can be obtained from equation (9) below.)

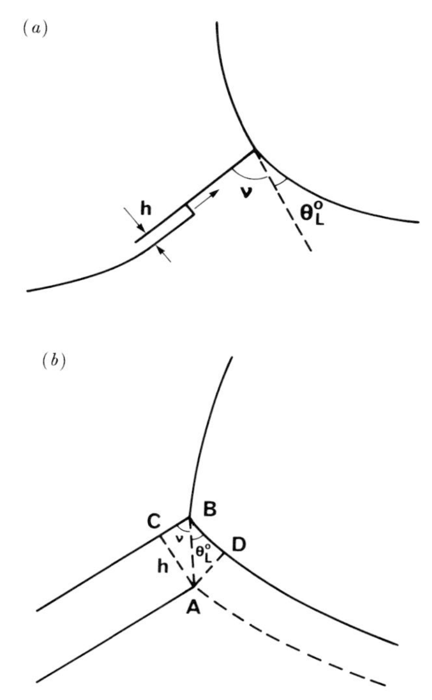


**Figure 4** Three-phase boundary having a faceted solid–liquid interface: (*a*) laterally advancing step; (*b*) geometric factors used in deriving equation (7).

The inequality expressed by equation (6) will be satisfied for some limited range of values of the angle $\nu$. The case where the facet extends to the triple boundary and the case where it does not are shown schematically in figures 5(*a*) and (*b*), respectively. To evaluate the inequality account must be taken of the anisotropy of $\sigma_{SG}$. At the high temperatures involved (close to melting point) we assume, following Voronkov, that the surface energies are isotropic except in the vicinity of the $\langle 111 \rangle$ directions, where a sharp minimum is exhibited (figure 6). These minimum values corresponding to the surface energy on the facet are denoted by the superscript 0.

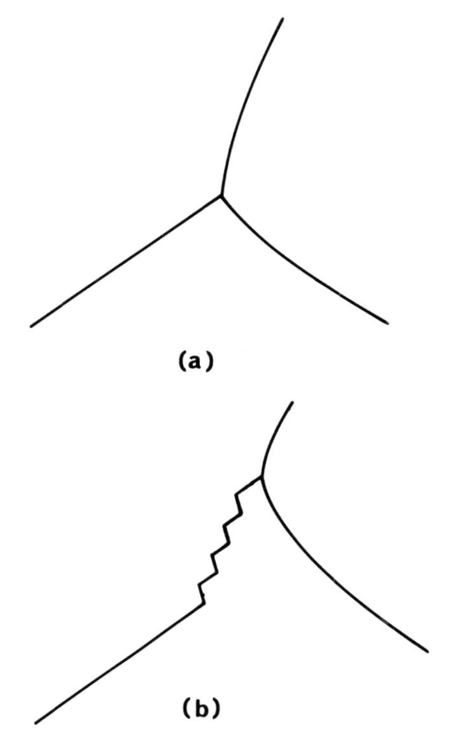


**Figure 5** Schematic representation of possible forms of the three-phase boundary: (*a*) case where laterally flowing steps on the facet are absorbed at the boundary, (*b*) case where the steps are not absorbed and the facet does not extend to the triple interface.

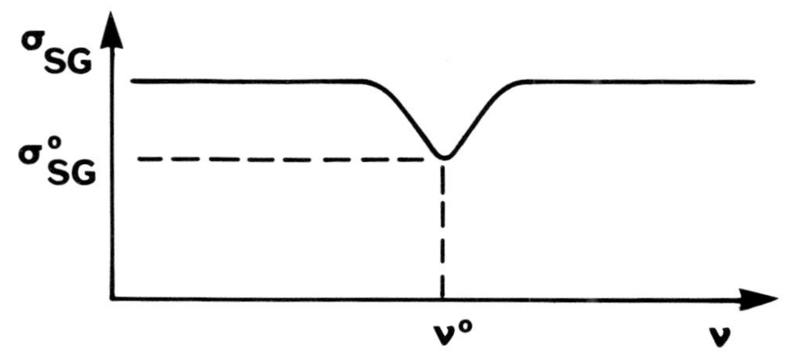


**Figure 6** Representation of the assumed anisotropy of the solid–gas interfacial energy. A sharp minimum occurs at the facet orientation $\nu = \nu^0$.

The three-phase boundary can nucleate steps as well as absorb them. The energy necessary to form a critical nucleus which is a truncated cylinder ($\Delta F_t^*$) at the three-phase boundary is less than that to form a cylindrical nucleus on the body of the facet ($\Delta F_c^*$) if

$$(\gamma_e - \gamma)h < 0.$$

This reduction in free energy has been calculated by Voronkov (1975) from geometrical factors:

$$E = \frac{\Delta F_c^* - \Delta F_t^*}{\Delta F_c^*} = \frac{1}{\pi} \left[\cos^{-1}(\gamma_e/\gamma) - (\gamma_e/\gamma)\left(1 - (\gamma_e/\gamma)^2\right)^{1/2}\right]. \qquad (8)$$

This is the fractional change in free energy and is plotted against the ratio $\gamma_e/\gamma$ in figure 7.

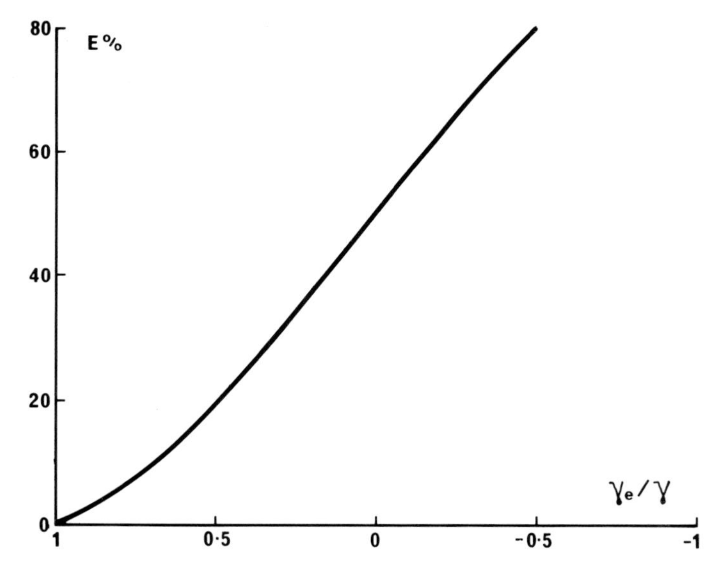


**Figure 7** Fractional reduction in the energy of formation of a critical nucleus when it occurs at the three-phase boundary, as compared with formation on an extended facet surface ($E$) against normalised free energy of absorption of a step (see text).

The derivation of the equation can be deduced from figure 8. Comparison is made between the free energy of formation of a cylindrical nucleus of height $h$ and radius $AC = r^*$ and a nucleus truncated by the section AMB of external surface. The free-energy difference between the two configurations is proportional to the difference in the energy of the two steps, ANB and AMB. Noting that, from Wulff's theorem, the ratio CM/CA is equal to $\gamma_e/\gamma$, equation (8) follows.

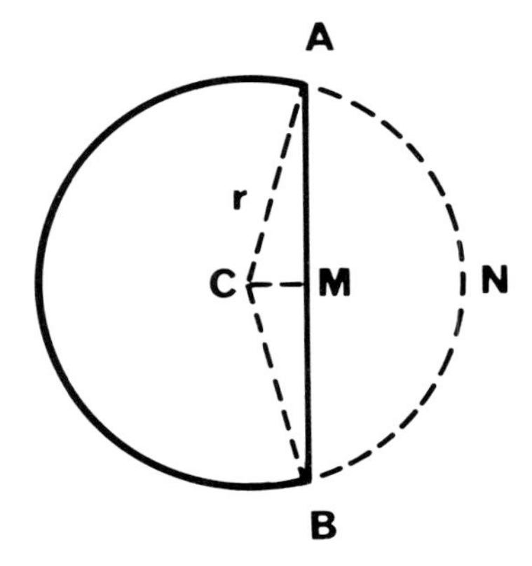


**Figure 8** Formation of a truncated nucleus of radius $r$.

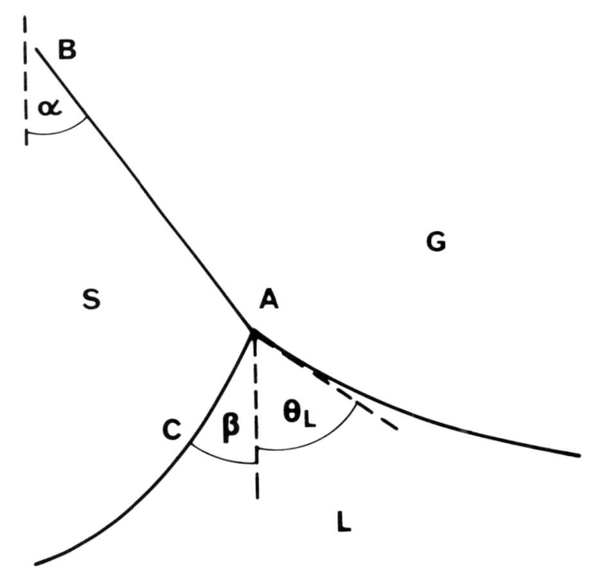


**Figure 9** Illustrating conical growth from a seed with semi-cone angle $\alpha$.

For $\gamma_e/\gamma < 1$ there will be a tendency for nucleation to occur on the facet at the three-phase boundary. We make the hypothesis that twinning is prone to occur in these circumstances.

Conditions might arise for which the facet becomes de-wetted, i.e. the three-phase boundary retreats along the facet in the direction AC shown in figure 9. This will occur when $\varphi = \beta + \theta_L$ is less than some critical value $\varphi^0$

(the wetting angle) given by

$$\sigma^0_{\mathrm{SL}} + \sigma_{\mathrm{LG}} \cos \varphi^0 = \sigma^0_{\mathrm{SG}}. \tag{9}$$

As we have already seen, $\theta^0_{\mathrm{L}} = 0$ for metals at their melting points but not for the {111} planes of the diamond cubic and zinc-blende semiconductors.

## 3 APPLICATION TO TWINNING

### 3.1 Introduction

It has been established in the previous section that edge facets will intersect the three-phase boundary if both of the following conditions are satisfied.

(1) $\gamma_{\mathrm{e}} < \gamma$. This occurs for a range of cone angles $\alpha$ less than some maximum value ($\alpha_{\max}$).
(2) The angle which the meniscus makes to the facet is greater than the wetting angle $\varphi^0$. This occurs for a range of values of $\alpha$ greater than some minimum value ($\alpha_{\min}$).

From equation (7) we see that $\alpha_{\max}$ is given by solution of the equation

$$\gamma \sin(\alpha_{\max} + \beta) = \sigma_{\mathrm{SG}} - \sigma^0_{\mathrm{SL}} \cos(\alpha_{\max} + \beta) - \sigma_{\mathrm{LG}} \cos \theta^0_{\mathrm{L}}. \tag{10}$$

$\alpha_{\min}$ is given by

$$\alpha_{\min} = \varphi^0 - \theta^0_{\mathrm{L}} - \beta \tag{11}$$

where $\beta$ is shown in figure 9. $\beta$ is the angle made by the {111} plane to the growth axis and is of course dependent on the orientation of the seed crystal.

Values of the required parameters for germanium, silicon, gallium arsenide and indium phosphide are listed in table 1. The step energy is calculated according to the expression given by Voronkov (1973) for a Kossel crystal:

$$\gamma = \sqrt{1.5}\, q^{-1/3} \{\tfrac{1}{3}W - kT_{\mathrm{m}} \ln[1 + 2\exp(-W/kT_{\mathrm{m}})]\} \tag{12}$$

where $q$ is the atomic density and $W$ is the excess energy of an interface bond. For germanium and silicon the ratio of $W$ to the latent heat per atom is $0.45 \pm 0.01$ (Voronkov 1973). We assume this ratio also holds true for the compound semiconductors.

On the basis of the data in table 1 values of $\alpha_{\min}$ and $\alpha_{\max}$ for growth on the two major orientations ⟨111⟩ and ⟨100⟩ have been calculated. These are listed in table 2. We infer from this that a facet can exist at the three-phase boundary over a wide range of grow-out conditions.

We suppose the above condition ($\alpha_{\min} < \alpha < \alpha_{\max}$) to be a necessary one for an enhanced probability of twinning. Note that very rapid grow-out ($\alpha > \alpha_{\max}$) causes the facet to pull away from the three-phase boundary and suggests that under these conditions twinning should be avoidable. This broadly accords with experimental information.

We also see that, for growth on the ⟨111⟩ orientation, $\alpha_{min}$ is generally greater than zero, so that growth of the cylindrical section of the crystal should be free from the risk of twinning. This is not, however, true for growth on the ⟨100⟩ orientation where $\alpha_{min}$ is less than zero (except for indium phosphide, when it is just greater than 0). Note that, from figure 7, $E$ increases as $\alpha$ decreases from $\alpha_{max}$ (where $E=0$). This increase occurs steadily right down to $\alpha = \alpha_{min}$, at which point the facet de-wets. Twinning should therefore be particularly prone to occur near the point of de-wetting.

**Table 2** Range of cone angles for which a {111} facet can be located at the three-phase boundary.

| | Ge | Si | GaAs | InP |
|---|---|---|---|---|
| $\alpha_{min}^{(111)}$ | 12.5° | 2.5° | 15.5° | 16.5° |
| $\alpha_{max}^{(111)}$ | 30.5° | 37.5° | 67.5° | 63° |
| $\alpha_{min}^{(100)}$ | −3.5° | −13.5° | −0.5° | +0.5° |
| $\alpha_{max}^{(100)}$ | 14.5° | 21.5° | 51.5° | 47° |

### 3.2 Energy required to form a twin oriented nucleus

Having established the geometric conditions under which a twin could form at the three-phase boundary, we next address the question of the magnitude of the energy required to form the twinned nucleus as against that required to form a correctly oriented one. The energy of the twin nucleus will be increased by an amount equal to the twin plane energy over the surface AC (see figure 10) and will be changed (either increased or decreased) by the difference in the free energy of the surface presented at AB as against that of the surface presented for an untwinned nucleus.

Consider growth in a [111] direction with a cone angle of $\alpha = 51°$. A twin nucleus forming on the ⟨111⟩ edge facet will have a low index {111} face presented at AB, which will have a lower surface energy than the irrational face presented by the correctly oriented nucleus. This will result in a reduction in the free-energy contribution of the AB surface and if this reduction exceeds the energy of the twin plane (the section of which is AC), then the twin nucleus will actually be energetically favoured over that of the correct one. This is most likely to occur, of course, for materials which have a very low twin plane energy. A similar situation applies to growth in a [100] direction with a cone angle $\alpha = 35°$. (For both orientations $\nu = 70.5°$.) In table 3 we list, from the work of Gottschalk *et al* (1978), values for the stacking fault energies of the considered materials. These will be roughly twice the twin plane energy (van Bueren 1960). We see that these are lowest for the indium compounds—indium phosphide, indium arsenide and indium antimonide—which is in accord with the experimental observations that it is these materials which twin most readily.

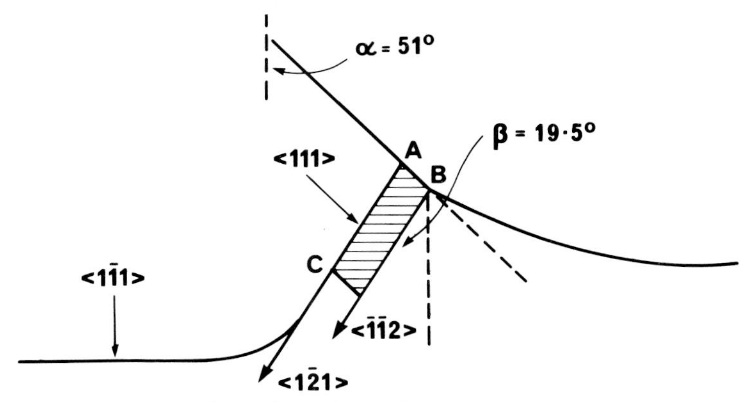


**Figure 10** Formation of a twin nucleus at the three-phase boundary for the case that the solid–vapour interface (AB) is also facetted.

**Table 3** Stacking fault energies: units $10^{-7}$ J cm$^{-2}$ (from Gottschalk *et al* 1978).

| GaAs | GaSb | GaP | InSb | InAs | InP |
|---|---|---|---|---|---|
| 55 ± 5 | 53 ± 7 | 41 ± 4 | 38 ± 4 | 30 ± 3 | 18 ± 3 |

These arguments suggest that, for a ⟨111⟩ oriented crystal, grow-out at an angle of 51° (and for a ⟨100⟩ crystal grow-out at 35°) is the most dangerous from the point of view of initiating twinning in those materials for which $\alpha_t = 51°$ (35°) lies in the range $\alpha_{max} > \alpha_t > \alpha_{min}$. Inspection of table 2 shows that $\alpha_t$ is encompassed by $\alpha_{min}$ and $\alpha_{max}$ for the cases of indium phosphide and gallium arsenide (which twin very readily) but not for germanium and silicon (which do not twin readily). Unfortunately, sufficient relevant data are not available for any other semiconductor. Of course, in any real crystal growth situation the meniscus angle is not exactly constant during the growth of the cone but can fluctuate over an angular range as a result of disturbances to the process. These can include changes in meniscus height due to thermal fluctuations in the melt, and also to changes in the meniscus angle due to mechanical infelicities (lack of alignment, etc) in the pulling mechanism. There is thus a range of $\alpha$ around $\alpha_t$, the extent of which depends on these disturbances, for which formation of this particularly favoured twin nucleus can occur.

We proceed now to calculate the conditions for growth at a cone angle $\alpha_t$ under which a twin nucleus requires a lower free energy than that for a correctly oriented one.

The critical radius $r_c^*$ of a cylindrical nucleus is

$$r_c^* = \frac{\sigma_{SL}^0}{\Delta F_v} = \frac{\sigma_{SL}^0 T_m}{\Delta H\,\delta T} \tag{13}$$

where $\Delta F_v$ is the volume-free energy change, $\Delta H$ is the latent heat and $\delta T$ the undercooling. The change in free energy on forming the critical nucleus is then

$$\Delta F_c^* = \frac{\pi h \sigma_{SL}^{0\,2} T_m}{\Delta H\;\delta T}. \tag{14}$$

The critical radius of a truncated nucleus is the same as that for a cylindrical one ($r_c^*$) but the free energy of formation of the truncated nucleus intersecting the three-phase boundary is reduced by an amount given by equation (8).

If now that nucleus is in twin orientation to the matrix and presents a {111} facet at the truncation surface intersecting the three-phase boundary, its free energy of formation will be

$$\Delta F_{tw} = (2r_{tw}\sigma_{SL} - r_{tw}^2\,\Delta F_v + \sigma_T r_{tw}^2/h)A^0 h \tag{15}$$

where

$$A^0 = \pi - \cos^{-1}(\gamma_e^0/\gamma) + (\gamma_e^0/\gamma)[1 - (\gamma_e^0/\gamma)^2]^{1/2}$$

where $\gamma_e^0$ is given by equation (7) but with $\sigma_{SG}$ replaced by $\sigma_{SG}^0$. $\sigma_T$ is the specific twin plane energy and $r_{tw}$ the radius of the twin nucleus.

Differentiating equation (15), we see that the critical radius, corresponding to the extremum in $\Delta F_{tw}$, is now

$$r_{tw}^* = \frac{\sigma_{SL}^0}{\Delta F_v - \sigma_T/h} \tag{16}$$

which is to be compared with equation (13) for a correctly oriented one. From equations (8), (14), (15) and (16) we obtain for the difference in free energy of formation for correct and twin-oriented nuclei:

$$\Delta F_{tw}^* - \Delta F_t^* = [A^0/(1 - \sigma_T T_m/h\;\Delta H\;\delta T) - A]\;\Delta F_c^*/\pi \tag{17}$$

where

$$A = \pi - \cos^{-1}(\gamma_e/\gamma) + (\gamma_e/\gamma)\;[1 - (\gamma_e/\gamma)^2]^{1/2}.$$

The twin critical nucleus will be favoured (i.e. will have a lower free energy of formation) if $\delta T$ is greater than some critical value ($\delta T_c$) where

$$\delta T_c = \frac{\sigma_T T_m}{h\;\Delta H}\frac{A}{(A - A_0)}. \tag{18}$$

This can be understood as follows.

At very small undercooling the size of the critical nucleus is very large and the energy contributed by the twin plane also relatively very large, being roughly proportional to $r_{tw}^{*2}$ (ignoring the truncation of the cylinder). At large undercooling, when the critical radius is small, the edge energy (roughly proportional to $r_{tw}^{*}$) dominates and the reduction in this quantity by the formation of a low index surface at the three-phase boundary serves to overcome the extra energy needed to form the twin plane. The critical undercooling at which the two effects balance is given by equation (18).

Taking $\sigma_T = \frac{1}{2}\sigma_{SF}$, where $\sigma_{SF}$ is the stacking fault energy, we calculate values of $\delta T_c$ for gallium arsenide and indium phosphide to be respectively 30 and 7 K. The actual steady-state undercooling on the facet can be inferred from a knowledge of the facet size and the temperature gradient in the melt. Voronkov (1973) has determined this to be 3.7 K for dislocation-free silicon. However, the attachment of particulate matter having a high emissivity ('scum') can in principle locally lower the temperature significantly, as can fluctuations in the growth process due to fluctuations in growth speed and/or crucible temperature. The required critical undercooling for indium phosphide (7 °C) is clearly much more readily attainable than that for gallium arsenide (30 °C), commensurate with the experimental observation that twinning is much more frequent in the former material.

### 3.3 Patch twins

During the growth of [111] axis crystals, $\{\bar{1}11\}$ facets can form on the exterior of the shoulder (see figure 11, taken from the paper by Bonner (1981)). Repeated twinning can sometimes occur under such conditions as seen in figure 11. For some range of $\alpha$, depending on the material, facets inclined at 109.5° to a $\langle 111 \rangle$ growth direction can be present at the three-phase boundary (at AB in figure 12). This is visible on the crystals in figure 11. If the cone angle $\alpha$ is encompassed by $\alpha_{min}$ and $\alpha_{max}$, then repeated twinning can also be anticipated on these facets but such twins will be confined to the peripheral areas of the crystal. In the author's laboratory they are referred to as 'patch' twins and are viewed as benign curiosities since they disappear as growth proceeds by 'growing out' of the crystal, unlike the single twin which traverses the whole crystal interface, giving rise to a new crystal of differing orientation. Patch twins are prone to occur in large diameter crystals of indium phosphide and have the beneficial effect of preventing the formation of the more deleterious single twins which form on the {111} facets inclined at 70.5° discussed above. Suppression of the 70.5° twins is thought to occur when the crystal has increased in diameter to the point that the twins on the 109.5° facets have spread radially until they have intersected each other in the angular region of the 70.5° facets and, by presenting a new growth direction there locally, have removed the possibility that 70.5° facets—and hence deleterious twins—could form at these points. One can readily show, by simple geometry, that if twins form simultaneously on all three 109.5° facets at the point that the crystal has a radius $R$, then they will have extended and intersected each other once the final crystal radius has exceeded a value $R_{\min}$ given by

$$R_{\min} = 2R\left(1 - \frac{\tan 19.5^\circ}{\tan \alpha}\right)\left(1 - \frac{2\tan 19.5^\circ}{\tan \alpha}\right)^{-1}.$$

Taking, for example, $R = 0.5$ cm and $\alpha = 55^\circ$, we have $R_{\min} = 1.5$ cm.

**Figure 11** 'Patch' twins on ⟨111⟩ LEC-grown InP crystals (courtesy of W Bonner (Bonner 1981)).

### 3.4 Polarity effects

Finally, we note that the twinning probability on the (111)B antimony-terminating faces of indium antimonide is much greater than that on the (111)A (indium-terminating) faces (Hulme and Mullin 1962). We suppose that this reflects the different interfacial energies of these two faces. $\delta T_c$ is sensitive to small differences in $\sigma^0_{SG}$. Oscherin (1976) calculates that $\sigma^0_{SG}$ is 7% higher on the indium-terminating facet as compared with the antimony-terminating one. Unfortunately data do not exist to permit calculation of $\delta T_c$ in the case of indium antimonide but, for illustration, an increase of 7% in $\sigma^0_{SG}$ for indium phosphide increases $\delta T_c$ from 7 to 16 °C. Further, the observation that impurities can induce twinning may be related to the effect of the impurities on the solid/liquid interfacial energy and on the melt surface tension.

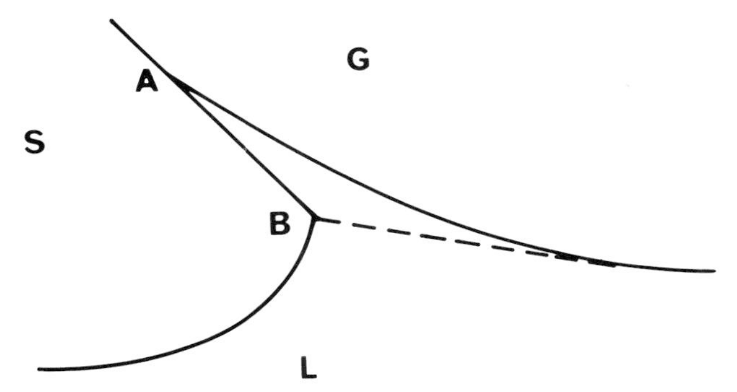


**Figure 12** Formation of an external facet AB on the cone section of a crystal.

## 4 SUMMARY

It has been hypothesised that single growth twins are likely to occur only when an edge facet is present at the three-phase boundary. We have shown that, for any given growth axis, this is possible over a limited range of grow-out angles ($\alpha$), that range depending on the material.

It has been shown that, for a specific value of grow-out angle $\alpha = \alpha_t = 51^\circ$ for the case of a $\langle 111 \rangle$ axis crystal and $\alpha_t = 35^\circ$ for a $\langle 100 \rangle$ one, a twinned nucleus can have a lower free energy than a correctly oriented one by presenting a low-surface-energy facet at the solid–vapour surface adjacent to the three-phase boundary provided that the undercooling on the facet exceeds a critical value. For this critical value to be low enough to be attained in practice the twin plane energy must be low. Further, $\alpha_t$ must lie in the range $\alpha_{min} < \alpha_t < \alpha_{max}$ which defines conditions under which a facet is present at the three-phase boundary. These conditions are shown to be well satisfied for indium phosphide (which twins readily), less well satisfied for gallium arsenide (which has an intermediate twinning incidence) and not satisfied for germanium and silicon (which seldom display twinning). It is also noted that the lowest twin plane energies occur for the indium-based compounds in agreement with the empirical observation that twinning occurs most readily in these materials. The model therefore provides a rationale of the observed behaviour.

It is suggested that the effects of impurities and facet polarity on twinning incidence are related to differences in interfacial energies.

## REFERENCES

Antonov P I 1968 *Growth of Crystals* vol 14, ed N N Sheftal (New York: Consultants Bureau) p 145
Bardsley W, Frank F C, Green G W and Hurle D T J 1974 *J. Crystal Growth* **23** 341
Bardsley W, Hurle D T J and Joyce G C 1977 *J. Crystal Growth* **40** 13
Billig E 1954/5 *J. Inst. Metals* **83** 53
Bonner W A 1981 *J. Crystal Growth* **54** 21
Brown G T, Cockayne B and McEwan W R 1980 *J. Mater. Sci.* **15** 1469
Buzynin A N, Antonov V A, Osiko V V and Tatarintsev V M 1988 *Izv. Akad Nauk SSSR Ser. Fiz.* **52** 1889
Churchman A T, Geach G A and Winton J 1956 *Proc. R. Soc.* A **238** 194
Frank F C 1963 *Metal Surfaces: Structure, Energetics and Kinetics* (Metals Park, OH: ASM) p 1
Gottschalk H, Patzner G and Alexander H 1978 *phys. status solidi* a **45** 207
Green G W 1970 Private communication
Hardy S C 1984 *J. Crystal Growth* **69** 456
Herring C 1951 *The Physics of Powder Metallurgy* ed W E Kingston (New York: McGraw-Hill) ch 8
Hulme K F and Mullin J B 1962 *Solid State Electron.* **5** 211
Karataev V V, Milvidskii M G and Zakharova N Ya 1966 *Izv. Akad Nauk SSSR Neorg. Mater.* **2** 833
Kashkooli I Y, Munier Z A and Williams L 1974 *J. Mater. Sci.* **9** 538
Kristensen J K and Cotterill R M J 1977 *Phil. Mag.* **36** 437
Mullin J B 1962 *Preparation and Properties of Indium Antimonide* ed R K Willardson and H L Goering (New York: Reinhold)
Oscherin B N 1956 *phys. status solidi* a **34** K181
Popov A S and Demberel L 1977 *Kristall und Technik* **12** 1167
Potemski R M and Small M B 1983 *J. Crystal Growth* **62** 317
Schultheiss P J and Rowland M C 1973 Unpublished work
Simon B and Kern R 1967 *Kristall und Technik* **2** 475
Surek T 1976 *Scripta Met.* **10** 425
van Bueren H G 1960 *Imperfections in Crystals* (Amsterdam: North Holland) p 164
Voronkov V V 1973 *Sov. Phys.–Crystallog.* **17** 807
—— 1974 *Sov. Phys.–Crystallog.* **19** 137
—— 1975 *Sov. Phys.–Crystallog.* **19** 573
Zadumkin S N 1960 *Fiz. Tverdogo Tela* **2** 878
—— 1961 *Izv. Akad Nauk SSSR, Met. i Topl.* **1** 55

# Bristol Anholonomy Calendar

Michael Berry

## 1 INTRODUCTION

For nearly half a century Charles Frank has set the intellectual tone of the Bristol physics department, with contributions—his own, and those he has inspired, directly and indirectly—whose unique character springs from a deep understanding of geometry. Here I will tease out a hidden thread that links seven of these contributions. The word 'hidden' is important: this will not be history but its *post hoc* reconstruction from a particular point of view, by the light of present knowledge.

Our thread is the concept of *anholonomy*. Abstractly stated, this concerns the behaviour of quantities $s$ when slaved to variables $X = X_1, X_2, \ldots$ that are taken round a circuit C. Anholonomy is the failure of $s$ to return when the $X$ do, as the result of non-integrability. A concrete example is the Foucault pendulum, whose direction of swing, $s$ (reckoned for example relative to north), is slaved to the local vertical $X$ (labelled for example by two polar angles) and does not return with $X$ after a day, when $X$ has completed a circular path C on its sphere (here the Earth). The slaving law here is that the direction of swing is parallel-transported by the local vertical. 'Parallel' means that the angular velocity of the direction of swing has no component along the vertical (i.e. along the instantaneous radius). A consequence of parallel transport is that $s$ at the end and beginning of C differ by the solid angle subtended by C at the centre of the sphere (this gives the familiar latitude dependence for the turning of the pendulum on the Earth's surface). In figure 1, parallel transport of a vector on a sphere is illustrated for a (non-diurnal) circuit made of great circles.

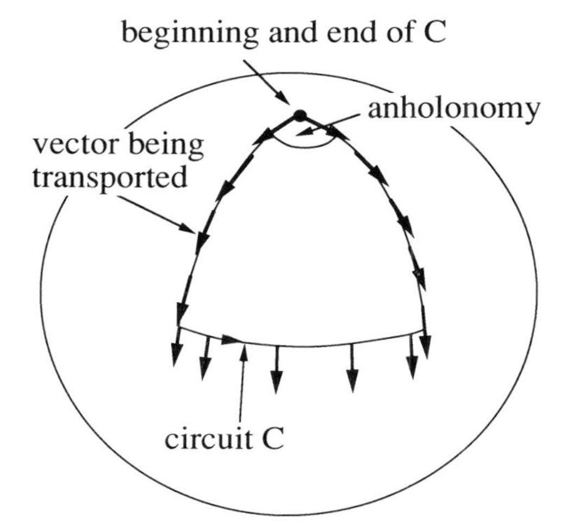


**Figure 1** Parallel transport of a vector on a sphere.

A note about terminology. The concept of anholonomy is familiar to geometers, who call it 'holonomy'. Physicists usually encounter it first in the study of mechanics in the presence of constraints: a constraint is holonomic if it can be integrated and thereby reduce the number of freedoms, and non-holonomic otherwise. According to the OED (1989), the word was first used by Hertz in 1894.

## 2 DISLOCATIONS IN CRYSTALS (1951)

The paper by Frank (1951) was a 'formal collection' of ideas developed during the preceding few years in terms of 'a private language and a number of special concepts, each simple enough in itself but together

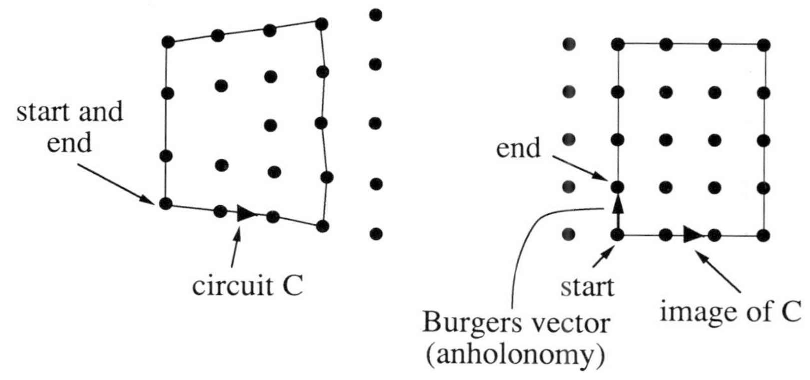


**Figure 2** Crystal dislocation as anholonomy.

forming a substantial self-contained body of theory'. Anholonomy lies at the heart of his definition of a dislocation.

In Frank's procedure (based on Burgers (1939)), C is a 'Burgers circuit' round the ions of the real—possibly dislocated—crystal (figure 2), whose positions are labelled by the (discrete) variables $X$. He shows how C has as image an 'associated path' in an imagined perfect reference lattice. Thus the slaved variables $s$ label the points of this reference lattice. If C encloses a dislocation, its image does not close. The closure failure is a lattice vector $s$—the 'Burgers vector'—which is thus the anholonomy in this case.

## 3 DISCLINATIONS IN LIQUID CRYSTALS (1958)

In his study of uniaxial liquid crystals, which Frank (1958) intended 'to urge the revival of experimental interest in its subject', he emphasised the importance of the line singularities of the swarms of unsigned directions (line fields) that represent these crystals. These are the disclinations (which he originally called 'disinclinations'), around which 'the cardinal direction of the preferred axis changes by a multiple of $\pi$'.

Disclinations exemplify anholonomy in the 'cardinal direction' $s$, which is slaved to position $X$ in the crystal. Because the line field depends continuously on $X$ away from singularities, it must return to itself after a circuit, but the change in $s$ can be a multiple of $\pi$ rather than $2\pi$ because the vectors representing molecular directions are unsigned. One such $\pi$ anholonomy, in a line field configuration originally calculated by Oseen (1933), is shown in figure 3.

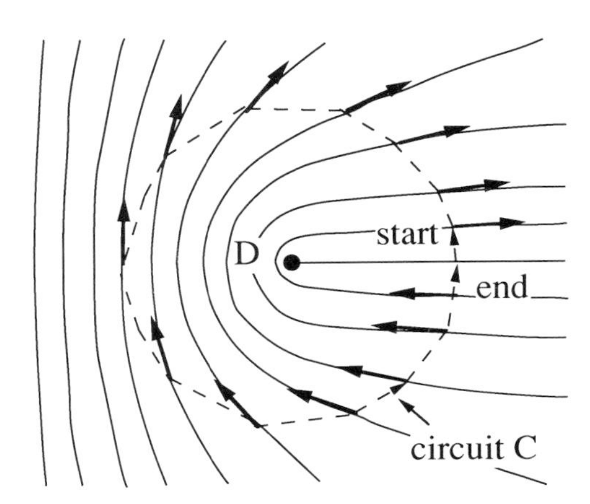


**Figure 3** Reversal ($\pi$ anholonomy) of liquid crystal direction around a disclination D.

There is a quite different context in which the same singularities occur: the line field consisting of the (orthogonal) directions of curvature of a surface in the neighbourhood of an umbilic point, where the two principal curvatures are equal (i.e. the surface is locally spherical). According to Darboux (1896), each of the two orthogonal direction fields rotates through $\pi$ in a circuit of the umbilic point. There are three typical 'disclination singularity' patterns of curvature lines. One of them is figure 3, complemented by superposition of the orthogonal set; for the others, see Berry and Hannay (1977).

For later reference it will help to know the mathematical origin of this $\pi$ anholonomy. Locally a curved surface can be described by its deviation $z(X)$ from the plane $X=(X_1,X_2)$ which is tangent to it at (say) $X=0$. From $z(X)$ can be formed the curvature matrix

$$K(X)=\begin{pmatrix} \dfrac{\partial^2 z}{\partial X_1^2} & \dfrac{\partial^2 z}{\partial X_1 \partial X_2} \\ \dfrac{\partial^2 z}{\partial X_1 \partial X_2} & \dfrac{\partial^2 z}{\partial X_2^2} \end{pmatrix}. \tag{1}$$

The principal curvatures at $X$ are the eigenvalues of $K(X)$. An umbilic point is therefore an $X$ at which $K(X)$ (regarded as depending parametrically on $X$) has degenerate eigenvalues. The curvature directions at $X$ are the (orthogonal) eigenvectors of $K(X)$. Therefore the umbilic/disclination $\pi$ anholonomy is equivalent to the following theorem: an eigenvector of a real symmetric matrix changes sign when smoothly continued round a circuit of parameters on which the matrix depends, provided the circuit encloses a matrix for which the corresponding eigenvalue is degenerate. Until recently this theorem was not widely known, but was of the type that is 'well known to those who know well' (e.g. Arnold 1978).

## 4 MOLECULAR ELECTRONIC DEGENERACIES (1958)

One of the authors of a paper by Longuet-Higgins *et al* (1958) was M H L Pryce, then the head of department at Bristol. The paper was about the dynamics of vibrations and rotations of molecules whose nuclear configuration is close to symmetric.

In the Born–Oppenheimer (adiabatic) approximation, where the quantum states of the electrons are slaved to the coordinates $X$ of the nuclei (regarded as frozen in place), nuclear symmetry gives rise to electronic degeneracy. This is illustrated in figure 4 for a circuit in the shape space of triatomic molecules, surrounding the equilateral molecule. From the solution of a particular model they noticed that when the nuclei $X$ make a cycle surrounding the symmetric configuration, the electronic wavefunctions change sign. This $\pi$-phase anholonomy is a particular case of the theorem stated at the end of the last section, and so is a general phenomenon associated with degeneracy where the Hamiltonian matrix is real symmetric (Herzberg and Longuet-Higgins 1963, Longuet-Higgins 1975).

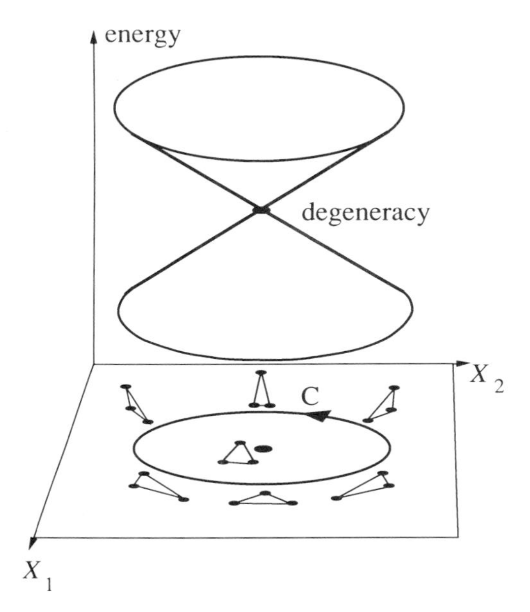


**Figure 4** Circuit in shape space $X$ of nuclei surrounding the equilateral molecule for which there is an electron energy degeneracy (conical intersection).

Moreover, they found that quantising the nuclear motion (i.e. the $X$) associated with such a circuit gave vibration–rotation energies with half-odd integer quantum number, rather than the usual integer. They remarked: 'This half-oddness is at first sight strange, but may be understood by noting that [around a circuit] the electronic factor in the wave function will be multiplied by $-1$, so that the angular part of the nuclear factor must do likewise if the total wave function is to be single-valued.' This is the phenomenon of pseudorotation, which has been of considerable interest more recently (Mead 1979, Zygelman 1987) and has been observed (Delacrétaz *et al* 1986) in the spectrum of $Na_3$.

## 5 QUANTUM EFFECTS OF ELECTROMAGNETIC POTENTIALS (1959)

Aharonov and Bohm (1959), in their paper from Bristol announcing the effect that now bears their name, showed that in quantum mechanics, in contrast to classical mechanics, the behaviour of a charged particle in a region where there are no electromagnetic fields can nevertheless be affected by potentials in the region. The effect is on the phase of the wave-function, and is anholonomy because the phase is a closure failure accumulated when the particle is taken round a circuit C in spacetime $X = \{X_\mu\}$ $(1 \leqslant \mu \leqslant 4)$. If $A_\mu(X)$ is the electromagnetic 4-potential in the region, and $q$ the charge, then

$$\text{phase change} = \frac{q}{\hbar} \oint_{\mathrm{C}} A_\mu(X)\, \mathrm{d}X_\mu. \tag{2}$$

They showed how the effect could be observed in an interference experiment: an electron beam is split into two; the beams pass on either side of a region from which electrons are excluded and which contains magnetic flux $\phi$, and are then recombined. The exclusion (for example by a metal cylinder enclosing a solenoid producing the flux) diffracts the electrons, and produces an interference pattern even without flux, but when the flux is switched on the fringes are shifted by the additional anholonomic phase. This is because

$$\begin{aligned}\text{fringe shift} &= (\text{phase above flux}) - (\text{phase below flux}) \\ &= (\text{phase above flux}) + (\text{phase below flux for backward path}) \\ &= \text{phase round circuit enclosing flux}\end{aligned} \tag{3}$$

which from (2) is $-e\phi/\hbar$. Aharonov told me that originally he thought a real (as opposed to gedanken) experiment would not be feasible, but that Charles Frank suggested using a magnetised iron whisker to produce the flux. This suggestion was taken up by Chambers (1960), who observed the fringe shift in the electron microscope. Since then, the effect has been seen in increasingly refined experiments (Olariu and Popescu 1985, Peshkin and Tonomura 1989).

Under a gauge transformation of the electromagnetic gauge transformations, the phase (2) is invariant, as it must be in order to be observable. Because of this fact, the Aharonov–Bohm effect played an important part in the development of gauge theories (Wu and Yang 1975).

## 6 DISLOCATIONS IN WAVES (1975)

Nye and Berry (1974) found that freely propagating waves can and typically do have phase singularities in the form of lines, closely analogous to dislocations in crystals. Again the circuit C is in spacetime $X = X_\mu$, and the anholonomy is in the phase $\chi$ of the wave $\psi$ when expressed as the complex function

$$\psi(X) = \rho(X)\exp\{\mathrm{i}\chi(X)\}. \tag{4}$$

Because $\psi$ must be single valued, the phase anholonomy is restricted to a multiple of $2\pi$, so that

$$\frac{1}{2\pi}\oint_{\mathrm{C}} \mathrm{d}\chi \tag{5}$$

must be an integer. Alternatively stated, the image of the 'Burgers circuit' C in a reference perfect plane wave fails to close by a whole number of wavecrests (defined as the contour surfaces $\chi = 0 \bmod 2\pi$). Figure 5 shows some wavefronts of a plane wave with an edge dislocation. In the analogy with crystals, wavefronts correspond to planes of atoms.

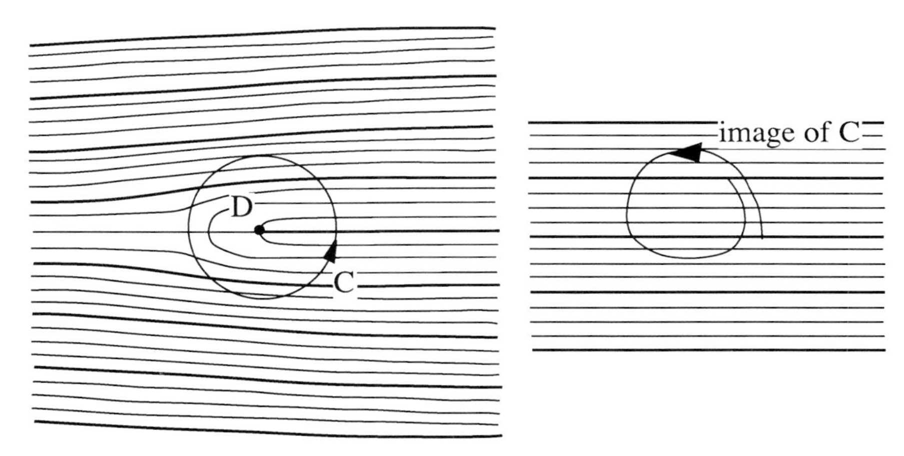


**Figure 5** Wavefronts (at intervals of $\frac{1}{2}\pi$ with crests bold) of a dislocated wave with a Burgers circuit surrounding the dislocation D, and its image in a plane wave.

If the integral (5) is not zero, C encloses a dislocation and can be shrunk down onto it, so that the phase anholonomy is accumulated round an infinitesimal circuit. Because $\psi$ is a smooth function of $X$, it follows that on the dislocation the modulus $\rho$ in (4) must vanish, so that another characterisation of dislocations is that they are nodal lines.

We found exact solutions of the wave equation with edge, screw and mixed edge–screw dislocations, curved or straight, infinite or looped, which could move by climb or glide, and interact by annihilation or bounce. Thus the analogy with crystal dislocations is very close, at least morphologically (as opposed to energetically).

The theory of wave dislocations has been extensively developed by Wright (1977), Nye (1981) and Wright and Nye (1982). Dislocations occur in ultrasonic pulses (Wright and Berry 1984, Nicholls and Nye 1986), in light (Berry *et al* 1979), in the tides (Berry 1981, Nye *et al* 1988) and in quantum scattering (Hirschfelder *et al* 1974a,b, Hirschfelder and Tang

1976) and bound states (Riess 1970a,b, 1976, Berry and Robnik 1986, Mondragon and Berry 1989).

The Aharonov–Bohm wavefunction also possesses a dislocation line. This differs from the previous cases in that the wave singularity does not propagate freely but is pinned to the flux line. Its strength (wavecrest anholonomy) is the closest integer to the quantum flux $e\phi/h$. The quantum wavefronts cannot be observed directly, but were seen in an analogous experiment with water waves encountering a bathtub vortex (Berry *et al* 1980) with the correct strength (proportional to the circulation of the vortex).

## 7 DISCLINATIONS IN ELECTROMAGNETIC WAVES (1983)

The vectorial nature of an electromagnetic wave gives rise to a structure of line singularities considerably richer than the dislocations of scalar waves (or of electromagnetism in the scalar-wave approximation). A fully general theory would consider time-dependent waves, and incorporate the intimate linkages between the electric field $\boldsymbol{E}(X)$ and the magnetic field $\boldsymbol{H}(X)$. This theory does not yet exist, but considerable progress has been made for monochromatic waves with $\boldsymbol{E}$ and $\boldsymbol{H}$ regarded as separate, in a series of papers by Nye and Hajnal.

Nye (1983a) considered waves for which a propagation direction can be defined, at least locally (for example paraxial propagation), and studied the lines on which the transverse field $\boldsymbol{E}_t$ vanishes. These are disclination lines of the transverse vector field, around which anholonomy of the vector direction is a multiple of $2\pi$ (rather than $\pi$ as for the disclinations of the line fields of liquid crystals). Typically the $\boldsymbol{E}_t$ disclinations form moving helices whose width is of the same order of magnitude as the wavelength. Intertwined with these are similar disclination helices of the magnetic field. In the limit when polarisation can be neglected, the electromagnetic double helices collapse onto the dislocation lines of the corresponding scalar wave.

Subsequently, Nye (1983b) considered not the strength but the state of polarisation of the transverse field, and was thereby led to study the quite different set of stationary lines on which $\boldsymbol{E}_t$ is circularly polarised (i.e. on which the two components of $\boldsymbol{E}_t$ have equal amplitude). He called these $\mathscr{C}$ lines. The associated anholonomy is in the polarisation ellipse, which typically turns through $\pi$ in a circuit of a $\mathscr{C}$ line. This is because polarisation is a tensorial property, so the principal axes of an ellipse are analogous to the directions of molecules in a liquid crystal near a disclination, or the curvature directions of a surface near an umbilic point.

In a generalisation, Nye and Hajnal (1987) (see also Hajnal 1987a) removed the paraxiality restriction by studying the three-dimensional pattern of polarisation in the full electric field $\boldsymbol{E}$, rather than the transverse field $\boldsymbol{E}_t$. They defined two sets of singular lines: $\mathscr{C}^{\mathrm{T}}$ lines on which $\boldsymbol{E}$ is circularly polarised (T stands for 'true' rather than transverse), and $\mathscr{L}^{\mathrm{T}}$ lines on which $\boldsymbol{E}$ is linearly polarised. They showed that the anholonomy of the polarisation ellipses is a $\pi$ rotation round a $\mathscr{C}^{\mathrm{T}}$ line, and a $2\pi$ rotation round an $\mathscr{L}^{\mathrm{T}}$ line.

The very complicated—but structurally stable—field patterns near these singular lines (and the surfaces swept out over time by the disclination lines, even for monochromatic waves) have been created and observed in microwaves, in beautiful experiments by Hajnal (1987b).

As Nye (1983a) pointed out, these line singularities are 'a useful way of describing the spatial and temporal geometrical features of electromagnetic diffraction fields that would otherwise be far too complicated to visualise. The disclination carries with it a certain local field structure; therefore, a description of the arrangement and motion of the disclinations contains much of the essential geometrical information about the field itself. They constitute elements of structure in the field.'

## 8 GEOMETRIC QUANTUM PHASES (1984)

Another kind of anholonomy, with several of the previous ones as special cases, was found by Berry (1984), initially in the context of quantum mechanics. In mathematical terms, this concerns eigenvectors $\boldsymbol{e}_n$ of general complex Hermitian (rather than $2\times 2$ real symmetric) matrices $\mathsf{H}(X)$, depending continuously on parameters $X$. Since $\mathsf{H}$ is complex, so is $\boldsymbol{e}_n$. This vector satisfies

$$\mathsf{H}(X)\boldsymbol{e}_n = E_n(X)\boldsymbol{e}_n \tag{6}$$

(where $E_n(X)$ is the corresponding eigenvalue), which defines $\boldsymbol{e}_n$ for each $X$, up to an arbitrary phase factor.

The question that now arises is: how can this phase factor be disambiguated? Mere continuity, which uniquely defines the evolution of real eigenvectors, does not do so for these complex $\boldsymbol{e}_n$. A natural continuation law which does define the phase is the stipulation that in a displacement $\mathrm{d}X$ the change in $\boldsymbol{e}_n$ is as small as possible, in the sense of being orthogonal to $\boldsymbol{e}_n$ itself, i.e.

$$\boldsymbol{e}_n^* \cdot \mathrm{d}\boldsymbol{e}_n = 0. \tag{7}$$

The surprise was that this innocent law (a generalisation of the parallel transport of ordinary vectors) is non-integrable: where $X$ makes a circuit C, $\boldsymbol{e}_n$, which is slaved to $X$, acquires (figure 6) a phase factor

$$\exp(\mathrm{i}\gamma_n(\mathrm{C})). \tag{8}$$

The phase anholonomy can be expressed in terms of $X$ derivatives of reference eigenvectors $\boldsymbol{u}_n(X)$, also solutions of (6), which are chosen to be single valued on C and across a surface S spanning C. Then $\gamma_n(\mathrm{C})$ is the flux

$$\gamma_n(\mathrm{C}) = -\mathrm{Im} \iint_{\mathrm{S}} \mathrm{d}\boldsymbol{u}_n^*(X) \wedge \cdot \mathrm{d}\boldsymbol{u}_n(X)$$

$$\equiv -\mathrm{Im} \iint_{\mathrm{S}} \left( \frac{\partial \boldsymbol{u}_n^*(X)}{\partial X_1} \cdot \frac{\partial \boldsymbol{u}_n(X)}{\partial X_2} - \frac{\partial \boldsymbol{u}_n^*(X)}{\partial X_2} \cdot \frac{\partial \boldsymbol{u}_n(X)}{\partial X_1} \right) \quad (9)$$

where in the second member the $\cdot$ links the $\boldsymbol{u}$ and the $\wedge$ links the d, and in the third member $X_1$ and $X_2$ are any coordinates in S.

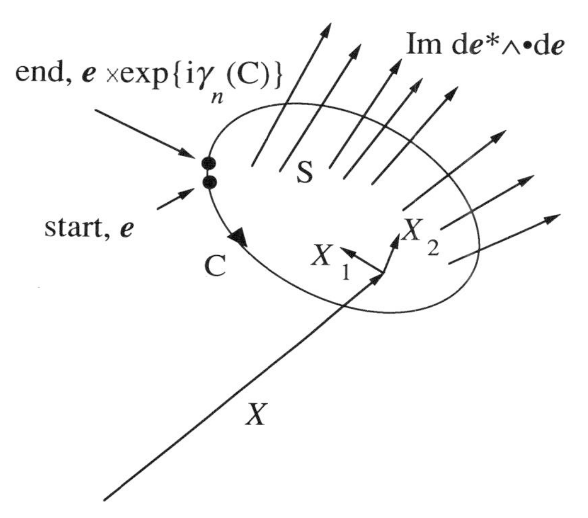


**Figure 6** Geometric phase as flux through circuit in parameter space.

The continuation law (7) is equivalent to the Schrödinger equation for a Hamiltonian that is changed slowly, so the geometric phase is ubiquitous in adiabatic quantum mechanics. We have already seen two special cases.

(i) The Aharonov–Bohm effect is the geometric phase for electrons transported round a line of magnetic flux. Here the mathematical abstractions in (9) become concrete: C is a circuit in ordinary position space $X$, and the flux is magnetic.

(ii) Electrons in molecules have interactions with time-reversal symmetry; thus **H** is real symmetric and the $\boldsymbol{u}_n$ are real, so any phase anholonomy must be a multiple of $\pi$. As we have seen, this is associated with degeneracies of the eigenvalue $E_n$. In the general complex case, degeneracies are (monopole) singularities of the integrands in (9), but $\gamma_n$ can be non-zero even in the absence of degeneracy (the Aharonov–Bohm effect is an example). For a molecule in a magnetic field, time-reversal symmetry is broken and $\gamma_n$ can have arbitrary values, as can the reaction of this anholonomy on the quantised nuclear rotation (Cottingham and Hassan 1990).

Another special case is a two-state system, such as a spin-$\frac{1}{2}$ particle in a magnetic field. Here $\mathbf{H}(X)$ is the complex Hermitian $2 \times 2$ matrix

$$\mathbf{H}(X) = \begin{pmatrix} X_3 & X_1 - \mathrm{i}X_2 \\ X_1 + \mathrm{i}X_2 & -X_3 \end{pmatrix} = \boldsymbol{X} \cdot \boldsymbol{\sigma} \qquad (10)$$

where $\boldsymbol{\sigma}$ is the vector of Pauli matrices and $X \equiv (X_1, X_2, X_3)$. It can be shown from (9) that the phase anholonomy is half the solid angle subtended by C at the origin $X = 0$. The matrix (10) also describes the state of polarisation of a light beam ($X$ is the vector of Stokes parameters), so light should acquire the same phase anholonomy in a cycle of polarisation changes. This was shown by Pancharatnam (1956) with very different arguments, whose equivalence to those underlying the geometric phase was demonstrated by Berry (1987). The relation between Pancharatnam's ideas and a natural phase convention for electromagnetic wavefields has been established by Nye (1991).

There have been many applications and generalisations of the geometric phase. The subject has been comprehensively reviewed by Zwanziger *et al* (1990) and there is also a reprint collection (Shapere and Wilczek 1989) and a volume of conference proceedings (Markovski and Vinitsky 1989).

The classical limit of the geometric phase provides yet another sort of anholonomy, discovered by Hannay (1985), in the angle variables of periodic (or multiply periodic) mechanical systems. One example is the Foucault pendulum, which thus completes this circuit in the anholonomy space.

## 9 CONCLUDING REMARKS

It is clear, then, that anholonomy has haunted Bristol physics for many years. This mathematical concept helps us to see relations between what at first sight seem very different areas of physics. Thus it can be regarded as an embodiment—albeit at a meta-level—of something Charles Frank said in his speech of retirement from Bristol in 1976: 'Physics is not just Concerning the Nature of Things, but Concerning the Interconnectedness of all the Natures of Things'.

## REFERENCES

Aharonov Y and Bohm D 1959 *Phys. Rev.* **115** 485–91
Arnold V I 1978 *Mathematical Methods of Classical Mechanics* (Berlin: Springer)

Berry M V 1981 *Les Houches Lecture Notes for Session XXXV* ed R Balian, M Kléman and J-P Poirier (Amsterdam: North-Holland) pp 453–543
—— 1984 *Proc. R. Soc.* A **392** 45–57
—— 1987 *J. Modern Opt.* **34** 1401–7
Berry M V, Chambers R G, Large M D, Upstill C and Walmsley J C 1980 *Eur. J. Phys* **1** 145–62
Berry M V and Hannay J H 1977 *J. Phys. A: Math. Gen.* **10** 1809–21
Berry M V, Nye J F and Wright F J 1979 *Phil. Trans. R. Soc.* A **291** 453–84
Berry M V and Robnik M 1986 *J. Phys. A: Math. Gen.* **19** 1365–72
Burgers J M 1939 *Proc. Phys. Soc.* **52** 23
Chambers R G 1960 *Phys. Rev. Lett.* **5** 3–5
Cottingham W N and Hassan N 1990 *J. Phys. B: At. Mol. Opt. Phys.* **23** 323–7
Darboux G 1896 *Leçons sur la Théorie Générale des Surfaces* vol 4 (Paris: Gauthier-Villars) note VII
Delacrétaz G, Grant E R, Whetten R L, Wöste L and Zwanziger J W 1986 *Phys. Rev. Lett.* **56** 2598–601
Frank F C 1951 *Phil. Mag.* **42** 809–19
—— 1958 *Discuss. Faraday Soc.* **25** 19–28
Hajnal J V 1987a *Proc. R. Soc.* A **414** 433–46
—— 1987b *Proc. R. Soc.* A **414** 447–68
Hannay J H 1985 *J Phys. A: Math. Gen.* **18** 221–30
Herzberg G and Longuet-Higgins H C 1963 *Discuss. Faraday Soc.* **35** 77–82
Hirschfelder J O, Christoph A C and Palke W E 1974a *J. Chem. Phys.* **61** 5435–55
Hirschfelder J O, Goebel C J and Bruch L W 1974b *J. Chem. Phys.* **61** 5456–9
Hirschfelder J O and Tang K T 1976 *J. Chem. Phys.* **64** 760–85, **65** 470–86
Longuet-Higgins H C 1975 *Proc. R. Soc* A **344** 147–56
Longuet-Higgins H C, Öpik U, Pryce M H L and Sack R A 1958 *Proc. R. Soc.* A **244** 1–16
Markovski B and Vinitsky S I (ed) 1989 *Topological Phases in Quantum Theory* (Singapore: World Scientific)
Mead C A 1979 *Chem Phys.* **49** 23–32, 33–8
Mondragon R J and Berry M V 1989 *Proc. R. Soc.* A **424** 263–78
Nicholls K W and Nye J F 1986 *J. Phys. A: Math. Gen.* **19** 375–83
Nye J F 1981 *Proc. R. Soc.* A **378** 219–39
—— 1983a *Proc. R. Soc.* A **387** 105–32
—— 1983b *Proc. R. Soc.* A **389** 279–90
—— 1991 this volume pp 220–31
Nye J F and Berry M V 1974 *Proc. R. Soc.* A **336** 165–90
Nye J F and Hajnal J V 1987 *Proc. R. Soc.* A **409** 21–36
Nye J F, Hajnal J V and Hannay J H 1988 *Proc. R. Soc.* A **417** 7–20
OED 1989 *Oxford English Dictionary* vol 7, 2nd edn (Oxford: Clarendon) p 314
Olariu S and Popescu I I 1985 *Rev. Mod. Phys.* **57** 339–436
Oseen C W 1933 *Trans. Faraday. Soc.* **29** 883
Pancharatnam S 1956 *Proc. Ind. Acad. Sci.* A **44** 247–62
Peshkin M and Tonomura A 1989 *The Aharonov–Bohm Effect* (Lecture Notes in Physics **340**) (Berlin: Springer)
Riess J 1970a *Ann Phys., NY* **57** 310–21
—— 1970b *Phys. Rev.* D **2** 647–53

—— 1976 *Phys. Rev.* B **13** 3862–9
Shapere A and Wilczek F (ed) 1989 *Geometric Phases in Physics* (Singapore: World Scientific)
Wright F J 1977 *PhD thesis* University of Bristol
Wright F J and Berry M V 1984 *J. Acoust. Soc. Am.* **75** 733–48
Wright F J and Nye J F 1982 *Phil. Trans. R. Soc.* A **305** 339–82
Wu T T and Yang C N 1975 *Phys. Rev.* D **12** 3845–57
Zwanziger J W, Koenig M and Pines A 1990 *Ann. Rev. Phys. Chem.* in press
Zygelman B 1987 *Phys. Lett.* **125A** 476–81

# Phase Gradient and Crystal-like Geometry in Electromagnetic and Elastic Wavefields

J F Nye

## 1 INTRODUCTION

A general field of scalar waves of a single frequency contains wave dislocations that are analogous to dislocations in crystals. Vector waves are more complicated, but a general field of electromagnetic or elastic waves of a single frequency has a structure that is in some ways similar to that of a liquid crystal. The spatially varying field of polarisation ellipses, oriented in different directions, provides the counterpart of the field of directors in the liquid crystal. Thus, the spirit of Charles Frank's geometrical approach to dislocations in crystals and disclinations in liquid crystals is also applicable to wavefields.

This idea has been worked out for the electromagnetic wavefield in a recent paper (Nye and Hajnal 1987). I thought it might be appropriate in this *Festschrift* to extend the treatment to cover elastic waves and at the same time to make an addition. This concerns the question of how to define a propagation direction at each point of a wavefield in terms of a spatial gradient of phase. One needs a definition that is equally suitable at all points, in order to treat the matter of chirality, and the paper referred to showed how this can be achieved. I show here that one reaches precisely the same definition by a quite different, and more satisfactory, route:

namely, by starting from the work of Pancharatnam (1956), who considered how one should treat the relative phase of two differently polarised beams of light. Berry (1987) has pointed out the equivalence between the Pancharatnam phase difference and that derived from the idea of the parallel transport of a quantum system during an adiabatic change. Thus this parallel-transport notion (Berry 1984, 1989) provides an equivalent starting point for the present problem.

## 2 THE FIELD OF POLARISATION ELLIPSES

We wish to consider a general field of electromagnetic or elastic waves of fixed frequency in a continuous medium, which may be anisotropic and inhomogeneous, but must be smoothly varying.

### 2.1 The electromagnetic wavefield

Taking the electromagnetic field first, it could be thought of as produced by the superposition of any number of waves of different amplitudes and polarisations propagating in different directions. Because the field is of fixed frequency its sources cannot be in relative motion, and therefore there is a preferred frame of reference: the one where the sources are at rest. This means that it is permissible to treat the electric and magnetic fields separately.

We represent either $\boldsymbol{E}$, $\boldsymbol{H}$, $\boldsymbol{D}$ or the magnetic induction, according to which is being considered, as the real part of a complex vector field $\boldsymbol{V}$. If $\omega$ is the angular frequency, $\boldsymbol{V}$ at any point may be written quite generally (Born and Wolf 1970) as

$$\boldsymbol{V} = (\boldsymbol{P} + \mathrm{i}\boldsymbol{Q})\mathrm{e}^{-\mathrm{i}\omega t} \tag{1}$$

where $\boldsymbol{P}$ and $\boldsymbol{Q}$ are fixed real vectors. By suitable choice of the scalar $\varepsilon$, this can be expressed as

$$\boldsymbol{V} = (\boldsymbol{A} + \mathrm{i}\boldsymbol{B})\mathrm{e}^{\mathrm{i}\varepsilon}\mathrm{e}^{-\mathrm{i}\omega t} \tag{2}$$

where $\boldsymbol{A}$ and $\boldsymbol{B}$ are real orthogonal vectors with $|\boldsymbol{A}| \geqslant |\boldsymbol{B}|$. Thus, the real physical field, $\boldsymbol{E}$, for example, is

$$\mathrm{Re}\ \boldsymbol{V} = \boldsymbol{A}\ \cos(\omega t - \varepsilon) + \boldsymbol{B}\ \sin(\omega t - \varepsilon). \tag{3}$$

The end of the $\boldsymbol{E}$ vector describes an ellipse, the polarisation ellipse, with $\boldsymbol{A}$ and $\boldsymbol{B}$ as the major and minor semi-axes, respectively. The meaning of $\varepsilon$ is that when $\omega t = \varepsilon$ the end of $\boldsymbol{E}$ is at one of the two extremities of the major axis. The polarisation ellipses for $\boldsymbol{E}$ and $\boldsymbol{H}$ are in general quite different.

It is useful to specify the plane of the ellipse by defining the normal vector $\boldsymbol{n} = \boldsymbol{A} \times \boldsymbol{B}$. The essential point is that the polarisation ellipse varies continuously with position in the field, and so does $\varepsilon$, which specifies how it is being executed in time. Thus, one has to imagine a continuous field of polarisation ellipses, each one specified by the orthogonal right-handed triad of vectors $\boldsymbol{A}$, $\boldsymbol{B}$, $\boldsymbol{n}$ and by the phase $\varepsilon$.

### 2.2 The elastic wavefield

The same formalism can be used to describe a general elastic wavefield of frequency $\omega$ taking $\boldsymbol{V}$ to be the complex displacement $\boldsymbol{s}$ of a particle (the physical displacement being Re $\boldsymbol{s}$). The polarisation ellipse is then simply the orbit of the particle. Since the velocity of the particle is $\dot{\boldsymbol{s}} = -\mathrm{i}\omega\boldsymbol{s}$, the same ellipse also represents the motion of the end of the velocity vector, scaled and changed in phase by $\frac{1}{2}\pi$.

If we wish, we can split $\boldsymbol{s}$ uniquely into two components, $\boldsymbol{s}_1$ and $\boldsymbol{s}_2$, of which one is curl free and the other divergence free, thus:

$$\boldsymbol{s} = \boldsymbol{s}_1 + \boldsymbol{s}_2$$
$$\mathrm{curl}\ \boldsymbol{s}_1 = 0 \qquad \mathrm{div}\ \boldsymbol{s}_2 = 0.$$

The complex vector $\boldsymbol{V}$ could represent either $\boldsymbol{s}$ or $\boldsymbol{s}_1$ or $\boldsymbol{s}_2$, but for the curl-free component $\boldsymbol{s}_1$ (which in a homogeneous isotropic medium corresponds to longitudinal waves) it is possible, as an alternative, to proceed more simply by defining a (complex) scalar potential $\psi$, say, such that

$$\boldsymbol{s}_1 = \mathrm{grad}\ \psi$$

and to use $\psi$ to represent the wavefield rather than the vector $\boldsymbol{s}_1$. (This would also be the case for a sound wavefield in a fluid, where $\boldsymbol{s}_2 \equiv 0$ and $\boldsymbol{s} = \boldsymbol{s}_1$.) But for $\boldsymbol{s}$ or $\boldsymbol{s}_2$ (as for $\boldsymbol{E}$ and $\boldsymbol{H}$) there is no escape from considering a vector field.

Any real elastic wavefield will not consist of either $\boldsymbol{s}_1$ or $\boldsymbol{s}_2$ alone, if only because the boundaries will impose conditions on $\boldsymbol{s}$, rather than on $\boldsymbol{s}_1$ and $\boldsymbol{s}_2$ separately. For example, a Rayleigh surface wave consists of a mixture of $\boldsymbol{s}_1$ and $\boldsymbol{s}_2$, and in general there is mode conversion at boundaries. For this reason it is not necessarily advantageous to split $\boldsymbol{s}$ into its two components.

## 3 THE DIRECTION OF PROPAGATION

At any point the polarisation ellipse is always executed, according to equation (3), in the sense from $\boldsymbol{A}$ to $\boldsymbol{B}$. The two major semi-axes give two choices for $\boldsymbol{A}$. This choice being made, the sense in which the ellipse is being executed decides which of the two minor semi-axes is $\boldsymbol{B}$. To decide the chirality at the point in question the sense of circulation of the ellipse has to be combined with a direction of propagation—and this needs definition. The direction of the flow of energy is a possible choice, but in this paper we shall explore the idea of using the direction of the spatial gradient of phase. For a scalar wavefield (pressure waves in a fluid, for example) the phase of the vibration at each point is well defined, except on dislocation lines (section 4), and so is its spatial gradient. However, for a vector wave the matter is less clear. When it is a uniform transverse elliptically polarised plane wave it propagates along the direction of $\pm \boldsymbol{n}$, depending on its chirality, but this is not true in general. Indeed for a plane linear polarised wave $\boldsymbol{n}$ is undefined, and even in a general wavefield there are line loci (see section 4) where the polarisation is linear and so where $\boldsymbol{n}$ is undefined.

As an alternative to $\boldsymbol{n}$ one might choose the vector $\nabla\varepsilon$ for the direction of propagation, as is done, for example, by Born and Wolf (1970). However, in a uniform circular polarised plane wave the direction of $\boldsymbol{A}$ becomes undefined; therefore so also does $\varepsilon$ (because it relates to the time at which the vibration reaches the end of the $\boldsymbol{A}$ axis) and $\nabla\varepsilon$ could not be used. Even in a general wavefield there are line loci (see section 4) where the polarisation ellipse becomes locally circular, so rendering the choice of $\nabla\varepsilon$ inappropriate.

Let us therefore proceed in the following way. We wish to find a spatial gradient of phase, and the problem, in essence, is to relate the 'phase' of an elliptical vibration at a given point to the 'phase' at a neighbouring point where the vibration ellipse has a slightly different size, shape and orientation in space. Now Pancharatnam (1956) has considered the problem of comparing the phases of two differently polarised beams of light by using a very simple approach. Let the two beams interfere and consider the resulting intensity. If the two fields are $\boldsymbol{V}_1$ and $\boldsymbol{V}_2$ we have for the resultant

$$\boldsymbol{V} = \boldsymbol{V}_1 + \boldsymbol{V}_2$$

and for the intensity $I$

$$\begin{aligned} I \propto \boldsymbol{V}\cdot\boldsymbol{V}^* &= (\boldsymbol{V}_1 + \boldsymbol{V}_2)\cdot(\boldsymbol{V}_1^* + \boldsymbol{V}_2^*) = \boldsymbol{V}_1\cdot\boldsymbol{V}_1^* + \boldsymbol{V}_2\cdot\boldsymbol{V}_2^* + 2\ \mathrm{Re}\ (\boldsymbol{V}_1^*\cdot\boldsymbol{V}_2) \\ &= |\boldsymbol{V}_1|^2 + |\boldsymbol{V}_2|^2 + 2|\boldsymbol{V}_1^*\cdot\boldsymbol{V}_2|\cos\arg(\boldsymbol{V}_1^*\cdot\boldsymbol{V}_2). \end{aligned}$$

Thus, the two vibrations are 'in phase' with maximum resultant intensity when $\arg(\boldsymbol{V}_1^*\cdot\boldsymbol{V}_2) = 0$, and 'out of phase' with minimum resultant intensity when $\arg(\boldsymbol{V}_1^*\cdot\boldsymbol{V}_2) = \pi$. We therefore take the phase difference to be $\arg(\boldsymbol{V}_1^*\cdot\boldsymbol{V}_2)$.

To apply this to our spatial continuum of polarisation ellipses we translate it into differential form, writing for the infinitesimal phase difference

$$\begin{aligned} \mathrm{d}\delta = \arg\{\boldsymbol{V}^*\cdot(\boldsymbol{V}+\mathrm{d}\boldsymbol{V})\} &= \arg(|\boldsymbol{V}|^2 + \boldsymbol{V}^*\cdot\mathrm{d}\boldsymbol{V}) \\ &= \frac{\mathrm{Im}(\boldsymbol{V}^*\cdot\mathrm{d}\boldsymbol{V})}{|\boldsymbol{V}|^2}. \end{aligned} \tag{4}$$

This is identical in form with the phase connection exploited by Berry (1984, 1987, 1989) in quantum mechanics. In that context it refers to the phase change of a single quantum system as certain parameters slowly change. Here it refers to the phase of the field of polarisation ellipses as one moves in three-dimensional space.

Using the representation (2) in equation (4) leads to

$$\mathrm{d}\delta = \frac{\boldsymbol{A}\cdot \mathrm{d}\boldsymbol{B} - \boldsymbol{B}\cdot \mathrm{d}\boldsymbol{A}}{A^2 + B^2} + \mathrm{d}\varepsilon \qquad A = |\boldsymbol{A}|,\ B = |\boldsymbol{B}|.$$

Since $\boldsymbol{A}\cdot\boldsymbol{B} = 0$, $\mathrm{d}(\boldsymbol{A}\cdot\boldsymbol{B}) = \boldsymbol{A}\cdot \mathrm{d}\boldsymbol{B} + \boldsymbol{B}\cdot \mathrm{d}\boldsymbol{A} = 0$. Hence

$$\mathrm{d}\delta = -\frac{2\boldsymbol{B}\cdot \mathrm{d}\boldsymbol{A}}{A^2 + B^2} + \mathrm{d}\varepsilon.$$

Referring to figure 1, let $\mathrm{d}\alpha$ be the (right-handed) angle of rotation of the axes $\boldsymbol{A}$, $\boldsymbol{B}$ about $\boldsymbol{n}$, and denote the component of $\mathrm{d}\boldsymbol{A}$ parallel to $\boldsymbol{B}$ by $\mathrm{d}A_B$. Then

$$\boldsymbol{B}\cdot \mathrm{d}\boldsymbol{A} = B\ \mathrm{d}A_B = AB\ \mathrm{d}\alpha$$

and hence

$$\mathrm{d}\delta = -\frac{2AB}{A^2 + B^2}\ \mathrm{d}\alpha + \mathrm{d}\varepsilon.$$

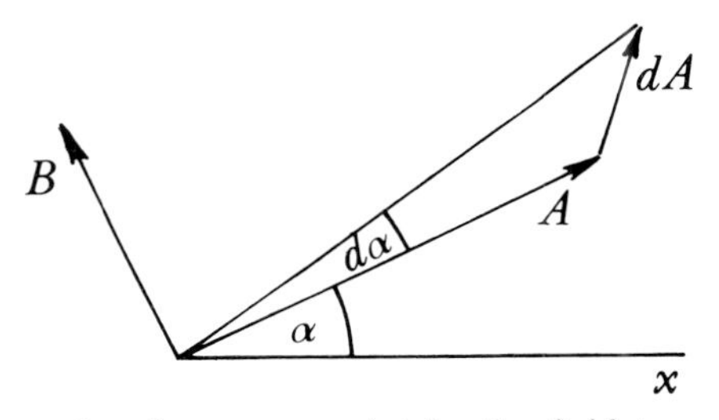


**Figure 1** In passing from one point in the field to a neighbouring point the major and minor semi-axes $\boldsymbol{A}$, $\boldsymbol{B}$ of the polarisation ellipse rotate by $\mathrm{d}\alpha$ about the normal $\boldsymbol{n}$ to the ellipse.

We should remark that $\mathrm{d}\delta$ is not integrable, as was shown by Pancharatnam (1956); nor is $\mathrm{d}\alpha$, because it is impossible to carry through three-dimensional space a reference azimuth against which to measure $\alpha$. Therefore, if we define $\boldsymbol{k}_\delta = \partial\delta/\partial\boldsymbol{r}$ and $\boldsymbol{k}_\alpha = \partial\alpha/\partial\boldsymbol{r}$, these vectors are not gradients of functions. In contrast, since $\varepsilon$ is well defined (modulo $\pi$) at each point (except on line singularities which we discuss later) $\partial\varepsilon/\partial\boldsymbol{r} = \nabla\varepsilon$ is (almost everywhere) the gradient of a function. With these definitions

$$\boldsymbol{k}_\delta = -\frac{2AB}{A^2 + B^2}\ \boldsymbol{k}_\alpha + \nabla\varepsilon. \tag{5}$$

The significance of this equation is that the 'phase gradient' $\boldsymbol{k}_\delta$ is thereby connected with the gradient of the phase $\varepsilon$ and also with the spatial rate of change of the ellipse orientation, or of the triad $\boldsymbol{A}$, $\boldsymbol{B}$, $\boldsymbol{n}$. The component of $\boldsymbol{k}_\alpha$ parallel to $\boldsymbol{n}$ represents twist of the ellipse about $\boldsymbol{n}$, while the components of $\boldsymbol{k}_\alpha$ transverse to $\boldsymbol{n}$ represent bend of the axial directions of the ellipse in its own plane.

Expression (5) takes a more symmetrical form if we resolve the elliptical vibration at each point into the sum of two counter-rotating circular vibrations in the same plane (figure 2). If these have amplitudes $\rho_1$ and $\rho_2$ ($\rho_1 \geqslant \rho_2$) and phases $\phi_1$ and $\phi_2$ with respect to a local azimuth O$x$ in the plane of vibration, it is easily shown that

$$\begin{aligned} A &= \rho_1 + \rho_2 & \qquad B &= \rho_1 - \rho_2 \\ \varepsilon &= \tfrac{1}{2}(\phi_2 + \phi_1) & \qquad \alpha &= \tfrac{1}{2}(\phi_2 - \phi_1) \end{aligned} \tag{6}$$

where $\alpha$ is the angle between $\boldsymbol{A}$ and O$x$ (positive when right-handed about $\boldsymbol{n}$). Because the reference azimuth O$x$ cannot be carried consistently throughout three-dimensional space, the two equations involving $\phi_1$ and $\phi_2$ have only local meaning; $\phi_1$ and $\phi_2$ do not exist as functions of $\boldsymbol{r}$. However, we can write $\mathrm{d}\varepsilon = \frac{1}{2}(\mathrm{d}\phi_2 + \mathrm{d}\phi_1)$ and $\mathrm{d}\alpha = \frac{1}{2}(\mathrm{d}\phi_2 - \mathrm{d}\phi_1)$ and, if we define $\boldsymbol{k}_1 = \partial\phi_1/\partial\boldsymbol{r}$ and $\boldsymbol{k}_2 = \partial\phi_2/\partial\boldsymbol{r}$, equation (5) becomes

$$\boldsymbol{k}_\delta = \frac{\rho_1^2}{\rho_1^2 + \rho_2^2}\,\boldsymbol{k}_1 + \frac{\rho_2^2}{\rho_1^2 + \rho_2^2}\,\boldsymbol{k}_2 \tag{7}$$

which is identical to the expression for $\boldsymbol{k}_\delta$ chosen for other reasons in Nye and Hajnal (1987). This is the main result of the present paper.

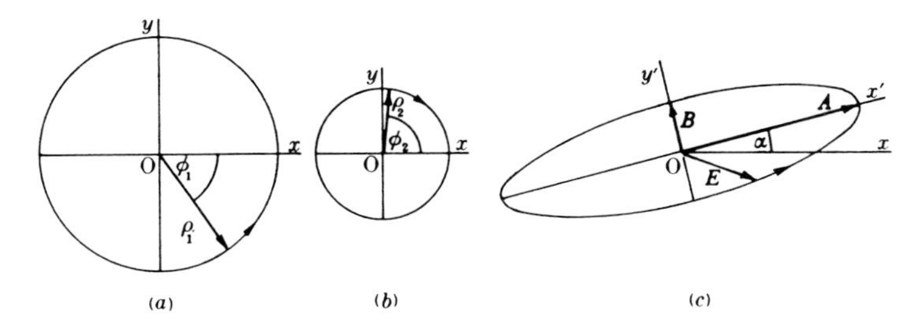


**Figure 2** The elliptical vibration in ($c$) is the sum of the two circular vibrations in ($a$) and ($b$). The vectors are drawn at $t = 0$.

It may alternatively be derived directly from (4) by first expressing $\boldsymbol{V}$ as the sum of two orthogonal complex vectors, $\boldsymbol{C}_1$ and $\boldsymbol{C}_2$, representing the two counter-rotating circular vibrations, so that $\boldsymbol{C}_1 \cdot \boldsymbol{C}_2^* = 0$. Since we wish to consider changes in $\boldsymbol{C}_1$ (that is, in $\rho_1$ and $\phi_1$) that keep it circular and so orthogonal to $\boldsymbol{C}_2$, it follows that $\mathrm{d}\boldsymbol{C}_1 \cdot \boldsymbol{C}_2^* = 0$. Similarly for changes in $\boldsymbol{C}_2$ we have $\mathrm{d}\boldsymbol{C}_2 \cdot \boldsymbol{C}_1^* = 0$. Hence

$$\begin{aligned} \mathrm{Im}(\boldsymbol{V}^* \cdot \mathrm{d}\boldsymbol{V}) &= \mathrm{Im}\{(\boldsymbol{C}_1^* + \boldsymbol{C}_2^*) \cdot (\mathrm{d}\boldsymbol{C}_1 + \mathrm{d}\boldsymbol{C}_2)\} \\ &= \mathrm{Im}(\boldsymbol{C}_1^* \cdot \mathrm{d}\boldsymbol{C}_1 + \boldsymbol{C}_2^* \cdot \mathrm{d}\boldsymbol{C}_2). \end{aligned}$$

Now the $x$, $y$ components of $\boldsymbol{C}_1$ are

$$\boldsymbol{C}_1 = \rho_1 e^{i\phi_1} e^{-i\omega t}(1, i)$$

and so

$$\mathrm{Im}(\boldsymbol{C}_1^* \cdot d\boldsymbol{C}_1) = 2\rho_1^2 \, d\phi_1.$$

Similarly

$$\boldsymbol{C}_2 = \rho_2 e^{i\phi_2} e^{-i\omega t}(1, -i)$$

and so

$$\mathrm{Im}(\boldsymbol{C}_2^* \cdot d\boldsymbol{C}_2) = 2\rho_2^2 \, d\phi_2.$$

Thus, using (4),

$$d\delta = \frac{\mathrm{Im}(\boldsymbol{V}^* \cdot d\boldsymbol{V})}{|\boldsymbol{V}|^2} = \frac{\rho_1^2 \, d\phi_1 + \rho_2^2 \, d\phi_2}{\rho_1^2 + \rho_2^2}$$

and the result (7) follows.

## 4 WHY CHOOSE THE PANCHARATNAM PHASE DIFFERENCE?

None of the four gradient quantities $\boldsymbol{k}_1$, $\boldsymbol{k}_2$, $\nabla\varepsilon$ or $\boldsymbol{k}_\alpha$ is suitable as an indicator of propagation direction over the whole field. The difficulty is that what is the natural definition at one place is unsuitable at another. For example, as we have already remarked, $\nabla\varepsilon$ fails, because it is undefined, at places where the polarisation ellipse is circular. At such places $\boldsymbol{k}_1$ is the natural choice, but $\boldsymbol{k}_1$ cannot be used over the whole field because it is indeterminate (see below) at places where the vibration is linear. However, $\boldsymbol{k}_\delta$ defined by (5) or (7) has the merit, as we shall now see, that it is well defined everywhere in a generic field; moreover, at places where there is a natural choice $\boldsymbol{k}_\delta$ coincides with it. To see that this is so we must look more closely at the singular places in the field where the vibration is circular or linear.

As a preliminary to doing so, let us recall briefly the singularities in a complex scalar monochromatic wavefield. Let the complex scalar at each point be denoted by $\psi = \rho e^{i\phi} e^{-i\omega t}$. Then there exist stationary line singularities, called wave dislocations (Nye and Berry 1974, Berry 1981, Nye 1981), where the magnitude $\rho$ is zero and the phase $\phi$ is correspondingly indeterminate. They are perfect interference fringes. The structure of a travelling plane (but dislocated) wave in the vicinity of such a line (figure 3) is similar to that around a crystal dislocation; the Burgers vector, so named for crystal dislocations by Frank (1951), is along the wave normal in this case of the travelling wave. But, in general, where there is no discernible plane wave, the characteristic feature of the wave dislocation is that, on a closed circuit around it, the phase $\phi$ changes by $2\pi$. Consequently $\phi$ at a given point is multi-valued, the values differing by an integer multiple of $2\pi$, and there are some points (on the dislocations) where it is undefined.

Coming now to the vector wavefield, there are two main kinds of singularity to examine. First, there are loci called $\mathscr{C}^{\mathrm{T}}$ lines (Nye and Hajnal 1987) where the wave vibration is perfectly circular, although the plane of the circle varies along the line. The structure of the wave near a $\mathscr{C}^{\mathrm{T}}$ line is shown in figure 4. The pattern of vibration ellipses, thought of as similar to a disclination (Frank 1958) in a liquid crystal, has a singularity index of either $-\frac{1}{2}$ or $+\frac{1}{2}$. On a circuit around a $\mathscr{C}^{\mathrm{T}}$ line both $\alpha$ and $\varepsilon$ change by $\pi$. On a cross section through a $\mathscr{C}^{\mathrm{T}}$ line the lines of equal $\varepsilon$ are radial, like the spokes of a wheel, and on the $\mathscr{C}^{\mathrm{T}}$ line itself $\varepsilon$ is indeterminate. At other points $\varepsilon$ is well defined, but only to within an arbitrary integer multiple of $\pi$.

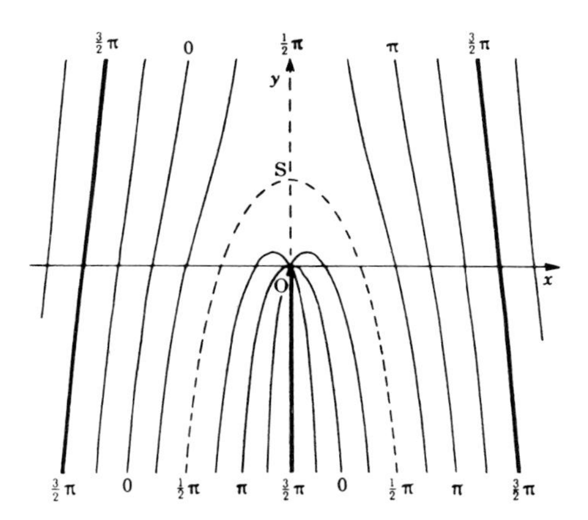


**Figure 3** A section through a dislocation line in a plane travelling scalar wave. The lines of equal phase $\phi$ are drawn. Any fixed phase value (e.g. $\phi = \frac{3}{2}\pi$) may be taken to denote a wave crest. The extra half-plane may be seen extending downwards from the dislocation at O. S is a saddle point for phase.

$\nabla\varepsilon$ is clearly singular on the $\mathscr{C}^{\mathrm{T}}$ line. So likewise is $\boldsymbol{k}_\alpha$. To see the behaviour of $\boldsymbol{k}_2$ notice that this measures the change of phase of a circular scalar component whose magnitude vanishes at the point in question. So the point lies on a dislocation of this scalar component, and we know that at such points the corresponding phase gradient $\boldsymbol{k}_2$ is singular. Thus $\nabla\varepsilon, \boldsymbol{k}_\alpha$ and $\boldsymbol{k}_2$ are all singular. Equation (5) is inapplicable but equation (7) reduces simply to $\boldsymbol{k}_\delta = \boldsymbol{k}_1$ (because at a small distance $r$ from the singularity $|\boldsymbol{k}_2| \propto r^{-1}$ while $\rho_2 \propto r$) as one would hope, because the circular component of amplitude $\rho_1$ is the only one present.

The second kind of singular locus consists of points where the vibration is linear. This is the $\mathscr{L}^{\mathrm{T}}$ line where the wave vibration is perfectly linear, although the direction of the vibration varies along the line. Near the line the vibration is a long thin ellipse and when viewed along the major axis the possible patterns made by the very short minor axes are characteristic of disclinations of index $\pm 1$ (figure 5). Since $\boldsymbol{B} = 0$ on the $\mathscr{L}^{\mathrm{T}}$ line the direction of $\boldsymbol{n}$ is undefined. This means that $\boldsymbol{k}_\alpha$ is undefined, because it measures the rate of rotation about $\boldsymbol{n}$. $\boldsymbol{k}_1$ and $\boldsymbol{k}_2$ are badly behaved because they also depend upon the direction of $\boldsymbol{n}$. Equation (7) is now inapplicable, but equation (5) as $B \to 0$ gives simply $\boldsymbol{k}_\delta = \nabla\varepsilon$, again as one would hope.

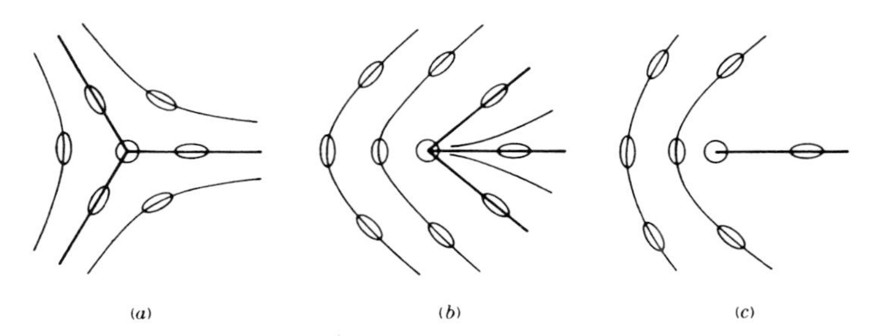


**Figure 4** The three possible patterns of polarisation ellipses surrounding a $\mathscr{C}^{\mathrm{T}}$ line. The $\mathscr{C}^{\mathrm{T}}$ line cuts through the centre of each diagram, where the polarisation ellipse is a circle. The lines, which show the directions of the major axes of the polarisation ellipses, make patterns whose singularity index is $-\frac{1}{2}$ for ($a$) and $+\frac{1}{2}$ for ($b$) and ($c$). (For example, in pattern ($a$), if one makes a clockwise circuit around the singularity, the direction of the major axis of the ellipse rotates by $\pi$ anticlockwise.)

In terms of the behaviour of the triad ($\boldsymbol{A}$, $\boldsymbol{B}$, $\boldsymbol{n}$) both $\mathscr{C}^{\mathrm{T}}$ and $\mathscr{L}^{\mathrm{T}}$ mark singular behaviour: $\mathscr{C}^{\mathrm{T}}$ is the centre of a vortex of rotation about $\boldsymbol{n}$, while $\mathscr{L}^{\mathrm{T}}$ is the centre of a vortex about $\boldsymbol{A}$. The difference of index, $\pm\frac{1}{2}$ for $\mathscr{C}^{\mathrm{T}}$ but $\pm 1$ for $\mathscr{L}^{\mathrm{T}}$, arises because the triad has inherent two-fold symmetry about $\boldsymbol{n}$ but not (because of the circulation around the ellipse) about $\boldsymbol{A}$.

A third kind of singular behaviour would be a vortex about $\boldsymbol{B}$, which would entail $\boldsymbol{A} = \boldsymbol{B} = 0$. However, this would be non-generic (of codimension 6) and structurally unstable. It would be an intersection point of an $\mathscr{L}^{\mathrm{T}}$ line with one or more $\mathscr{C}^{\mathrm{T}}$ lines and on perturbation they would separate. We may note that, since $\rho_1 = \rho_2 = 0$ at such a point, equation (7) shows that $\boldsymbol{k}_\delta$ would be singular. We should also note that in a generic field there is a set of isolated points where $\boldsymbol{k}_\delta = 0$; these can be thought of as standing-wave points where the wave has no direction of travel.

The conclusion is that, while no one of the quantities $\nabla\varepsilon, \boldsymbol{k}_\alpha, \boldsymbol{k}_1, \boldsymbol{k}_2$ is regular everywhere, in a generic field, the weighted mean of $\nabla\varepsilon$ and $\boldsymbol{k}_\alpha$

represented by (5) or of $\boldsymbol{k}_1$, $\boldsymbol{k}_2$ represented by (7) is indeed regular. This, in fact, was the reason for suggesting equation (7) in the earlier paper.

In summary, the primary reasons for choosing the phase connection (4) are that it is mathematically natural and has a physical interpretation in terms of interference. It is not integrable, but its gradient has no singularities (other than point zeros) in the generic fields we are considering.

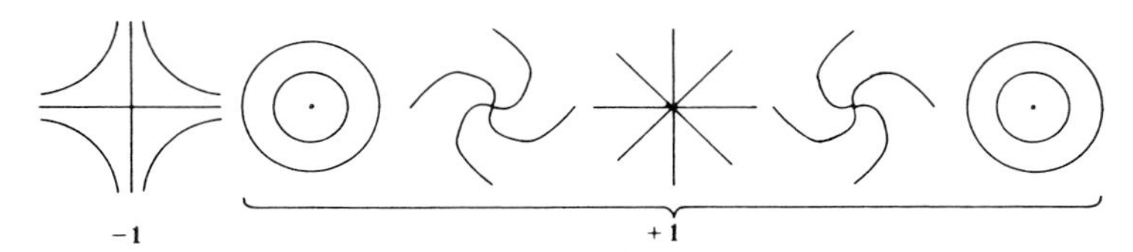


**Figure 5** Patterns of polarisation ellipses surrounding an $\mathscr{L}^{\mathrm{T}}$ line, which cuts through the centre of each diagram. The ellipses are long and thin with their major axes nearly normal to the page. The lines indicate the directions of the very short minor axes. There is only one pattern with singularity index $-1$ but a variety of patterns with index $+1$, as shown.

Having thus defined $\boldsymbol{k}_\delta$ we can regard it as the propagation direction and use it, in conjunction with the circulation sense of the polarisation ellipse, to establish a handedness at each point of the wavefield. The hand is decided by the sign of the triple product $T = [\boldsymbol{A}\ \boldsymbol{B}\ \boldsymbol{k}_\delta] = \boldsymbol{n} \cdot \boldsymbol{k}_\delta$: negative for right hand, positive for left hand (if we follow the usual optical convention, which is opposite to the IEEE convention). Surfaces on which $T = 0$ separate right- and left-handed regions of space.

For curl-free vector waves, as we have remarked, one has the alternative of representing them by a complex scalar $\psi$ and then it is straightforward to define the phase as $\phi$. However, $\nabla\phi$ is not the same as $\boldsymbol{k}_\delta$, derived from the associated vector field. One can see that this must be so, because whereas $\mathrm{d}\delta$ is not integrable, $\mathrm{d}\phi$ is integrable in the sense that it yields a multi-valued function $\phi$ which is well defined except at line singularities and except for an additive integer multiple of $2\pi$. Thus here the choice is between a non-integrable phase gradient without singularities (other than point zeros), or an integrable, but multi-valued, phase with singularities.

## 5 THE TRANSVERSE FIELD

All these considerations have applied to a general wavefield where the waves are travelling in all directions. If electromagnetic waves are all travelling in free space in approximately the same direction, paraxially, $\boldsymbol{E}$ and $\boldsymbol{H}$

will have only small longitudinal components, and it is then more appropriate to consider the transverse components $\boldsymbol{E}_t$ and $\boldsymbol{H}_t$ rather than the full fields. In this case the singularities are different (Nye 1983a,b, Hajnal 1987a,b). First, there are $\mathscr{C}$ lines where the transverse field is circularly polarised (while the full field, with its small additional longitudinal component, is not). These have structures very similar to those around $\mathscr{C}^T$ lines (the superscript T, for 'true', indicates that the vibration on $\mathscr{C}^T$ is truly circular). Second, there are $\mathscr{S}$ loci where the transverse field is linear polarised (while the full field is not). The $\mathscr{S}$ loci are surfaces rather than lines, and they separate space into right- and left-handed regions, the counterparts in this respect of the $T=0$ surfaces in the general case.

## 6 COMPARISON WITH CRYSTAL DEFECTS

In all this it is notable that the wave equation satisfied by the vector field, which for an inhomogeneous anisotropic medium will be quite complicated, plays no part. It does not, for example, restrict the existence of any of the singularities mentioned. They are topological objects, like dislocations in crystals and disclinations in liquid crystals. But they also have notable differences from crystal defects. For example, wave defects have no cores where a continuum description breaks down. More significantly, there is nothing in the wave singularities corresponding to line tension; and with wave defects, although one can superpose solutions of the linear differential wave equations, each solution destroys the defects of the other. This is because wave defects are places where a magnitude is zero, in contrast to a crystal dislocation where the stress and strain are infinite.

## ACKNOWLEDGMENT

I should like to thank Professor M V Berry for his helpful comments.

## REFERENCES

Berry M V 1981 *Physics of Defects, Les Houches Session XXXV* ed R Balian, M Kléman and J-P Poirier (Amsterdam: North-Holland) pp 453–543

—— 1984 *Proc. R. Soc.* A **392** 45–57

—— 1987 *J. Modern Opt.* **34** 1401–7

—— 1989 *Geometric Phases in Physics* ed A Shapere and F Wilczek (Singapore: World Scientific) pp 7–28

Born M and Wolf E 1970 *Principles of Optics* 4th edn (Oxford: Pergamon)

Frank F C 1951 *Phil. Mag.* **42** 809–19
—— 1958 *Discuss. Faraday Soc.* **25** 19–28
Hajnal J V 1987a *Proc R. Soc.* A **414** 433–46
—— 1987b *Proc. R. Soc.* A **414** 447–68
Nye J F 1981 *Proc. R. Soc.* A **378** 219–39
—— 1983a *Proc. R. Soc.* A **387** 105–32
—— 1983b *Proc. R. Soc.* A **389** 279–90
Nye J F and Berry M V 1974 *Proc. R. Soc.* A **336** 165–90
Nye J F and Hajnal J V 1987 *Proc. R. Soc.* A **409** 21–36
Pancharatnam S 1956 *Proc. Ind. Acad. Sci.* A **44** 247–62 (reprinted 1975 *Collected Works of S Pancharatnam* (Oxford: Oxford University Press)

# Aperiodic Crystals: Structures and Structural Defects

M Kléman

*Lapides crescunt; vegetabilia crescunt et vivunt; animadia crescunt, vivunt et sentiunt.*

Linné (cited in Romé de l'Isle, 1783)

## 1 INTRODUCTION

Aperiodic crystals or, as they are more commonly called, quasi-crystals, came to birth in 1984 when Shechtman *et al* discovered that certain metallic alloys (they studied mostly an alloy whose composition is close to $Al_4Mn$) produced diffraction patterns consisting of sharp peaks belonging to a 3D icosahedral pattern. Icosahedral symmetry does not appear in any of the 230 Schoenflies space groups, since five-fold symmetry is not crystallographic (an icosahedron has six five-fold axes), a result that can be inferred easily from the fact that the peaks located in reciprocal space at points

$$\boldsymbol{k} = \sum_{i=1}^{6} n_i \boldsymbol{k}_i \tag{1}$$

where the $\boldsymbol{k}_i$ point towards six independent vertices of an icosahedron, and the $n_i$ are integers, form a dense set in space. Crystallographically speaking, such a situation is intolerable, even if the intensities are significantly different from zero only on a small number of peaks, those corresponding to small $n_i$ in the experiments of Shechtman *et al*. Many more metallic alloys were soon discovered with the same symmetry; although some controversies arose about the origin of the diffraction patterns (Pauling (1985) explained the presence of a five-fold symmetry as resulting from multiple twinning), it was soon evident that icosahedral order had its place as a central concept in crystallography.

Theoretical research on aperiodicity started simultaneously, with the demonstration by Levine *et al* (1985), that highly ordered-like crystals, but quasi-periodic rather than periodic, were possible. This result of a very general nature had been preceded by a number of efforts to construct 2D or 3D quasi-lattices, especially by Penrose (1974, 1979), Mackay (1962, 1981) and Kramer (1982). The experimental discovery of quasi-crystals arrived on a soil which was already fertilised. Penrose had devised the famous 2D tiling which bears his name (figure 1). This tiling can be built in a number of fashions (cf Gardner 1977), the simplest one from the point of view of the crystallography of quasi-crystals being a tiling of two types of rhombuses with angles that are multiples of 36°, as in figure 1.

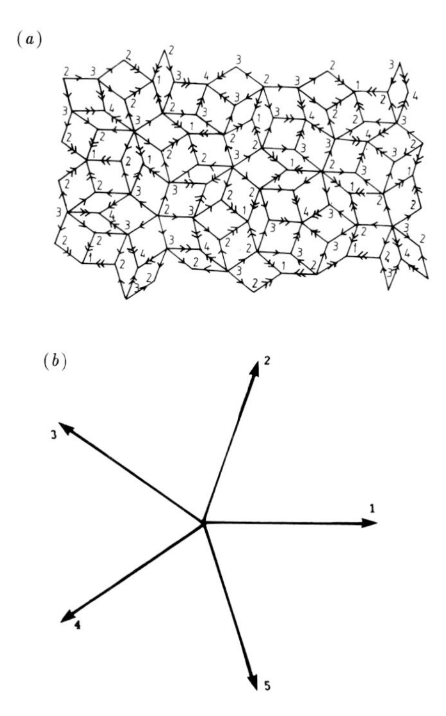


**Figure 1** A piece of Penrose lattice.

It is the name of de Bruijn (1981) which must be attached to the discovery of the essential mathematical features of quasi-crystals, long before their experimental demonstration, since he proved that Penrose tilings with rhombuses can be obtained as projections of a selected number of vertices of a 5D hypercubic lattice on a 2D planar irrational cut; the rhombuses are themselves the projections of the faces of the hypercubes. This method was generalised by Kramer and Neri (1984) and Duneau and Katz (1985) for the purpose of constructing a quasi-crystal in any dimension, soon after the discovery of Shechtman *et al.* Since then, we have seen a remarkable outburst of theoretical works on this new crystallogaphy (see Steinhardt and Ostlund (1987) for a collection of papers up to 1987). Even if it appears in the long run that *perfect aperiodic crystallinity* is a view of the mind, an asymptotical concept, and not a physical reality, the importance of these works justify entirely the present belief in their existence.

Another reason why the experimental discovery of Shechtman *et al* has arrived right on time is the current interest in structural models of disorder based on local icosahedral symmetry. Frank (1952) was certainly the first to stress the importance of local icosahedral arrangements of atoms; he explained the (meta)-stability of supercooled liquids by showing that the stable configuration of a cluster of 13 atoms bound by a Lennard-Jones potential is icosahedral, and not FCC or HCP. In fact, an icosahedral cluster is more compact than a FCC or HCP cluster, but cannot of course extend with an even density through all space. Soon after Frank's short but pioneering 1952 paper, the icosahedral model was adopted for simple liquids (Bernal 1964) and amorphous metals. In 1979, Kléman and Sadoc showed that icosahedral order, which is not crystallographic in our common Euclidean space, tiles a 3D sphere, and that disorder can be described as the result of introducing disclinations (a line defect which was first conceptualised by Frank (1958) for the purpose of explaining the nature of singularities in liquid crystals) in a $\{3,3,5\}$ spherical 'crystal'†, tiled with regular tetrahedra $\{3,3\}$, 5 along each edge, i.e. with centred icosahedra whose edge length is equal to the radius. A disclination line of the right sign introduces a wedge of extra matter along the line, which has the effect of decurving the spherical crystal (figure 2); with a suitable density of such lines, the spherical crystal approaches flatness, in such a way that finally a disordered monatomic medium can be described as a matrix made of $Z = 12$ coordinated atoms, pierced by a *random* 3D array of disclinations located in their large majority along lines of $Z > 12$ atoms (Kléman and Sadoc 1979, Kléman 1990a).

A few years after his 1952 paper, Frank produced with Kasper a series of two papers (Frank and Kasper 1958, 1959) devoted to polytetrahedral order, a structure which is met in many transition-metal complex alloys (Laves phases, $\sigma$ phases, etc). Their main result is that these so-called 'Frank and Kasper' phases can be described, as above, as a matrix of $Z = 12$ coordinated atoms, pierced by a *periodic* array of lines of $Z \neq 12$ coordinated atoms. They did not recognise that these lines are disclinations (this new concept due to Frank had just been applied to nematics, but we

† The notation $\{p,q,r\}$ is due to Schläfli; $\{p\}$ stands for a regular polygon with $p$ edges; $\{p,q\}$ stands for a Platonic polyhedron with $q\{p\}$s at each vertex; $r$ such polyhedra merge along each edge in the $\{p,q,r\}$ tiling.

had to wait for Nelson (1983) and Sadoc (1983) to identify these lines as disclinations, in the wake of the Kléman and Sadoc (1979) paper) and, even more surprisingly, did not mention the relationship with the question of the stability of local icosahedral order (Frank 1952). It is indeed remarkable that these polytetrahedral alloys provide an example of structures which possess all the characteristics of disorder inferred from the Kléman–Sadoc model and at the same time are periodic. Many of the known quasi-crystals possess near-by neighbours in the phase diagram which are Frank and Kasper alloys with large unit cells. This effect of proximity, akin in a sense to polytypism, has been studied experimentally on many examples by Kuo and his group (see, for example, Kuo 1986). In fact, these Frank and Kasper phases fall in the class of the rational approximants of quasi-crystals.

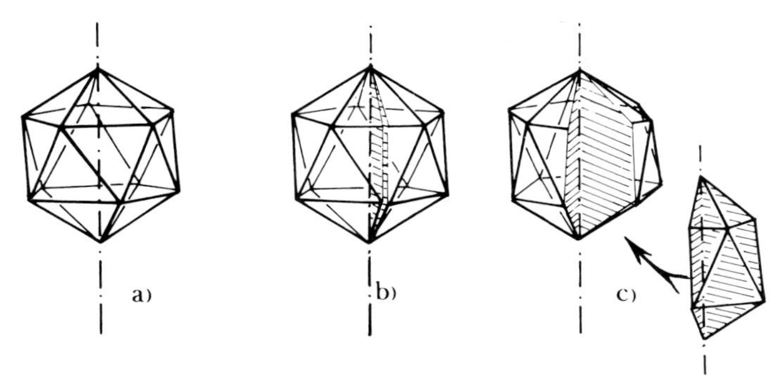


**Figure 2** Construction of a disclination in a icosahedral lattice; local aspect. (*a*) Perfect icosahedron; (*b*) cut surface; (*c*) introduction of extra matter.

These relationships between icosahedral local order, Frank and Kasper phases, and quasi-crystals put in a new perspective the questions of the structure of quasi-crystals and their thermodynamic stability. Concerning the question of their structure, it should certainly be possible to describe it as a regularly (but not periodically) disclinated version of a curved icosahedral crystal; an attempt has been made in this direction by Mosseri and Sadoc (1985). But in this paper, which shall deal mostly with structural defects of quasi-crystals, we shall use the more natural geometrical approach to QC starting from a high-dimensional space. Concerning their thermodynamic stability, it has been suggested by various authors (Henley 1987, Friedel 1988) that quasi-crystals might be Hume–Rothery phases; i.e. their electronic stability is related to a definite number of valence electrons per atom. More trivially, one can also think that the fact that those quasi-crystals *grow* in the shape of definite polyhedra (for example, large single crystals with a morphology of rhombic triacontahedra have been observed in Al–Li–Cu alloys (Dubost *et al* 1986)) is an argument in favour of their thermodynamic stability. *Lapides crescunt...*.

Let us here make a point of history and terminology. Frank proposed the name 'non-haüyan crystallography' for the new area of structural science opened by the discovery of quasi-crystals. According to Romé de l'Isle (1783), who coined the name, and Haüy (1784), *crystallography* is the science of minerals which take polyhedral shapes under growth, with constant dihedral angles. In that sense quasi-crystals belong to the realm of crystals, but with one fundamental difference, which would have been perceived at once by Haüy: the indexing of the facets does not obey the famous *loi des troncatures rationnelles* (law of rational truncations or of rational indices). Donnay in the early thirties studied a natural crystal, calaverite, whose facets indexing necessitated unreasonably large integral numbers; according to Occam's razor, nature should be more simple. Therefore Donnay (1935) tried to explain this effect which looked like a breaking of Haüy's law, in terms of twins (as Pauling did 50 years later for quasi-crystals); in fact, the solution to this mystery was given by Dam *et al* (1985), in a paper to which Donnay contributed himself; they showed that calaverite is an *incommensurate* crystal, whose facets can be indexed with small integers in a 4D Euclidean crystal which contains the crystal as a three-dimensional subspace. The law of rational truncations is still valid, but in four dimensions. Quasi-crystals are of that sort; their incommensurability originates in the inherent five-fold symmetry and does not vary continuously with temperature as it does in usual incommensurate crystals, but the same crystallographic approach is at work: aperiodic crystals can be described as projections of high-dimensional Euclidean crystals, namely five-dimensional hypercubic crystals for the decagonal phases which have 1D translational periodicity and 2D quasi-crystalline character and six-dimensional hypercubic crystals for the icosahedral phases (figure 3) on a Euclidean subset $P_{\parallel}$ of dimension 2 or 3 according to the case, which cuts the hypercubic lattice irrationally. Rational cuts yield the various rational approximants.

The golden ratio $(1+\sqrt{5})/2$ plays the most important role in quasi-crystals of pentagonal and icosahedral symmetry. Quasi-crystals involving other symmetries than five-fold symmetries are not known. Five-fold symmetry is also extremely frequent in biology (see Rivier 1986). But the golden ratio has essentially long been known for its prevalence in aesthetics (the famous book of Fra Luca Pacioli di Borgo, 'Divine Proportion' illustrated by Leonardo da Vinci, was published in 1598; see the French translation and facsimile (1980)), in music (where the most harmonious intervals involve successive integers of the Fibonacci series) and in other domains of art. In connection with this universality, it has been recently claimed (Rosolato 1985) that the number 5 is present in the formation and the organisation of the mind, in particular in the metaphor, which is the central element present in the emergence of meaning and harmony from the unknown and the irrational. It is impossible not to give an astonished look to this comeback of the pythagorean views on nature.

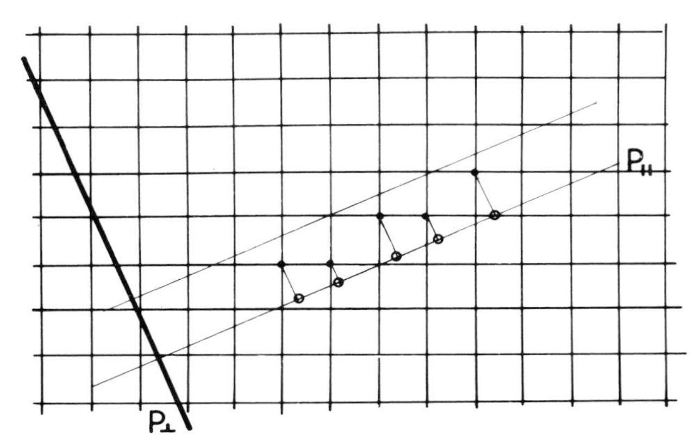


**Figure 3** Illustration of the cut-and-projection method for the case $d = 2$, $d_{\|} = 1$. The 1D aperiodic crystal is made of a sequence of short and long segments which reproduce the sequence of 0s and 1s of the binary representation of the Fibonacci sequence.

## 2 THE CRYSTALLOGRAPHY OF APERIODIC CRYSTALS

Continuous models and discrete models have been used to construct quasicrystals.

Continuous models start from a $d = 5$ dimensional (pentagonal case) or $d = 6$ dimensional (icosahedral case) Fourier expansion of the density $\rho(\boldsymbol{x})$

$$\rho(\boldsymbol{x}) = \sum \rho_k \exp(\mathrm{i}\boldsymbol{k}\boldsymbol{x}) \qquad \boldsymbol{x} \in E_d \tag{2}$$

which represents a cubic structure in $E_d$, and consider its restriction $\rho_{\mathrm{r}}(\boldsymbol{x})$ to a $d_{\|} = 2$ $(d = 5)$ or $d_{\|} = 3$ $(d = 6)$ subspace $P_{\|}$, which is the physical space $\boldsymbol{x} \in P_{\|}$. $P_{\|}$ is so chosen that in the $d = 5$ case (resp $d = 6$) the basic reciprocal vectors $\boldsymbol{k}_i$ $(i = 1, 2, 3, \ldots, d)$ project along the edges $\boldsymbol{k}_{\|i}$ of a regular pentagon (resp along the radii joining its centre to the vertices of a regular icosahedron), or to any other star of $\boldsymbol{k}$ vectors deduced from the former. Replacing now $P_{\|}$ by a vector $\gamma \in E_d$ parallel to itself, we get

$$\rho_{\mathrm{r}}(\boldsymbol{x} + \boldsymbol{\gamma}) = \sum \rho_k \exp[\mathrm{i}(\boldsymbol{k}_{\|} \cdot \boldsymbol{x} + \boldsymbol{k} \cdot \boldsymbol{\gamma})] \tag{3}$$

where $\boldsymbol{k} \cdot \boldsymbol{\gamma} = \varphi_k$ is a phase. Writing $\boldsymbol{k} \cdot \boldsymbol{\gamma} = \boldsymbol{k}_{\|} \cdot \boldsymbol{\gamma}_{\|} + \boldsymbol{k}_{\perp} \cdot \boldsymbol{\gamma}_{\perp}$, we see that the density wave depends therefore on $d$ degrees of freedom: $d_{\|}$ degrees of displacement which span all the values of $\boldsymbol{\gamma}_{\|} \in P_{\|}$, and $d_{\perp} = d - d_{\|}$

degrees of phase which span all the values of $\gamma_\perp \in P_\perp$, the subspace perpendicular to $P_\parallel$. The generalised Fourier coefficients of $\rho_r$ therefore give $\rho_k \exp(\mathrm{i}\,\boldsymbol{k}\cdot\boldsymbol{\gamma})$.

Inserting now these coefficients in a Landau expansion

$$F = \tfrac{1}{2}\sum_k A\,|\rho_k|^2 + B\sum_{k_1+k_2+k_3=0}\rho_{k_1}\rho_{k_2}\rho_{k_3} + C\sum_{k_1+k_2+k_3+k_4=0}\rho_{k_1k_2k_3k_4} + \ldots \quad (4)$$

a standard calculation shows that $F$, restricted to fifth- and tenth-order terms, does not depend on the phase variables $\varphi_k$ in the icosahedral case, while $F$ depends only on the scalar variable $\varphi = \Sigma\,\varphi_{k_i}$ in the pentagonal case, where the $\boldsymbol{k}_i$ are the five reciprocal vectors which project along the regular pentagon. In this latter case $\varphi$ is the (unnormalised) component of $\boldsymbol{\gamma}$ along the large diagonal $\Delta$ of the hypercube $\langle 1, 1, 1, 1, 1\rangle$, and measures the component of the parallel displacements of $P_\parallel$ along $\Delta$. The parallel displacement of $P_\parallel$ perpendicular to $\Delta$ does yield phase shifts which do not change the Landau free energy, but changes the configuration of the density wave. Many papers deal at length with the continuous structural properties and phase transitions of quasi-crystals in the Landau picture (see for example Jaric (1985) and Kalugin *et al* (1985)). It is worth citing at this stage the pioneering work of Alexander and McTague (1978), who discussed long ago the stability of an hypothetical icosahedral phase versus BCC.

The continuous models described above are energetical models. Discrete models are crystallographic in nature; they discuss in detail the construction of the quasi-crystal starting from a $d$-dimensional hypercubic lattice (whose symmetry groups have been classified by various authors; see for example Janssen (1985)), and how the projection of a selected number of vertices of the hypercubic lattice on the physical space $P_\parallel$ yields a quasi-periodic tiling of $P_\parallel$, whose elementary tiles are two rhombuses with matching rules indicated by the arrows carried by the edges (figure 1) in the 5D case and two rhombohedra in the 6D case. This is the so-called cut-and-projection method, which was developed after de Bruijn's fundamental paper (1981) by Kramer and Neri (1984) and Duneau and Katz (1985) (figure 3). The selected vertices belong to the interior of a strip in $E_\mathrm{d}$, whose $d_\parallel$ generatrices are parallel to $P_\parallel$, and in which a unit cell is inscribed. In this method, the phason degrees of freedom are *discontinuous*. The phase changes are indeed obtained in this model by moving the strip parallel to itself; when a new vertex $N$ enters the strip, an old one $M$, which is diagonally opposed to $N$ on the unit cube, leaves it simultaneously, and the set of tiles projected from the cube in question flips in a manner which is illustrated in figure 4 for the 2D case ($d_\parallel = 2$; $d = 5$). A global phase change of the perfect quasi-crystal involves such changes along an aperiodic set of lines ($d_\parallel = 2$) or planes ($d_\parallel = 3$). Figure 5 visualises such a set for $d_\parallel = 2$. We can refer to the $d_\parallel$ tiling of the quasi-crystal obtained by the cut-and-

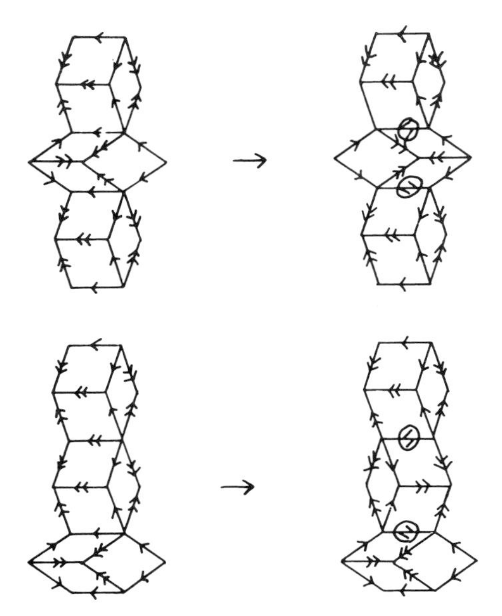

**Figure 4** Local effect of a phase change on the Bravais lattice of a Penrose tiling. From Pavlovitch *et al* (1989) with permission.

projection method as a Bravais lattice; this lattice varies discontinuously under a change of phase, but its reciprocal lattice, which is the starting point of the continuous method summarised above, does not depend on the phase.

The decoration of the Bravais lattice of a quasi-crystal is a difficult problem, which has received no completely satisfactory solution in real cases. The decoration can be introduced directly in the hypercubic lattice, as a motif $\sigma_i$ in each hypercubic cell $i$, and the construction of the quasi-crystal has to be modified as follows. The motif is a $d_\perp$-dimensional manifold, which cuts $P_\parallel$ (which is $d_\parallel$ dimensional) generically in a point or a set of points. When $P_\parallel$ moves parallel to itself, its intersection set with the motifs move, and the quasi-crystal suffers a phase change. If the motif is a copy of a piece of $P_\perp$, restricted to the interior of each unit cell, the quasi-crystal so constructed is the Bravais lattice of the cut-and-projection method. Moving $P_\parallel$ yields the same discontinuous phase changes we have discussed above. Levitov (1989) has shown that it is possible to choose this motif (which is called usually an atomic surface) in such a way that the phase changes are continuous. This extremely important result, which came as a surprise to most of the experts in the field, who were much accustomed to the Bravais lattice model, puts the physics of quasi-crystals in a new perspective. However, as we shall see later, the topological properties of phasons depends only on the symmetry properties of the quasi-crystals, i.e. on the Bravais lattice.

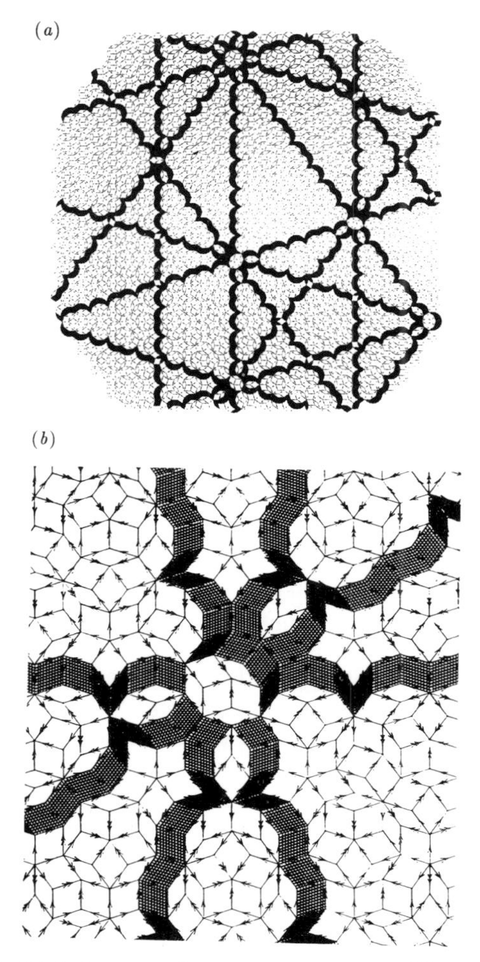


**Figure 5** (*a*) Global phase change on a Penrose tiling. The regions which are affected by the phase changes are along 'worms', which appear shaded (from Pavlovitch *et al* (1989) with permission). (*b*) Enlargement of a piece of (*a*).

## 3 THE VOLTERRA CLASSIFICATION OF DEFECTS IN QUASI-CRYSTALS

The Volterra classification of defects starts from the so-called Volterra process, which consists, in a normal 3D crystal, in displacing rigidly the two lips of some 'cut surface' $\Sigma$ with respect to each other in such a way that the void which is so created (if it is void; the results apply *mutatis mutandis* if the rigid displacement creates extra matter in some region) can be filled by a volume of perfect, unstrained, crystal whose boundaries fit perfectly along the two displaced lips (for details, see Friedel (1964)). It is easy to realise that this condition imposes a special rigid displacement: it must be a symmetry of the lattice, i.e. the sum of a translation $\boldsymbol{b}$ and a rotation $\bar{\bar{\Omega}}$ belonging to the Schoenflies group of the crystal. The line which bounds $\Sigma$ becomes singular in this process; it is called a *dislocation* of the Burgers vector $\boldsymbol{b}$ if $\bar{\bar{\Omega}} = 0$, a *disclination* of strength $\bar{\bar{\Omega}}$ if $\boldsymbol{b} = 0$, a *dispiration* in the general case $\boldsymbol{b} \neq 0$, $\bar{\bar{\Omega}} \neq 0$.

The essence of the Volterra process is the utilisation of the displacements of the group of symmetry. Any element of the group of symmetry of the crystal is either a direct displacement or a displacement with an inversion. The Volterra process does not employ these latter elements. The group of symmetry of a hypercrystal in $d$ Euclidean dimensions contains the same types of elements, and no other. Therefore the Volterra process can be defined in the same words. But now $\Sigma$ is a *hypersurface* of dimension $(d-1)$, and the singularity which bounds $\Sigma$ is a *hyperline* of dimension $(d-2)$. Consequently, a dislocation or a disclination in a six-dimensional hypercubic lattice is a four-dimensional object.

In this section, we will study the properties of a line of singularity $L_\parallel$ in a quasi-crystal obtained as the intersection of the hyperline $L_\mathrm{d}$ with $P_\parallel$. This intersection is, generically, in $d = 6$, $d_\parallel = 3$, a true 1D line, and in $d = 5$, $d_\parallel = 2$, a point, as expected. The intersection can be of higher dimension in non-generic cases, yielding intriguing objects like walls or volumes ($d_\parallel = 3$), or lines or walls ($d_\parallel = 2$), but we shall not consider these cases.

### 3.1 Dislocations; the analytical approach (Kléman 1988, Kléman and Sommers 1990)

We assume that $L_d$ is a dislocation hyperline of the Burgers vector $\boldsymbol{b}$ which is a $d$-vector. We write $\boldsymbol{b} = \boldsymbol{b}_\parallel + \boldsymbol{b}_\perp$, where $\boldsymbol{b}_\parallel$ is the $P_\parallel$ component of $\boldsymbol{b}$, and $\boldsymbol{b}_\perp$ the $P_\perp$ component. Both are non-vanishing vectors, necessarily, since $P_\parallel$ is an irrational plane. The difficult question is how to determine the strain field (which contains a phonon part and a phason part) carried by $L_\parallel$ in $P_\parallel$. Consider, as an heuristic guide, the very simple case $d = 3$, $d_\parallel = 2$, $d_\perp = 1$ (figure 6). The atoms in $P_\parallel$ are the intersections of the motifs (the atomic surfaces) $\sigma_i$; the 'bad crystal' in $P_\parallel$ is made of the intersections with $P_\parallel$ of the motifs belonging to the bad crystal in $E_3$. The bad crystal in $P_\parallel$ clearly has the shape of a $d_\perp$-dimensional furrow (in figure 6, $d_\perp = 1$) attached to $L_\parallel$. Physically, this singular furrow is considerably 'phonon' strained; it has a large elastic energy. The stability, if any, of the line $L_\parallel$ in the presence of such a furrow can henceforth be due only to some large anisotropy of the elastic constants in $P_\parallel$. This, however, is an unreasonable assumption since quasi-crystals are more symmetrical than cubic crystals. Therefore we need to reduce the furrow to zero, which requires $L_3$ ($L_d$) to be perpendicular to $P_{\|}$. We have therefore

$$L_d = L_{\|} \times P_{\perp} \tag{5}$$

$L_d$ is a cyclinder with basis $L_{\|}$ and generators $P_{\perp}$.

There is no difficulty in analytically calculating the strain field (in $d$-space) of the hyperline $L_d$, at least its singular contribution (Kléman 1988). Locally in the vicinity of the hyperline, the strain field created by $L_{\|} \times P_{\perp}$ is not much different from the strain field of a straight hyperline tangent to $L_{\|}$ in $P_{\|}$. Let $r$, $\varphi$, $z$ be cylindrical coordinates in the coordinate frame attached to the tangent to $L_{\|}$ whose axes are along $\boldsymbol{b}_e$ (the edge part of $\boldsymbol{b}_{\|}$), $\boldsymbol{b}_s \times \boldsymbol{b}_e$, and $\boldsymbol{b}_s$ (the screw part of $\boldsymbol{b}_{\|} = \boldsymbol{b}_e + \boldsymbol{b}_s$). The total displacement field generated by $\boldsymbol{b}_s$ and $\boldsymbol{b}_{\perp}$ has cylindrical symmetry about $L_d$, since $\boldsymbol{b}_s + \boldsymbol{b}_{\perp}$ is the total screw component of $\boldsymbol{b}$. Hence any 2-plane belonging to $P_{\|}$ and perpendicular to $L_{\|}$ (and consequently to $P_{\perp}$, i.e. to $L_d$) is invariant. Thus

$$\boldsymbol{u}_{s\perp} = \boldsymbol{u}_s + \boldsymbol{u}_{\perp} = \frac{\boldsymbol{b}_s + \boldsymbol{b}_{\perp}}{2\pi}\,\varphi + \boldsymbol{u}'(x, y) \tag{6}$$

where $\boldsymbol{u}'(x, y)$ is non-singular. All the vectors considered in equation (6) are $d$-vectors.

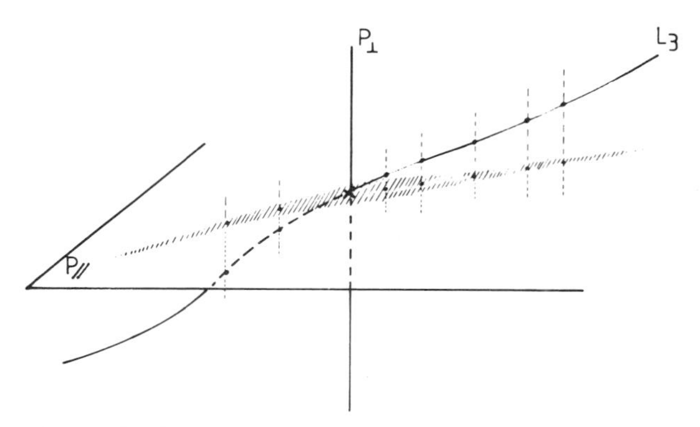


**Figure 6** Projecting a dislocation from $d$-space on the $d_{\|}$ physical space. Here $d = 3$, $d_{\|} = 2$, $d_{\perp} = 1$. Formation of a furrow.

The edge component $\boldsymbol{b}_e$ generates a field which is also translationally invariant along $P_{\|}$ and $L_{\|}$, but not rotationally invariant about them. However, the singular part is still the same, and we have

$$\boldsymbol{u}_e = \frac{\boldsymbol{b}_e}{2\pi}\,\varphi + \boldsymbol{u}''(x, y) \tag{7}$$

leading to a total displacement field in $d$-space

$$\boldsymbol{u} = \frac{\boldsymbol{b}}{2\pi}\,\varphi + \boldsymbol{u}_0(x, y) \tag{8}$$

where $\boldsymbol{u}_0(x, y)$ is non-singular. This expression is a valid approximation, at least in the vicinity of $L_d$, even if $L_d$ is curved.

Coming now to the dislocated physical space, we construct it as the intersection set of the displaced $\sigma_i$ with the same $P_\parallel$, fixed in $E_d$. Note that the $\sigma_i$ are not deformed along $P_\perp$, since $P_\perp$ is along a set of directions of translational symmetry for the dislocation. Therefore, if $\sigma_i$ is a piece of $P_\perp$, limited to the interior of each unit cell, the construction of the dislocated quasi-crystal becomes very easy: $P_\parallel$ and the strip are conserved, the hypercubic lattice is dislocated, and the vertices of the quasi-crystal are the projections of the vertices of the hypercubic lattice, which are inside the strip. Our choice of keeping $P_\parallel$ fixed spreads the phason field all over the sample. If we had made the opposite choice by considering that $P_\parallel$ is deformed with the dislocated hypercubic lattice the phason field would have been concentrated along the intersection of the cut hypersurface $\Sigma$ with $P_\parallel$, which is a 2D surface in $d_\parallel = 3$, a line in $d_\parallel = 2$. These concentrated phasons would have appeared much as a stacking fault; in this sense, the dispersed phasons look like a dispersed 'stacking' fault. But we prefer at this stage to use the term 'mismatch', which implies that the elementary tiles do not match rightly along the fault (as in figure 4), making a kind of singularity which is stronger than the usual stacking fault (where the local matching is correct, the order being broken only at some distance).

Other analytical approaches to the construction of dislocations can be found in Levine *et al* (1985), Socolar *et al* (1986) and Bohsung and Trebin (1987). See also Levine *et al* (1984) and Kléman *et al* (1986).

### 3.2 Dislocations in quasi-crystals are partials

Figure 7 represents the result of a numerical simulation done along the lines discussed above for $d_\parallel = 2$, $\boldsymbol{b} = (1, \bar{1}, 0, 0, 0)$. The core region has been left blank and is bounded by a loop made up of a set of edges which can be easily labelled along the directions of the edges of a regular pentagon representing the projections in $P_\parallel$ of the base vectors in $E_5$. When mapping the loop on the pentagon, the image loop fails to close precisely by a vector $\boldsymbol{b}_\parallel$. The localised phasons, which are unit mismatches, are marked.

Figure 8 represents a dislocation with a very small Burgers' physical component $\boldsymbol{b}_\parallel$ and a large component $\boldsymbol{b}_\perp$. Such dislocations are a sum of 'elementary' perfect dislocations. Take for example

$$\boldsymbol{b} = p(1, 0, \bar{1}, 0, 0) + q(0, 0, 0, 1, \bar{1}) \tag{9}$$

with $p, q$ integers. We find

$$b_\parallel = 2 \mid p\tau - q \mid \cos 54. \tag{10}$$

$b_\parallel$ can be made very small, in fact as close as one wishes to any small value fixed in advance, if $p = f_n$, $q = f_{n+1}$, $n \to \infty$, $f_n$ being the $n$th term of the Fibonacci sequence $\{1, 1, 2, 3, 5, 8, 13, 21, \ldots\}$. The Burgers vector of figure 8 is of that sort, with $p = 2$, $q = 3$.

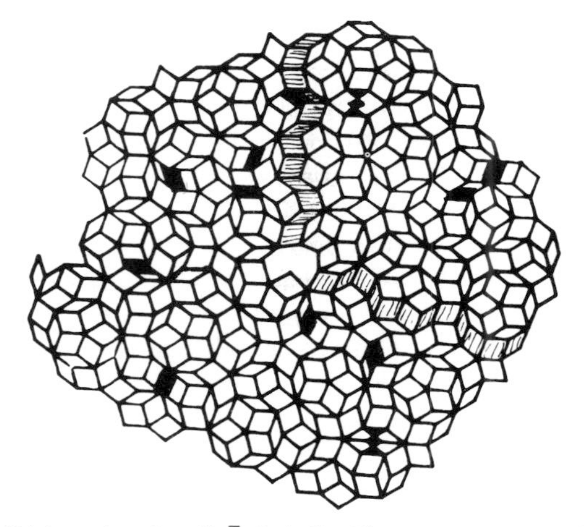

**Figure 7** Dislocation $\boldsymbol{b} = (1, \bar{1}, 0, 0, 0)$. The extra sectors of matter are marked, as well as the localised phasons. The components of $\boldsymbol{b}$ are ordered as in the drawing of the numbering of the five pentagonal directions.

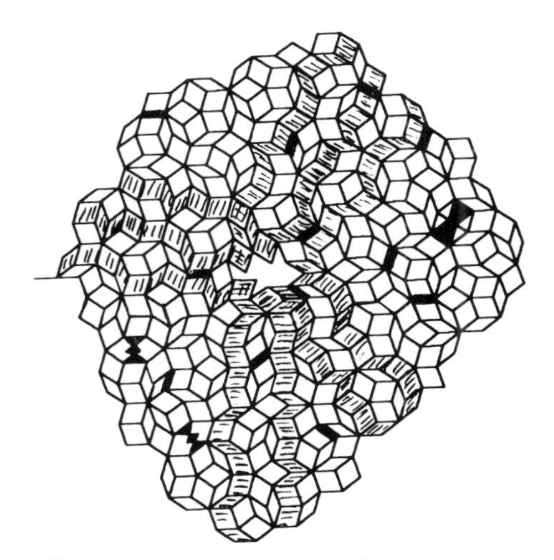

**Figure 8** Dislocation $\boldsymbol{b} = (\bar{2}, 3, \bar{3}, 2, 0)$.

The possibility of making physical Burgers vectors $\boldsymbol{b}_{\parallel}$ extremely small, if not vanishingly small, is due to the irrationality, and suggests that the dislocations of a quasi-crystal should be considered as *partial dislocations*. A direct way of convincing oneself of this status is to remark that the $\boldsymbol{b}_{\parallel}$ Burgers vectors of quasi-crystals become a true partial of their approximants, which can be described as periodic tilings made of the same tiles as the quasi-crystal itself (Entin-Wohlman *et al* 1988). A perfect dislocation of an approximant has a finite Burgers vector, which becomes infinite in the quasi-crystal. Garg (1988) has also argued that dislocations in quasi-crystals are surface rather than line defects.

A simple way of looking at this phenomenon is to construct directly a dislocation in a Penrose tiling by removing trails of constant thickness according to the non-zero components of $\boldsymbol{b}_{\parallel}$. Glueing back the lips of the trails shows that the edges do not match correctly any longer. The result of the Volterra process is clearly a stacking fault. Now, the various mismatches along the cut surface can diffuse away individually and the stacking fault be dispersed. This process presents a great analogy with the reshuffling process devised by Krönberg (1961) to describe the complex atomic rearrangements which occur in materials with more than one atomic species, the standard example being sapphire $Al_2O_3$, which crystallises in the HC system. Oxygen occupies the close-packed planes of the hexagonal structure, while the smaller ions of Al occupy the octahedral interstices. When a stacking fault $\boldsymbol{b}_1$ is introduced along a close-packed plane, octahedral interstices become tetrahedral interstices; the Al atoms have to move by an amount $\boldsymbol{b}_2$ to other interstices, a motion which is described by Krönberg as the motion of a dipole. Note that the analogy goes far: $\boldsymbol{b}_1 + \boldsymbol{b}_2 = \boldsymbol{b}$ is a perfect Burgers vector of the HC lattice, as well as $\boldsymbol{b}_{\parallel} + \boldsymbol{b}_{\perp}$ ($\boldsymbol{b}_{\parallel}$ identical to $\boldsymbol{b}_1$, $\boldsymbol{b}_{\perp}$ identical to $\boldsymbol{b}_2$). Furthermore the motion of a mismatch involves a rearrangement along a worm which can be truly represented by the motion of a dipole of dislocations (see section 4). It is also interesting to remark that, when the mismatches are dispersed, the stacking fault is no longer visible as a surface. But by making a loop around the dislocation, this loop sitting in the good quasi-crystal everywhere, and lifting this loop in $E_d$, one finds $\boldsymbol{b}_{\perp}$, which is the measure of a phase shift. Similarly, $\boldsymbol{b}_2$ can be thought of as a measure of the phase shift of the crystal due to a stacking fault.

### 3.3 Motion of a dislocation

The motion of a dislocation in a quasi-crystal is attended by a complex motion (analogous to Krönberg's synchroshear) of the cloud of localised phasons which surround the dislocation line and, probably, by a rearrangement of the core. 2D simulations (Bohsung and Trebin 1987, Kléman and Sommers 1990) show up clearly the collapse of mismatches of different signs, and creation of pairs, and the diffusion-like motion of these mismatches. However, these simulations, which are geometrical, cannot tell us anything about the properties of the plastic deformation of quasi-crystals. It has been advanced that, due to the probably small diffusivity constants of the mismatches (Lubensky 1988), the motion of dislocation should be difficult, the yield stress high, and consequently quasi-crystals should be brittle. It is known that Frank and Kasper phases are brittle (Frank 1978); similarly quasi-crystals seem indeed to break easily at room temperature. But nothing is known about their high-temperature behaviour (in their thermodynamic domain of existence). Recent annealing experiments in Al–Cu–Fe quasi-crystal, which result in x-ray peaks sharpening (Bancel 1989), seem to indicate a larger phason diffusivity constant than expected.

The categories of glide and climb are still true in quasi-crystals. Consider *glide*. We introduce the concept of a 'glide manifold' $G$ in $E_d$, a $(d-1)$-dimensional hyperplane which contains the Burgers vector $\boldsymbol{b}$ and the dislocation hyperline. $G$ intersects $P_\parallel$ along a $(d_\parallel - 1)$-manifold, i.e. a line when $d = 5$ and a plane when $d = 6$, as expected. This $(d_\parallel - 1)$-manifold $g$ is the usual glide line (or plane), since it has most of the properties attached to the usual glide plane: it contains $\boldsymbol{b}_\parallel$ and the dislocation line $L_\parallel$ itself, in the case where $L_d$ is defined as above as the direct product $L_d = L_\parallel \times P_\perp$ (in this case $g$ is at the same time the projection of $G$ on $P_\parallel$ and its intersection). $g$ does not contain $\boldsymbol{b}_\perp$, so that, even if the motion of $L_\parallel$ in the direction of $\boldsymbol{b}_\parallel$ in $g$ is conservative, it is accompanied by a phason rearrangement.

*Pure climb* is easily defined as a displacement along a unique direction $\boldsymbol{C}$ in $E_d$ which is perpendicular both to $L_d$ and to $\boldsymbol{b}$. Since $L_d$ contains $P_\perp$, $\boldsymbol{C}$ is perpendicular to $P_\perp$, and consequently belongs to the physical space $P_\parallel$. In fact it is easy to see that $\boldsymbol{C}$ is perpendicular to both $\boldsymbol{b}_\parallel$ and $L_\parallel$ in $P_\parallel$; pure climb in $P_\parallel$ is pure climb in the hypercubic lattice and equivalent to the usual notion of climb. Note that since the displacement of $L_d$ is along $P_\parallel$ in pure climb, no matter leaves or enters the strip (in the cut-and-projection method), except if due to elastic deformation, and we expect smaller phason rearrangements than in glide. A complete analysis remains to be done. If $L_d$ is not a direct product $L_\parallel \times P_\perp$, one expects rather more complex geometrical properties of glide and climb than in the case just discussed. In particular the glide plane $g$ would be replaced by a 'thick' plane, related to a core of a more complex nature as in figure 6.

Finally, aside this discussion which is of a pure geometrical nature, the character of the atomic motif should play an important role (Peierls lattice friction for the core, continuous or discrete phasons for the reshuffling).

### 3.4 Disclinations

As in solids, we expect disclinations (projected from $E_d$) to carry a very large elastic strain, and consequently a large elastic energy; they are therefore forbidden in principle, except perhaps those whose dihedral angle measured in $P_\parallel$ is very small (in $E_d$ this angle is an angle of symmetry of the hypercubic lattice), and which can be split into dislocations of small $\boldsymbol{b}_\parallel$ Burgers vector. See Bohsung and Trebin (1989) for a simulation of a disclination in a 2D Penrose tiling.

## 4 THE TOPOLOGICAL CLASSIFICATION OF DEFECTS IN QUASI-CRYSTALS

In this section we underline the topological classification of defects in quasi-crystals, by making use of the general principles of this classification (Toulouse and Kléman 1976, Michel 1980), which we do not develop here. Our discussion is in two steps. First we apply this method to the classification of defects in the $d$-dimensional hypercubic lattice; the corresponding defects with $P_{\parallel}$, as above. Apart from the dislocations and the disclinations which have already been discussed, and of walls in non-symmorphic lattices, we shall find no other defects of real interest, except those related to pathological intersections which are hard to understand at the present stage of our experimental knowledge of quasi-crystals. The second step consists in employing directly the topological method to $P_{\parallel}$ considered as a $d_{\parallel}$-dimensional boundary of a $d$-dimensional crystal; this method, which was suggested to us by Michel (1989), is a generalisation of Volovik's (1978) for surface singularities in usual ordered media. We find dislocations and disclinations, as above, but also another class of line defects (in $d = 6$) and point defects in ($d = 5$) which we identify as the topological defects of the phase, in fact the localised phasons which have been repeatedly alluded to above.

### 4.1 Topological classification of defects in the hyperlattice

The topologically stable (TS) defects of the hypercubic lattice in dimension $d$ are classified by the successive groups of homotopy of the order parameter space $V_d$, which is the quotient of the full Euclidean group in $E^d$

$$G_d = O(d) \square R_d \tag{11}$$

by the discrete group of symmetry of the hypercubic lattice

$$H_d = \Omega(d) \square Z^d \tag{12}$$

where $O(d)$ is the group of rotations in $d$ dimensions, $R_d$ the group of continuous translations, $\Omega(d)$ the hyperoctahedral group in $d$ dimensions, and $Z^d$ the Abelian group of discrete translations with $d$ generators. We have $V_d = G_d/H_d$. Kléman and Michel (1978) have given the standard methods to calculate all the homotopy groups $\pi_n(V_d)$, which classify the defects of dimension $d - n - 1$ in the hypercubic lattice, i.e. which cut $P_{\parallel}$ along defects of dimension $d_{\parallel} - n - 1$. We limit ourselves to the discussion of $n = 0$ (topological walls), $n = 1$ (line or point dislocations and disclinations), and $n = 2$ (point defects for $d_{\parallel} = 3$). Let us remark incidentally that the topological classification enables us to enumerate more types of defects than the Volterra process.

*n = 0*. $\pi_0(V_d)$ is trivial for all the hypercubic lattices whose group contains a centre of symmetry (no topologically stable walls) and is isomorphic to $Z_2$ if it does not. These walls annihilate by pairs and are hypersurfaces in $E^d$ and the analogues of the twins by reticular merihedry classified long ago by Friedel (1929).

*n = 1*. $\pi_1(V_d)$ is the lift in $\bar{G}_{d_0}$ (the double universal covering of $G_{d_0}$, which is the connected subgroup of $G_d$, i.e. $G_d$ without reflections) of $H'_d = H_d \cap G_{d_0}$. This yields

$$\pi_1(V_d) = \overline{\Omega_0(d)} \,\square\, Z^d \tag{13}$$

where $\Omega_0(d)$ is the subgroup of the hyperoctahedral group without reflections, and $\overline{\Omega_0(d)}$ is the lift of $\Omega_0(d)$ in $\bar{G}_{d_0}$. In practice, we are only interested in $Z^d$, which classifies the dislocations, and which is also the fundamental group of the $d$-torus $T_d$ (i.e. the unit cell of the hypercubic lattice with $(d-1)$-dimensional faces identified by the translations). $\overline{\Omega_0(d)}$ classifies the disclinations.

*n = 2*. The group $\pi_2(V_d)$ is always trivial, hence there are no corresponding topologically stable defects.

*n = 3*. We get $\pi_3(V_d) \simeq Z$ in any dimension $d$. Those defects are in $d = 5$ line defects, and in $d = 6$ surface defects; in both cases, they do not cut a $d_\parallel$ manifold generically. Therefore they would deform physical space, but in a manner which can be elastically relaxed, even if these defects are topologically stable in $E_d$. But we should bear in mind that they could generate defects of an unknown type in $P_\parallel$, if they cut it non-generically.

*n > 3*. The corresponding defects do not cut the physical space generically, and comments similar to those just above are in order. Note the case $n = d$, corresponding in $E_d$ to configurations; they have been discussed by Trebin (1983) for $d = 3$, and Bohsung and Trebin (1989) for quasi-crystals.

### 4.2 Topological classification of defects in the quasi-crystal

The foregoing discussion deals with the order parameter space of the hypercrystal in $E_d$. In fact, $P_\parallel$ carries a *reduced* order parameter space $V'_d$. We discuss first its nature. Consider the projection $U$ in $P_\perp$ of the unit cell of the hypercubic lattice ($U$ is a triacontahedron TR in the case $d = 6$ and a rhombic icosahedron RI in the case $d = 5$). Each of the points $A_\perp$ of $U$ represents a quasi-crystal which is made of the set of atoms at the intersection of $P_\parallel$ and of the set of motifs $\sigma_i$ ($P_\parallel$ intersects $U$ in $A_\perp$). All the points of $U$ are in correspondence with different $P_\parallel$, which differ one from the other by a phase shift and a translation. Since $U$ is the projection of the unit cell, all the inequivalent translations of the hyperlattice relevant for the quasi-crystal are represented in $U$, as well as all the inequivalent phase shifts which do not change the class of local isomorphism of the quasi-crystal (at least in the case $d = 6$; in the case $d = 5$, $A_\perp$ has to be restricted to some 2D subset of RI for each class of local isomorphism). Therefore $U$ is a part of the space of the reduced order parameter, with rotations excluded. We shall here restrict ourselves to this space smaller than the total $V_d$, since it contains most of the information we require to classify the current phase singularities. Now, since $U$ is the projection of the unit cell, it can be lifted back inside it, after the manner of Frenkel *et al* (1986), and be considered as a part of the unit cell, which is the order parameter space of the hypercubic crystal, rotations excluded. Of course, the equivalent faces of the unit cell have to be identified to make it a valid order parameter space $T_d$ (the $d$-dimensional torus), and similarly the equivalent faces of $U$ have to be identified.

$\pi_1(U)$ contains not only the dislocations (not the disclinations since rotations are excluded) of the physical space (in fact, those defects which are the intersections of the hyperline defects in $E_d$ with $P_\parallel$) but also other line defects in $P_\parallel$, as we shall see now. In fact, the most general method consists in employing the concepts of relative homotopy, since $U$ is a manifold which is included in $T_d$. This method has been used with success in comparing the classification of defects in the bulk and in the boundaries of 3D media (Volovik 1978). $P_\parallel$ is analogously the boundary of a domain in $E_d$. We have the exact sequence of homomorphisms between homotopy groups (Steenrod 1974):

$$\to \pi_2(U) \xrightarrow{i_2} \pi_2(T_d) \xrightarrow{j_2} \pi_2(T_d, U) \xrightarrow{\partial_2} \pi_1(U) \xrightarrow{i_1} \pi_1(T_d). \qquad (14)$$

In this sequence $i$ denotes the operator of inclusion of $U$ and $j$ is the operation of inclusion which consists in considering any 2D circuit in $U$ as belonging to $T_d$. As regards $\pi_2$ $(T_d, U)$, it is the group of the classes of equivalence (homotopically speaking) of loops $\Gamma$ in $P_\parallel$ capped by a 2D surface $\sigma$ belonging to $E_d$ (but not to $P_\parallel$) and homotopic to a half-sphere. $\sigma$ is on one side of $P_\parallel$. $\pi_2(T_d, U)$ therefore classifies the topologically stable line singularities of the quasi-crystal (with a 'Burgers' circuit $\Gamma$) which belong to a $(d-3)$-dimensional object in $E_d$. As mentioned above, such objects are *not* topologically stable in the hypercubic lattice. Finally, $\partial$ is the operation which consists in restricting $\sigma$ to its boundary $\partial\sigma = \Gamma$. The property of exactness means that the image of any homomorphism belonging to the sequence is the kernel of the following homomorphism; for example: im $\partial_2 =$ ker $i_1$ (cf figure 9). As an illustration, consider the icosahedral case $d = 6$; then $\pi_2(T_d, U)$ classifies $d_\perp$ $(= d - n - 1)$-dimensional non-TS defects in $E_d$ which cut $P_\parallel$ (non-generically) along a line, i.e. classifies some special line defects in $P_\parallel$ which are not dislocations or disclinations. These defects were not foreseen by the above topological classification which uses $T_d$ only; neither could they be discovered with the help of the restricted order parameter space $U$ alone. Since $\pi_2(T_d) = 1$, the exactness of the sequence implies that $\pi(T_d, U) = 1$, and the exactness of the sequence implies that $\pi_2(T_d, U)$ is a subgroup of $\pi_1(U)$:

$$\pi_2(T_d, U) = \text{im } \partial_2 = \ker i_1. \tag{15}$$

Therefore $\pi_1(U)$ contains as a subgroup the topologically stable line defects of the quasi-crystal classified by the relative homotopy group $\pi_2(T_d, U)$, as well as its dislocations. But ker $i_1$ (by definition) projects on the unit element of $\pi_1(T_d)$. Therefore the elements of $\pi_1(U)$ which classify dislocation and disclination lines map on an invariant subgroup im $i_1$ of $\pi_1(T_d)$.

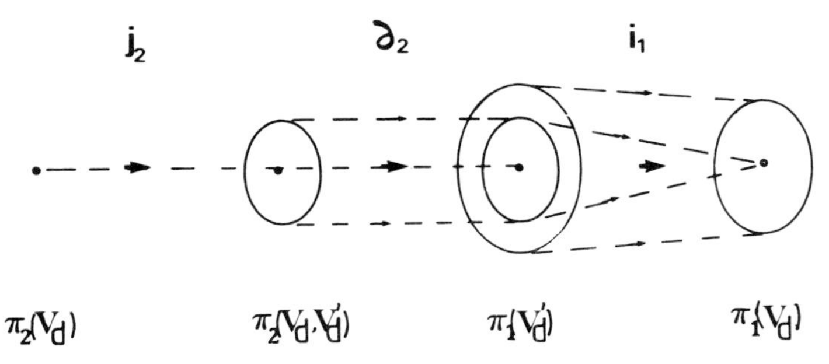


**Figure 9** A pictorial representation of the exact homotopy sequence of equation (14).

The line defects of equation (15) cannot be anything else than phase singularities. A first indication of this result is the fact that their Burgers vector is identically vanishing (this is the meaning of the homomorphism $i_1$). The various groups which appear in equation (14) for the case $d = 6$ and for the case $d = 5$ (which, as mentioned above, requires some special care) have been calculated (Kléman 1990b).

We have in figure 4 an illustration of some phase singularities in a Penrose tiling. Each mismatch is a topologically stable defect which can move from one site to another along a worm (by discrete jumps in the present Bravais model of the quasi-crystal) but cannot disappear. It might anneal by collapse with another mismatch of opposite sign. It is easy to visualise each mismatch as a dislocation dipole with a small stacking fault along the edge which carries the mismatch.

The exact homotopy sequence of equation (14) can be used, *mutatis* mutandis, with $V_d$ in place of $T_d$, and $V'_d$ in place of $U$. Then disclinations and phase singularities related to rotations should also appear. Bohsung and Trebin (1987, 1989) have defined $V'_d$ as a coset space $V'_d = G'_d/H'_d$, where $G'_d = \mathrm{SO}(d_\parallel) \times \mathrm{SO}(d_\perp) \,\square\, R_d$ is the product of the group of rotations in $P_\parallel$ by the group of rotations in $P_\perp$ by the group of translations $R_d$, and $H'_d = H_d \cap G'_d$ is the intersection of $G'_d$ with $H'_d$. Subsequently, they restrict their considerations to defects defined by the homotopy groups of $V_d$ only. We have not studied the relationship between their $V_d'$ and our $U$, which incidentally is not defined as a coset space of groups (an order parameter space is not necessarily a coset space (Michel 1989).

As I complete this paper, it is a real pleasure to recognise the scientific influence Sir Charles Frank's ideas and physical concepts have had on most of the aspects of my own work, and in particular on my approach to the theory of defects. His stay in Orsay twelve years ago was for me a time of high scientific friendly excitation.

## ACKNOWLEDGMENTS

I thank Dr C Sommers for permission to publish the computer images of the dislocations.

## REFERENCES

Alexander S and McTague J 1978 *Phys. Rev. Lett.* **41** 702–6
Bancel P A 1989 *Phys. Rev. Lett.* **63** 2741–4
Bernal J D 1964 *Proc. R. Soc.* A **280** 299–322
Bohsung J and Trebin H-R 1987 *Phys. Rev. Lett.* **58** 2277–80
—— 1989 *Aperiodicity and Order* vol 2, ed M V Jaric (New York: Academic) pp 183–221
de Bruijn N G 1981 *Kon. Nederl. Akad. Weterasch. Proc.* A **84** (*Indigat. Math* **43**) 39–66
Dam B, Janner A and Donnay J H D 1985 *Phys. Rev. Lett.* **55** 2301–4
Donnay J D H 1935 *Ann. Soc. Geol. Belg.* **18** B222
Dubost B, Lang J-M, Tanaka M, Sainfort P and Audier M 1986 *Nature* **324** 48
Duneau M and Katz A 1985 *Phys. Rev. Lett.* **54** 2688–91
Entin-Wohlmann O, Kléman M and Pavlovitch A 1988 *J. Physique* **49** 587–98
Frank F C 1952 *Proc. R. Soc.* A **215** 43–6
—— 1958 *Discuss. Faraday Soc.* **25** 19–28
—— 1978 Private communication
Frank F C and Kasper J S 1958 *Acta Crystallogr.* **11** 184–90
—— 1959 *Acta Crystallogr.* **12** 483–99
Frenkel D M, Henley C L and Siggia E D 1986 *Phys. Rev.* B **34** 3649–69
Friedel G 1929 *Leçons de Cristallographie* (Paris: Berger-Levrault)
Friedel J 1964 *Dislocations* (Oxford: Pergamon)
—— 1988 *Helv. Phys. Acta* **61** 538–56

Gardner M 1977 *Sci. Am.* **236** 110–21
Garg A 1988 *NSF-ITP preprint* 88–97
Haüy R-J 1784 *Essai d'une Théorie sur la Structure de Cristaux Appliquée à Plusieurs Genres de Substances Cristallines* Paris
Henley C L 1987 *Commun. Condens. Matter Phys.* **13** 59–117
Janssen T 1985 *Acta Crystallogr.* A **42** 261–71
Jaric M V 1985 *Phys. Rev. Lett.* **55** 607–10
Kalugin P A, Kitaey A and Levitov L 1985 *JETP Lett.* **41** 145–9
Kléman M 1988 *Quasi-crystalline Materials. Proc. ILL/CODEST Workshop* ed C H Janot and J M Dubois (Singapore: World Scientific) pp 318–26
—— 1990a *Adv. Phys.* **38** 605–68
—— 1990b *J. Physique* in press
Kléman M, Gefen Y and Pavlovitch A 1986 *Europhys. Lett.* **1** 61–9
Kléman M and Michel L 1978 *Phys. Rev. Lett.* **40** 1387–90
Kléman M and Sadoc J-F 1979 *J. Physique Lett.* **40** L569–74
Kléman M and Sommers C 1990 *Acta Metall.* in press
Kramer P 1982 *Acta. Crystallogr.* A **38** 257–64
Kramer P and Neri R 1984 *Acta Crystallogr.* A **40** 580–7
Krönberg M L 1961 *Acta Metall.* **9** 970–1
Kuo K H 1986 *J. Physique* **47** C3 425–36
Levine D, Lubensky T C, Ostlund S, Ramaswamy S, Steinhardt P J and Toner J 1985 *Phys. Rev. Lett.* **54** 1520–3
Levine D and Steinhardt P J 1984 *Phys. Rev. Lett.* **53** 2477–80
Levitov L S 1989 *J. Physique* **50** 3181–90
Lubensky T C 1988 *Aperiodicity and Order* vol 2, ed M V Jaric (New York: Academic) pp 199–280
Mackay A L 1962 *Acta Crystallogr.* **15** 916–18
—— 1981 *Kristallografiya* **26** 910–19 (*Sov. Phys–Cryst.* **26** 517–22)
Michel L 1980 *Rev. Mod. Phys.* **52** 617–51
—— 1989 Private communication
Mosseri R and Sadoc J-F 1985 *J. Non-Cryst. Solids* **75** 115–23
Nelson D R 1983 *Phys. Rev. Lett.* **28** 5515–35
Pacioli, Fra Luca 1980 *Divine Proportion* (Paris: Librairie du Compagnonnage)
Pauling L 1985 *Nature* **317** 512–14
Pavlovitch A, Gefen Y and Kléman M 1989 *J. Phys. A: Math. Gen* **22** 4347–73
Penrose R 1974 *Bull. Inst. Math. Appl.* **10** 266–71
—— 1979 *Math. Intell.* **2** 32–8
Rivier N 1986 *J. Physique* **47** C3 299, 309
Romé de lsle J-B 1783 *Cristallographie ou Description des Formes Propres à tous les Corps du Règne Minéral* (Paris: Imprimerie de Monsieur)
Rosolato G 1985 *Eléments de l'interprétation* (Paris: Gallimard) pp 133–65
Sadoc J-F 1983 *J. Physique Lett.* **44** L707–15
Shechtman D, Blech I, Gratias D and Cahn J W 1984 *Phys. Rev. Lett.* **53** 1951–4
Socolar J E S, Lubensky T C and Steinhardt P J 1986 *Phys. Rev.* B **34** 3345–59
Steenrod N 1974 *The Topology of Fiber Bundles* (Princeton: Princeton University Press)
Steinhardt P J and Ostlund S (ed) 1987 *The Physics of Quasicrystals* (Singapore: World Scientific)

Toulouse G and Kléman M 1976 *J. Physique Lett.* **37** L149–51
Trebin H R 1983 *Phys. Rev. Lett.* **50** 1381–4
Volovik G E 1978 *JETP Lett.* **28** 59

# The Biaxial Nematic Liquid Crystal

S Chandrasekhar

## 1 INTRODUCTION

As is well known, Sir Charles Frank's paper 'On the Theory of Liquid Crystals' (Frank 1958), which he presented at a Discussion of the Faraday Society in 1958, was a landmark in the development of the subject. The paper is remarkable for its prescience and contains ideas that have a bearing even today on problems of current interest. We give just one example.

The ferroelectric liquid crystal, which was discovered by Meyer *et al* (1975), is now attracting considerable attention largely because of its enormous practical importance as a material for fast electro-optical switching (Clark and Lagerwall 1980; see, for example, *Proc. 1st* and *2nd Int. Conf. on Ferroelectric Liquid Crystals* 1988, 1990). R B Meyer (1977) has given the following account of the logical steps that led him to look for spontaneous polarisation in the chiral smectic C phase:

> In the early spring of 1974, I realized that by an unusual combination of symmetry properties, smectic C liquid crystals composed of chiral molecules ought to be ferroelectric. This discovery grew out of my attempt, at that time, to review the most general possible coupling between polar ordering and spontaneous curvature in liquid crystals. F. C. Frank had pointed out in 1958 that molecular chirality was responsible for the spontaneous torsion of cholesteric liquid crystals, and that a state of uniform torsion of the nematic director field could be achieved by a simple helicoidal structure, filling three-dimensional space without the necessity of any systematic defect structures. In the same article Frank explored the possible consequences of a spon-

taneous polarization parallel to the director. In that case, a spontaneous splay curvature is induced, but as Frank indicated, a three-dimensional structure containing uniform splay and filling space without defects does not exist. The spontaneous splay could only be achieved in combination with other curvature, in an inhomogeneous structure. The energy associated with these structural complexities could be high enough to prevent the appearance of a spontaneously splayed state as evidence of the presence of polar symmetry along the director. The contrast with a cholesteric is significant; even the slightest chiral perturbation of a nematic produces an observable torsion.

The third case of interest is that of spontaneous bending curvature coupled to a polarization normal to the director, which has been considered in connection with the flexoelectric effect. Is there a space filling structure containing uniform bending of the director field? There is; if one starts with a cholesteric structure, and adds a constant component of the director parallel to the helix axis, the result is a state of uniform torsion and bending, the relative amounts of bend and twist being determined by the tilt angle of the director relative to the helix axis. Since this is the structure of the helix in a chiral smectic C, I realized that in fact there must be a polarization associated with that helix.

Because the spontaneous torsion and bending in the helix, and the associated polarization, are all perturbations of the same order of magnitude, it seemed there must be some fundamental coupling of molecular chirality and polarization in a smectic C system. The idea of preferential orientation of molecules in the smectic C structure, suggested by W. L. McMillan, led me to realize that this coupling of chirality and polarization occurs at the molecular level, and is independent of the presence of the helix. Although the final result can be stated most clearly in molecular terms, it was Frank's ideas on macroscopic symmetry that stimulated the initial discovery.

In the present paper we consider another newly discovered phase of fundamental interest, namely the biaxial nematic liquid crystal, the possibility of which was in fact envisaged by Frank in 1958.

## 2 THE BIAXIAL NEMATIC PHASE

Fluidity (in the sense that no shear stress can persist in the absence of flow) is in principle compatible with biaxial orientational order, with or without translational order in one dimension.

F C Frank (1958)

The biaxial nematic ($N_b$) phase was first identified by Yu and Saupe (1980) as recently as 1980 in a ternary amphiphilic system composed of potassium laurate, 1-decanol and $D_2O$. In such systems the constituent units are molecular aggregates, called micelles, whose size and shape are sensitive to the temperature and concentration; the $N_b$ phase was found to occur over a range of temperature and concentration. There are obvious advantages in having a single-component, low-molar-mass thermotropic $N_b$

phase. The suggestion was made (Chandrasekhar 1985) that a convenient way of achieving this would be by preparing a hybrid mesogen that combines the features of the rod and the disc. This has proved to be efficacious and the $N_b$ phase has been observed in relatively simple compounds (Chandrasekhar *et al* 1988a,b, 1990, Galerne 1988 and references therein, Praefcke *et al* 1990).

### 2.1 Identification of the $N_b$ phase

We first describe the experimental evidence establishing the biaxiality of the thermotropic nematic phase of bis[1-(p-n-decylbiphenyl)3-(p-ethoxyphenyl) propane-1,3-dionato] copper(II), a paramagnetic complex that was prepared in this laboratory and studied in some detail (Chandrasekhar *et al* 1986, 1988a,b, 1990). The structural formula of this complex (hereafter referred to as complex A) is given in figure 1. Its transition temperatures are: melting transition 186.6 °C and (monotropic) isotropic–nematic transition 168.5 °C.

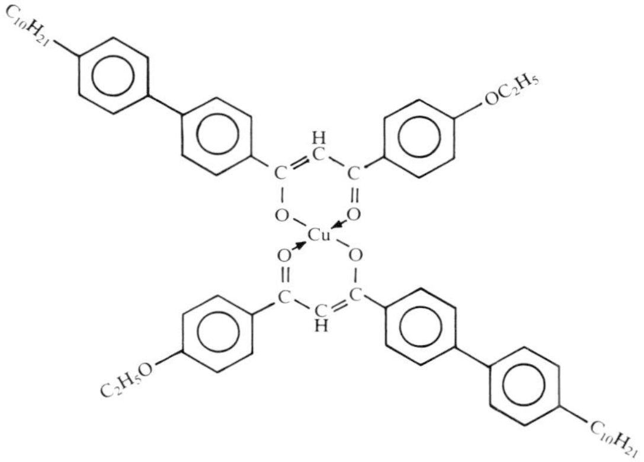

**Figure 1** Structural formula of complex A.

The mesophase shows the usual schlieren texture of the nematic, except that often the pattern consists entirely of $|s| = \frac{1}{2}$ (two-brush) disclinations (figure 2). Zig-zag disclinations are also seen, but only very rarely. Conoscopic observations were made on thick samples ($\sim 125$ μm), aligned homeotropically by the combined effect of silane coating and a 3 kHz AC electric field. The sample was sandwiched between two cover slips (each of thickness $\sim 100$ μm), the *external* surfaces of which were coated with tin oxide, which served as electrodes, and the *internal* surfaces with silane. The alignment was checked by visual observation as well as by measuring the intensity of light transmitted by the sample between crossed polaroids under orthoscopic conditions using a He–Ne laser and a photo-diode. For 'perfect' alignment there was almost complete extinction and the

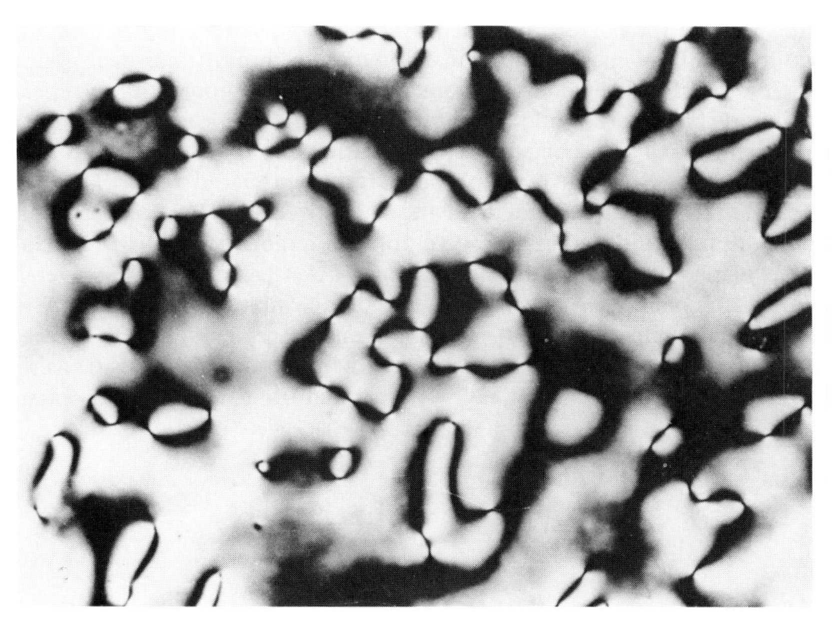

**Figure 2** Schlieren texture of the nematic phase of complex A.

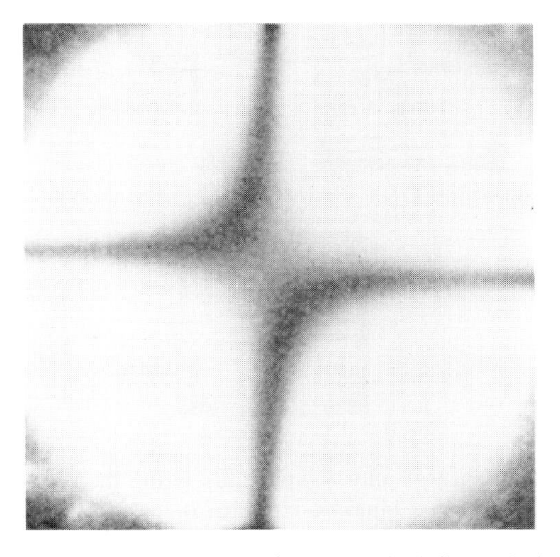

**Figure 3** Conoscopic figure showing the biaxiality of the nematic phase of complex A. $T_{NI} - T = 1.5\,^{\circ}C$. Film thickness $\sim 125\ \mu m$, homeotropic alignment. Numerical aperture of the objective $= 0.40$.

transmitted intensity was equal to that for the isotropic phase. The saturation voltage for perfect alignment was usually above 200 V across a film of thickness 125 $\mu$m. No electrohydrodynamic motion was seen in pure samples. There was evidence of some chemical decomposition on repeated heating of the material, and therefore only fresh samples were used for the experiments. All the conoscopic observations were found to be reproducible with well-aligned samples in freshly prepared cells. The conoscopic pattern was independent of the applied voltage for voltages greater than the saturation value. We were able to demonstrate (i) the biaxiality of the nematic phase of the pure complex A (figure 3), (ii) a reversible uniaxial–biaxial ($N_u$–$N_b$) transition in a binary mixture of complex A with the uniaxial nematogen 4′-n-pentyl-4-cyano-p-terphenyl (5CT, figure 4) and (iii) the variation of the biaxiality with temperature near this transition. We also obtained the $I$–$N_u$–$N_b$ phase diagram for this binary system (figure 5).

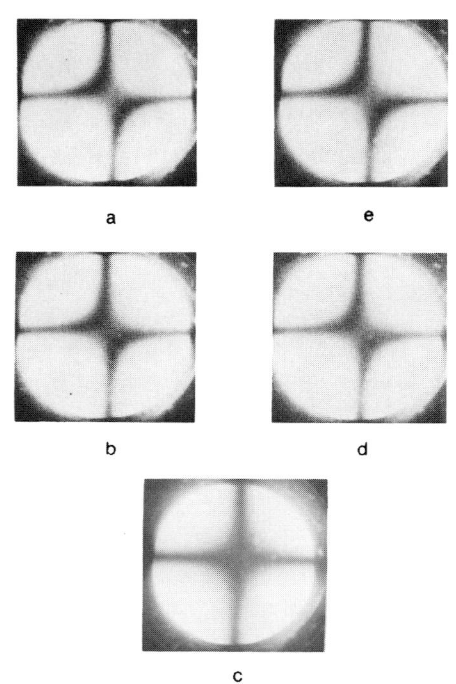


**Figure 4** Sequence of photographs illustrating the reversibility of the $N_b$–$N_u$ transition in a binary mixture of 0.25% (by weight) of 5CT in A. Heating mode: (*a*) $N_b$ at 165.8 °C, (*b*) $N_b$ at 167.0 °C, (*c*) $N_u$ at 167.8 °C. Cooling mode: (*d*) $N_b$ at 167.2 °C, (*e*) $N_b$ at 166.3 °C. The biaxiality can be seen to decrease on approaching the $N_b$–$N_u$ transition at 167.5 °C. Film thickness ~125 $\mu$m, homeotropic alignment (from Chandrasekhar *et al* 1988a).

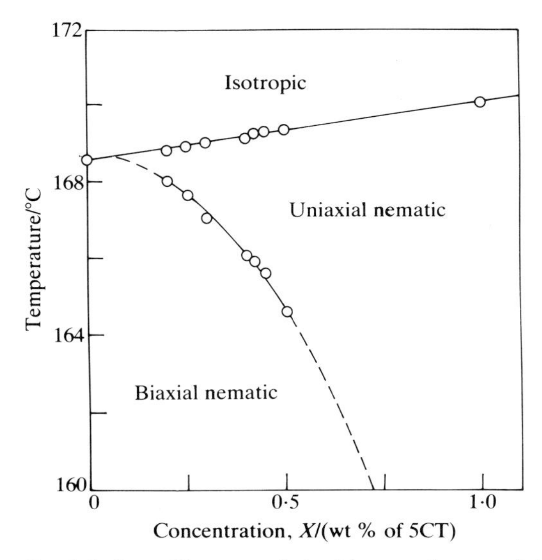


**Figure 5** Partial phase diagram of the binary mixture of A and 5CT in the concentration range 0–1% 5CT (by weight), showing the $I$–$N_u$ and $N_u$–$N_b$ phase boundaries. The broken portions of the curve represent an interpolation or an extrapolation (from Chandrasekhar *et al* 1988a).

Confirmatory evidence was obtained by x-ray studies (Chandrasekhar *et al* 1990). X-ray diffraction photographs were taken of aligned samples (aligned in a magnetic field of strength 1.8 T) using filtered $CuK_\alpha$ radiation and a flat film, and intensity scans were obtained with a Joyce–Loebl microdensitometer. For comparison, we first present in figure 6 (top) the intensity scans along the meridional and equatorial directions (parallel and perpendicular to the magnetic field, respectively) for the uniaxial nematic phase of 4′-n-octyloxy-4-cyanobiphenyl (80CB). As expected there are two pairs of diffuse (liquid-like) peaks, the low-angle meridional peaks corresponding to the mean repeat distance along the director axis, and the high-angle equatorial ones to the mean lateral intermolecular spacing. Figure 6 (bottom) gives the scans for the $N_b$ phase of complex A. There is now an additional pair of diffuse peaks in the equatorial scan, as would be expected say from an orthorhombic fluid. (It may be recalled that the sample was unaligned in the equatorial plane.) Thus the x-ray evidence supports the conclusions drawn from optical observations. Very similar results have been obtained by Ebert *et al* (1988) in their x-ray study of the $N_b$ phase of a sanidic discotic polymer. (Evidence of biaxiality in certain nematic polymers has also been presented by Windle *et al* (1985) and Hessel and Finkelmann (1986).)

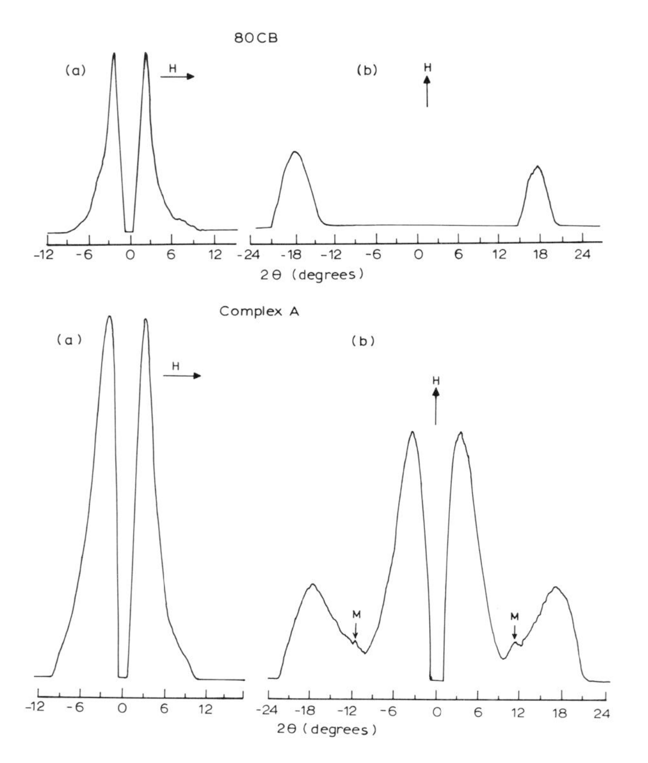


**Figure 6** Raw microdensitometer scans of the x-ray intensity plotted against diffracting angle ($2\theta$) for magnetically aligned nematic samples: top, the uniaxial nematic phase of 8OCB at 77 °C; bottom, the biaxial nematic phase of complex A at 166.5 °C. (*a*) Meridional scan (parallel to $H$), (*b*) equatorial scan (perpendicular to $H$). M represents the diffraction peaks from the mylar film which covered the windows of the sample holder and heater assembly (from Chandrasekhar *et al* 1990).

## 2.2 Elastic constants of an orthorhombic nematic

Several continuum theories have been developed for biaxial nematics (see Kini and Chandrasekhar (1989) for a full list of references on continuum theories) but the one formulated by Saupe (1981) and rederived by Kini (1984a,b) appears to us to be the most satisfactory. We write down the basic equations for an incompressible orthorhombic nematic and indicate how some of the material parameters may be determined experimentally (Kini and Chandrasekhar (1989, 1990).

The elastic free energy density of the classical uniaxial nematic ($N_u$) is given by (Frank 1958)

$$W_{\text{elast}} = [k_{11}(\nabla \cdot \boldsymbol{n})^2 + k_{22}(\boldsymbol{n} \cdot \nabla \times \boldsymbol{n})^2 + k_{33}(\boldsymbol{n} \times \nabla \times \boldsymbol{n})^2] \qquad (1)$$

where $\boldsymbol{n}$ in the director and $k_{11}$, $k_{22}$ and $k_{33}$ are the Frank constants for splay, twist and bend, respectively. In the orthorhombic case, the preferred molecular orientation is described by an orthonormal triad of director fields $\boldsymbol{a}, \boldsymbol{b}, \boldsymbol{c}$ and (1) takes the form

$$W_{\text{elast}} = \sum_{a} [k_{aa} \mid (\boldsymbol{a} \cdot \nabla \boldsymbol{b}) \cdot \boldsymbol{c} \mid^2 + k_{\text{ab}} \mid (\boldsymbol{a} \cdot \nabla \boldsymbol{a}) \cdot \boldsymbol{b} \mid^2 + k_{ac} \mid (\boldsymbol{a} \cdot \nabla \boldsymbol{a}) \cdot \boldsymbol{c} \mid^2$$

$$+ 2C_{ab} (\boldsymbol{a} \cdot \nabla \boldsymbol{a}) \cdot (\boldsymbol{b} \cdot \nabla \boldsymbol{b}) + 2k_{0a} \nabla \cdot (\boldsymbol{a} \cdot \nabla \boldsymbol{a} - \boldsymbol{a} \nabla \cdot \boldsymbol{a})] . \quad (2)$$

There are 12 $k$ (of which $k_{0a}$, $k_{0b}$ and $k_{0c}$ contribute only to the surface torque) and 3 $C$. Assuming that the initial alignment of the undistorted nematic is given by $\boldsymbol{a}_0 = (1, 0, 0)$, $\boldsymbol{b}_0 = (0, 1, 0)$ and $\boldsymbol{c}_0 = (0, 0, 1)$ in cartesian coordinates, and that the distortions are small, the 9 $k$ may be interpreted as follows:

$k_{aa}$ : twist of $(\boldsymbol{b}, \boldsymbol{c})$ about $x$
$k_{bb}$ : twist of $(\boldsymbol{c}, \boldsymbol{a})$ about $y$
$k_{cc}$ : twist of $(\boldsymbol{a}, \boldsymbol{b})$ about $z$
$k_{ab}$ : bend of $\boldsymbol{a}$ in the $xy$ plane
$k_{bc}$ : bend of $\boldsymbol{b}$ in the $yz$ plane
$k_{ca}$ : bend of $\boldsymbol{c}$ in the $xz$ plane
$k_{ba}$ : splay of $\boldsymbol{a}$ in the $xy$ plane
$k_{cb}$ : splay of $\boldsymbol{b}$ in the $yz$ plane
$k_{ac}$ : splay of $\boldsymbol{c}$ in the $xz$ plane.

The constants $C_{ab}, C_{bc}, C_{ca}$ are somewhat more difficult to describe as they involve simultaneous distortions of $\boldsymbol{a}$ and $\boldsymbol{b}$, or $\boldsymbol{b}$ and $\boldsymbol{c}$, or $\boldsymbol{c}$ and $\boldsymbol{a}$. For the enantiomorphic (cholesteric) phase, there are three additional terms

$$\sum_{a} k_a (\boldsymbol{a} \cdot \nabla \boldsymbol{b}) \cdot \boldsymbol{c}$$

where $k_a, k_b, k_c$ can be related to the equilibrium pitch; for example, for a cholesteric with its helical axis along $\boldsymbol{a}$, the pitch $P_a = 2\pi k_{aa}/k_a$.

In the presence of magnetic and electric fields,

$$W_H = -\tfrac{1}{2} \sum_{a} \chi_a (\boldsymbol{H} \cdot \boldsymbol{a})^2$$

$$W_E = -\sum_{a} \varepsilon_a (\boldsymbol{E} \cdot \boldsymbol{a})^2 / 8\pi$$

where $\chi_a, \chi_b, \chi_c$ are the principal dia- or paramagnetic susceptibilities along $\boldsymbol{a}, \boldsymbol{b}, \boldsymbol{c}$ and $\varepsilon_a, \varepsilon_b, \varepsilon_c$ are the corresponding dielectric susceptibilities. The total free energy density is then

$$W = W_{\text{elast}} + W_H + W_E.$$

Minimisation of the free energy with respect to $\boldsymbol{a}, \boldsymbol{b}, \boldsymbol{c}$ leads to the three

Euler–Lagrange equations

$$\sum_{a} \{a_i[(\partial W/\partial a_{k,j}),_j - (\partial W/\partial a_k)] - a_k[(\partial W/\partial a_{i,j}),_j - (\partial W/\partial a_i)]\} = 0$$

where the comma denotes partial differentiation.

In principle, some of the elastic constants can be determined by studying the response of the system to external magnetic and electric fields. As an example, let us consider one case in some detail. To simplify the discussion we suppose that $\boldsymbol{a}$ defines the preferred direction of orientation of the long molecular axis. We suppose further that by using standard techniques of surface treatment $\boldsymbol{a}$ can be firmly anchored homeotropically (i.e. normal to the bounding surfaces) and that the interactions of $\boldsymbol{b}$ and $\boldsymbol{c}$ with the boundaries are comparatively small. We also assume, for convenience, that the magnetic and dielectric susceptibilities satisfy the inequalities

$$\chi_a > \chi_b > \chi_c \qquad \varepsilon_a > \varepsilon_b > \varepsilon_c.$$

Consider the sample between glass plates $x = \pm h$, with $\boldsymbol{a}$ aligned homeotropically. When a magnetic field of moderate strength is applied along $y$, $\boldsymbol{b}$ will initially orient itself along $y$, since $\boldsymbol{b}$ and $\boldsymbol{c}$ are assumed not to have much interaction with the boundaries, and $\chi_b > \chi_c$. On increasing the field a Freedericksz transition will take place with the director field distorted in the $(\boldsymbol{a}, \boldsymbol{b})$ plane. The threshold field is given by

$$H_{y1} = (\pi/2h)\,(k_{ab}/\chi_{ab})^{1/2}$$

where $\chi_{ab} = \chi_a - \chi_b$.

If, on the other hand, $H = (H_x, 0, 0)$ so that the initial homeotropic orientation of $\boldsymbol{a}$ is stabilised, and a destabilising electric field $E = (0, E_y, 0)$ is applied in the $(\boldsymbol{b}, \boldsymbol{c})$ plane, $\boldsymbol{b}$ will align itself along $E_y$ since $\varepsilon_b > \varepsilon_c$. On increasing $E$, a Freedericksz transition will take place. The threshold is now given by

$$(E_{y1}h)^2 = \pi^3(k_{ab}/\varepsilon_{ab}) + 4\pi(H_xh)^2(\chi_{ab}/\varepsilon_{ab})$$

where $\varepsilon_{ab} = \varepsilon_a - \varepsilon_b$. A plot of $(E_{y1}h)^2$ (which is one-quarter of the square of the threshold voltage) against $4\pi(H_xh)^2$ results in a straight line of slope $(\chi_{ab}/\varepsilon_{ab})$. The intercept of this line with the $(E_{y1}h)^2$ axis gives an estimate of $k_{ab}/\varepsilon_{ab}$. Further, by studying the deformation above the threshold one can determine the ratio $k_{ba}/k_{ab}$.

Similarly, other sample geometries, and other techniques (such as light scattering), can be used to measure the $k$. However, the determination of the $C$ is not so straightforward, but estimates of their magnitudes may be possible by employing tilted boundary conditions (Kini and Chandrasekhar 1989).

### 2.3 Viscosity coefficients of an orthorhombic nematic

The $N_u$ phase has six viscosity (or Leslie) coefficients, which reduce to five if one assumes Onsager's reciprocal relations, whereas the orthorhombic nematic has twelve (Saupe 1981, Kini 1984a). The viscous stress tensor $\sigma'_{ji}$ in the latter case is given by

$$2\sigma'_{ji} = \sum_a [\eta_{aaaa}(a_i d_{ij} a_j)\,\delta_{ij} + a_i a_j(\bar{\eta}_{aabb} b_i d_{ij} b_j + \bar{\eta}_{ccaa} c_i d_{ij} c_j) + a_i b_j\,(\bar{\eta}_{abab} a_i d_{ij} b_j + \bar{\gamma}_{cab} N_1 c_1) + a_j b_i(\eta'_{abab} a_i d_{ij} b_j + \gamma'_{cab} N_1 c_1)]$$

$$\bar{\eta}_{aabb} = 2\eta_{aabb} - \eta_{aaaa} - \eta_{bbbb} \qquad \bar{\eta}_{abab} = 4\eta_{abab} + \gamma_{cab}$$

$$\eta'_{abab} = 4\eta_{abab} - \gamma_{cab} \qquad \bar{\gamma}_{cab} = \gamma_{cab} + \gamma_{cc}$$

$$\gamma'_{cab} = \gamma_{cab} - \gamma_{cc} \qquad d_{ij} = (v_{i,j} + v_{j,i})/2$$

$$N_i = \sum_a (\dot{b}_k - \omega_{kl} b_l) c_k a_i \qquad \omega_{ij} = (v_{i,j} - v_{j,i})/2$$

where the $\bar{\eta}_{aabb}$, $\eta_{abab}$, $\gamma_{abc}$ and $\gamma_{aa}$ are twelve viscosity coefficients ($\gamma_{aa}$, $\gamma_{bb}$, $\gamma_{cc}$ being the twist viscosities about $\boldsymbol{a}, \boldsymbol{b}$, and $\boldsymbol{c}$ respectively), a comma denotes partial differentiation and a superposed dot the material time derivative.

Different techniques can be used to determine the viscosity coefficients, namely measurements of (i) the effective viscosity under shear, (ii) ultrasonic reflection, (iii) damped oscillations in a magnetic field, (iv) flow alignment and (v) the dynamics above the Freedericksz threshold, and it has been shown that, in principle, all 12 coefficients of $N_b$ can be determined (Kini and Chandrasekhar 1990).

An an illustration let us consider the simplest case, namely the Miesowicz experiment. While $N_u$ has only three Miesowicz coefficients, $N_b$ has six, which are set out in table 1. In all of them $\boldsymbol{a}$, $\boldsymbol{b}$, $\boldsymbol{c}$ have been assumed to be aligned along $x$, $y$, $z$, respectively: $\boldsymbol{a}$ is aligned first by the application of a strong $H$ (or $E$) along $x$ and also by surface treatment. Then $\boldsymbol{b}$ is aligned with $E$ (or $H$) of moderate strength along $y$. A measurement of the viscous response to an imposed low shear rate by the oscillating plate method (Miesowicz 1946) or by flow in a rectangular capillary (Gähwiller 1973, O'Neill 1986 and references therein) yields the effective viscosity for the given geometry (ignoring the effects of director gradients).

**Table 1** The Miesowicz coefficients for an orthorhombic nematic.

| Velocity | Plates | Velocity gradient, $\nabla v$ | Viscous shear stress, $\sigma'$ | $\eta = \sigma' / \nabla v$ |
|---|---|---|---|---|
| $x$ | $y = \pm h$ | $v_{x,y}$ | $\sigma'_{yx}$ | $\tau_{abab}$ |
| $y$ | $x = \pm h$ | $v_{y,x}$ | $\sigma'_{xy}$ | $\zeta_{abab}$ |
| $y$ | $z = \pm h$ | $v_{y,z}$ | $\sigma'_{zy}$ | $\tau_{bcbc}$ |
| $z$ | $y = \pm h$ | $v_{z,y}$ | $\sigma'_{yz}$ | $\zeta_{bcbc}$ |
| $z$ | $x = \pm h$ | $v_{z,x}$ | $\sigma'_{xz}$ | $\tau_{caca}$ |
| $x$ | $z = \pm h$ | $v_{x,z}$ | $\sigma'_{zx}$ | $\zeta_{caca}$ |

$\tau_{abab} = \frac{1}{4}(4\eta_{abab} + \gamma_{cc} + 2\gamma_{cab})$; $\zeta_{abab} = \frac{1}{4}(4\eta_{abab} + \gamma_{cc} - 2\gamma_{cab})$.

## ACKNOWLEDGMENTS

I am very grateful to Dr U D Kini and Mr V N Raja for their valuable help in the preparation of this paper.

## REFERENCES

Chandrasekhar S 1985 *Plenary Lecture, 10th Int. Liquid Crystal Conf., York, July 1984. Mol. Cryst. Liquid Cryst.* **124** 1

Chandrasekhar S, Raja V N and Sadashiva B K 1990 *Mol. Cryst. Liquid Cryst. Lett.* **7** 65

Chandrasekhar S, Ratna B R, Sadashiva B K and Raja V N 1988a *Mol. Cryst. Liquid Cryst.* **165** 123

Chandrasekhar S, Sadashiva B K, Ramesha S and Srikanta B S 1986 *Pramana* **27** L713

—— 1988b *Pramana* **30** L491

Clark N A and Lagerwall S T 1980 *Appl. Phys. Lett.* **36** 899

Ebert M, Hermann-Schönherr O, Wendorff J H, Ringsdorf H and Tschirner P 1988 *Makromol. Chem. Rapid Commun.* **9** 445

Frank F C 1958 *Discuss. Faraday Soc.* **25** 19

Gähwiller Ch 1973 *Mol. Cryst. Liquid Cryst.* **20** 301

Galerne Y 1988 *Mol. Cryst. Liquid Cryst.* **165** 147

Hessel F and Finkelmann H 1986 *Polym. Bull.* **15** 349

Kini U D 1984a *Mol. Cryst. Liquid Cryst.* **108** 71

—— 1984b *Mol. Cryst. Liquid Cryst.* **112** 265

Kini U D and Chandrasekhar S 1989 *Physica* **156A** 364

—— 1990 *Mol. Cryst. Liquid Cryst.* **179** 27

Malthete J, Liébert L, Levelut A M and Galerne Y 1986 *C. R. Acad. Sci. Paris* **303** 1073

Meyer R B 1977 *Plenary Lecture, 6th Int. Liquid Crystal Conf., Kent State University, August 1976. Mol. Cryst. Liquid Cryst.* **40** 33

Meyer R B, Liébert L, Strzelecki L and Keller P 1975 *J. Physique Lett.* **36** L69

Miesowicz M 1946 *Nature* **158** 27

O'Neill G J 1986 *Liquid Cryst.* **1** 271

Praefcke K, Kohne B, Gündoğan B, Demus D, Diele S and Pelzl G 1990 *Mol. Cryst. Liquid Cryst. Lett.* **7** 27

Saupe A 1981 *J. Chem. Phys.* **75** 5118

Windle A H, Viney C, Golombok R, Donald A M and Mitchell G R 1985 *Faraday Discuss. Chem. Soc.* **879** 55

Yu L J and Saupe A 1980 *Phys. Rev. Lett.* **45** 1000

1988 *Proc. 1st Int. Conf. on Ferroelectric Liquid Crystals Bordeaux-Arcachon, September 1987 Ferroelectrics* **84/85**

1990 *Proc. 2nd Int. Conf. on Ferroelectric Liquid Crystals, Goteborg, June 1989, Ferroelectrics* in press

# Chain-Folded Crystallisation of Polymers From Discovery to Present Day: A Personalised Journey

A Keller

## 1 INTRODUCTION: RECOGNITION OF CHAIN FOLDING

When writing about polymer crystallisation for the present volume I have been facing the choice between reminiscing on personal experiences, particularly on those relating to Charles Frank's part in the subject, between reviewing the subject and between presenting new material at the forefront of the field. Not having been able to choose, I am now attempting to do all three in one article in the belief that only this can do justice both to the subject and Sir Charles' part in it.

Over more than three decades polymer crystals have presented us with never-ending surprises and they still continue to do so today. The continual stream of new features emerging have repeatedly defied *a priori* expectations, yet at the same time they remained within one aspect or other parts of the established discipline of crystal growth. This dual feature has been singularly appropriate to attract Sir Charles, always eager for new intellectual challenges, while keeping to the solid foundations of exact science. It is this dual feature, the surprise at the unforeseen and links to the main body of crystal studies, which I choose as the guide theme for the scientific part of this paper.

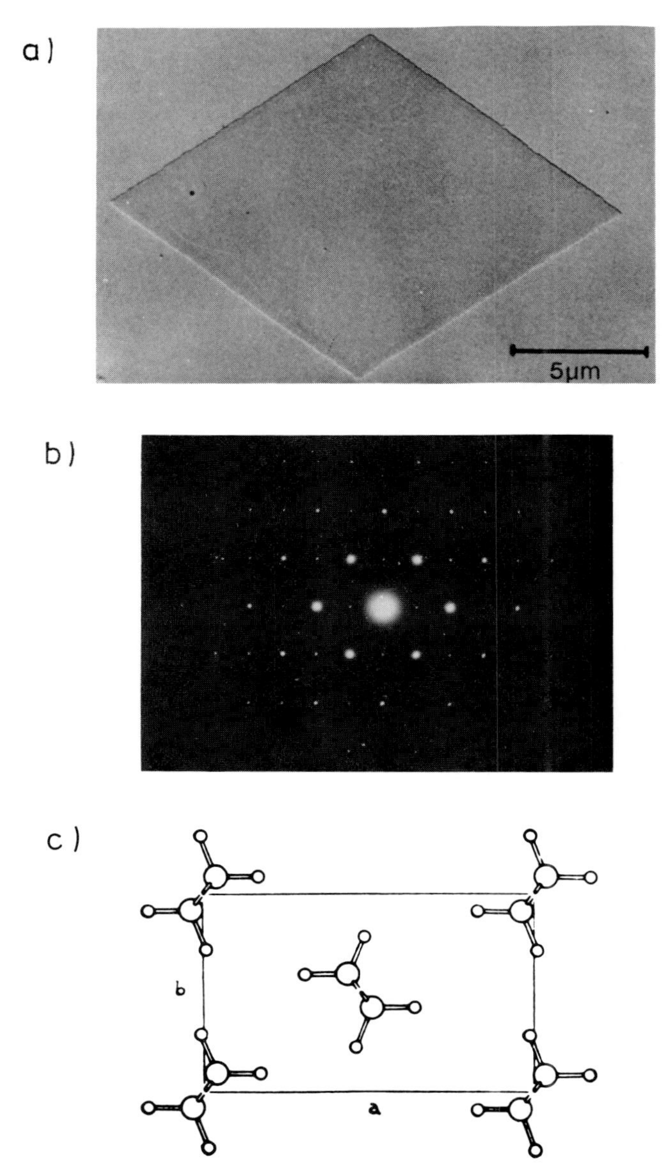


**Figure 1** The basic facts leading to the recognition of chain folding. (*a*) A monolayer single crystal of polyethylene (crystallised from solution) with uniform layer thickness of ~10 nm (electron micrograph). (*b*) Electron diffraction pattern given by same crystal as in (*a*). (*c*) Unit cell of polyethylene as seen along the chain direction corresponding to (*b*); hence chains are perpendicular to layer in (*a*). (These particular illustrations, as due to Dr Sally Organ, originate from a later date.)

First, to introduce the central subject of this article, consider the chain-folded lamellar polymer crystal. For this, out of a variety of polymeric matter, we focus on the regularly repeating, chemically unspecific, highly flexible chain, of which polyethylene (PE) is possibly the most representative example, with chains sufficiently long for ends to be neglected to a good approximation. It was an astounding experimental recognition in 1957 (Till 1957, Keller 1957, Fischer 1957) that such chains were found to crystallise, not by forming fibres as might have been expected from filamentous molecules, but by forming thin platelets with regular facets, with the chain direction perpendicular (or at a specifiable large angle) to the lamellar surface (figure 1). As the lamellae were very much thinner than the length of the chains there was no other conceivable option but to envisage the chains as folding up and down within the confines of the lamella (figure 2), the basic evidence for chain folding (Keller 1957).

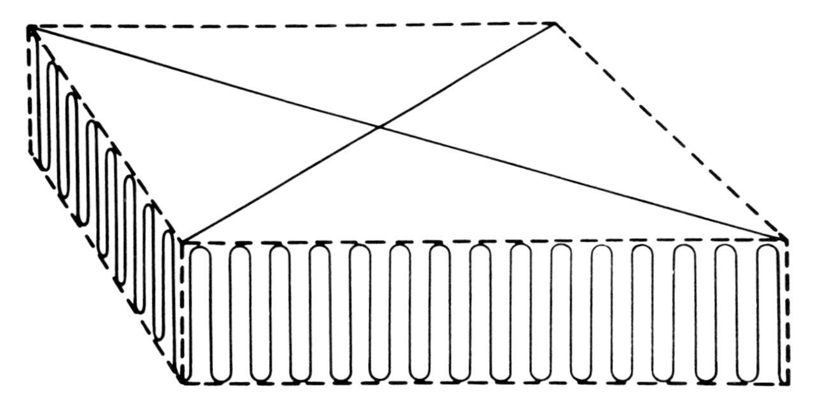

**Figure 2** Model of chain folding as consistent with facts in figure 1 in the idealised representation of complete adjacency of fold re-entry and uniform sharpness of folds. (For sectorisation represented by diagonals see section 2.)

In retrospect, possibly nowhere else could this discovery have been made, neither by myself nor anybody else, except in Bristol as it was then, in the environment of Charles Frank; the observation of lamellar crystals themselves yes, but not the facing-up to the consequences. This is apparent from the several independent announcements of the observational facts which, important as they are in themselves, provided nothing more. In fact, Jaccodine (1955), one of the less frequently quoted authors, told me subsequently (at a chance meeting, as he was not in the field anymore) that he himself thought of chain folding but was dissuaded (or even forbidden?) by his superiors to announce it so as not to expose himself to ridicule and bring discredit to himself and the laboratory. So he eventually published an ineffectual note which did not offend conservative sensibilities but at the same time did not advance the subject either.

How different it was with myself! Let me retell from the beginning. I came to Bristol in 1955 as a Research Assistant on a fellowship from the Ministry of Supply to pursue and exploit some recognitions on crystallising polymers due to myself while in Industry previously (ICI, Manchester). The latter can be summed up as follows: crystallising polymers display a range of morphologies from the submicroscopic (few tens of nms as revealed elsewhere by low-angle x-ray scattering (SAX) and by some rudimentary electron microscopy (EM)) to the (polarising) light optical dimensions (spherulites) observed by myself (see review by Keller (1958)). I maintained that we cannot hope to understand polymer crystallisation, nor the structure of crystalline polymers, neither properties depending on these structures until the nature of the newly recognised structure hierarchy has been unravelled. However, the idea of a morphological hierarchy was alien to the scientific establishment in polymer science at that time. The authorities believed that everything worth knowing can be accounted for by simply considering the statistical behaviour of chain molecules. Accordingly, the chains are like heaps of spaghetti in permanent Brownian motion from which alone all follows. Crystallisation in particular was seen as a chance coming together of adjacent chain portions forming little micellar bundles but no larger entities. The latter seems the more incomprehensible in retrospect, as large-scale organisations (spherulites) were readily visible even at that time, if one cared to look through a polarising microscope. Fortunately, however, Mr Warburton Hall at the Ministry of Supply thought otherwise, and so did Professor Charles Frank, who recognised the underlying scientific potential when the appropriate facts and ideas were brought to his notice. So it happened that I was given a chance to work in Bristol.

On entering the H H Wills Physics Laboratory, I was rather stunned by what I saw and by what was awaiting me, both in a positive and in a negative sense (where, however, the pluses eventually outweighed the minuses, hence my staying here over 35 years). As I shall show, the negative experience eventually turned out to have positive value.

The most positive aspect was the extraordinary intellectual ferment coupled with open-mindedness which permeated the whole place. Physics was in the air, was discussed everywhere: on the stairs, over tea, in the doors (in the process of leaving the building—which could become protracted to the despair of spouses waiting with the dinner at home); passage of time was simply forgotten or ignored. The topics themselves were as wide as the universe. There was no distinction between high and low brow, it was all one intellectual adventure. That is how polymers eventually slotted in between quantum mechanics, dislocations, particle physics, liquid helium, design of new optical instruments and much else. Here I saw science in action, not as fragmented into specialities but as an indivisible whole, a single enterprise of the human mind. It is this spirit that has guided me ever since in my research and in my associated educational activities and which, in my own way, I am still trying to perpetuate. And Sir Charles was central to it all! Like a chess virtuoso playing several games simultaneously he was conducting these unforgettable tea-time discussions on virtually all subjects in science. While a protagonist in one subject was pausing to think for a reply Charles was turning to something quite different with somebody else only to return to the previous subject when the reply arrived. In all this, polymers acquired first a foothold then gradually a more substantial part, to the inestimable benefit of the whole subject, not only through myself but also through many others who happened to pass by.

Amongst my 'negative' experiences was first and foremost the nearly total lack of equipment; and I relied on some when coming here. Spherulites, the observation of which set me on my path, required a polarising microscope which I was assured to be available. In fact, it was Charles' personal schoolboy microscope he specially brought in for me. It was to be used with a reading lamp to provide the light which was to bounce off a mirror into the illuminating system (just as well that Abbe was not there any longer to tell me that I could not hope to see much; in fact I did see quite a lot!). When it came to polarisation I was given two small sheets of polaroid to hold with my hands one below and above the specimen. Undoubtedly I was free to cross or uncross them as I pleased but I had no more hands left for focusing.

Also when I came I planned to reveal the connection between the electron density fluctuations in a crystalline polymer on the 10–30 nm scale as revealed by small-angle x-ray (SAX) evidence with electron-microscopic image features, which I thought should be possible in view of the ~2.5 nm resolution of electron microscopes at the time. Also I intended to exploit the diffraction capabilities of the EM technique, hardly utilised at that time anywhere, so as to connect standard crystallographic x-ray diffraction with electron microscopy. The eventual accomplishment of these objectives culminated in a way which nobody would foresee: the recognition of single crystals and chain folding. But I needed the instruments first. True, an electron microscope was available (not common in those days) but under operational conditions which are better left unsaid: eventually we got it right. As regards x-ray source, I was given a sealed-off tube with a window encrusted with tungsten which nevertheless still allowed some x-rays to get through, once these had been generated. The generator itself needed to be assembled first from ex Army–Navy discarded miscellaneous stock. When it worked at last, the generator was all enveloped in sparks of a corona discharge at the high-tension air terminal (there were no funds for an oil-sealed terminal) and the x-rays were freely scattering all over the same room which was also my office. Lead for shielding, so I was told, was expensive. Further, there was no fume cupboard, and the drains were open ducts of the kind you see in Pompeii, so a mixture of boiling organic solvents was constantly filling the room. (Once Charles made me purge it from fumes properly, which meant stop experimenting, before a visit by Sir Lawrence Bragg, the only occasion I remember when Charles showed awareness of the 'air' we were breathing.)

I cannot deny that the above experimental conditions were frustrating to the extreme, yet they were inducive to make best use of the little there was and always to concentrate on the essentials, lessons well worth learning.

Another 'negative' experience was the total absence of anybody knowledgeable in polymers. As I was still unknown in the field nobody visited me and I had no funds to visit anybody else. Also my access to the polymer literature was highly limited. So I lived and worked for two full years in near complete isolation from the relevant scientific community. But if there was no polymer science and scientist to consult there was Charles Frank (after all it is only basic physics and chemistry!) and that proved worth more than anything else. So it happened that when I showed my first polymer single crystal such as figure 1, he was not incredulous, in fact just the contrary! Further, when I told him that I cannot see how long chains, which I found to lie perpendicular to the basal surface (by combined electron microscopy and diffraction (figure 1)), of layers much thinner than the molecules are long (thickness assessed by EM shadowing and SAX technique), can do anything else but fold, he said 'of course' and encouraged me to publish immediately. What a contrast to the case of Jaccodine ! That is how in 1957 in an 'office' filled with fumes, sparks and scattered x-rays, amidst total isolation from, in fact in ignorance of the rest of polymer science, single crystals and chain folding were recognised†.

## 2 CONSEQUENCES OF CHAIN FOLDING

The recognition of single crystals and chain folding was followed by exciting times. Not only were new facts coming to light thick and fast but with unprecedented definiteness and considerable aesthetic appeal (e.g.

† Actually chain folding had been invoked much earlier (Storks 1938), based on a similar reasoning not through single crystals but thin films and under much less definitive circumstances. While in a main stream journal it was totally unnoticed at the time. I came across it accidentally, just when weighing up the pros and cons for my own model, and it gave me added confidence to publish. I acknowledged the precedent and was amply rewarded by the ensuing amicability when fate brought us together afterwards, while in the same laboratory (Bell, Murray Hill) in 1960, i.e. twenty-two years after the original discovery, by which time Ken Storks was in a different field.

figure 3). This was particularly remarkable for a subject area which shortly before had been inaccessible to any direct approach and was even believed to be intrinsically immersed in a kind of mist of meandering random chains where the order inherent to the crystalline state was only a local departure from randomness never to become apparent by direct viewing. And now suddenly a new world has emerged with clearly defined crystals, where faces can be identified, dihedral angles clearly measured, twinning modes, often fascinatingly complex (Kovacs *et al* 1967), can be determined with undisputed clarity just like in any other crystal. (See reviews: Keller 1968 (main work), 1984 (teaching text) and other interim ones by myself like Keller 1962, 1964, 1969, 1972, 1979, 1981 (emphasis on methodology), 1986; some reviews by other authors: Khoury and Passaglia 1975; Magill 1977; books: Geil 1963, Wunderlich 1973, Bassett 1981; atlas (remarkable collection of illustrations): Woodward 1988.)

**Figure 3** Multilayer polymer crystal growth through spiral terraces. In this instance, there is a regular rotation of consecutive terraces, a very special even if not an exceptional occurrence. The polymer of this example is poly(ethylene oxide) (as one block in a polystyrene–poly(ethylene oxide) two block copolymer). Electron micrograph (Lotz *et al* 1966).

But even more was to come. The folding of the chain, so inescapably derived from the most straightforward observation of crystal and diffraction pattern (see last section), was bringing with it quite new, yet straightforward laws adding to those of standard crystals, correspondingly enriching rather than complicating the study of the subject. In this new kind of crystallography everything seemed to fit into a single pattern, one fact following from the other, all arising from the apparently overwhelming inclination of the chains to fold. Here the layer thickness became identifiable with the fold length, hence, for the first time in the crystal world, a measurable crystal dimension became identifiable with a molecular feature. In other words, morphology, the preoccupation of the nineteenth century crystallographers, and atomic arrangement, the theme of modern crystal studies, had become closely linked in a wider subject we may call, after Bernal (1967), generalised crystallography. And this layer thickness, hence fold length, was not an immutable property of the chain, but was found to be determined by crystallisation conditions to an astonishing accuracy and reproducibility (within 1% or so), to be described specifically in the next section.

But the layer and its thickness was not the only link between crystal habit (and dimension) and molecular structure. Even by the simplest picture (only 'man of the street' common sense and not 'high science' was needed at this stage!) the chains were expected to fold along the lateral faces of the growing lamellae, which, if true, would subdivide a crystal into as many sectors as there are prism faces, as for example sketched in figure 2 for a four-sided lozenge. Within such a crystal the lattice, as conventionally defined by the atomic periodicity along the chain, would be identical throughout the sectors, becoming distinct only when considering the fold as part of the crystal. Hardly was this, from the conventional point of view, unlikely possibility recognised when it was actually observed in a wide variety of manifestations. Figure 4 shows an electron micrograph of a truncated lozenge with six sides and, correspondingly, six sectors. Here the sectors appear through diffraction contrast, but in fact the underlying sectorisation was recognised, with the contribution of Charles Frank, a few days earlier, even without these diffraction features, merely by seeing the crystals through the phase contrast optical microscope. The focus of his attention was the central pleat which he attributed immediately to the collapse of a hollow pyramid when sedimenting from the solvent from which it had been crystallised (Bassett *et al* 1959) like that of a tent, which, as is familiar from common experience, cannot be laid flat without at least one crease; here the sectors correspond to the tent panels. The proof (as far as it was still needed) that such crystals had in fact been tent-like before sedimentation was directly forthcoming by viewing the crystal light optically while floating in their mother liquor (figure 5) (Bassett *et al* 1963a,b).

**Figure 4** Truncated lozenge single crystal of polyethylene showing six-fold sectorisation through diffraction contrast; central pleat results from collapse of hollow pyramid. Electron micrograph (after Bassett *et al* 1959; from Keller 1964).

The effect of sectorisation became apparent by a variety of other physical effects. For example, a crack, accidentally or deliberately induced, will cleave a crystal clearly when running parallel to the assumed fold plane, yet it will become crossed by fibrils when entering into an adjacent sector where, by the argument underlying figure 2, the crack should cut across the fold plane, which it will do by pulling threads. In the case when sets of sectors are crystallographically non-equivalent, such as are in the six-sided crystals, four associated with lozenge faces and two with the truncating faces, the two sets are observed to melt at different temperatures; thus, there are two different melting points within the same crystal! Further, the non-planar shapes could be quantitatively defined: e.g. the facets of the hollow pyramids (which turned out to be non-planar based) could be given simple crystallographic indices (see, e.g., review by Keller (1968)).

All the above implied complete crystallographic definition of the crystal entity, including the fold which, as was just shown, subdivides the crystal itself, in a unique, crystallographically specifiable manner. It was natural therefore to attempt to define the fold itself in crystallographic terms. Following the guidance of Charles Frank, qualitatively the existence of pyramids and their non-planar base were attributed to the packing requirements of the folds. Namely, the folds were visualised as asymmetric, both with respect to the plane of folding and within this plane, where the nestling of consecutive fold planes gives rise to the obliquity of the pyramidal facets, and the staggered arrangement within the fold plane to the inclination of the lateral edges connecting the basal apices lying at different levels (figures 6(*a*)(i) and (ii), respectively). Beyond this, Charles Frank made a more quantitative attempt to map out the fold in terms of carbon–carbon bond pathways in a diamond lattice, such as would most closely approach the trajectory of a polyethylene chain within a polyethylene crystal. Allowing for differences between lattices of diamond and polyethylene, and for the corresponding distortions of the diamond pathways, the optimum fold configurations derived did indeed have the required two-fold asymmetries while containing no more than 3–4 C–C bonds (figure 6(*b*)).

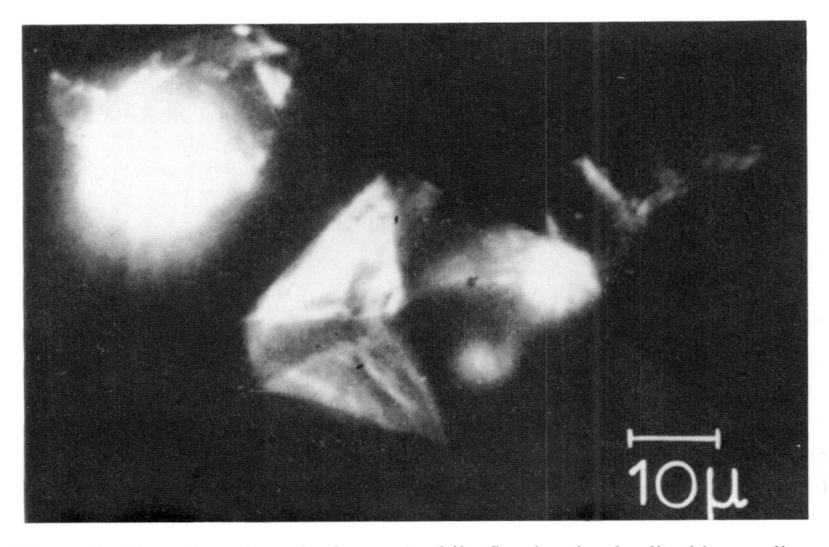


**Figure 5** Crystal such as in figure 4 while floating in the liquid revealing hollow pyramidal form. Dark-field optical micrograph (from Bassett *et al* 1963b).

As a next step, the fold itself was introduced as a crystallographic element in the polyethylene lattice. Accordingly, the true unit cell was to incorporate the full fold length, including the folded-over portion, with the conventionally defined unit cell constituted by the monomer repeat being relegated to the role of a subcell. This new formulation created, amongst others, an unexpected continuity with the familiar crystallography of paraffin, the fold stem length replacing the extended chain length (see also section 4) (Bassett *et al* 1963b).

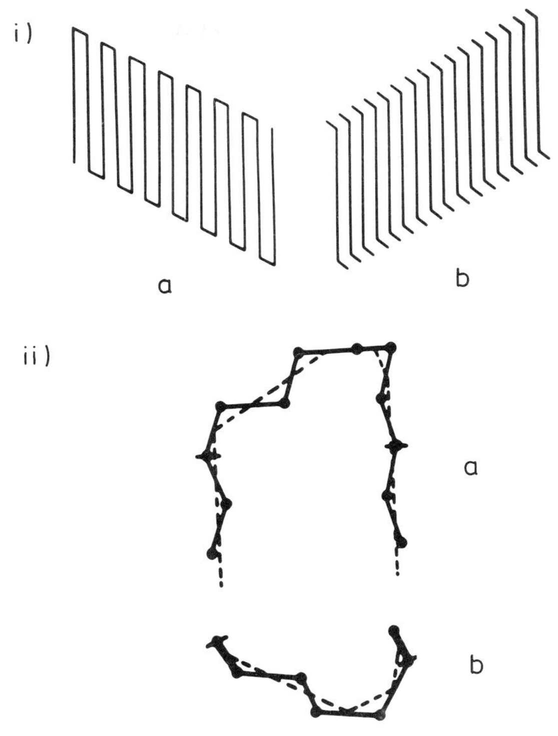


**Figure 6** Attempts to define the fold structure. (*a*) Schematised asymmetries envisaged from the pyramidal crystals. (i) Seen normal to the fold plane, (ii) folded ribbons seen edge-on (after Bassett *et al* 1963b, from Keller 1964). (*b*) One of the simple and energetically most favourable carbon–carbon paths in a diamond lattice giving rise to a fold in polyethylene. (i) Normal to plane of folding, (ii) seen along the straight segment direction (after Frank 1962, from Keller 1962, 1968).

At this stage the potential of this newly created 'generalised crystallography' as centred on the regularly folded chain, seemed boundless. However, at this point our new delights were rudely interrupted. In some other laboratories the degree of crystallinity of single-crystal preparations was measured and was found to be too low to be readily compatible with such regular crystals as in our models. The 20–25% deficiency (even if it could be argued that according to measurement techniques it need not be quite as high) was attributed to amorphous content in the sense of the traditional amorphous crystalline two-phase model, and by various reasonings this amorphous material was envisaged as constituting the fold surface.

Accordingly, there would be no regularity along the fold surface, merely an amorphous mush, all the actually observed regularity on the morphological level notwithstanding. Chain folding was considered merely as a statistical consequence of the random chain and the fold surface of an isolated layer—obtainable only from solutions at the time—as a random telephone switchboard†. Those subscribing to this view as the only truth remained unimpressed by subsequent discoveries of similarly regular lamellar crystals arising also through crystallisation from the molten mass (see e.g. Bassett 1981, 1982, Keller 1986) and of a range of further effects indicating pronounced regularities along the fold surfaces themselves (such as spectacular dislocation networks between mutually twisted layers (Holland and Lindenmeyer 1965, Sadler and Keller 1970)) and, later, decoration effects relating to the direction of the fold planes such as shown in figure 7 (Wittman and Lotz 1985).

The above divergences in facts and views formed the origin of the by now notorious fold surface controversy which, as I see it presently, has deflected the community, including ourselves, from its natural path of enquiry. Thus, instead of pursuing the fruitful enquiries into our newly opened world we, like everybody else, were seeking the 'decisive' evidence as to who is right or wrong. This controversy raged for more than 15 years with ever-increasing acrimony. It reached its peak with the advent of the application of neutron scattering which, by some, was seen as the ultimate arbiter in the dispute. Without entering into detail, through the introduction of isotopically distinct 'guest molecules' (deuterated polyethylene in this case) neutron scattering should identify the trajectory of a chain within its own environment as before as after crystallisation; thus in principle it should tell us all. However, in reality there were limits on crystallisation conditions which could be employed (the deuterated material tended to segregate) and on angular resolution that was practically achievable. So while it did provide a certain amount of valuable information (of which I myself

† Here I err by putting the dispute in its extreme form into which it eventually developed. In fact, the necessity of having some sharp, adjacent re-entry was recognised early on due to the simple fact that the same amount of material requires more space in its random than in its crystalline state which makes molecular continuity between crystal core and a random basal surface layer impossible (a fact first pointed out by Charles Frank at a Conference at Cooperstown in an argument with Flory (Frank 1958, Flory 1958). The fact that there has to be a reduced 'chain flux' at the interface, as a consequence, thus at least some adjacency in the folding, was in fact built into Flory's picture of a lamellar crystal as early as 1962 (Flory 1962) from which the switchboard terminology eventually evolved. While many of the specific arguments turned around the specific amount of adjacency required by pure geometry, or permitted by supposedly compelling experiment (neutron scattering), the overall dispute polarised into the two sweeping extremes of complete order and complete randomness, the 'switchboard' standing for the latter.

benefitted at one stage, through my association with the late David Sadler) it did not provide the ultimate answer once for all by a single decisive experiment as it was then believed, raising expectations accordingly. In fact, when the first experiments revealed that the chains did nothing at all on crystallisation, i.e. on their own they would not even have revealed that they had crystallised, this was hailed as evidence against chain folding and against the whole edifice which ourselves and like minded colleagues had been building so convincingly over the preceding two decades. At this stage the situation got out of hand. Scientific news spread, by what I would call pamphleteering, even by articles in *Nature*, one entitled 'Folded Chains Last Stand?'. (I refrain from selecting references on this issue; these can be found in *Faraday Discussions* no 68, 1979.)

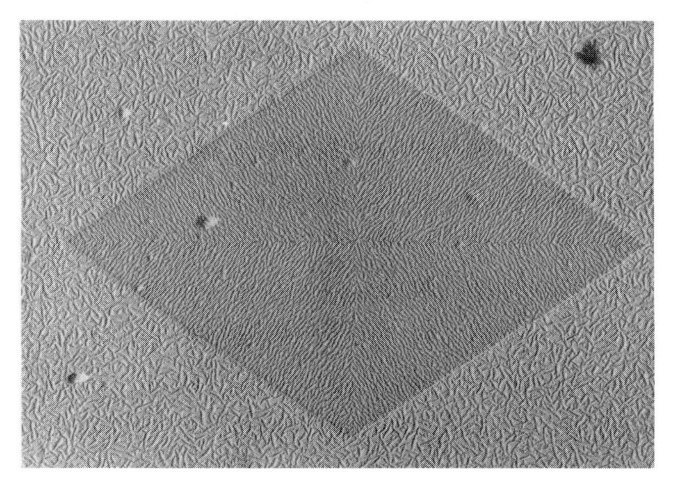

**Figure 7** Monolayer polyethylene crystal with four-fold sectorisation revealed by decoration. The decorating particles are paraffin crystals resulting from vapour deposition of evaporated (and degraded) polyethylene. There is a crystallographic relation between decorating particles and substrate including the fold plane (Wittman and Lotz 1985).

Charles Frank took it all in his stride. I remember his comment after a seminar on the then latest results from neutron scattering: 'what a tremendous waste of money'. (I hasten to emphasise that the technique has proved itself in many ways in a less heated atmosphere since, ourselves having also been amongst its practitioners.)

This whole irrational turmoil came to an end suddenly with the Faraday Discussion in 1979. This meeting was remarkable not by advancing knowledge through any particular discovery, but by allowing all steam to be vented until there was nothing more left to say, so that the participants could return to their desks or benches and proceed with their work in an orderly and productive fashion thereafter. As Chairman of the Polymer Physics Group of the Institute of Physics at the time, I admit to having been one of its initiators and planners in response to an approach of the Faraday Division of the Royal Society of Chemistry to have a joint meeting. This turned out to be a dubious distinction in view of the condemnation by some of the tone of the meeting after the event. Yet the near impossible was achieved: all of the protagonists were present, all delivered their piece also in writing which is now on record (except for a few extreme excesses removed by the editor) for posterity to read (*Faraday Discussion* 1979). It was like a storm on a sultry summer afternoon: it cleared the air. And there was plenty of lightning and thunder with Charles Frank in the midst of it. Thus, in an introductory lecture, he put the 'switchboard model' in its place as a topological impossibility which, at least in the form then held, violated basic space requirements. He referred to the unfortunate telephone switchboard girl who was to create these connections, impossible to achieve with the dimensions of cables and sockets provided: she would not have persisted for twenty years (reference to Frank (1958), see preceding footnote) said Charles Frank, but would have long gone on strike. He disposed of other argument points in a similar manner. Thus, he dismissed the alleged impossibility of random, intermingled chains ever to form a regularly chain-folded structure by statistical motion. For this he invoked experience in cooking spaghetti; while true that the disentangling of an individual spaghetti by mere shaking in a ladle (the culinary equivalent of Brownian motion) would take an infinitely long time, this would occur readily when taking one filament between one's lips and sucking it out, the equivalent of what the crystallisation force would do.

And so, after all was said, the storm blew itself out. In retrospect it may be difficult to apprehend how these polarised and entrenched positions had ever arisen. It had become more a matter of belief than of arguments. Not infrequently during this period I was asked questions such as 'do you believe in the sharp, adjacently reentrant fold?' by interested but puzzled bystanders, as if it were a matter of religious faith which the issue had in fact started to resemble. From the present standpoint it seems apparent to me that the fault lay not with one view or the other but in the way the questions were asked and the issues weighted. The possibility and existence of sharp and adjacent folds could not have been in doubt from the very beginning, and current experiments on ultra-long model paraffins strikingly bear this out (see section 4): the issue merely boils down to what extent this idealised situation of ordering can be achieved under the specific conditions of solidification of a particular material. On the other hand, it must also be true that there will be many departures from the most regular chain deposition, the more, the longer the chain, with chances of imparting amorphous characteristics (and intermediate states of ordering) particularly along the fold surface, as currently is being revealed by several admirable studies (e.g. Mandelkern 1986, Fatou 1989). In my opinion the fault lay with the laying out of the alternatives in a mutually exclusive manner and in the search for a single decisive experiment. When a few days after the Faraday meeting I was asked to give an account of the whole hullabaloo at the Medical Research Council (Cambridge) at the invitation of a puzzled Aaron Klug and Max Perutz, and I concluded by saying that 'nature bypasses the irrelevant question' I felt immediately from the response that I got it right, and so I still feel today. I only regret that we did not pursue further the crystallographic fold concept, true unit cell–subcell, etc, following Charles Frank's guidance in the early days, but allowed ourselves to be distracted. Highly idealised as the afore-mentioned systems may appear, as compared to the 'real' material, I feel we would all have gained by persisting with them at the time. After all, there are ideal gases, ideal lattices, and departures from them in real systems do not negate the essence of the idealised model, so why should it be different for polymer crystals?

## 3 CRYSTALLISATION THEORIES

A remarkable phenomenon such as chain folding, of course, requires explanation which leads to the theories. The subject of theories is also punctuated by disputes, but these are different in kind compared to those surrounding the fold surface controversy. Namely, they are confined to the camp of 'believers' in chain folding, as only they are prepared to take the trouble of modelling the phenomenon quantitatively and then argue amongst themselves, sometimes most heatedly, about the best way of doing it. The 'non-believers' dismissing chain folding, or that there is anything special about it beyond random coil chain statistics (even if such statistics failed to predict it!), not unnaturally leave the subject alone. Encouragingly, since the fold surface controversy subsided, the boundaries between the two 'camps' are becoming increasingly blurred, which also applies to the attitudes to theory. Unfortunately, however, at the same time the theoretical scene in its own has become more obscure and the literature nearly intractable, a situation not helped by heated disputes where sometimes, so it seems, only the parties involved know what they are arguing about. This is having the consequence that, except for a handful of people directly concerned, the scientific public is left uninstructed, inevitably leading to general indifference and loss of interest, even without any *a priori* prejudice as regards chain folding. It is this latter situation I am trying to remedy in this theoretical part of the present article, even if only in a verbalised manner. I believe that the simple sounding yet loaded statements to follow, and particularly their critical juxtaposition, are not readily extracted from the existing literature without great labour and long-standing immersion in the subject. For providing the latter two I wish to acknowledge the input of my colleague Gerhard Goldbeck-Wood without whom I could not have written this section in its present form.

As will be seen, I focus much of the presentation on Charles Frank's input. Of course, this is in the spirit of the '*Festschrift*'; but the occasion aside, I hope it will become apparent that, without any unfairness to the all important and much more ubiquitous contributions by other authors more frequently quoted, at least in the polymer field, this is not far off the mark by any criterion.

### 3.1 General background

Charles Frank has two publications relating to the theory of polymer crystallisation of seminal importance which may not be fully apparent from the usual references to them in the literature. Here I shall try to place the role of these contributions in what I believe appropriate perspective. Even so, Charles Frank's true contribution to polymer crystallisation theory is not confined to the two published papers, important as they are; for a full appreciation we need to go back to basic models of crystal growth predating subsequent interest in polymers. This I shall do in what follows.

Crystallisation, like most phase transformations, consists of two consecutive stages: primary nucleation and growth. Here we shall be concerned with crystal growth only which, in the case of polymers, is the factor of consequence for the final crystal (i.e. determining the fold length—see below).

There are three distinct modes for crystal growth. Historically the first, due to Stranski and Kaischev (1935), is the secondary nucleation mechanism. Here the rate determining step is the formation of a secondary nucleus on an atomically smooth substrate of the same crystal. This occurs through deposition and clustering of the crystallising entities onto the substrate within a supercooled system. When for the prevailing supercooling the size of critical stability is reached, the two-dimensional cluster will spread until it covers the full substrate layer, the process repeating itself for each consecutive layer (figure 8(*a*)(i)). However, as is familiar, growth was observed to occur at much lower supercoolings and much faster than permitted by the above model with secondary nucleation as the rate determining step (Volmer 1939). A way out was provided by the screw-dislocation-based growth mechanism (Frank 1949). Here the growing face contains a step centred on an emerging screw dislocation, the step height corresponding to the Burgers vector of the dislocation. This step obviates the need for a secondary nucleus and can propagate, with the concomitant growth of the crystal, at much lower supercoolings. In the course of this propagation the step winds itself into a spiral, centred on the emerging dislocation, giving rise to spiral terraces (figure 8(*b*)). Such spiral terraces were first predicted theoretically, to be observed experimentally in innumerable cases ever since. About concurrently with the screw-dislocation-based growth another mechanism obviating the need for secondary nucleation was also postulated theoretically (Burton *et al* 1951), later termed roughening-based growth (Jackson 1968). Here the starting point is the

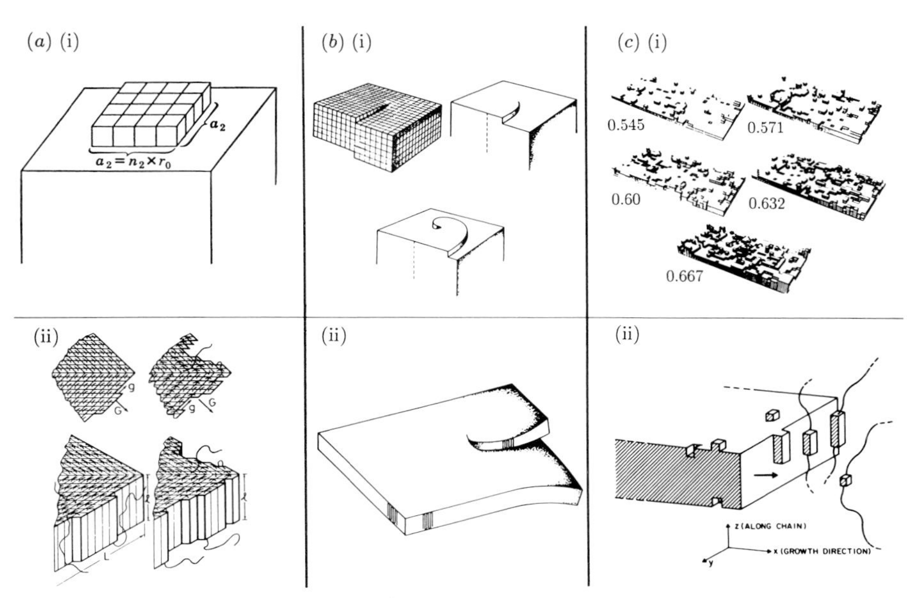


**Figure 8** Models of crystal growth as constructed for simple substances (upper row) and as applied to polymers (lower row). (*a*) Growth by secondary nucleation. (i) Classical model by Stranski and Kaischev (1935). (ii) As applied to chain-folded polymers, in this particular representation by Hoffman *et al* (1975). Nucleation (at a rate $i$) occurs by successive fold stem depositions where the nuclei spread out along the lateral face at a rate $g$ up to a limiting substrate length $L$ producing an overall advance $G$. In the left-hand pair there is, on average, only one nucleus per $L$, i.e. $g \gg i$ (regime I); in the right-hand pair there are more, i.e. $g$ and $i$ compete (regime II). In this particular model the stems are all of uniform length: to this Frank and Tosi (1961) added fluctuations. Also, stem deposition is in a single step. (Frank and Tosi suggested the more realistic finer grain process.) (*b*) Growth by screw dislocations. (i) The classical model by Frank (1949). (ii) As applied to polymers. Here the Burgers vector equals the thickness of the chain-folded layer, the consecutive layers themselves not necessarily being in crystallographic register (they may splay apart). While it accounts for crystal thickening by layer multiplication it is not directly relevant to the origin of chain folding. (*c*) Growth by roughening. (i) Classical roughening. Simulation by Leamy *et al* (1975) of roughening theory by Jackson (1968), itself based on Burton *et al* (1951). (ii) Application to polymer crystal growth by Sadler. Here the depositing (roughening) entities are short segments of the chain; their consecutive deposition probability coupled with their connectedness results in the final finite layer thickness (Sadler and Gilmer 1986).

self-evident fact that for entropic reasons at all temperatures above 0 K the surface is not ideally smooth. By the Burton–Cabrera–Frank theory this equilibrium surface roughness increases dramatically at a certain temperature, later termed the roughening transition. At and above this transition the growth face will advance, hence the crystal will grow, as atoms (or small molecules) can now attach at pre-existing niche sites with the equilibrium roughness conformation statistically being maintained in the process (figure 8(*c*)). Crystals growing through such a mechanism (at low supercooling) acquire rounded contours (or rather lose crystallographic faceting).

All the three basic mechanisms in figure 8 feature in polymer crystals and/or in treatments intended to account for their growth. Or, conversely, models invoked for polymer crystal growth are all in essence based on mechanisms previously established in simple crystal growth. Franks' explicit contribution to polymer crystal growth theory was in the class of secondary nucleation (figure 8(*a*)(ii)), while the other two mechanisms (figures 8(*b*)(ii), 8(*c*)) are in direct lineage from his seminal works on basic crystal growth, as will be apparent from figure 8 and its caption.

The applicability of the schemes in figure 8 to polymers is most apparent in the case of the screw-dislocation-based mechanism (figure 8(*b*)) as growth spirals are directly displayed by the polymer crystals (figure 3). While obviously pertinent, this has not been the subject of the theories themselves. The reason for this lies in the particular relation between crystal structure, Burgers vector and morphology in the case of chain-folded polymer crystals. Namely, the Burgers vector here is the thickness of the chain-folded crystal layer itself. It is this layer which winds up into the growth spiral, thus leading to multilayer development (figure 8(*b*)(ii)). As discussed in the previous section, in the prevailing view, thickening of the crystal by layer multiplication is presently not considered to represent an accretion of the crystal by a crystallographically defined periodicity (i.e. except for our own attempts in this direction (Bassett *et al* 1963b), the layer thickness is not considered as a lattice period). The focus in the subject is rather on the lateral growth of the individual layer with constant thickness, as this embodies the origin of the fold and it is to this that the quantitative theories are directed. Most of them, including the Frank contributions, take the secondary nucleation model as their basis considering the formation of the (secondary) nucleus along the growing edge as the rate determining step (figure 8(*a*)(ii)). A new, comparatively recent departure applies the roughening concept to this same growing edge (figure 8(*c*)(ii)). Even if Charles Frank has made no explicit contribution to the latter within the confines of the polymer field, nor has he expressed views in the resulting controversies, nevertheless in essence the whole approach, as already stated, goes back to him (and to Burton and Cabrera).

### 3.2 Experimental basis of theories and its influence on historical development

All theories have to account for certain salient experimental facts which I shall recapitulate and/or newly introduce in the listing that follows.

(1) The thin lamellar nature of the crystals, implying the existence of chain folding with uniform fold length under given growth conditions.

(2) The fold length and its variations. The fold length ($l$) is directly measurable as the lamellar thickness (by electron microscopy, small-angle x-ray scattering, low-frequency Raman spectroscopy) and is found to be inversely related to the crystallisation temperature ($T_c$) and, more explicitly, to the supercooling ($\Delta T \equiv T_m^0 - T_c$, where $T_m^0$ is the melting, or in solutions, dissolution, temperature of the stablest infinite-size extended chain crystal) (figure 9). This functional dependence is expressable as

$$1 \propto \frac{\text{constant}}{\Delta T} \tag{1}$$

except for the highest $\Delta T$ (achievable in some slowly crystallisable polymers, not in polyethylene) where $l$ becomes a constant (invariant with $\Delta T$).

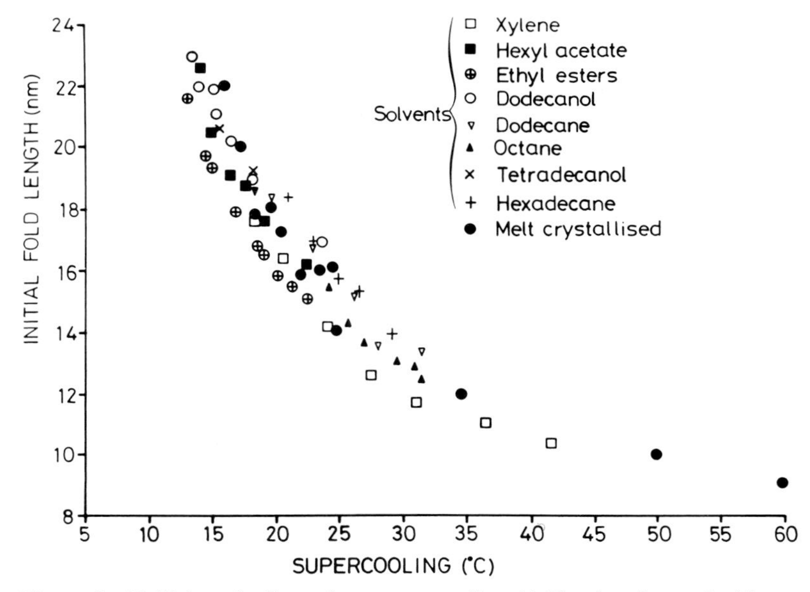


**Figure 9** Fold length ($l$) against supercooling ($\Delta T$), showing coincidence between curves for a range of solvents and the melt. Such dependence on $\Delta T$, as distinct from crystallisation temperature *per se*, was established for solutions in 1965 (Kawai and Keller 1965) but establishment of full correspondence with the melt (for a long time hindered by secondary thickenings) was only achieved in 1985 (Barham *et al* 1985).

(3) Relation (1) applies at any stage of crystal growth: a change in $\Delta T$ in the course of growth leads to a corresponding change in crystal thickness: an increase in thickness on decrease of $\Delta T$ and vice versa. Thus a stepwise change in crystallisation temperature produces steps in either direction within the observable single lamellar crystal (figure 10 is an example of recent origin).

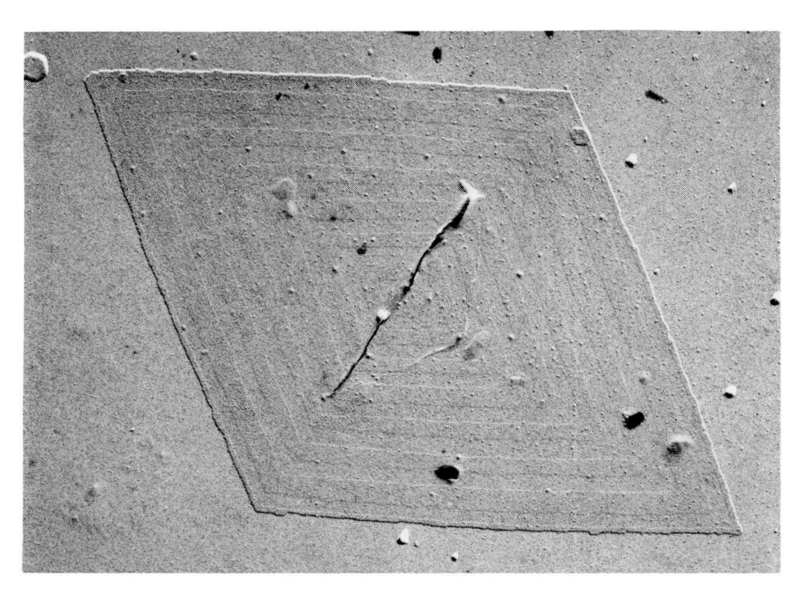

**Figure 10** Isochronously decorated polyethylene crystal. The up and down steps correspond to the response of the fold length to the respective raising and lowering of the temperature during crystallisation. Electron micrograph (Dosière *et al* 1986).

(4) The trend to refold irreversibly to higher fold length subsequent to primary crystallisation. This was first observed on heating (annealing) crystals, formed already, to above the crystallisation temperature, but later also during isothermal crystallisation itself as a secondary perfectioning process.

(5) Crystals grow laterally at a constant rate uniquely dependent on $\Delta T$ according to the relation

$$G \propto e^{-K/\Delta T} \tag{2}$$

$G$ is the lateral growth rate and $K$ is a constant.

Here, according to accumulation of experience, $K$ may have different values in different $\Delta T$ ranges (growth 'regimes').

(6) Where crystals are morphologically well defined, the plane of folding is parallel to the growth face (sectorisation, see preceding section).

(7) Curved-edged crystals can be observed in the same system which normally gives faceted growth, the former occurring usually at the highest temperatures of crystallisation (figure 11).

The evidence underlying points 1–7 arose at different dates, correspondingly affecting the historical development of the theoretical works. Thus, in the early stages the possibility of chain folding being due to a favoured equilibrium state was still open. Charles Frank (together with Maurice Pryce) disqualified an early attempt at equilibrium theory through an unpublished message I was authorised to read out at the IUPAC Conference at Wiesbaden in 1959. The same equilibrium theory was then considerably elaborated (Peterlin *et al* 1962). However, at this stage it had become too intricate for a ready assessment of its intrinsic validity and Charles Frank had other priorities for his attention. The equilibrium approach in its final form has in fact never been disproved; it was overtaken by events primarily through the consolidation of point 4 in the above list of experimental recognitions. As a result, there is now consensus that chain folding is the consequence of kinetic factors and accordingly is kinetically controlled. By all evidence, therefore, the equilibrium state is the fully chain extended infinite crystal; chain folding sets in merely because in that way crystallisation can proceed faster.

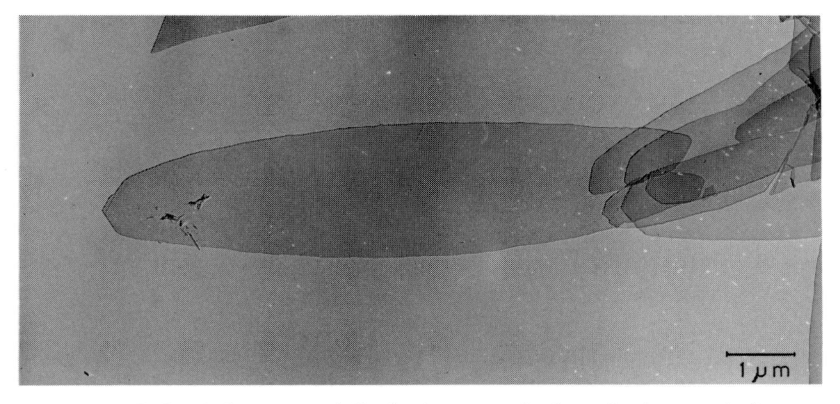


**Figure 11** Polyethylene crystal displaying curved edges. Such crystals form at the highest growth temperatures. Electron micrograph (Organ and Keller 1985).

Within the various kinetic possibilities that of control by secondary nucleation arose even before the recognition of single crystals. Namely, spherulite growth was observed to obey equation (2) (point 5), which from the onset was attributed to secondary nucleation (Flory and McIntyre 1955, Mandelkern 1958). Since by subsequent knowledge polymer spherulites are known to consist of lamellae this is pertinent also for such crystals. The same applies to equation (1) (point 2) for the fold length, which I myself identified first as reflecting control by initial nucleus size (Keller and O'Connor 1958—addendum). The conforming of fold length and growth rate with the functional relations in equations (1) and (2) has become the cornerstone of nearly all subsequent theories which (with the exception of the later approach by Sadler) all took nucleation as a rate controlling step. Sadler, however, pointed out that equations (1) and (2) need not necessarily imply nucleation, but merely kinetically controlled growth with an activation barrier. Whether it is nucleation or roughening control (see later) will then depend on the theory itself.

Still within the framework of nucleation-based theories, points 3 and aspects of points 4 and 5, as they emerged experimentally, have had major impact. The recognition that a pre-existing fold length (when different) had no influence on subsequent fold length and growth (as follows from point 3) led at an early stage to the dismissal of primary nucleation as a relevant factor for the purposes of theories and to the focusing on crystal growth alone. The recognition of the invariance of $l$ (with $\Delta T$) at the highest $\Delta T$, i.e. the breakdown of the relation in equation (1) at the highest supercoolings, achievable only with some slowly crystallising polymers (see under point 2) has led to new initiatives (still within the framework of the nucleation theories), and so has more recently the recognition that rearrangement subsequent to chain deposition (see point 4) may contribute to the rate determining step. Much earlier, the experimental findings of variations in the constant $K$ with $\Delta T$ by discrete amounts (under point 5) led to the extension of the nucleation approach to incorporate 'regimes', a step to which Charles Frank made a contribution.,

In all the above, lateral habits received little attention *per se*. Irregular and degenerate lamellar habits were disregarded as mere complications, and so were the curvi-linear lamellar outlines seen, even if occasionally, for a long time. The increasing prominence of such curved habits (point 7), however, prompted the late David Sadler to explore the possibility of applying roughening theory, as originating from Burton *et al* (1951), to polymer crystal growth leading to an altogether new departure in the theoretical works. This has led to a revival of theoretical efforts, somewhat staid by that time in general. Thus only quite recently this new thrust has received a riposte from the adherents of the nucleation theory falling back, amongst others, on one of Charles Frank's works relating to the 'regimes'.

It will be apparent that the subject has had a tortuous history, potentially difficult to follow up from its inception; on the other hand plunging into it midstream could also be a perplexing experience. It is hoped that the above 'sketch-map', and its sequel below, will be able to provide some guidance. In what follows I shall choose a few selected items which highlight Charles Frank's input. It then happens (not unexpectedly as it is coming from Charles Frank) that these few items have a special part to play both in shaping our historic perspective and in providing pointers to the future.

### 3.3 The fold length

The first kinetic theory was that by Lauritzen and Hoffman (1960) setting a line which has been followed in numerous variants ever since by the same authors and their associates (see e.g. Hoffman *et al* 1979). It is based on the secondary nucleation approach and has, amongst others, the great merit of providing simple analytical expressions for the primary variables for ready comparison with experiment, with meaningful physical parameters which are derivable from the experimental data. Here, as already implied (figure 8(*a*)(ii)), the rate determining step is the formation of the secondary nucleus along the growing edge which then propagates rapidly along that face. Folding, hence the limiting layer thickness, arises through the agency of two opposing trends: an activation barrier to the deposition of the first straight stem along an otherwise smooth face, a barrier which increases (thus opposing growth) with increasing stem length, and the thermodynamic driving force which increases (thus promoting growth) with increasing stem length. The full stem is considered as depositing in a single step as one unit, after which it either detaches itself or stays on with the chain folding over so as to deposit a second segment of the same length adjacently. This, folded over, segment again may either detach itself or be followed by a second fold plus straight segment, and so on. This process can lead to growth if at each step the free energy of folding is more than being paid for by the gain in free energy through new segment deposition. The latter is ensured if the depositing stem length exceeds the stability requirement of the initial two-dimensional secondary nucleus appropriate for that supercooling. For such growth the overall flux is the resultant of the respective attachment and detachment rates which are calculated through the appropriate rate equations. In the theory such calculations are carried out with the stem length ($l$) as a variable, when (in view of the two opposing factors indicated above) a favoured length, $l_g^*$, is obtained for which the overall flux is largest, hence growth fastest. Analytically this gives

$$l_g^* = \frac{2\sigma_e T_m^0}{\Delta H \Delta T} + \delta l \tag{3}$$

where $T_m^0$ is the equilibrium melting point of the fully extended chain crystal, $\sigma_e$ is the fold surface free energy and $\Delta H$ is the heat of fusion. The first term on the right-hand side represents the size of the critical nucleus; this is exceeded by a small factor $\delta l$. Subsequent elaborations of the theory then provide analytical expressions for $\delta l$ to provide $l_g^*$.

As is apparent, equation (3) embodies the observed dependence on $\Delta T$ as expressed by equation (1). In addition to this functional dependence reasonable numerical fit with experiment can be obtained for realistic values of the parameters $\sigma_e$ and $\Delta H$.

It was at this stage that Frank (with Tosi) made his input (Frank and Tosi

1961). First they set out to improve on one unrealistic feature of the original model, namely that the $l$ values are taken as all identical in a given crystal where the spread in $l$ about $l_g^*$, as given by the flux equations, is provided by the distribution of crystal thicknesses, each crystal in the total population representing a particular $l$ value. Frank and Tosi improved on this model by considering a distribution in $l$ within each crystal, which in turn they generated through fluctuations in $l$ during the deposition itself. The usual attribution to Frank and Tosi in the relevant review and textbook literature is this introduction of fluctuations into the Lauritzen and Hoffman theory. However, this conventional reference hides the deeper significance of the Frank and Tosi theory, which I shall attempt to convey in what follows.

Here I refer first to point 3 in the previous section, namely that $l_g^*$ (in the above notation) only depends on the prevailing $\Delta T$ and not on the pre-existing substrate. This was first revealed in our laboratory by steps arising, both downwards and upwards, in the crystals as the temperature was lowered or raised respectively during the growth of the crystal (Bassett and Keller 1962) and demonstrated by spectacular examples later by Point and collaborators (figure 10) through a study to be referred to again further below. An even more prominent example is provided by the subsequently discovered 'shish-kebab' crystals (for illustration see article by Mackley in this volume, p 309) where the lamellae are nucleated by extended chain fibres (products of crystallisation of chains stretched out by flow first) but nevertheless crystallise by chain folding with the appropriate $l_g^*$, taking no note of the thermodynamically much more favourable substrate already present. Amongst others, it is these experiences which justify the usual practice in theoretical modelling which consider the deposition of the chain along an infinite substrate. At first sight it appeared that introduction of fluctuations would wipe out any such step (referring to the case of stepped crystals) as fluctuations might be expected to drive the crystal thickness upwards. I well remember those discussions around the tea table, an interim state lost for the literature, where the possible incompatibility of the prevailing theoretical framework with experimental evidence was seriously raised. It was then a major surprise when the fluctuations were in fact found to converge to a final limiting thickness determined by $\Delta T$ alone irrespective whether approached from $l$ above or below $l_g^*$. It is seldom remembered nowadays that without this accomplishment the Lauritzen–Hoffman approach would hardly have remained viable in its initial form.

The Frank and Tosi paper has sown some seeds of major consequence for further developments such as are not generally appreciated. Two are contained in their appendix. One is the drawing of attention to the importance of the apportioning of the free energy change between the forward and backward steps in the reaction path. First Lauritzen and Hoffman apportioned this free energy change equally between the two (expressed by a factor $\psi$ where $\psi = \frac{1}{2}$) but Frank and Tosi draw attention to the physical significance of biasing this apportioning, in their case, in the forward direction, an issue which has acquired prominence subsequently (see below). The second forward-looking point was their introduction of a 'fine grain' representation of the reaction path (figure 12). This replaces the unrealistic single-step deposition of the full chain segment with the sequential attachment of much smaller flexible subunits simulating the true behaviour of a flexible macromolecule. This multistage deposition, followed by that of a fold portion, which was shown to be equivalent to the overall single-stage deposition in the case of a high barrier to the fold, was not in fact used in their quantitative formulation, but the scheme, as laid out in their appendix, was taken up profitably by others, as will be seen below.

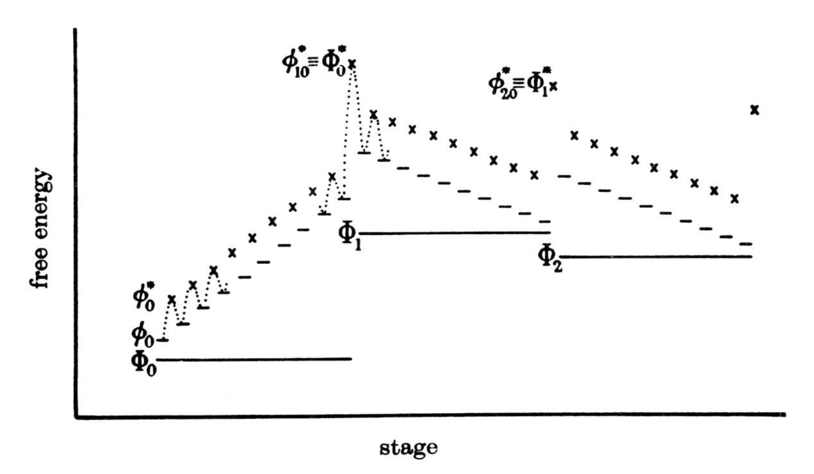


**Figure 12** Schematic representation of the free energy levels in the process of consecutive attachment of fold segments. The fine-scale oscillations correspond to attachment of flexible subunit, the 'fine grain' representation suggested first by Frank and Tosi (1961).

The Frank–Tosi paper, however, has led to one apparently unrealistic consequence (which in fact turned out to be a common feature also of other nucleation-based theories). Namely, that $l_g^*$ did not continue to decrease with increasing $\Delta T$ in accordance with equation (1) but increased catastrophically beyond a certain $\Delta T$ (referred to subsequently as '$\delta l$ catastrophe' because of its origin in the $\delta l$ term).

The prediction of such a 'blow-up' of $l$ on decreasing temperature seems obviously strange. While conflict with experiment did not arise at that stage, because polyethylene could not be supercooled to such an extent where the blow-up effect could become observable, it did arise a decade later when the much more slowly crystallisable isotactic polystyrene was examined (Jones *et al* 1973). Here the $\Delta T$ range of the anticipated '$\delta l$ catastrophe' was reached and, perhaps not surprisingly, the blow-up was not observed. Instead $l$ became invariant with $\Delta T$, curves like figure 9 levelling off to a plateau.

The matter of the absence of the '$\delta l$ catastrophe' was attended by Lauritzen and Hoffman (see Hoffman *et al* 1975). (Frank and Tosi considered the behaviour of a polymer at such high supercooling outside the range of $\Delta T$ for the kind of theory in question from the very beginning, and themselves refrained from providing a physical interpretation of the effect.) The reason for a blow-up in $l$ is nevertheless obvious: it is due to the removal of the activation barrier to the deposition of the first segment. Accordingly, beyond a certain $\Delta T$ a segment would deposit with decreasing free energy *ab initio* (the increase due to creation of surface is compensated by decrease due to new bulk). Therefore, to retain correspondence with experience a barrier would need to be created by the theory. It was at this point that attention became focused on the apportioning factor $\psi$, the importance of which had been highlighted by Frank and Tosi previously. I do not know how far Frank and Tosi would have concurred with what followed, yet undoubtedly the apportioning of only a small fraction of the change to the first deposition could shift the 'blow-up' to below any arbitrarily chosen crystallisation temperature. The associated physical picture involved was an adsorption of the still largely 'amorphous' chain portion, and thus the creation of a barrier, prior to its fitting into the lattice, a picture with some appeal. However, this success to comply with experiment was at a price; namely the all important fluctuations first introduced by Frank and Tosi could, due to analytical complexities arising, no longer be incorporated. This situation pertains up to the present day for all developments along the lines of the original nucleation-based theories. Thus an analytical treatment which both eliminates the 'blow-up' and retains fluctuations still needs constructing.

Another line of approach for the elimination of the '$\delta l$ catastrophe', while still adhering to the nucleation theory, was taken by Point (1979a,b). This again goes back to the Frank and Tosi paper as its source, in particular by adopting the fine grain representation therein, but this time in a quantitative manner. It considers flexible segments depositing consecutively as in figure 12 but at each stage giving the chain a two-fold chance: either to continue its straight progression or fold back, in the latter case, for sake of simplicity, retaining the fold length thus acquired thereafter. At conventional supercooolings, the theory converges to the preceding nucleation theories, but at high supercoolings the chains are deprived of the chance of extending sufficiently, so as to reach the 'blow-up', by folding up beforehand as a consequence of pure statistical probability. (In fact detailed calculations claim to lead to a plateau in $l$.) The significant new element here is the probabilistic restriction to chain extension even when this would be provided by the energetics otherwise. In other words, the barrier to deposition acquires an entropic character in contrast to the enthalpic nature in the preceding version of the nucleation-based theories. In this latter respect at least it approaches the conceptual framework of the later roughening-based theory of Sadler (Sadler 1983, 1987a,b, Sadler and Gilmer 1984) and conforms with some latest inferences from experiment (see section 4).

The roughening theories themselves will not be pursued in any detail at this stage. It will merely be reiterated that they were first motivated by the observation of curvilinear crystal faces appearing systematically at the highest crystallisation temperatures (figure 11). It was argued that, as such faces must be covered by niches, crystal growth cannot be niche (hence nucleation) controlled. Again a 'fine grain' deposition of flexible segments was adopted to provide the equilibrium roughness in the sense of Burton, Cabrera and Frank, which then ensures continuing growth. By due consideration of the connectedness of the segments and by incorporation of certain 'pinning rules' the formation of lamellae of finite thickness (thus incorporating chain folding) could be ensured with the obeying of experiments as expressed by equations (1) and (2). A significant feature of this approach was the nature of the activation barrier to the chain deposition: here this was explicitly entropic, i.e. the achievement of the thermodynamically stablest extended chain is frustrated by diminishing probability rather than by excess energy of surfaces created (a point which it has in common with Point's otherwise nucleation-based theory).

The latest attempt by the nucleation theory (Hoffman and Miller 1988, 1989a,b) to account for curvilinear crystals, and thus meet the challenge posed by the roughening theories, will be referred to in the next section as it relies on further concepts, again due to Frank as its source, to be introduced first in what follows.

All the theories mentioned so far considered the attaching chain as either staying on or detaching, according to the prevailing energetics and/or statistics, but not the possibility that an attaching chain may improve its chances of staying by subsequent rearrangement, still as part of the rate determining step. The latter possibility is incorporated in the recent theories of Hikosaka (1987, 1990). This author, taking his lead from frequently observed refolding behaviour, considers that the chain can increase its initial deposition length by sliding diffusion while in contact with the substrate, thus biasing it against detachment. Otherwise, Hikosaka's approach follows the framework of the Lauritzen–Hoffman nucleation theory. This approach may be particularly pertinent at the lowest $\Delta T$ and especially at elevated pressures where the chains in the crystal phase can acquire high mobility (see section 4).

### 3.4 The growth rates

Next we turn to the issue of the lateral growth rate. Here, the most basic observation is the linear growth with time, allowing the growth to be described by a single growth rate ($G$). It will be obvious, although it took better part of a decade to formulate explicitly, that this is inconsistent with nucleation-controlled growth without the introduction of an additional factor: namely, as the growing surface increases, so does the nucleation density which, for a spreading rate that is very much faster in comparison, should lead to an increase of the lateral crystal growth rate with size, hence time. In order to preserve the constancy of the growth rate, the extent of spreading has to be confined, which is the origin of the introduction of a 'substrate' length $L$ (see figure 8($a$)(ii)), limiting the lateral spread from a given growth nucleus in cases where the growth face exceeds $L$, which is always the situation in observable practice (for qualification see later). The physical meaning and numerical value of $L$ is still a matter of dispute today.

As already stated, experimentally $G$ is found to depend on $\Delta T$ as expressed by equation (2), a basic functional relation with which all theories need to comply. Further scrutinies focus on the constant $K$. Again, experiment revealed, first for crystallisation from the melt, and much later from solutions, that $K$ can have different values in different $\Delta T$ regimes, where, in most cases, the change-over is fairly sharp in a particular $\Delta T$ region. This observation is the origin of the subject of 'growth regimes' and of the underlying regime theories (Hoffman *et al* 1975). Specifically, on increasing $\Delta T, K$ reduces fairly suddenly by a factor of two at a particular $\Delta T$ below which (i.e. at the highest $T$) growth is termed to be in regime I, and above which (i.e. lower $T$) regime II. Where a sufficiently high $\Delta T$ can be achieved there is a further sudden change in $K$, now an increase by a factor of two, i.e. a return to the regime I value, the corresponding $\Delta T$ region being termed regime III (e.g. Hoffman 1983, Phillips 1990). In what follows we confine ourselves to regimes I and II.

As derived by Hillig (1966) for the three-dimensional situations in simple substances, growth regimes of the above kind can be interpreted in terms of a competition between the nucleation ($g$) and spreading ($i$) rates. The approach was first introduced to the polymer case, i.e. to the situation of a two-dimensional growing edge of substrate length $L$, by Sanchez and Di Marzio (1972), and subsequently extensively applied by Lauritzen (1973). Explicitly, in this model edge-wise growth of a crystal lamella is controlled by the nucleation of stems on the growth face occurring at a rate of $i$ events per unit length and time. Each nucleation event initiates locally a new layer of thickness $b$ which then grows in both directions along the edge with a constant velocity $g$ until it either reaches the end of its 'substrate' of length $L$ or meets an independently nucleated growth layer at the same level. Predominant nucleation control ($i \ll g$) yields regime I, in average one nucleus per $L$, and comparable nucleation and spreading yields regime II (many nuclei per $L$) (left- and right-side pair in figure 8(a)(ii) represent regimes

I and II respectively). Lauritzen derived $G$ for the former case exactly leading to the expression:

$$G = bLi \tag{4}$$

corresponding to regime I. In the other extreme of very large $i$ (and/or $L$) Lauritzen obtained

$$G = Cb(ig)^{1/2}. \tag{5}$$

This corresponds to regime II behaviour where $G$ becomes independent of $L$ with $C \geqslant 1/\sqrt{\pi}$. The often quoted dimensionless parameter

$$Z = \frac{iL^2}{4g} \tag{6}$$

is introduced where a value $\sim 1$ defines the $\Delta T$ (through the corresponding expressions giving $i$ and $g$) region where regime I changes into regime II.

Lauritzen's treatment only provides upper and lower bounds for $G$. It was Charles Frank (1974) who solved the full problem exactly within the framework of a mean field approach for all $P$ where

$$P = (2Z)^{1/2}. \tag{7}$$

The rather elaborate relations simplify to the above regime growth laws (equations (4) and (5)) for the asymptotes $P < 0.1$ (regime I) and $P > 10$ (regime II) with a precisely defined constant $C$ for the latter. Here $G_{\mathrm{I}}$ and $G_{\mathrm{II}}$ intersect at $P = 1$ (subscripts refer to the regimes).

More recently there have been other analytical and numerical solutions without the constraints of the mean field approach (Bennet *et al* 1981, Goldenfeld 1984, Toda and Tanazawa 1986) in the light of which the Frank solution was still found to be very satisfactory for all $Z$ concerned (within an error of 4%).

It is to be noted, however, that Frank's exact solution predicts a smooth, rather than a sharp, transition, in spite of the rather sharp transition more often observed (or is such a transition due to a different cause altogether as queried in an unpublished seminar by the late David Sadler?). Also the value of intersection differs by a factor of $\sqrt{2}$ from the more empirical criterion of Lauritzen.

So far, more satisfying as it is in comparison with its predecessors, the virtues of the Frank treatment may appear more cosmetic than decisive for the physical issue in question. Yet, as a closer scrutiny will show, particularly that of the appendix (again the importance of the appendix!), its implications go much deeper and its impact is more forward looking than it may perhaps appear at first sight. In particular, the treatment gives attention to the crystal profile, considering its possible curvature. Actually, in the situation taken the curvature is small, but nevertheless this possibility, together with an expression for the mean growth front profile, is included.

This is of great topical interest in the light of current debates about the origin of rounded habits observed in PE single crystals and their implications for the validity of current theoretical approaches already referred to. In fact, Frank's equations form the starting point for attempts to account for such curved-edge crystals within the framework of nucleation theories. Toda (1986) invoking the additional effect of impurities, in a calculation which is based on the Frank treatment, claims to be able to provide the curvature required by experiment, while Mansfield (1988) claims to derive the experimentally observed curvatures from the Frank equations when applied to moving, rather than to the previously considered, fixed substrate length. The latter is now being incorporated into the most recent version of the nucleation-based theories along the original lines of Lauritzen and Hoffman, as already mentioned, in response to the challenge of the roughening approach to the same problem (Hoffman and Miller 1988, 1989a,b). Here the reduction in $g$ required by the Mansfield treatment is attributed to strains within the lattice due to the folds along the basal surface.

Having pursued the subject so far, a few comments on the values of $L$ are appropriate. The original quotes, derived from fits to growth rate measurements, were in the range of 1 $\mu$m (Hoffman 1985). Currently, the estimates have reduced to a few tens of nm. The latter is mainly in response to the salient experimental work of Point and associates (Point *et al* 1986). To appreciate the issue we recall that constancy of lateral growth is only expected for crystal facets with lengths which are larger than $L$; for lengths which are smaller growth should be exponential with size (hence time). Thus, identification of crystal size below which $G$ becomes size dependent would provide $L$. The above authors using an ingenious morphological method were unable to detect any departure from constancy of $G$ down to the smallest sizes that could be observed, which limits $L$ (if such exists) to within about 200 nm.

However, $L$ must have a lower limit ($L_s$) defined by the stability limit of the secondary nucleus itself with a value which is dependent on the modelling adopted. There have been some arguments as to whether possible $L$ values, which are small enough not to be incompatible with the current status of the experiment, are above or below the stability limit required by the individual theories (if below this would make theory untenable). Thus at one stage a value of 21 nm featured in theoretical works (Hoffman and Miller 1988) which, as was pointed out subsequently (Point and Dosière 1989), was below $L_s$ (51 nm) as arising from the same model. An adjustment which followed achieved the more tenable value of 50 nm (against a calculation $L_s \leqslant 46$ nm) (Hoffman and Miller 1989b).

Whatever the true case may be, it will be clear that a continuing search to find $L$ experimentally is imperative for the subject; to our knowledge no such work is being currently pursued. In this context the limitations of existing theoretical treatments may also need to be borne in mind; for example the spreading rate $g$ is everywhere taken as independent of temperature, which need not be so (Point 1989). Whatever the ultimate answer, the Frank paper provides both a base and signpost for further work in this particular direction, the importance of which, due to the sophistication and subtleness of the issues, is in my opinion not widely enough appreciated.

### 3.5 Concluding remarks: theory and experiment

It may appear from the above sketch and somewhat selective survey that there are both wide and complicated divergences in the theories trying to cope with the intricacies of highly complex phenomena. In order to dispel any possible despondency this may create, it is salutary to recall the positive aspects of the situation, namely the definitive, clear-cut and reproducible nature of the experimental evidence to be accounted for. This is that we have an intrinsic molecular quantity, the fold length ($l$), which determines a readily measurable crystal dimension (the layer thickness), a situation which has no parallel in the field of conventional crystal growth. This $l$ is uniquely and reproducibly determined by $\Delta T$, and by this alone (in fact to such an extent that when a fresh student finds departure from established values I can confidently advise him to calibrate his thermometers). Thus the crystals can be grown with strictly uniform thickness and (with suitable seeding procedures) also with uniform lateral habit and size. The latter, amongst others, enables precise growth rate determinations where the results again give, not only precisely defined functional relations, but also reproducible numerical values. Thus, complex as polymers may be thought of, the habits, sizes (including $l$) and growth rates of their crystals can be better defined and are more reproducible than of many simpler substances. In fact, in the above respects polymers seem to be insensitive to the vagaries of undefinable external circumstances (like impurities) which usually beset conventional crystallisation studies. It appears that complex as they may be the chains know precisely what they 'want' to do and do not readily let themselves be deflected from it. This means that they must have strong and specific motivations which is up to theory to uncover, which as far as it has not yet been adequately achieved, should provide encouragement for continuing efforts. One would think that clear-cut effects must have correspondingly clear-cut and hopefully simple causes.

## 4 SOME LATEST TRENDS

In this final section I shall point to two latest lines in polymer crystallisation to provide some up-to-date information and forward-looking perspectives. Not unnaturally both are from our Bristol laboratory being pursued in continuing consultation, even if not any longer in active collaboration, with Charles Frank.

### 4.1 From paraffins to polyethylene: onset of chain folding

The first is work on strictly uniform ultralong paraffins in a quest for the onset of chain folding, an item briefly hinted on in the second section of this article. Here we need to recall that polyethylene is nothing but a long *n*-alkane, the crystal characteristics of which were known well before that of polymers. It was known that the crystals are lamellae with the chains perpendicular to the basal planes, but here fully extended, with the extended chain length defining the true lattice period. The analogy between polyethylene is complete except that the conventionally defined unit cell in polyethylene is only a subcell in the paraffin (it is the gap we tried to bridge under the guidance of Charles Frank early on by considering the fold length as the true unit cell, an approach from which we became deflected by events (see section 2)). The question clearly arises of how long need an *n*-alkane be before chain folding sets in. In order to answer this question, together with ensuing issues about the structures and mechanisms involved, we would need to pass on to *n*-alkanes of increasing chain lengths where the chains are strictly uniform, as done in the previously studied shorter paraffins. This would enable correlation to be made between lamellar thickness and the related true crystallographic unit cell and molecular length or, alternatively, to establish the point of cessation of such correlation. However, materials of sufficiently long chains, whereby the precedent of polyethylene chain folding could be expected, have not been available till recently, at least of sufficient uniformity.

At this point we need to step back and reflect upon the fact that polymer synthesis involves the random coming together of monomers, hence the statistical accretion of reacting species. The result is inevitably a distribution of chain lengths, and for this reason polymer science remains necessarily a statistical science as opposed to the more exact science of small molecules†.

It is this basic boundary between statistical and exact molecular science which we set out to cross with the works to be introduced next.

The key to these developments is the strictly controlled step-by-step synthesis of long chains by methods of traditional organic chemistry. We were extremely fortunate that Professor Mark Whiting, Chemistry Department in Bristol, has undertaken this exceedingly difficult and lengthy work in response to our needs. Chains of up to 394 carbon atom lengths were

† This is not to depreciate the beautiful studies on polyethylene oxides of sharp, yet still not quite uniform, distribution by Kovacs (following preceding works by Skoulios) which I cannot find space to describe here (see e.g. Buckley and Kovacs 1984).

reached in mg quantities on which the physical investigations were carried out by my colleague Goran Ungar, joined later by Sally Organ, as part of our wider programme of polymer crystallisation. This work, while placing extreme demands both on the preceding chemical and subsequent physical procedures, has proved to bring new revelations in terms of precision, definiteness and reproducibility of the results (for up to date reference list see Keller (1989) and for an interim review see Ungar (1988)).

Very briefly, the chains indeed were found to fold, and this at around 100–150 C atom lengths, the precise value depending on crystallisation temperature. The fold lengths in the final stages of crystallisation were found to be integer fractions of the chain length changing by such fractional lengths with changes in crystallisation temperatures and/or annealing conditions. Amongst many other points of interest, in view of the strict uniformity of the chains and in the exact knowledge of their lengths, the portion of the chain within the fold region could be assessed. This, in the most regular cases at least, amounted to no more than 2–4 C atoms. Such cases also revealed specific vibrational bands in the infrared spectrum providing access to the study of the structure of the fold itself. A similar and coordinated approach by x-ray crystal structure analysis is now clearly invited to bridge the gap between the crystallography of paraffins and polyethylene.

However, it was recognised by synchrotron-based x-ray work that the folds were not always equally sharp. In fact, it was observed that crystallisation starts with rather imperfect crystals with significant disorder along the fold surface, where the fold length itself is not closely of integer fractional value. However, during isothermal crystal growth the crystal can still perfect itself, in the course of which the fold length either increases or, rather surprisingly, decreases to its nearest lying integer fractional value with a concommitant regularisation of the fold surface. The process, as envisaged presently for a once folded chain, is illustrated schematically in figure 13, the meaning of which, with the help of the caption, should be self-explanatory.

At this point the following features deserve highlighting, particularly with reference to the previous sections.

First consider the fold surface problem. It seems that the folds can be sharp with identifiable structure features and, at least for up to four folds per chain (the maximum number we could get with our compounds), must be adjacently re-entrant. But, as just seen, there can be disorder even for adjacent re-entry and even for such a short chain. Here the nature of the final product will depend on the stage at which crystallisation and rearrangement is interrupted by cooling. Thus the practically achieved fold surface is not unique and can span a spectrum of order from near amorphousness to crystallographic. The most important point, however, is the pronounced drive for the chains to fold and the trend towards maximum regularity in the folded state which, depending on the degree of rearrangement allowed, they may achieve to varying degrees.

The second and third points relate to the mode of deposition and subsequent rearrangements, the two being in a way interrelated. The scheme in figure 13 will reveal the emphasis on probability-controlled deposition, and hence on the entropic nature of the initial act of growth, a concept already alluded to in the theoretical section in a different, yet not quite unrelated, context. The chain will attach itself according to laws of probability and will 'stick' provided that the minimum thermodynamic stability criterion is satisfied. Optimisation, hence the resulting maximum rate, will not correspond to the most perfect structure even within the limits of a particular fold number (one fold in figure 13). Perfectioning will then be a further stage which may be subsequent to growth, or may still be part of the rate determining step. The reality of the probabilistic nature of the deposition will be highlighted by the next (and last) example to be quoted from the *n*-alkane work, while the importance of rearrangement will be brought out further by a different mode of crystallisation of polyethylene with which I shall finish this article.

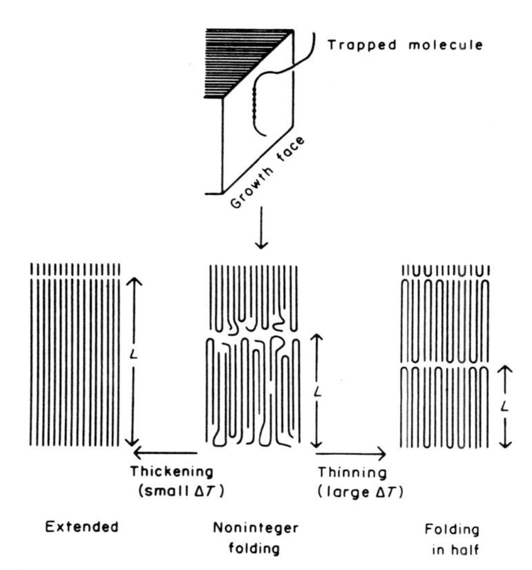


**Figure 13** Schematic representation of crystallisation of a long *n*-alkane of uniform length. Top: initial deposition, in general unfavourable by probability, for formation of either fully extended or exactly once folded conformation. The optimum kinetic pathway leads to a once folded crystal with disordered fold and internal defects (bottom row middle). The latter can then rearrange subsequently in two ways: either fully chain extend by thickening (bottom left) or achieve a precisely once folded state with ends at surface and folds regular through thinning (bottom right) (Ungar and Keller 1987).

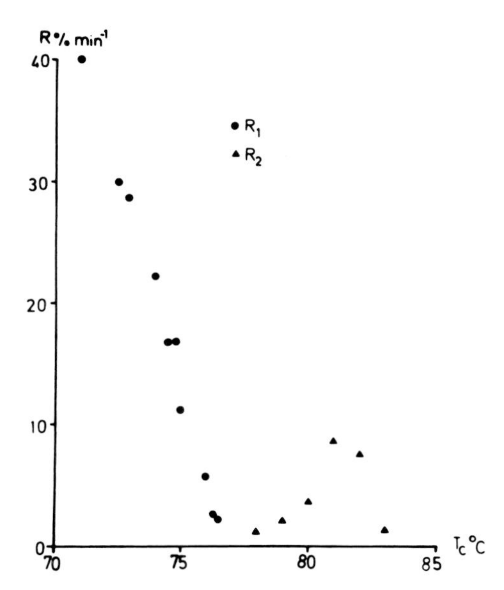


**Figure 14** Rate of crystallisation of *n*-alkane $C_{198}H_{398}$ from solution as a function of crystallisation temperature ($T_c$) displaying pronounced minimum. $R_1$ and $R_2$ refer to folded and extended chain crystallisation respectively (Organ *et al* 1989).

Possibly the most unique feature of the *n*-alkane work emerged from the study of crystallisation rates as a function of temperature, including both primary nucleation and growth, both in combination and individually. It needs recalling that in these alkanes on lowering the crystallisation temperature $T$ (i.e. raising $\Delta T$) crystallisation will occur consecutively in extended, once folded, twice folded, etc, configurations. The overall rates, $G$, would (in the absence of transport terms which are negligible in the vicinity of the melting point, the $T$ range here concerned) be expected to increase exponentially with increasing $\Delta T$. In fact, $G$, after an initial increase, was found to reach a maximum, then to dip into a sharp minimum to rise again thereafter (figure 14). This was observed to apply both for growth and primary nucleation, both from melts and solutions, as assessed by x-rays, calorimetry and by visual microscopy. At growth temperatures above the minimum the chains were extended, below they were in a once folded conformation, the change-over occurring at the start of the second rise. This effect of a minimum is unprecedented in any crystal growth; it is extremely sharp and strictly reproducible. It appears as if the chain 'hesitated' whether to extend or to fold, and in fact does neither, with growth nearly stopping as a consequence.

An explanation was suggested by my colleague Goran Ungar, which is becoming increasingly substantiated by ongoing experiments. This is that downwards (in terms of $T$) till the minimum, only the extended chain deposition is stable but not that of the once folded chains. However, even if transiently, once folded depositions will still occur and their probability will increase along the growth face with decreasing $T$. Being subcritical, these once folded nuclei cannot grow, yet they block the deposition of the chains in their stable extended conformation, hence the growth of the crystal (and also of the primary nucleation) itself. Ungar termed this novel phenomenon, seemingly characteristic of polymers alone, 'self-poisoning'. Clearly it is a specific and striking manifestation of the general fact that the chain has to deposit in the 'right' conformation to perpetuate the crystal, and hence of the probabilistic control of chain-folded crystallisation. Concurrent computer simulations of the late David Sadler (Sadler and Gilmer 1987) as based on his roughening theory (Sadler and Gilmer 1987), in fact did contain the above growth minimum, and became fully appreciated only after the actual experimental observation itself. This seems to testify to the predictive power of roughening-based theories; however, whether this is exclusive to this class of theory, or whether others would also be able to display the same, has to my knowledge not been tested.

### 4.2 Metastable transient mobile phases: their role in chain folding

This final topic has a long background history. Here, I shall only try to convey essentials such as relevant to the mainstream issues, adding only one more illustration.

The origin of the subject lies in crystallisation studies of polyethylene under hydrostatic pressure. As Wunderlich and colleagues observed first (see e.g. Wunderlich 1973), polyethylene, when crystallised above 3.3 kbars (the precise value is of later origin) differed from the familiar plastic; it was hard and brittle and of near 100% crystallinity by conventional criteria. Morphologically, the samples had a highly stratified structure with chains perpendicular to the stratification. It was inferred from the scale of the stratification, which was 0.1–several $\mu$m, that the chains must be in an extended, or at least not too numerously folded conformation, which by the morphological evidence of tapering crystal edges must have been attained by refolding subsequent to an initially chain-folded crystallisation of more conventional fold length (few tens nm). It was then an important finding by Bassett (see e.g. Bassett 1982) that polyethylene exists in a hexagonal crystal phase ($h$) (as opposed to the conventional orthorhombic ($o$)) at elevated pressures ($P$), 3.3 kbar being a triple point. This hexagonal phase is highly mobile (we currently attribute liquid crystal characteristics to it) which facilitates refolding to longer fold lengths and ultimately to full chain extension while in this phase, which is the origin of the 'extended chain type' texture observed in crystallisation under pressure (figure 15, a recent version by Hikosaka, see below).

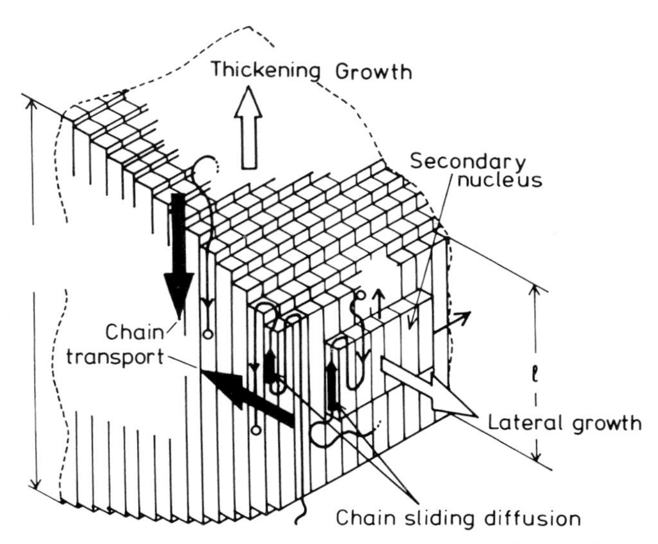


**Figure 15** Development of a crystal via chain folding to chain extension (and beyond) as visualised by Hikosaka (1989). This picture is based on experiments with polyethylene melts at elevated pressures and is most pertinent to the mobile hexagonal phase which allows long-range sliding diffusion.

In Bristol we became involved in high-pressure crystallisation studies through the working visit of Masa Hikosaka, who not only brought along his own apparatus but graciously allowed us to copy it on his departure. This apparatus, to our knowledge, is unique by preserving constant pressure during volume contraction on crystallisation (not ensured by most other facilities elsewhere) while also allowing direct viewing optically and recording x-ray diffraction patterns of visually selected crystals during crystallisation. These features allowed him, and subsequently Sanjay Rastogi and Hideo Kawabata in these laboratories, to observe crystals growing, to measure their growth rates (lateral *in situ* and thickness by electron microscopy after pressure quenching) and to identify crystal structure and changes therein at any preselected portion or along chosen pathways in the *PT* phase diagram. Amongst much else it was found that the crystals were only arising and growing in their *h* form, even within the stability regime of the *o* form of the phase diagram, where the *h* phase is metastable. What is more, when transforming into their stable *o* form, an event which could be identified *in situ* during growth, both lateral growth and thickening stopped (at least on the timescale of our observation) (Rastogi *et al* 1990).

It follows from the above that crystals grow both laterally and in the thickness direction (where the latter implies refolding) only while in the *h* phase, irrespective as to whether this is the stable or the metastable phase. In view of the fact that this applies also to the metastable regime the phase diagram has ceased to provide guidance as to which crystal phase to expect; in fact, by the above, the *h* phase is strongly favoured, if not the only form capable of growth. As the *h* phase is highly mobile, allowing substantial rearrangement, and refolding in particular, this suggests that the ability of refolding is a requirement or at least a promoting factor for growth.

The above is a statement of fact within our observational range, i.e. down to 1.8 kbar so far. The question arises as to how far and how deep into the *o* stability regime does this apply. In particular, does it apply to conventional conditions of crystallisation at atmospheric pressure where the *h* phase, being so transient there if it exists at all, may have remained undetected so far? Alternatively, is there a change in crystallisation behaviour from that via an *h* phase to one directly into the *o* phase at *P* and *T* which are unspecifiable by thermodynamics alone? In view of the fact that all theories so far have only considered chain deposition in the *o* phase and, with the exception of Hikosaka, have not considered the possibility of rearrangement subsequent to chain deposition as a growth controlling factor, the pertinence of the new findings and of the questions arising are obvious. Hikosata himself first only considered the *o* phase alone (1987a,b) with the incorporation of refolding by sliding diffusion (see section 3) but in the light of the joint recent recognition has taken cognisance of the role of the hexagonal phase as providing hugely enhanced scope for such sliding diffusion, and of the possible effect of $h \rightarrow o$ transformation during growth, in his latest, essentially nucleation-based theoretical approach, and this in a quantitative manner (Hikosata *et al* 1990), figure 15 being one way in which this can be visualised.

While of major influence on the present issue, the mere fact that a new phase state first passes through a metastable station in the course of change in matter of state (e.g. vapour → solid, liquid → solid) is in itself not quite unusual and is in fact expressed by Ostwald's 'Stage Rule' dating back to 1892. An intriguing possibility arises when considering the effect of phase size on the phase diagram. Namely, with appropriate (and reasonable) choice of surface parameters the situation can arise that the true thermodynamic stability conditions can invert with size. Specifically, for a polyethylene crystal that is thin enough (our case, the argument in principle is more general) the *h* phase could be the stable phase and the *o* phase the metastable one even in the *P–T* regime where normally the converse applies, i.e. stable *o* and metastable *h* phase for an infinite size crystal. If and when this is the case, true metastability need not even be involved to account for the observation of a metastable phase appearing first. In fact, here the metastable phase will, in its diminutive form, be the stable phase, with an inversion of phase stability on growth; an accompanying $h \rightarrow o$ transformation will then set in beyond a certain size (lamellar thickness, fold length) as the crystal grows (thickness for the present purpose). While the occurrence of the above is a general possibility (being one manifestation of the 'Ostwald Stage Rule', here rather founded on equilibrium thermodynamics) it acquires special significance for polymers such as polyethylene. Here, below the triple point, the initial (and for infinite sizes metastable) $h$ phase is mobile allowing for ready refolding to greater thicknesses thus leading up to a size-induced stability inversion, in the present case to $h \rightarrow o$ transformation, which then in turn arrests (or drastically reduces) the thickening which has brought the newly formed phase into being with the effect of limiting the final thickness of the lamellae. The latter (i.e. lamellae of uniform small thickness) is in fact the principal feature of a crystalline flexible chain polymer, hence the core issue of this article.

It follows that one may envisage crystallisation to start through a transient phase which is the stable one when small (thin), but metastable beyond a certain thickness. This means that in our polymer crystallisation could, within this transient phase, start with a much reduced length, for which the chain deposition probability, and hence consequent growth rate, is vastly (about exponentially) enhanced. The transient phase being mobile, thickening by chain sliding would follow, until terminated, or slowed down, by the transformation into the final, less mobile phase at or beyond a certain thickness which is characteristic of the final lamella thus created.

The fact that change in crystal size, in particular thickening, can induce phase transformation through chain refolding while in an initially mobile phase, into a less mobile phase which then terminates (or slows down) this refolding, has already been conclusively demonstrated (Rastogi and Ungar 1990). Here, the material was poly 1-4 transbutadiene which has two crystal forms, a mobile, and a less mobile even at atmospheric pressure, and the experiment was carried out not during crystallisation but on the fully crystalline material. The experimental conditions were chosen so that thickening induced a transition into the less mobile phase, thus demonstrating that the coexistence temperature between the two phases is indeed crystal size, specifically, thickness dependent. If follows that primary crystallisation, if it could be monitored near the phase coexistence temperature (which in fact cannot be done with poly 1-4 transbutadiene) would indeed involve a phase change with crystal growth. This would also include self-limiting lamellar thickening, hence a limiting specific lamellar thickness, here by a combination of thermodynamic (size-dependent phase transition) *and* kinetic (crystal phase determined chain mobility) reasoning. From here it is only one step to envisage the same for the situation where the mobile phase is virtual (for the $P$ and $T$ concerned) for the same to apply to the polyethylene and possibly other flexible polymers, a projection with which I am concluding this article.

## 5 CONCLUDING REMARKS

The tale, as here presented, starting from historical reminiscences and reaching out to some of the latest results, has in this particular form not been told before. I can only hope that it will be helpful for a retrospective appreciation of the subject on all levels of specialisation, for glimpsing new perspectives and possibly with lessons to be learnt on making choices and judgements. The inexhaustable wealth of the subject will hopefully have become apparent, and also that we are nowhere near the end of the road still studded with novelties and surprises. Some of the latest may well modify existing views, who knows, perhaps out of recognition. The excitement and challenge never ceases and I am sure, that as previously, this will continue to attract Sir Charles' interest, to whom, as I tried to show here, the subject owes so much both in an explicit and in an indirect manner.

## ACKNOWLEDGMENT

I am indebted to Mr Gerhard Goldbeck-Wood for providing invaluable guidance on the subtler aspects of theory as is also stated in the text.

## REFERENCES

Barham P J, Chivers R A, Keller A, Martinez-Salazar J and Organ S J 1985 *J. Mater. Sci.* **20** 1625

Bassett D C 1981 *Principles of Polymer Morphology* (Cambridge: Cambridge University Press)

—— 1982 *Developments in Crystalline Polymers* vol 1, ed D C Bassett (London: Applied Science Publishers) pp 115–50

Bassett D C, Frank F C and Keller A 1959 *Nature* **184** 810–11

—— 1963a *Phil. Mag.* **10** 1739–51

—— 1963b *Phil. Mag.* **12** 1753–87

Bassett D C and Keller A 1962 *Phil. Mag.* **7** 1553–84

Bernal J D 1967 *The Origin of Life* (London: Weidenfield and Nicholson)

Bennet C H, Buttiker M, Londoner R and Thomas H 1981 *J. Stat. Phys.* **24** 419

Buckley C P and Kovacs A J 1984 *Structure of Crystalline Polymers* ed I H Hall (Amsterdam: Elsevier) pp 261–307

Burton W K, Cabrera N and Frank F C 1951 *Phil. Trans. R. Soc.* **243** 299–358

Dosière M, Colet M Ch and Point J J 1986 *Morphology of Polymers* ed B Sedlacek (Berlin: de Gruyter) pp 171–8

Fatou J G 1989 *Polymer Preprints, Am. Chem. Soc. Symp. Miami Beach* **30** no 2, 286–7

Fischer E W 1957 *Z. Naturf.* **12a** 753–4

Flory P J 1958 *Growth and Perfection of Crystals; Proc. Int. Conf. on Crystal Growth, Cooperstown* (New York: Wiley) p 530

—— 1962 *J. Am. Chem. Soc.* **84** 2857–67
Flory P J and McIntyre A D 1955 *J. Polymer Sci.* **18** 592
Frank F C 1949 *Discuss. Faraday Soc.* **5** 48–54
—— 1958 *Growth and Perfection of Crystals*; *Proc. Int. Conf. on Crystal Growth, Cooperstown* (New York: Wiley) pp 529–30
—— 1962 unpublished
—— 1974 *J. Crystal Growth* **22** 233–6
Frank F C and Tosi M 1961 *Proc. R. Soc.* A **263** 323–9
Geil P H 1963 *Polymer Single Crystals* (New York: Interscience)
Goldenfeld N 1984 *J. Phys. A: Math. Gen.* **17** 2807
Hikosaka M 1987 *Polymer* **28** 1257–64
—— 1989 Private communication
—— 1990 *Polymer* **31** 458–68
Hillig W B 1966 *Acta Metall.* **14** 1868
Hoffman J D 1983 *Polymer* **24** 3–26
—— 1985 *Polymer* **26** 803
Hoffman J D, Davis G T and Lauritzen J I 1975 *Treatise on Solid State Chemistry* vol 3, ed N B Hannay (New York: Plenum) pp 497–614
Hoffman J D, Guttman C M and Di Marzio E A 1979 *Discuss. Faraday Soc.* **68** 177–95
Hoffman J D and Miller R 1988 *Macromol.* **21** 3038–51
—— 1989a *Macromol.* **22** 3038–54
—— 1989b *Macromol.* **22** 3502–5
Holland V F and Lindenmeyer P H 1965 *J. Appl. Phys.* **36** 3049–56
Jaccodine R 1955 *Nature* **176** 305
Jackson K A 1968 *J. Crystal Growth* **3/4** 507–17
Jones D H, Latham A J, Keller A and Girolamo M 1973 *J. Polymer Sci.* C **38** 237–50
Kawai T and Keller A 1965 *Phil. Mag.* **11** 1165–77
Keller A 1957 *Phil. Mag.* **2** 1171–5
—— 1958 *Growth and Perfection of Crystals*; *Proc. Int. Conf. on Crystal Growth, Cooperstown* (New York: Wiley) pp 449–528
—— 1962 *Polymer* **3** 393–421
—— 1964 *Kolloidzschr. Z. Polymere* **219** 118–31
—— 1969 *Kolloidzschr. Z. Polymere* **231** 386–418
—— 1968 *Rep. Prog. Phys.* **32** 623–704
—— 1972 *MTP International Review of Science, Macromolecular Science, Physical Chemistry* series I, vol 8, ed A D Buckingham and C E H Bawn (London: Butterworth) pp 105–56
—— 1979 *Discuss. Faraday Soc.* **68** 145–66
—— 1981 *Structural Order in Polymers; International Union of Pure and Appl. Chem.* ed F Ciardelli and P Giusti (Oxford: Pergamon) pp 135–80
—— 1984 *Polymers, Liquid Crystals and Low-Dimensional Solids* ed M March and M Tosi (New York: Plenum) pp 3–142
—— 1986 *Recent Developments in Morphology of Crystalline Polymers* ed B Sedlacek (Berlin: de Gruyter) pp 3–26
—— 1989 *Polymer Preprints, Am. Chem. Soc. Symp. Miami Beach* **30** no 2, 263–4
Keller A and O'Connor A 1958 *Discuss. Faraday Soc.* **25** 114–21

Khoury F and Passaglia E 1975 *Treatise on Solid State Chemistry* vol 3, ed N B Hannay (New York: Plenum) pp 335–496
Kovacs A J, Lotz B and Keller A 1969 *J. Macromol. Sci.* B **3** 385–425
Lauritzen J I 1973 *J. Appl. Phys.* **44** 4353–9
Lauritzen J I and Hoffman J D 1960 *J. Res. NBS* A **64** 73–102
Leamy H J, Gilmer G H and Jackson K A 1975 *Surface Physics of Materials* vol 1, ed J M Blackley (New York: Academic) pp 121–88
Lotz B, Kovacs A J, Bassett G A and Keller A 1966 *Kolloidztsch. Z. Polymere* **209** 115–28
Magill J H 1977 *Treatise on Materials Science and Technology* vol 10, ed J M Schultz (New York: Academic) pp 1–368
Mandelkern L 1958 *Growth and Perfection of Crystals, Proc. Int. Conf. on Crystal Growth, Cooperstown* (New York: Wiley) pp 467–95
—— 1986 *Polymer* **17** 337–50
Mansfield M L 1988 *Polymer* **29** 1755–9
Organ S J and Keller A 1985 *J. Mater. Sci.* **20** 1571–85
Organ S J, Ungar G and Keller A 1989 *Macromol.* **22** 1995–2000
Peterlin A, Fischer E W and Rheinhold C 1962 *J. Chem. Phys.* **37** 1403–8
Phillips P J 1989 *Rep. Prog. Phys.* **53** 549–604
Point J J 1979a *Discuss. Faraday Soc.* **25** 165–76
—— 1979b *Macromol.* **12** 770–5
—— 1989 Private communication
Point J J, Colet M Ch and Dosière M 1986 *J. Polymer Sci. Phys. Ed.* **24** 357–88
Point J J and Dosière M 1989 *Macromol.* **22** 3501–2
Rastogi S, Hikosaka M, Kawabata H and Keller A 1990 *Macromol. Chem. Symp.* (Sorrento: EPF) in press
Rastogi S and Ungar G 1990 *Macromol.* in press
Sadler D M 1983 *Polymer* **24** 1401–9
—— 1987a *Nature* **326** 174–6
—— 1987b *J. Chem. Phys.* **87** 1771–84
Sadler D M and Gilmer G H 1984 *Polymer* **25** 446–52
—— 1986 *Phys. Rev. Lett.* **56** 2708–11
—— 1987 *Polymer Commun.* **28** 241–6
Sadler D M and Keller A 1970 *Kolloidzschr. Z. Polymere* **239** 641–54
Sanchez I C and Di Marzio E A 1972 *J. Res. NBS* A **76** 213–23
Storks K H 1938 *J. Am. Chem. Soc.* **60** 1753–61
Stranski I and Kaischev R 1935 *Z. Phys.* **36** 393–403
Till P H 1957 *J. Polymer Sci.* **24** 301–6
Toda A 1986 *J. Phys. Soc. Japan* **55** 3419–27
Toda A and Tanazawa Y 1986 *J. Crystal Growth* **76** 462–8
Ungar G 1988 *Integration of Fundamental Polymer Science and Technology* vol 2, ed P J Lemstra and L A Kleintjens (Amsterdam: Elsevier) pp 346–62
Ungar G and Keller A 1987 *Polymer* **27** 1835–44
Volmer M 1939 *Kinetic der Phasenbildung* (Leipzig: Steinkopft Verlag)
Wittman J C and Lotz B 1985 *J. Polymer. Sci. Phys. Ed.* **23** 205–26
Woodward A E 1988 *Atlas of Polymer Morphology* (Munich: Hanser)
Wunderlich B 1973 *Macromolecular Physics* vol 1 (New York: Academic)

# Polymer Fluids

M R Mackley

## 1 FIRST ENCOUNTER

My first meeting with Charles Frank concerned Christmas trees and thermodynamics. I attended his lecture course on thermodynamics at Bristol in 1969 and 1972 and at the beginning of both he introduced the idea of Maxwell-type equations using conjugate force and extension vectors. This he did by decorating a 'Christmas tree' with individual force and extension units hanging from branches and then proceeding to consider the equilibrium of the whole system. Certainly in 1969 the whole experience was a little strange to the class and many of us were left somewhat mystified. In addition the historical journey he took us through concerning the birth of thermodynamics from Clapeyron to Le Chatelier and the Carnot family at that time seemed somewhat obscure to the course subject of Physics of Materials and comments that in the nineteenth century Anglo-Saxons seemed to have a 'broad and shallow approach to Science' in the field of thermodynamics and the French 'a narrower but deeper insight' were almost certainly lost on most of us. However, even as first year postgraduate students, there was an immediate appreciation by the class that Professor Frank was a scientist of great substance and what he said should be considered very carefully indeed.

In the early 1970s the MSc course on the Physics of Materials was enjoying a most productive period. Lectures were stimulating and interesting research projects abounded. These projects covered a very wide range of topics from polymer physics, dislocation theory and composites to grain-boundary studies and x-ray topography. Each project was supervised by a member of staff and in a large number of cases it seemed that many conceptual ideas or suggestions had come in some way from Professor Frank. Even if the project did not involve the Professor directly his general interest and enthusiasm for exploring uncharted scientific waters was infectious to us all.

At this early stage direct contact with Professor Frank was not often made outside lectures except on a number of social occasions. On reflection these events were extremely important to a number of us as they showed the very human side of him and later, when we had serious scientific problems, enabled us to approach him. The Bristol Physics Christmas party was always interesting and well patronised and this annual event was very much linked to the social awareness of Professor Frank's wife. Maita Frank was the perfect back-up for Professor Frank and her presence always ensured that the conversation was lively and that spirits were high. The Departmental cabaret was often rather good and I well remember one occasion when Maita greatly enjoyed secretly passing on one of the Professor's fine waistcoats to be used by students for a sketch on him.

These were exciting days; it was the age of the Beatles, free thinking and rapid scientific advances in the field of materials physics. Within the UK, Bristol Physics seemed well advanced in terms of a broad-ranging and fundamental team examining many aspects of solid state physics; this certainly did not happen at Bristol by accident and it can almost certainly be attributed initially to Professor Mott and then to Professor Frank's appreciation of the importance and scientific interest of the physics of solids.

## 2 FLUID MOTIONS

Charles Frank's breadth of knowledge and interest in physical science is enormous and one facet of this interest concerns polymers and the way polymers can crystallise. Charles was kept abreast of developments by Professor Andrew Keller's very active group at Bristol. In particular, the Keller group had fully established the morphological organisation of polymer single crystals (see, for example, Keller 1968) and in addition had carried out studies of the so-called 'shish kebab' fibrous crystals that could be induced from solutions grown during mechanical stirring. Micrographs and schematic structures of single crystal and shish kebabs are shown in figure 1. In 1970 the Bristol Group was sent a preprint of work carried out by Pennings and co-workers (1970a) on flow-induced crystallisation of polyethylene. Using a Couette rotating cylinder apparatus, they had shown in an elegant series of experiments that shish-kebab fibrous crystals were only formed when a form of secondary flow known as 'Taylor vortices' were present and if the usual 'simple shearing' flow alone was acting no

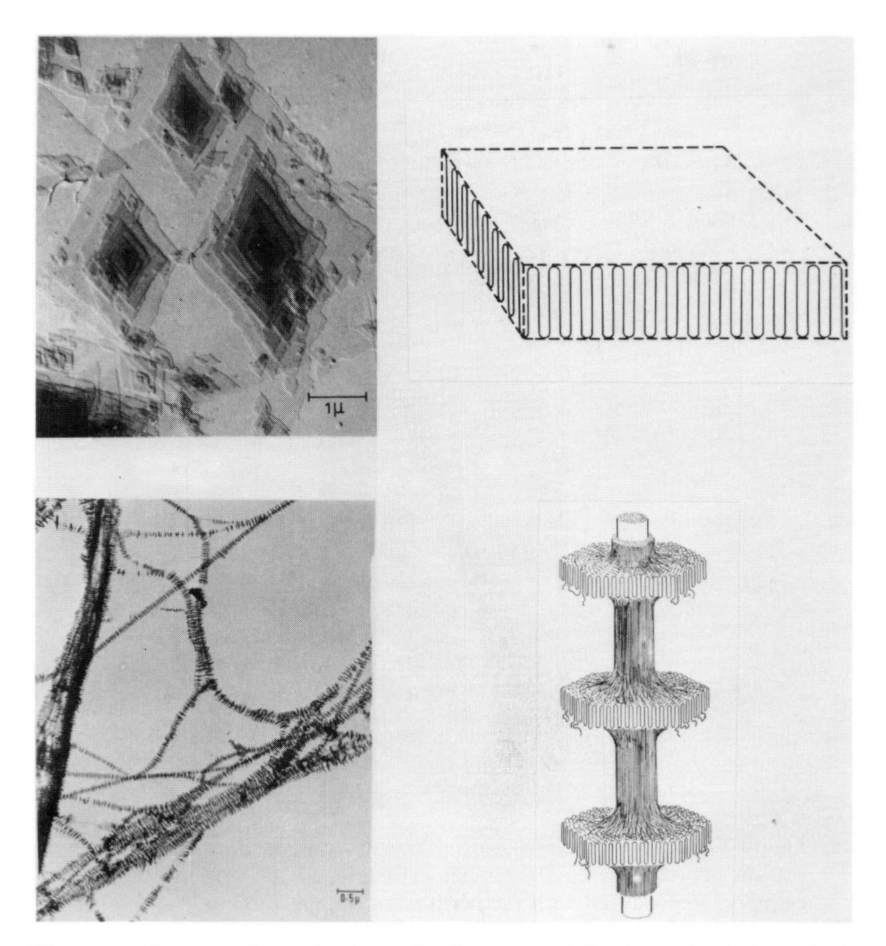


**Figure 1** Photographs and schematic diagrams of platelet polyethylene single crystals grown from a quiescent solution and 'shish-kebab' fibrous crystals grown from a flowing polyethylene solution. Parts reproduced with the permission of the Royal Society from Mackley and Keller (1975).

fibrous crystals would form. Professor Frank identified the 'extensional flow' components between adjacent vortices as the crucial 'chain stretching' element of the flow. Theoretical work by, amongst others, Ziabicki (1959), had shown that rotation-free extensional flows were more efficient at stretching chains than simple shearing flows and it was concluded that chain stretching in solution was a necessary requirement for fibrous crystallisation. A section of the Taylor vortex flow between rotating cylinders is shown in figure 2. The symmetry of this flow pattern is elegant and this is an aspect that Charles Frank appreciated.

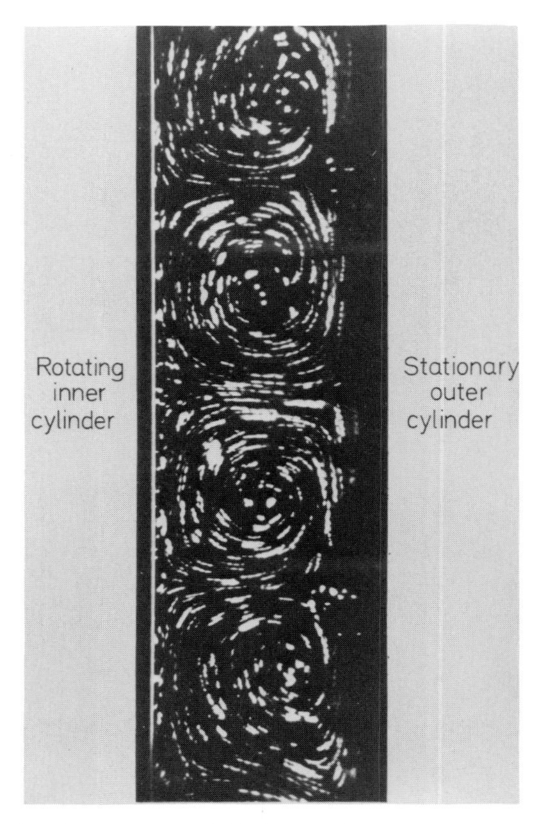


**Figure 2** The flow pattern of Taylor vortices, viewed as a transverse section between the rotating inner cylinder and the stationary outer cylinder. Reproduced with the permission of *Nature* from Keller *et al* (1975).

The theory of Ziabicki stated that extensional flows were required to obtain efficient chain stretching, but in order to achieve this condition $\dot{\gamma}\tau \geqslant 1$, where $\dot{\gamma}$ is the extensional deformation rate and $\tau$ the longest relaxation time of the chain. For polymers in solution relaxation times of order $10^{-2} - 10^{-3}$ s might be expected and consequently extensional deformation rates of $\dot{\gamma} \sim 10^2$–$10^3\,\mathrm{s}^{-1}$ are required. At this point Charles Frank made an incisive suggestion concerning an experimental way that these conditions might be met. He suggested that by firing two jets into the faces of each other uniaxial compression flow would be generated, as shown schematically in figure 3. By generating a stagnation point at the centre of symmetry the deformation rate tensor would be given by

$$\dot{\gamma} = \begin{bmatrix} -\dot{\gamma} & 0 & 0 \\ 0 & \frac{1}{2}\dot{\gamma} & 0 \\ 0 & 0 & \frac{1}{2}\dot{\gamma} \end{bmatrix}.$$

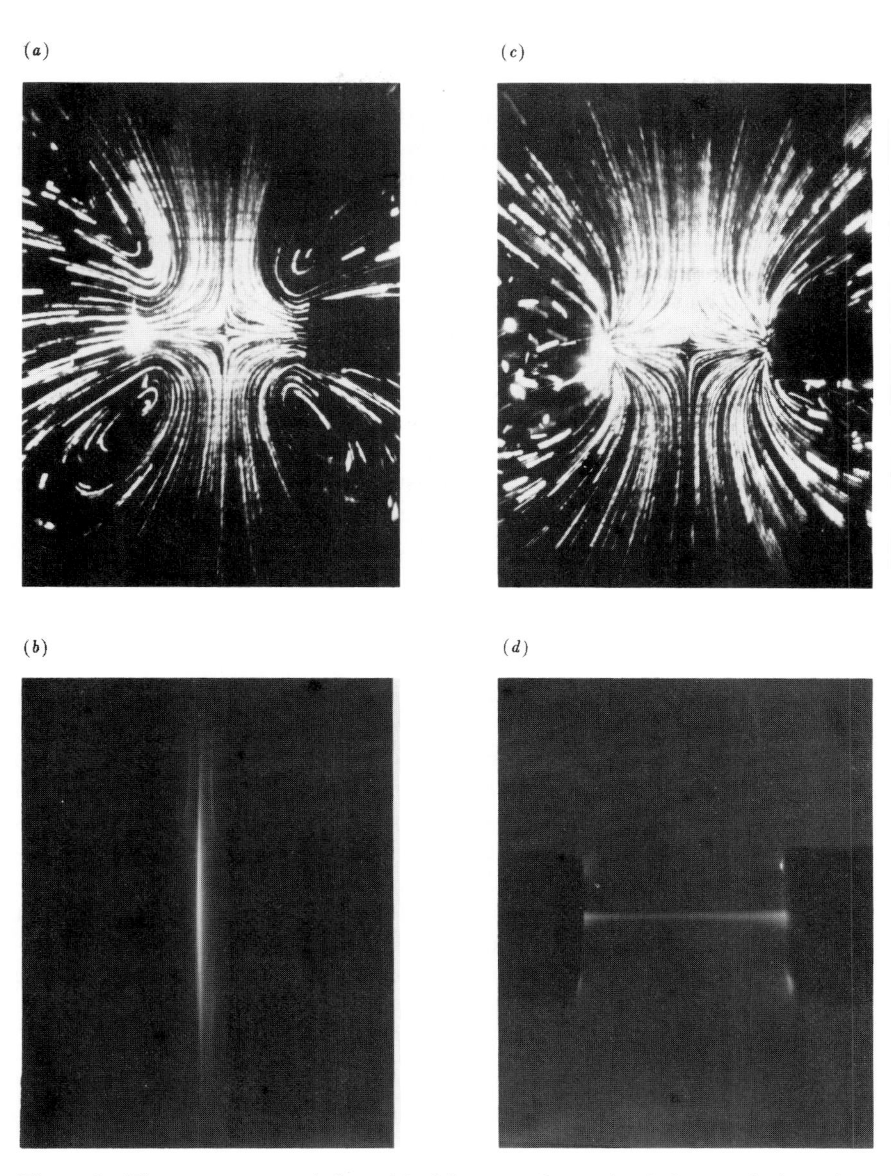


**Figure 3** Flow pattern and flow birefringence for polyethylene solutions in a double jet apparatus. (*a*) Flow pattern for outflow (axial compression), (*b*) flow birefringence for outflow, (*c*) flow pattern for inflow (uniaxial extension), (*d*) flow birefringence for inflow. Reproduced with the permission of John Wiley publishers from Crowley *et al* (1976).

The use of fine bore diameter jets of order 1 mm would enable velocities to change from the order of $1\ \mathrm{m\,s^{-1}}$ to zero in a distance of order 1 mm thereby producing the required deformation rates of $10^3\ \mathrm{s^{-1}}$.

I was given the task of trying this idea out. The polymer group's photographer 'Laurie' Robinson had told me that she had been involved in some early experimental work with Professor Frank and this had not been an astounding success; however, in this instance with the guidance of both Charles Frank and Andrew Keller a system was successfully conceived and tested (Frank *et al* 1971). We found that the reverse flow to uniaxial compression, i.e. uniaxial extension, was the more stable of the flows and this was achieved by sucking fluid into the jets rather than out of them as shown in figure 3. Using a flow birefringence technique we were able to observe *in situ* localised chain extension in extensional flow (Mackley 1972, Mackley and Keller 1975). The effect has since been seen and studied by many others and reviews on the topic are given for example by Keller and Odell (1985). Professor Andrew Keller and in particular Dr Geoff Odell are still highly active in the field at Bristol and have greatly extended the technique in a number of different areas.

## 3 FOUR, TWO AND SIX ROLL MILLS

The reason why localised flow birefringence was seen was at first not fully understood. However a further series of experiments soon made the effect seem obvious.

I do not think of Charles Frank as primarily a 'fluids man'. The fluids scientists he spoke most often about were Prandtl (and his ingenious tea pot) and G I Taylor. I suspect G I Taylor's mind and Charles Frank's worked on similar lines. They were both very good mathematicians, free-thinking scientists and often ready to exploit, wherever possible, the symmetry of the problem. In this respect G I Taylor had realised that efficient droplet deformation could be achieved in extensional flows and he had considered the rotation-free two-dimensional pure shear deformation rate given by

$$\dot{\gamma} = \begin{bmatrix} \dot{\gamma} & 0 & 0 \\ 0 & -\dot{\gamma} & 0 \\ 0 & 0 & 0 \end{bmatrix}.$$

In order to generate this flow he proposed using a four roll mill geometry and used it to study liquid droplet break-up (Taylor 1934). In 1974 I had a four roll mill built in the Physics Department and together with two final-year project students we observed localised flow birefringence for polymer solutions in this flow deformation, as shown in figure 4 (Crowley *et al* 1976). We felt confident that the flow was dominantly pure shear and of a

(*a*)

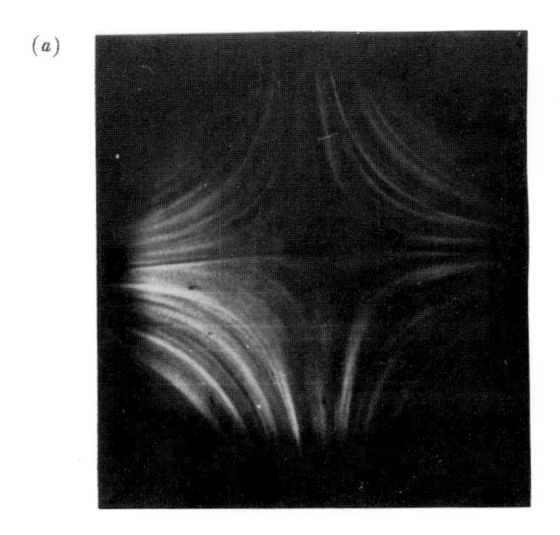

(*b*)

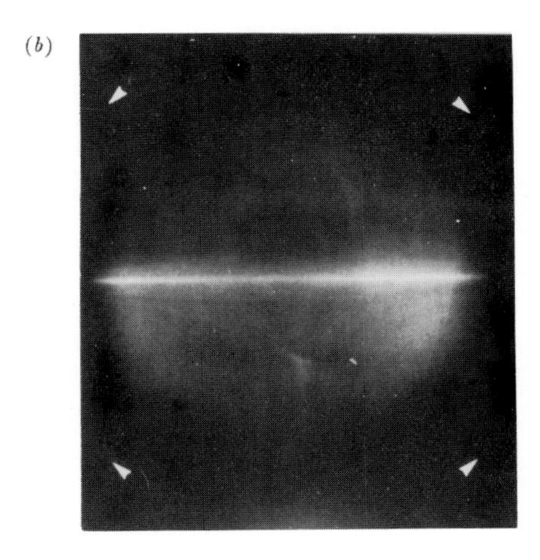

**Figure 4** Flow pattern (*a*) and flow birefringence (*b*) for a polyethylene oxide solution in a four roll mill. Reproduced with the permission of John Wiley Publishers from Crowley *et al* (1976).

uniform magnitude within the region bounded by the rollers. Realisation of the localised plane or orientation seen as a birefringent line in figure 4 became clear to us, in that not only must the $\dot{\gamma}\tau > 1$ condition be satisfied but also $\dot{\gamma}t \gg 1$. This second condition merely states that the chain must be subject to a deformation gradient $\dot{\gamma}$ for sufficient time in order that the chain becomes stretched. In the centrosymmetric flow field of the double jet and four roll mill this is only satisfied for fluid elements that pass close to the stagnation point. On reflection the conclusion was obvious but we all missed the point until the four roll mill photographs were in front of us.

(*a*)

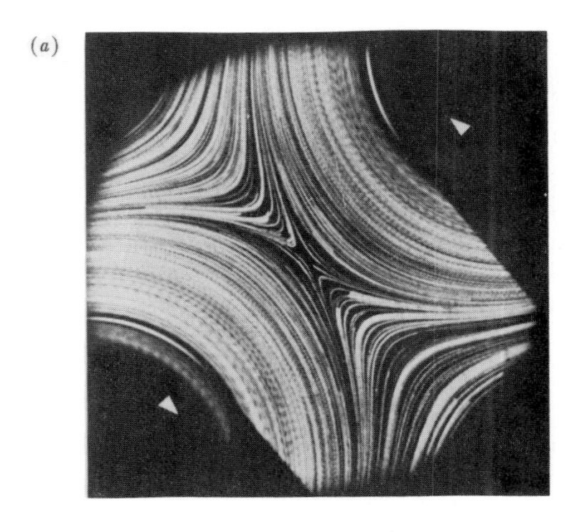

(*b*)

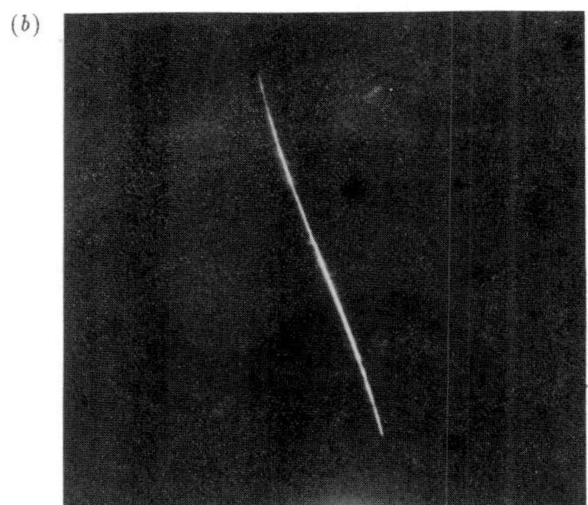

**Figure 5** Flow pattern (*a*) and flow birefringence (*b*) for a polyethylene oxide solution in a two roll mill. Reproduced with the permission of John Wiley Publishers from Frank and Mackley (1976).

In my opinion, during the period that I have known him, Charles is at his best when being supplied with experimental results which at the time cannot be explained. An example of this was the work I carried out on a two roll mill. Having worked with the Taylor four roll mill, the two roll mill seemed a possibly simpler variant. The flow pattern generated by two co-rotating rollers is shown in figure 5. In this case the orthogonal symmetry of the flow is broken and a rotational component to the flow is introduced. We again observed localised flow birefringence in this flow geometry (Frank and Mackley 1976) and from photographs such as these, Charles developed the idea of a 'persistent strain rate'. This way of classifying flow has not, as yet, become widely used or indeed recognised but it

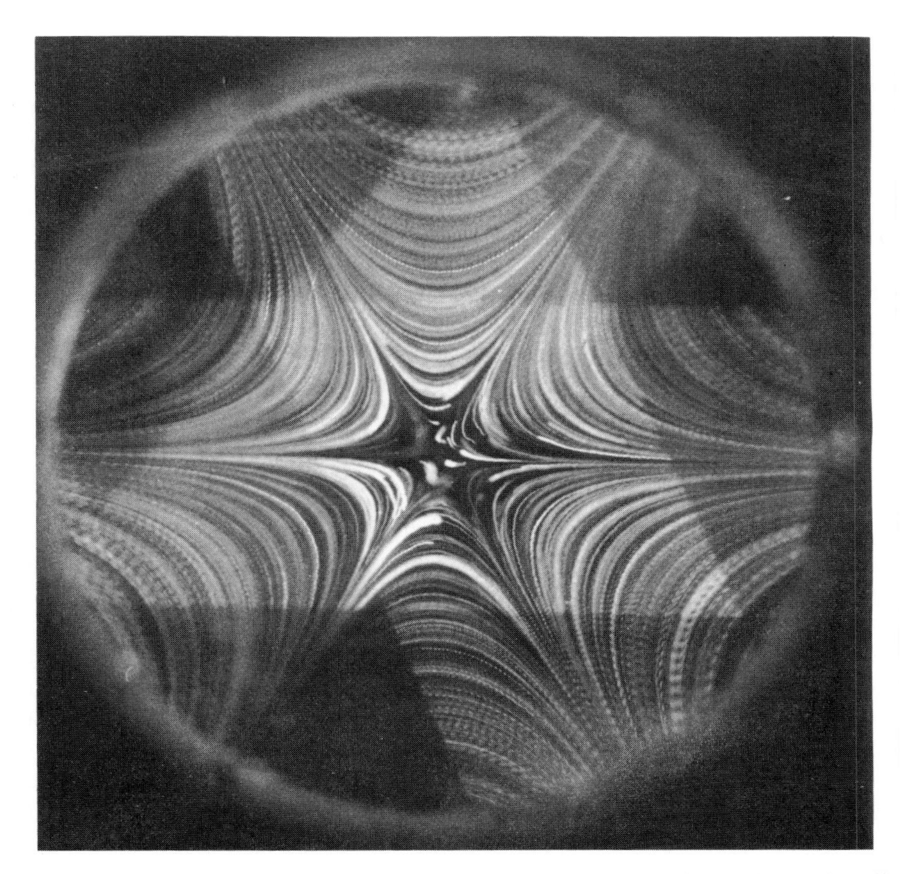

**Figure 6** Flow pattern of the germ of the elliptic umbilic obtained using a six roll mill. Reproduced with the permission of the Royal Society from Berry and Mackley (1977).

is one of many examples where Charles would think through the basics of a physical problem without necessarily first referring to other workers' formulations or indeed conforming to current trends in the literature.

In 1975 the two roll mill work eventually led to the six roll mill (Berry and Mackley 1977). At this point much of the theoretical input came from Michael Berry and his involvement with catastrophe theory. The symmetry of the two and four roll mills had led us to speculate whether stagnation point flows had to have two inflows and two outflows. The answer of course is no and in two dimensions the next level of complexity is the six roll mill as shown in figure 6 with three inflows and outflows. The stream function for this flow happened to be the germ of the elliptic umbilic catastrophe as classified by Thom (1975), and from this realisation there followed a wonderfully entertaining period following the emergence of degenerate critical points in control space. Charles took an amused interest in this work and Michael and I dedicated the paper on the work to him. As with so many pieces of work, Charles had acted as instigator of a research concept and encouraged collaboration between colleagues by his genius and ceaseless interest in unravelling the complexities of the physical world.

## 4 HIGH MODULUS POLYETHYLENE

1970 was a key year concerning the development of the notion that polyethylene could be physically manipulated to yield a high strength and stiffness material. The background to the ideas was centred at Bristol with Andrew Keller's work on crystal morphology and in particular the PhD work of Frank Willmouth (1968), the flow papers from Pennings' school (Pennings *et al* 1970a,b) and the solid state deformation studies carried out by Ian Ward's group (Andrews and Ward 1970). Charles Frank synthesised these findings and added to them his breadth of knowledge relating to diamond. He realised that the unit cell of the all-trans-polyethylene configuration had certain similarities to that of diamond and concluded that polyethylene should plausibly have an ultimate Youngs modulus of the order of 300 GPa in the direction of the chain axis. He recognised that in order to achieve this very high modulus it would be necessary to have high degrees of chain extension with little back folding. The 'shish-kebab' morphology offered hope that this could be achieved in the backbone of the fibre and it was the bringing together of these ideas in his *Proceedings of the Royal Society* paper in 1970 which started an international effort into realising this goal. Various routes were pursued by different groups. In the UK Ian Ward left Bristol to become professor at Leeds University and he vigorously developed variants of solid state drawing techniques. He soon found ways of increasing the modulus of polyethylene from its usual 1 GPa to around 50 GPa (Cappacio and Ward 1975). Zwinjnenburg and Pennings were the first to break the 100 GPa barrier, initially with their free growth technique (1976a) and then their surface growth method (1976b). In 1980 Dutch State Mines (DSM) announced the discovery of a 'gel spinning' route for the manufacture of high modulus polyethylene and Paul Smith and Piet Lemstra published findings of the process (Smith and Lemstra 1980). Before joining DSM Professor Piet Lemstra had worked in the Bristol Laboratory where there had been a strong scientific interest in both polyethylene solution processing and gels.

High modulus polyethylene is now a commercial reality with DSM, Allied Chemicals and Mitsui all producing the material in increasing tonnages. In my own mind there is absolutely no doubt that Charles Frank's speculations in his 1970 *Proc. R. Soc.* paper provided the catalyst and motivation for most if not all the work that followed in the next decade on that subject. Whilst Charles is first and foremost an academic, he took a considerable interest in the scientific developments within the commercial sector and this particular advance can certainly be attributed largely to his own ideas.

## 5 1976

The summer of 1976 was memorably hot in England and day after day was met with glorious blue skies. It was the year that Charles officially retired and many of us at Bristol had great difficulty visualising how such an intellectually active mind could contemplate retirement. 1976 was also the year that I left Bristol to take up an appointment in the Materials group with Professor Robert Cahn at Sussex University. I found saying goodbye to Charles very difficult as he had been such an inspiration to me over the past six years. The early evening dialogues were a persistent feature. I would report to him the latest experimental findings, and these conversations, often held on the staircase or in the tea room, might extend for a couple of hours. At the time, issues might not be resolved but more often than not on the following day Charles might well have come up with a plausible explanation. Much of his thinking and writing was done at Orchard Cottage in the early evening and late into the night. Talking to Charles requires particular skills because if you ask him a scientific question he may respond initially with what seems like a completely unrelated topic, but as the argument develops he will work the logic round to answering the original question. Charles would always find time to talk to someone with an interest in science and this was done even whilst he was Head of Department. He did not become bogged down with the bureaucracy of administration, and the department seemed to run efficiently and very smoothly while he was Head. Charles has his own pace and this was probably one of his great strengths in dealing with the everyday dramas of running a Department.

On the retirement of Charles from the H H Wills Physics Laboratory the staff presented him with a wooden garden seat. This seemed a most appropriate gift as the garden of Orchard Cottage is a very important part of Charles and Maita's home. Peaceful and tranquil surroundings which are conducive for deep thought are places not easily found in this ever-faster moving world. However, Orchard Cottage is one such place and both Maita and Charles have put many years of work into making it such a delightful home and garden. With Charles's retirement the centre of gravity of his thinking moved from the Physics Laboratory to his home. Within a short while it was clear that it was business as usual in terms of generating scientific ideas. Perhaps Charles seemed a little more relaxed but he was still ready and willing and able to discuss scientific points with scientific colleagues and friends.

## 6 LADY FRANK

After we had moved from Bristol to Sussex in 1976 and then in 1979 to Cambridge our friendship with the Franks strengthened and with it the recognition that Maita Frank has played a most significant role in supporting Charles through his scientific career. Charles came from a Suffolk farming background but from an early age always seemed to be destined to be an intellectual, although evidently before starting his academic career he was a 'hairs breadth' away from signing a contract to join ICI. Maita appears to have recognised very early in his career that Charles's intellect was special and has subsequently quite unselfishly supported Charles in every way possible. Maita is extremely able herself, and she has actively involved herself in enlivening the social side of both Bristol University and the H H Wills Physics Laboratory. She must have also arranged endless tea or dinner parties which, while nominally being social occasions, may have slipped partly into surreptitious scientific debate where the genesis of new ideas were created. Maita I think also played a significant part in their appreciation of the arts. Both are lovers of paintings and the theatre and have patronised Bath and Bristol functions for many years. The scientific community owes a deep level of gratitude to Maita for all her background support to Charles and it seemed most fitting that some public recognition

**Figure 7** Photographs of Sir Charles and Maita Frank relaxing in the garden of 2 Iford Manor Cottages, Iford, Sussex, in 1977.

came with the award of the knighthood in 1977 and her title of Lady Frank. Before being presented at Buckingham Palace Charles and Maita made one of their regular visits to my family and their photographs taken relaxing at Iford, Sussex, are shown in figure 7.

## 7 LIQUID CRYSTAL POLYMERS

Thermotropic liquid crystal polymers were first reported in 1976 and we were fortunate enough at Sussex to be supplied with a small quantity by the Eastman Kodak Company. These materials are a physicist's delight and when viewed in the optical microscope at elevated temperature they are highly birefringent and also dense with defects (Mackley *et al* 1981). It was only when we started to look into what kind of defects might be present that I discovered the key contribution Sir Charles had made in the field of small molecule liquid crystals. His paper of 1958 (Frank 1958) is a true classic and it has underpinned the understanding and classification of disclinations in liquid crystals since then. To me the paper exemplifies Sir Charles's mastery of physics, mathematics and symmetry. It also comes as no real surprise that the disclination concept introduced by Sir Charles can be extended from small molecules to liquid crystal polymers.

To a certain extent disclinations in small molecule liquid crystals were something of academic interest only although of course the 'Frank' elastic constants are the cornerstone of the Leslie–Erickson constitutive equation which describes the rheology of these materials (Leslie 1966). This rheology controls the speed of electrical and or magnetic switching which is important in any liquid crystal display device.

The position in the case of liquid crystal polymers is somewhat more complex as the defect texture within the material is very shear sensitive (Graziano and Mackley 1984a) and can also influence the solid state properties of the resulting material. Sir Charles took a strong interest in the forms of defect textures we observed in the optical microscope and helped us interpret the director trajectories around disclination loops that were formed as a consequence of shear in MBBA as shown in figure 8 (from Graziano and Mackley 1984b).

Sir Charles's interest in liquid crystal polymers (LCPs) prompted him to attend the Faraday Discussion Meeting on LCPs at Cambridge in 1985. He had had a preview of some most interesting electron micrographs prepared by Professor Ned Thomas and Miss Wood (1985) and this resulted in him rapidly grasping the significance of the crystal lamellae structure and making a presentation of his results to the meeting (Frank 1985).

At the time of writing Sir Charles and Lady Frank's most recent visit to Cambridge was in November 1989 to attend the R H Fowler memorial lecture. Charles claimed it was this man who found a job for him at

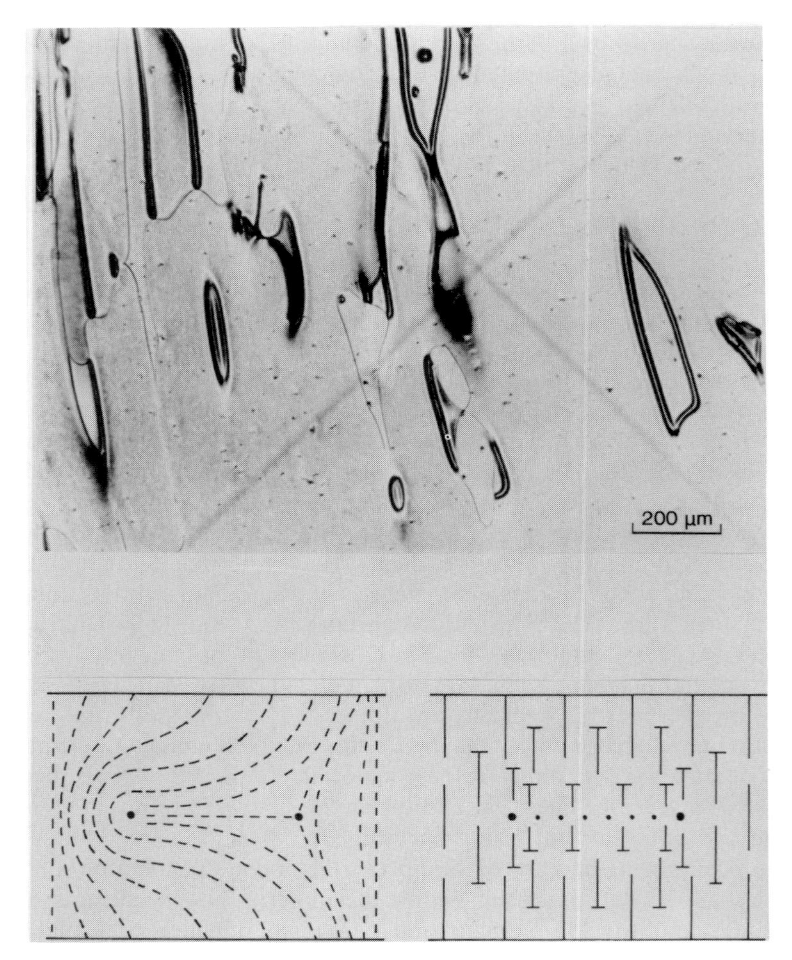


**Figure 8** Photograph and schematic diagram of disclination loops in MBBA observed to form as a consequence of shear. Reproduced with the permission of Gordon and Breach, from Graziano and Mackley (1984b).

Cambridge and saved him from signing up for ICI. During the visit, I took him into the laboratory to see a video of our latest shearing work on thermotropic LCPs. On seeing the video he immediately picked up a point missed by us relating to the way in which the banded texture seen on the cessation of shear subsequently relaxed to a defected texture.

Sir Charles's enthusiasm and appetite for science has not diminished since I first met him twenty years ago. His contribution to the physics community has been immense and it has been a great privilege and pleasure for me to share a small part of his life.

## REFERENCES

Andrews J A and Ward I M 1970 *J. Mater. Sci.* **5** 411–17
Berry M V and Mackley M R 1977 *Phil. Trans. R. Soc.* **287** 1–16
Cappacio G C and Ward I M 1975 *Polymer Eng. Sci.* **15** 219–24
Crowley D G, Frank F C, Mackley M R and Stephenson R C 1976 *J. Polymer Sci.* A **2** 1111–19
Frank F C 1958 *Discuss. Faraday Soc.* **25** 19–30
—— 1970 *Proc. R. Soc.* A **319** 127–36
—— 1985 *Faraday Discuss. Chem. Soc.* **79** 274–80
Frank F C and Mackley M R 1976 *J. Polymer Sci.* A **2** 1121–32
Frank F C, Mackley M R and Keller A 1971 *Polymer* **12** 467–73
Graziano D J and Mackley M R 1984a *Mol. Cryst. LIquid Cryst.* **106** 73–93
—— 1984b *Mol. Cryst. Liquid Cryst.* **106** 103–19
Keller A 1968 *Rep. Prog. Phys.* **30** 624–701
Keller A, Kiss G and Mackley M R 1975 *Nature* **257** 304–5
Keller A and Odell J A 1985 *Colloid Polymer Sci.* **263** 181–201
Leslie F M 1966 *Q. J. Mech. Appl. Mech.* **19** 358–69
Mackley M R 1972 *PhD thesis* Bristol University
Mackley M R and Keller A 1975 *Phil. Trans. R. Soc.* **278** 29–66
Mackley M R, Pinaud F and Siekmann G 1981 *Polymer* **22** 437–46
Pennings A J, van der Mark J M A A, Booij H C 1970a *Kolloid Z. Polymere* **236** 99–111
Pennings A J, van der Mark J M A A and Kiel A M 1970b *Kolloid Z. Polymere* **237** 336–58
Smith P and Lemstra P J 1980 *J. Mater. Sci.* **15** 505–14
Taylor G I 1934 *Proc. R. Soc.* A **146** 501–23
Thom R 1975 *Structural Stability and Morphogenesis* (San Francisco: Benjamin)
Thomas E L and Wood B A 1985 *Faraday Discuss. Chem. Soc.* **79** 229–39
Willmouth F M 1968 *PhD thesis* Bristol University
Ziabicki A 1959 *J. Appl. Polymer Sci.* **2** 14–23
Zwinjnenburg A and Pennings A J 1976a *Colloid Polymer Sci.* **253** 452–61
—— 1976b *Colloid Polymer Sci.* **254** 868–81

# The Anisotropic Mechanical Behaviour of Polymers at Low Strains

I M Ward

## 1 INTRODUCTION

Although it has always been appreciated that the introduction of molecular orientation usually produces enhanced mechanical properties in polymers, it is only since the advent of Kevlar and high modulus polyethylene fibres that materials with very high orientation have been available. These new materials have heightened our perception of what can be achieved in this way. Sir Charles Frank was one of the first persons to recognise the significance of these developments in an extremely prescient paper in the *Proceedings of the Royal Society* in 1970. It is therefore fitting that the *Festschrift* in his honour should contain a review of the anisotropic mechanical behaviour of oriented polymers.

This review will be concerned with the mechanical anisotropy of oriented polymers at low strains and will discuss the following major aspects:

(1) The determination of the elastic constants for a fully aligned polymer. For a crystalline polymer this requires the measurement or calculation of the so-called 'crystal elastic constants'.
(2) Experimental determination of mechanical anisotropy for oriented polymers.
(3) Interpretation of the mechanical behaviour of oriented polymers in terms of our understanding of their structure.

## 2 THEORY

The mechanical behaviour of an anisotropic *elastic* solid is described by a generalised Hooke's law with strains $\varepsilon_{ij}$ related to stresses $\sigma_{ij}$ by the relationships

$$\varepsilon_{ij} = S_{ijkl}\sigma_{kl}$$

$$\sigma_{ij} = C_{ijkl}\varepsilon_{kl}$$

where $S_{ijkl}$, $C_{ijkl}$ are the compliance constants and stiffness constants, respectively.

At low strains, an anisotropic polymer can be regarded as a linear viscoelastic solid. The compliance and stiffness constants are then time (or frequency) dependent and define creep compliances and stress relaxation moduli (or complex compliances or complex moduli in a dynamic mechanical experiment) (Ward 1983). It is customary to adopt an abbreviated notation

$$e_p = S_{pq}\sigma_q$$

$$\sigma_p = C_{pq}e_q$$

where $S_{pq}$ and $C_{pq}$ are the compliance constants and stiffness constants respectively and $e_1$, $e_2$, $e_3$ are the extensional strains, $e_4$, $e_5$, $e_6$ are the *engineering* shear strains, $\sigma_1$, $\sigma_2$, $\sigma_3$ are the normal stresses, and $\sigma_4$, $\sigma_5$, $\sigma_6$ are the shear stresses.

It is usually most straightforward to present the discussion in terms of the compliance constants, where we consider the strains produced by a simple loading system such as a single stress or shear stress. The most complex oriented polymer samples considered are films with three mutually perpendicular planes of symmetry, where there are nine independent compliance constants, represented by the matrix

$$\begin{pmatrix} S_{11} & S_{12} & S_{13} & 0 & 0 & 0 \\ S_{12} & S_{22} & S_{23} & 0 & 0 & 0 \\ S_{13} & S_{23} & S_{33} & 0 & 0 & 0 \\ 0 & 0 & 0 & S_{44} & 0 & 0 \\ 0 & 0 & 0 & 0 & S_{55} & 0 \\ 0 & 0 & 0 & 0 & 0 & S_{66} \end{pmatrix}.$$

These are illustrated in figure 1, where it is noted that the $z$ axis is chosen to be the initial drawing or rolling direction for a system of rectangular cartesian coordinates, the $x$ axis lies in the plane of the film and the $y$ axis normal to the plane of the film.

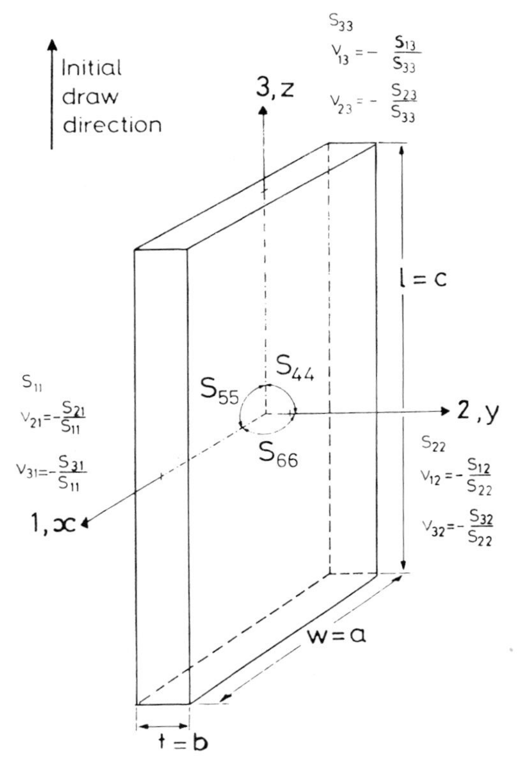


**Figure 1** The elastic constants for a sample with orthorhombic symmetry.

The so-called 'technical' elastic constants are the Young's moduli

$$E_1 = \frac{1}{S_{11}} \qquad E_2 = \frac{1}{S_{22}} \qquad E_3 = \frac{1}{S_{33}}$$

Poisson's ratios

$$\nu_{21} = -\frac{S_{12}}{S_{22}} \qquad \nu_{31} = -\frac{S_{13}}{S_{33}} \qquad \nu_{32} = -\frac{S_{23}}{S_{33}}$$

and shear moduli

$$G_1 = \frac{1}{S_{44}} \qquad G_2 = \frac{1}{S_{55}} \qquad G_3 = \frac{1}{S_{66}}.$$

For a fibre or a film with uniaxial symmetry, the $z$ axis is chosen to be the fibre axis, and $S_{11} = S_{22}$, $S_{13} = S_{23}$, $S_{44} = S_{55}$ and $S_{66} = 2(S_{11} - S_{12})$, so that the number of independent compliance constants reduces to five.

## 3 DETERMINATION OF ELASTIC CONSTANTS FOR FULLY ALIGNED POLYMERS

Comprehensive reviews of experimental and theoretical estimates for the elastic constants of the crystalline regions of crystalline polymers have been presented previously by Holliday (1975), Holliday and White (1971) and the present author (Ward 1982). In this paper, the overall situation will be reviewed, giving particular attention to recent developments, but it is not intended to be comprehensive.

### 3.1 Experimental determination of crystal elastic constants

#### *3.1.1 X-ray measurements*

X-ray measurements of the change in lattice dimensions under stress still make the largest single contribution to our knowledge of the elastic constants for the crystalline regions of a polymer. This type of measurement was initiated by Dulmage and Contois (1958) and by Sakurada and his co-workers (Sakurada *et al* 1966, Sakurada and Kaji 1970, Kaji *et al* 1974). During the period 1966–74 Sakurada and his collaborators measured the crystal chain moduli for all the major crystalline polymers on the assumption that the stress applied to the crystal is the external applied load divided by the cross-sectional area of the sample, i.e. homogeneous stress exists throughout the samples.

In general the results obtained for the chain moduli in this way were consistent with expectations in terms of the chain structure. Where the molecular chains were fully extended (e.g. polyethylene and nylon) chain moduli in the range 150–200 GPa were obtained, consistent with the modes of chain deformation being bond bending and bond stretching. Where the molecular chains were helical (e.g. polypropylene and polymethylene) chain moduli in the range 40–60 GPa or even lower were obtained, consistent with bond rotations becoming significant in chain deformation.

The availability of highly oriented polymers such as Kevlar, high modulus polyethylene and more recently the thermotropic liquid crystalline polymers has made the accurate measurement of crystal chain moduli of even greater importance. Consequently there has been renewed interest in the determination of these chain moduli by x-ray diffraction measurements. Recent research in this area will now be reviewed, and it will be shown that the assumption of homogeneous stress throughout the sample is not generally correct. In fact, the measurements of crystal strain can be of value in establishing relationships between structure and properties. This recent research does not, however, substantially detract from the earlier work of Sakurada and his collaborators, most of whose results have been confirmed. The low modulus samples examined by Sakurada *et al* were generally annealed at high temperatures to produce structures in which the crystalline and amorphous regions alternated (approximating to the Takayanagi series model) so that the crystalline regions experienced the macroscopic stress applied to the sample.

*3.1.1 (a) X-ray crystal strain measurements on flexible highly oriented polymers* Recent studies by Jakeways, Ward and collaborators have been concerned with measurements of crystal strain in a range of highly oriented polymers. The apparent lattice modulus (or crystal modulus) $E_c^{app}$ is defined from a plot of lattice strain against applied stress on the assumption of homogeneous stress in the sample.

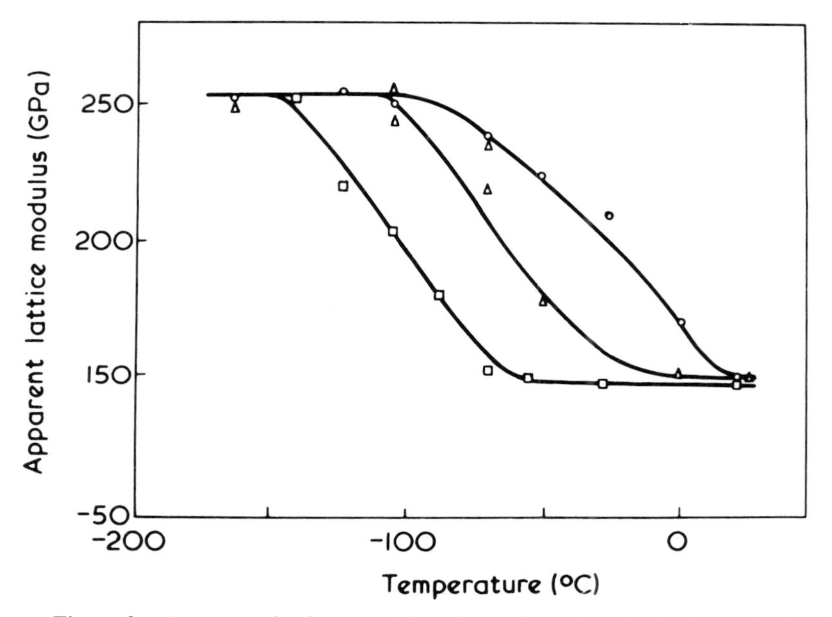


**Figure 2** Apparent lattice modulus for oriented polyethylene obtained from crystal strain measurements. Draw ratios: ○, 10; △, 15; □, 25. Reproduced from Clements *et al* (1978) by permission of Butterworth & Co, Publishers Ltd (c).

Figure 2 shows $E_c^{app}$ as a function of temperature for three samples of H020 grade linear polyethylene of differing draw ratio (Clements *et al* 1978). It can be seen that $E_c^{app}$ rises to a value of 255 GPa at low temperatures for all three samples, but at room temperature $E_c^{app} \sim 150$ GPa. It is not difficult to obtain a satisfactory explanation of these results. Although there is not complete agreement regarding the structure of these highly oriented polyethylenes, it is known that there is a considerable increase in the lengths of the crystalline regions with increasing draw ratio, so that the crystal lengths of a substantial fraction of the material exceed the long period. The Leeds view is that these long crystals can be regarded either as producing intercrystalline bridges between the periodic crystal blocks (as originally proposed by Fischer *et al* (1971)), or as akin to the reinforcing fibres in a short fibre composite (which is similar but not identical to a model originally proposed by Arridge *et al* (1977) at Bristol University).

The simplest theoretical analysis can be made in terms of a Takayanagi model where there is a continuous crystal fraction $b$, with a non-crystalline fraction defined by the quantity $a$ (figure 3). The mechanical response of this model can be evaluated in two ways, the 'series–parallel' and 'parallel–series' assumptions.

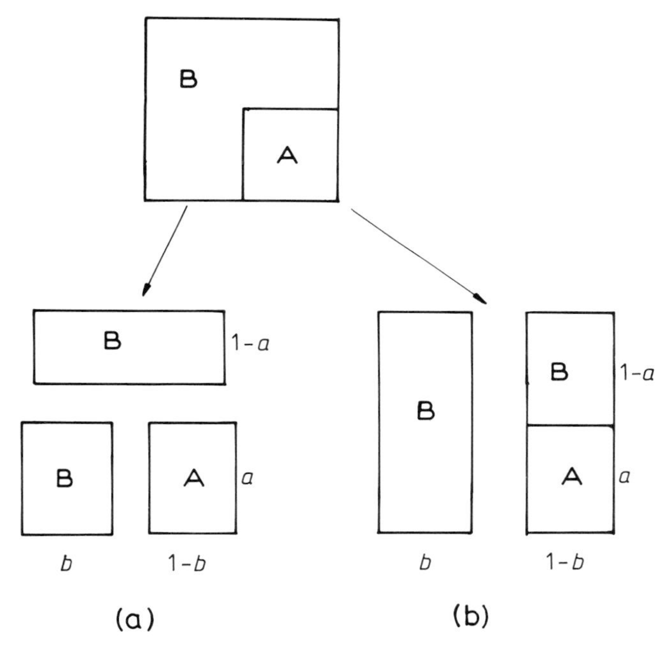


**Figure 3** The Takayanagi series–parallel and parallel–series models.

The 'series–parallel' assumption gives

$$E_c^{app} = \chi E_c(1-a) + \left(\frac{ab}{b+(1-b)E_a/E_c}\right)^{-1} \tag{1}$$

and the 'parallel–series' assumption gives

$$E_c^{app} = \chi E_c\left(b + \frac{(1-b)E_a/E_c}{a+(1-a)E_a/E_c}\right)\left(b + \frac{(1-a)(1-b)E_a/E_c}{a+(1-a)E_a/E_c}\right)^{-1} \tag{2}$$

where $\chi = 1 - a(1-b)$ is the volume of crystalline material and $E_c$ and $E_a$

are the elastic moduli of the crystalline and non-crystalline phases respectively.

Irrespective of the choice of assumption which leads to either equation (1) or (2) above, it can be seen that as

$$E_a \rightarrow 0 \qquad E_c^{app} \rightarrow \chi E_c$$

and as

$$b \rightarrow 0 \text{ or } E_a \rightarrow E_c \qquad E_c^{app} \rightarrow E_c.$$

The general features of the experimental results are therefore reproduced. At high temperatures $E_a$ falls and $E_c^{app}$ falls to a constant value $\chi E_c$. At low temperatures where $E_a$ approaches $E_c$, $E_c^{app}$ rises to approach $E_c$. Moreover, the higher draw ratio samples have higher values of $b$ (which represent the crystal continuity) and therefore require higher values of $E_a$ (i.e. lower temperatures) for $E_c^{app}$ to approach $E_c$.

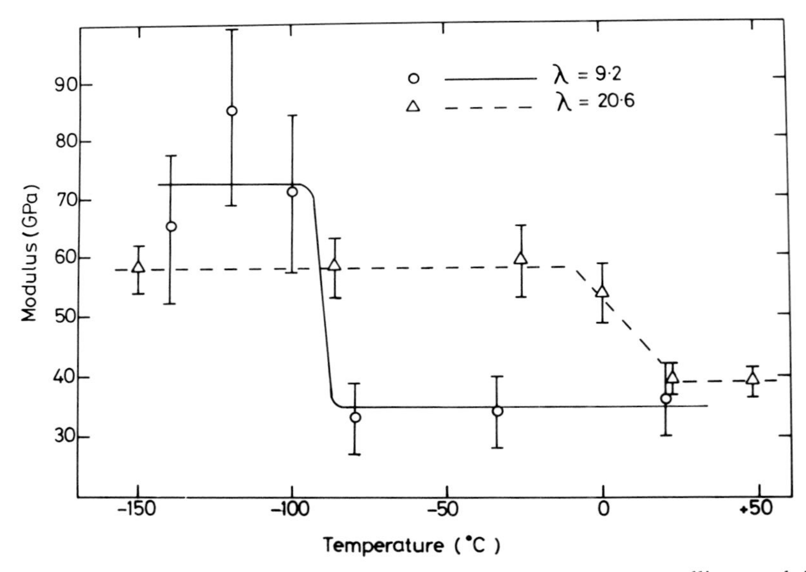


**Figure 4** Variation with temperature of the apparent crystalline modulus of two 'as drawn' POM tapes of different draw ratios. Reproduced from Jungnitz *et al* (1986) by permission of Butterworth & Co, Publishers Ltd (c).

The fibre composite model yields similar results, which is not surprising, as in algebraic terms it is identical to the 'parallel–series' model with the fibre fraction taking the role of the continuous crystal fraction $b$. There are, however, two final comments to make on these results. First it must be clear that even at low temperatures, where the results converge to a value for the crystal chain modulus of 255 GPa, this can only be regarded as a lower bound to the true crystal modulus, because the non-crystalline phase modulus $E_a$ can never reach $E_c$. Secondly, the value for $E_c^{app}$ at high temperatures would suggest a value for $\chi$, the volume fraction of crystalline material of about 0.6, which seems to be unrealistic, since values from density, DSC and NMR suggest at least 0.8. A possible conclusion, which must be tentative, is that the x-ray measurements are concerned only with this smaller fraction of the total material, and that a fraction $\sim 0.2$ of the molecular chains are aligned without possessing three-dimensional order over a sufficiently large volume.

Another polymer where high modulus has been achieved by high draw is polyoxymethylene (POM). In this case the apparent lattice modulus $E_c^{app}$ was also found to be markedly temperature dependent, but as shown in figures 4 and 5, the low-temperature value varies from specimen to specimen, and is dependent on both draw ratio and annealing (Jungnitz *et al* 1986). In POM there is no support for the type of model proposed for the highly drawn polyethylenes. X-ray diffraction data show that crystallite lengths do not significantly exceed the long period. Moreover such models predict that the rises in $E_c^{app}$ should occur at lower temperatures for higher draw samples, which is contrary to the results.

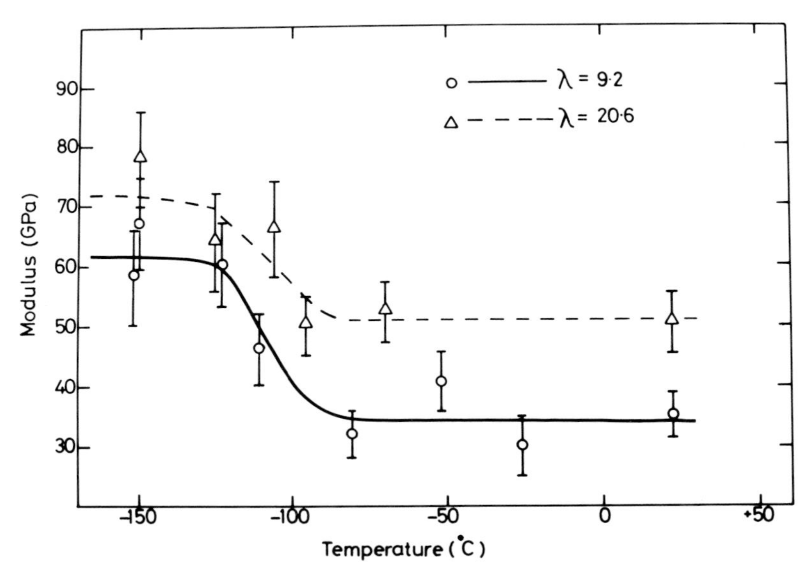


**Figure 5** Variation with temperature of the apparent crystalline modulus of two annealed POM tapes of different draw ratios. Reproduced from Jungnitz *et al* (1986) by permission of Butterworth & Co Publishers Ltd (c).

The starting point for the analysis of these results comes from examination of the two annealed tapes (figure 5). Comparison of measurements of density with the measurements of long period and crystal lengths suggests that these samples have an essentially series structure which is not quite as well developed in the draw ratio 9.2 sample as in the draw ratio 20.6 sample. It was therefore concluded that the upper curve in figure 5 is close to a true representation of the chain modulus and that this is truly temperature dependent, as indicated by previous x-ray diffraction and NMR studies (Chiba *et al* 1966, Olf and Peterlin 1964). The behaviour of the low draw unannealed tapes (figure 4) is similar to the annealed tapes, and is therefore explicable in terms of a series model and temperature-dependent crystal chain modulus. The high draw unannealed sample does, however, show distinctly different behaviour. It was proposed that this behaviour requires a third phase consisting of chains which are aligned in the draw direction but not packed in crystalline register. This third phase is considered to be in parallel with the remaining structure and to possess a stiffness which varies between the limits calculated by Sugeta and Miyazawa (1970) for 'free' chains in a 2/1 helical conformation. The lower limit, with freedom to rotate, is 19 GPa and this obtains at room temperature so that the stress is concentrated on the remaining material in parallel and $E_c^{app}$ falls to 35 GPa. At low temperatures the stiffness of this phase is 48 GPa and $E_c^{app}$ is 58 GPa, which is much closer to the value of 70 GPa observed in the high draw annealed sample at these temperatures.

It was shown that this model of a third phase in parallel with the crystalline regions and the remaining amorphous regions could give a reasonable fit to the experimental data for both $E_c^{app}$ and the macroscopic Young's modulus of the sample, on the basis of plausible values for the modulus of the remaining amorphous regions as a function of temperature and the parameters of the Takayanagi parallel–series model (20% of the third phase in parallel). The comparison between the calculated and measured modulus is shown in figure 6.

In a very recent investigation, Tashiro and co-workers (Wu *et al* 1989) have combined x-ray diffraction measurements on POM with Raman and infrared measurements under stress to obtain estimates of the true crystal modulus. They are in agreement with Jungnitz *et al* regarding the need to consider the temperature dependence of the chain modulus and the inhomogeneous stress distribution.

A further polymer which is of considerable technological importance is polyethylene terephthalate (PET), well known as a textile fibre under its several trademark names, such as Terylene, Dacron and Trevira. Recent studies of the crystal strain in oriented samples of this polymer (Thistlethwaite *et al* 1988) have revealed a further degree of complexity, together with further confirmation that it is unlikely that the assumption of homogeneous stress can be assumed to be universally valid.

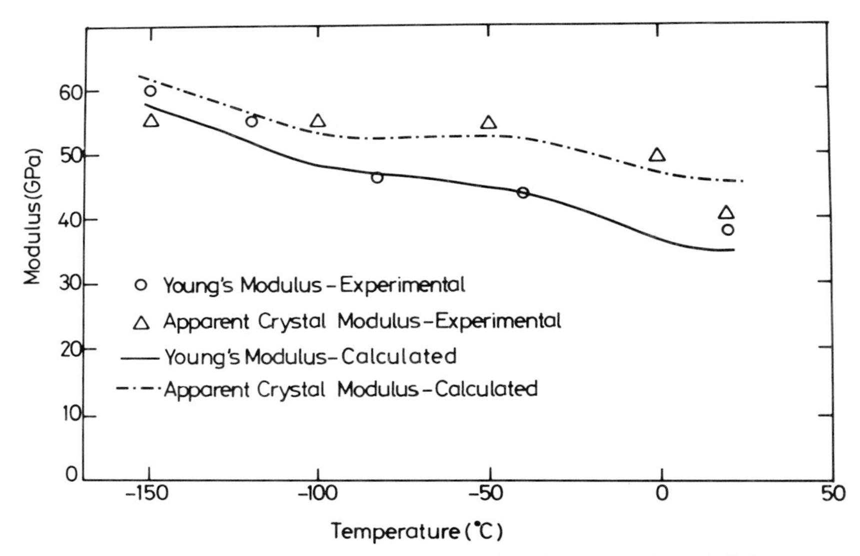


**Figure 6** Comparison between the calculated and measured moduli for POM tapes of draw ratio 20.6 : 1. Reproduced from Jungnitz *et al* (1986) by permission of Butterworth & Co, Publishers Ltd (c).

The crystal structure of PET is well known. The unit cell is triclinic and the *c* axis is usually found to be inclined by up to 6° or so to the draw direction in a drawn fibre or tape. A further complexity is that there is no (001) reflection available as in the case of PE or POM to give a direct measure of the change in chain length. The $(\bar{1}05)$ reflection has to be used and the normal to this plane lies at some 11.5° from the *c* axis. Measurements show that the degree of tilt decreases when the fibre is stressed. The inclined unit cell is clearly undergoing a change that might well include a change of shape, the interaxial angles changing by more than would occur for zero tilt, giving a different relationship between the changes in the $(\bar{1}05)$ plane spacing and changes in the *c* axis. There is then a systematic variation in $E_c^{app}$ as the measured tilt angle varies. In terms of the $(\bar{1}05)$ reflection $E_c^{app}$ varies as the azimuthal angle $\chi_{(\bar{1}05)}$ for this reflection varies (smaller tilt, larger values of $\chi_{(\bar{1}05)}$). The results (Thistlethwaite *et al* 1988) produced a value for $E_c^{app} = 2.31\chi_{(\bar{1}05)} + 57.6$ GPa.

From this equation we can calculate the value of $E_c^{app}$ for zero tilt ($\chi_{(\bar{1}05)} = 11.56°$) which is 84.3 ± 4 GPa. Although this can be regarded as a true measure of the *c*-axis extension under load, there is still the very real question of the distribution of stress.

It is generally accepted that in PET fibres (following the work of Prevorsek *et al* (1974)) a parallel–series model is applicable, with a continuous oriented amorphous phase in parallel with the crystalline material and the remaining amorphous material which are in series. Strong evidence from this model was confirmed by structural measurements which showed that the crystallinity is much less than that deduced from a comparison of the mean crystallite length and the long period (Thistlethwaite *et al* 1988). From their collected data, Thistlethwaite *et al* concluded that the true corrected value of the crystal chain modulus of PET is 110 ± 10 GPa. No temperature dependence was observed for this value (nor for $E_c^{app}$) over the temperature range − 164 to + 20 °C. It is interesting that this value is quite close to a recent calculated value of 95 GPa (Tashiro *et al* 1977), and very close to the revised value of 105 GPa given by Sakurada and Kaji (1970).

*3.1.2 Raman scattering and inelastic neutron scattering measurements*
It is interesting to note that the first experimental determination of the crystal chain modulus of polyethylene was made by determining the frequency of the longitudinal acoustic mode of vibration by Raman spectroscopy. Assuming that the chain behaves as an elastic rod the wave number shift $\Delta\tilde{\nu}$ is

$$\Delta\tilde{\nu} = \frac{m}{2Lc}\left(\frac{E}{\rho}\right)^{1/2} \tag{3}$$

where $m$ is the mode number (1, 3, 5, ...), $L$ is the chain length, $E$ is the chain modulus, $\rho$ is the density and $c$ is the velocity of light. Mizushima and Shimanouchi (1949), and later Shauffele and Shimanouchi (1967), found values of 340 and 358 GPa on the basis of equation (3) which is substantially higher than the values of 235–255 GPa found by x-ray diffraction. These differences have been variously associated with end effects due to methyl groups and interchain interactions. Ströbl and Eckel (1976) made a careful study of the frequency shifts obtained for both $m = 1$ and $m = 3$, concluded that coupling between chains does exist and obtained a value of 290 GPa.

The Raman method has been applied to several polymers, including polytetrafluorethylene, polyethyleneoxide, polyoxymethylene and polypropylene. Unfortunately, there are considerable difficulties in obtaining a satisfactory value of chain length $L$, partly due to the problems of end effects and partly due to the complication of obtaining this from estimates of the long period and the crystallinity in the presence of chain tilt. Although there has been much interest in the Raman method, it has therefore not emerged as a serious contender for accurate chain modulus data.

Another technique which appears to be extremely interesting for the determination of crystal elastic constants is inelastic neutron scattering (White 1976). The advantage of this method lies in the fact that the neutron wavelengths (~ 1 Å) are much shorter than the crystalline sequences (~ 100 Å). Only comparatively few measurements have been made, but the results are extremely encouraging. In polyethylene a value of 329 GPa was obtained for the chain modulus. Although this value is appreciably higher than either the x-ray crystal strain value or that from Raman measurements, it would be expected to be the most reliable value, because there are fewer complicating assumptions in its determination.

### 3.2 Theoretical calculation of elastic constants

The calculation of the chain moduli is comparatively straightforward. In some instances it is only necessary to consider two modes of deformation, bond stretching and bond angle opening, and force constants may be readily obtained from spectroscopic data. In most polymers, internal rotation around the bonds in the chain can also occur, but it is again straightforward to model this by a suitable potential energy function. Calculations of the elastic modulus of a single chain may also involve steric interactions within the chain, for example in the aromatic polyamides between the benzene ring and the hydrogen and oxygen atoms of the amide group.

In his 1970 review paper Sir Charles Frank showed how easily this type of calculation can be made for polyethylene, emphasising the close parallelism of the stiffness of a polyethylene chain with that of diamond, which has structural kinship to polyethylene.

Recent research in this area has been stimulated by the discovery of highly oriented liquid crystalline polymers. First, the lyotropic liquid crystalline polymers, the aramids such as Kevlar and poly(parabenzamide), where Tadokoro and his collaborators (Tashiro *et al* 1977) have calculated values for the chain modulus which are in reasonable agreement with those obtained by x-ray crystal strain measurements. Secondly there are the oriented thermotropic liquid crystalline polymers, such as the copolyesters of hydroxybenzoic acid (HBA) and hydroxynaphthoic acid (HNA) (marketed under the tradename of 'Vectra' by Hoechst-Celanese). In this case, the Leeds group has attempted to compare calculated chain moduli with those obtained from x-ray crystal strain measurements. The results are of some complexity and will therefore be considered in detail later, because there is an important extra ingredient required to make this comparison, a model for the aggregate of oriented chains.

In comparison with the many attempts to calculate chain moduli, there are very few cases where a calculation of all the elastic constants has been undertaken. Only in polyethylene have several determined efforts been reported, notably by Odajima and Maeda (1966) and more recently by Tadokoro and co-workers (Tashiro *et al* 1978). There are appreciable differences between the various estimates, which Tadokoro *et al* attribute to differences in the setting angle $\phi$ which defines the angle made by the planar zig-zag with the $b$ axis. In some instances, notably the value of 0.06 obtained by Tadokoro *et al* for the Poisson ratio $\nu_{13}$, the results are not in accord with physical expectation. It is, however, encouraging that their value of 315.5 GPa for the chain modulus is close to the value obtained by the neutron scattering technique (see table 1 for a summary of the elastic constants).

In a more recent publication Boyd and co-workers (Sorensen *et al* 1988) have calculated the stiffness matrices for polyethylene and POM. It is disappointing to note, however, that inversion of these matrices gives values for the Poisson's ratios which do not appear to be physically reasonable, for example $\nu_{13}$ for PE is zero and $\nu_{23}$ for the orthorhombic POM structure is negative. It can only be concluded that this is an area where much research is required.

**Table 1** Theoretical prediction for elastic constants of a polyethylene crystal (after Tashiro *et al* 1977).

$$C_{ij} = \begin{pmatrix} 7.99 & 3.28 & 1.13 & & & \\ 3.28 & 9.92 & 2.14 & & & \\ 1.13 & 2.14 & 315.9 & & & \\ & & & 3.19 & & \\ & & & & 1.62 & \\ & & & & & 3.62 \end{pmatrix} \text{GPa}$$

$$S_{ij} = \begin{pmatrix} 14.5 & -4.78 & -0.02 & & & \\ -4.78 & 11.67 & -0.06 & & & \\ -0.02 & -0.06 & 0.32 & & & \\ & & & 31.3 & & \\ & & & & 61.8 & \\ & & & & & 27.6 \end{pmatrix} (\text{GPa})^{-1} \times 100$$

## 4 EXPERIMENTAL STUDIES OF ANISOTROPIC BEHAVIOUR AND THEIR INTERPRETATION

The degree of mechanical anisotropy observed for oriented polymers is usually much less than that expected on the basis of theoretical considerations, and the Young's modulus takes a much lower value than that predicted.

In several instances, however, notably ultra high modulus polyethylene (especially that produced by the gel-spinning route) and for the polyaramids such as Kevlar, and more recently for the oriented thermotropic liquid crystalline polymers, experimental values for the Young's modulus are very close to the theoretical values. In figure 7 results are presented for the dependence of the elastic moduli of linear polyethylene (LPE) on draw ratio measured ultrasonically (Leung *et al* 1984). The collected data for draw ratio 20 are shown in table 2, for comparison with the theoretical predictions in table 1. It can be seen from figure 7 that all the stiffness constants appear to have reached limiting values by draw ratio 20, with the exception of $c_{33}$ which is still increasing. We can therefore conclude that there are genuine discrepancies between the measured and predicted values of the elastic constants, especially the lateral compliances $S_{12}$, $S_{13}$ and $S_{23}$, with the measured values being more in accord with intuitive physical expectation.

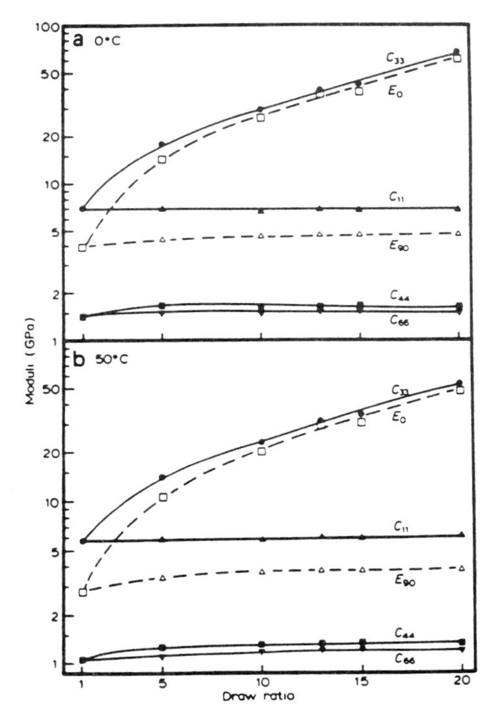


**Figure 7** Draw ratio dependence of the elastic moduli of LPE at (*a*) 0 °C and (*b*) 50 °C. Reproduced from Leung *et al* (1984) by permission of Butterworth & Co, Publishers Ltd (c).

Polyethylene is the only polymer for which such a detailed comparison can be made, and this is clearly an area which demands much more attention in terms of both theory and experiment.

For all other cases, the interpretation of the anisotropic mechanical behaviour rests on the recognition that even the most highly oriented samples available are very appreciably disordered. Quantitative calculation then requires the quantitative modelling of two major factors: (1) molecular orientation, and (2) the composite nature of crystalline polymers.

**Table 2** Elastic constants for highly oriented LPE (draw ratio 20) at 0 °C obtained from ultrasonic measurements.

$$C_{ij} = \begin{pmatrix} 6.9 & 3.9 & 4.4 & 0 & 0 & 0 \\ 3.9 & 6.9 & 4.4 & 0 & 0 & 0 \\ 4.4 & 4.4 & 66 & 0 & 0 & 0 \\ 0 & 0 & 0 & 1.6 & 0 & 0 \\ 0 & 0 & 0 & 0 & 1.6 & 0 \\ 0 & 0 & 0 & 0 & 0 & 1.5 \end{pmatrix} \text{GPa}$$

$$S_{ij} = \begin{pmatrix} 21 & -11 & -0.7 & 0 & 0 & 0 \\ -11 & 21 & -0.7 & 0 & 0 & 0 \\ -0.7 & -0.7 & 1.6 & 0 & 0 & 0 \\ 0 & 0 & 0 & 64 & 0 & 0 \\ 0 & 0 & 0 & 0 & 64 & 0 \\ 0 & 0 & 0 & 0 & 0 & 66 \end{pmatrix} (\text{GPa})^{-1} \times 100$$

### 4.1 Molecular orientation and the aggregate model

The aggregate model (Ward 1962) considers that the polymer consists of an aggregate of anisotropic units. The model originated as a single-phase model, and the mechanical properties of the units were considered to be those of a highly oriented polymer obtained from experimental data. As is clear from the above discussion these properties could be much different from those of the fully aligned polymer, especially with regard to the chain axis direction. In spite of its arbitrary nature, the aggregate model was successful in describing the development of mechanical anisotropy with molecular orientation (often defined by the draw ratio) in amorphous polymers, and also in PET (which is partly crystalline) and in low-density polyethylene where it fairly accurately modelled the influence of the *c*-shear relaxation.

More recently the aggregate model has proved to be useful in understanding the mechanical anisotropy of liquid crystalline polymers and oriented pressure-crystallised polyethylenes.

On the aggregate model, the Reuss averaging scheme (assuming homogeneous stress) gives the tensile compliance of the sample in the orientation direction ($S'_{33}$) as

$$S'_{33} = S_{11}\langle \sin^4 \theta \rangle + S_{33}\langle \cos^4 \theta \rangle + (2S_{13} + S_{44})\langle \sin^2 \theta \cos^2 \theta \rangle. \quad (4)$$

For simplicity we assume an aggregate of transversely isotropic units, with the orientation of the units being defined by the angle $\theta$ between the unique axis (the 3 axis) of the units and the unique axis of the macroscopic sample.

For samples with a high degree of molecular orientation $\langle \cos^2 \theta \rangle \sim 1$ and $\langle \sin^2 \theta \rangle \gg \langle \sin^4 \theta \rangle$. In addition, for oriented polymers, we have seen that

$S_{13} \sim S_{33} \ll S_{44}$. Equation (4) therefore reduces to

$$S'_{33} = S_{33} + S_{44}\langle \sin^2 \theta \rangle. \tag{5}$$

This equation encapsulates the physical meaning of the aggregate model. In polymers, the extensional compliance in the chain direction $S_{33} = 1/E_{33}$ is always at least one order of magnitude less than the shear compliance $S_{44}$ (and in LPE and liquid crystalline polymers two orders of magnitude less). This means that even a comparatively small degree of misorientation $\langle \sin^2 \theta \rangle \sim 0.1$ or even $\sim 0.01$ can increase the measured extensional compliance $S'_{33}$ by a very significant amount.

To a similar approximation it can be shown that the shear compliance of the sample

$$S'_{44} = S_{44} = 1/G \tag{6}$$

where $S_{44}$ is the shear compliance of the units of structure and $G$ is the shear modulus of the sample. Combining equations (5) and (6) we obtain

$$\frac{1}{E} = S_{33} + \frac{1}{G}\langle \sin^2 \theta \rangle. \tag{7}$$

In the oriented liquid crystalline polymers, Davies, Ward and their collaborators (Davies and Ward 1988, Troughton *et al* 1988, Green *et al* 1990) examined the applicability of equation (7) to dynamic mechanical measurements in tension and torsion for a series of polyesters based on hydroxybenzoic acid (HBA), hydroxynaphthoic acid (HNA), biphenyl, dihydroxynaphthalene and terephthalic acid (table 3). These mechanical measurements were combined with x-ray measurements under stress and theoretical calculations of the chain undertaken. The substituents in these

**Table 3** Chemical composition of liquid crystalline polymers. HBA = hydroxybenzoic acid, HNA = hydroxynaphthoic acid, TPA = terephthalic acid, DHN = dihydroxynaphthalene, BP = biphenyl (all para substituted).

| | Molar percentage of | | | | |
|---|---|---|---|---|---|
| Polymer | HBA | HNA | TPA | DHN | BP |
| 30/70 | 30 | 70 | | | |
| 73/27 | 73 | 27 | | | |
| CO2,6 | 60 | | 20 | 20 | |
| COTBP | 60 | 5 | 17.5 | | 17.5 |

polyesters are randomly substituted, and the x-ray measurements show distinctive meridional diffraction patterns with well defined but aperiodic peaks. The HBA/HNA copolyesters show a strong peak at $2\theta \sim 43^\circ$ ($CuK_\alpha$) which has been used to measure an apparent chain modulus. Results were obtained for a range of temperatures (figure 8), which showed that the apparent chain modulus in these materials falls substantially with temperature, with the exception of that containing biphenyl.

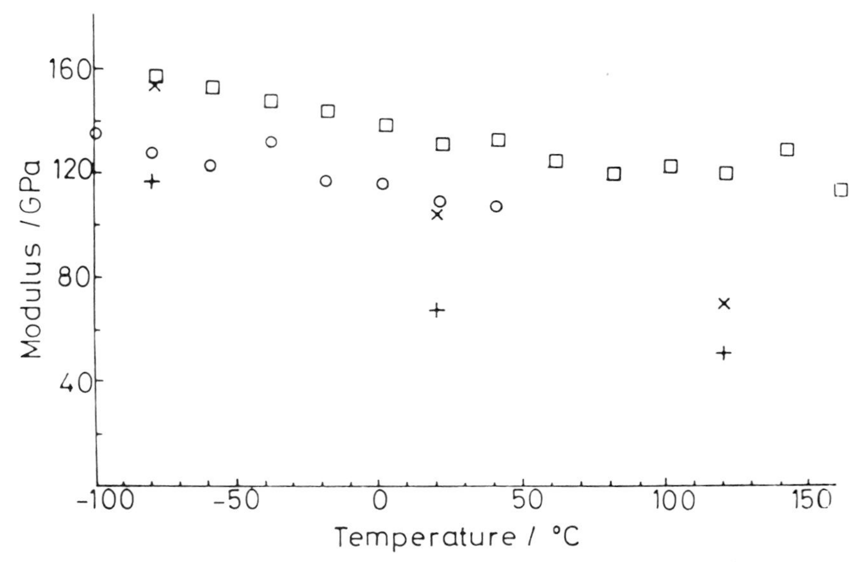


**Figure 8** Apparent crystal modulus of aromatic polyesters as a function of temperature. □, COTBP; ○, CO2,6; ×, CO73/27; +, CO30/70.

In the first evaluation of equation (7), the extensional compliance of the units was identified as $S_{33} = 1/E_c$ where $E_c$ is the apparent chain modulus. Figure 9 shows a graph of $(1/E - 1/E_c)$ against 1/G where some care has been taken to refer the individual measurements to a comparable frequency range for each temperature.

A very convincing straight-line plot has been obtained, passing through the origin. It is, however, also necessary to address the origin of the magnitudes of the apparent chain moduli and their temperature dependence. In these polymers, the diffraction pattern is sampling all the polymer, so that the fall in apparent chain modulus with temperature cannot be attributed to a stress concentration effect, as in highly oriented polyethylene. Consideration of theoretical estimates of the chain moduli (Troughton *et al* 1988) showed that for isolated chains these would be expected to be comparatively low, because of the implication of random substitution of the monomer units in the chains. The theoretical modelling showed that even to produce values of chain modulus comparable to those determined from high-temperature x-ray measurements, the chain has to be constructed with comparatively low degree of deviation from straightness, i.e. a low sinuosity. It was therefore concluded that the high value for the chain modulus, consistent with the fit to equation (7) and the low-temperature x-ray results, arises because intermolecular forces constrain the straightening of the molecular chains at low temperatures. This idea led to a reappraisal of equation (7), now identifying the unit of structure as a localised group of monomer units rather than the average over at least 10 units which is identified by the x-ray measurements. With this in mind, $1/E$ is now plotted directly against $1/G$ as shown in figure 10. Again a reasonable straight-line plot is obtained, but with an extrapolation to the $1/E$ axis corresponding to the compliance $S_{33}$ of the localised group of monomer units. This value corresponds well with both the low-temperature x-ray value for the chain modulus, and that calculated theoretically for a localised group of monomer units, i.e. relating to bond stretching and bond angle opening within a very short length of chain.

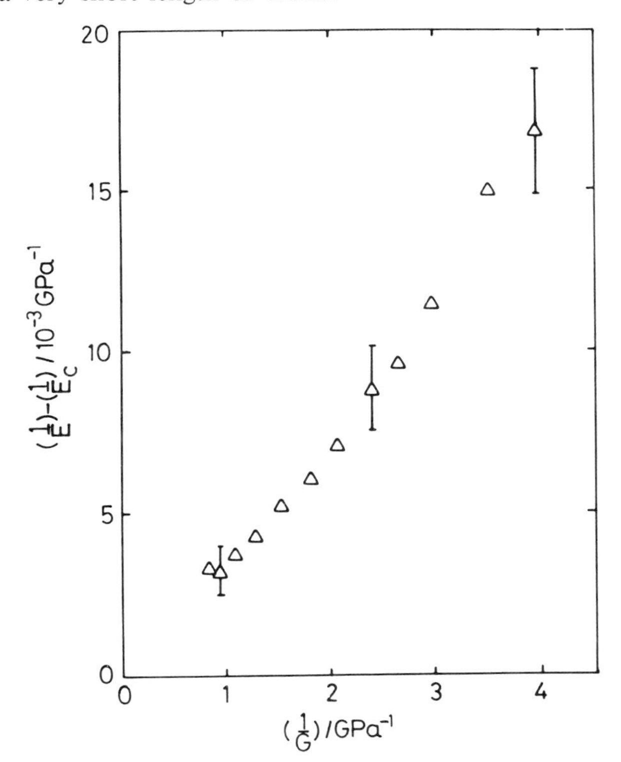


**Figure 9** The aggregate model for COTBP assuming temperature-dependent unit [ = *x*-ray] modulus.

A final point of some interest is the comparison of the values of $\langle \sin^2\theta \rangle$ found from the two different plots (i.e. figures 9 and 10, respectively). From figure 9 we obtain a value of $\langle \sin^2\theta \rangle$ which corresponds to overall chain orientation, whereas figure 10 gives a value of $\langle \sin^2\theta \rangle$ for the aggregate of localised units. As anticipated, the latter value is substantially greater than the former, reflecting the degree of sinuosity of the molecular chain. It has been found that this comparison does relate to the structural composition in a physically satisfactory fashion. For example, the chains for compositions incorporating the linear biphenyl unit appear to have a lower sinuosity than those for other compositions.

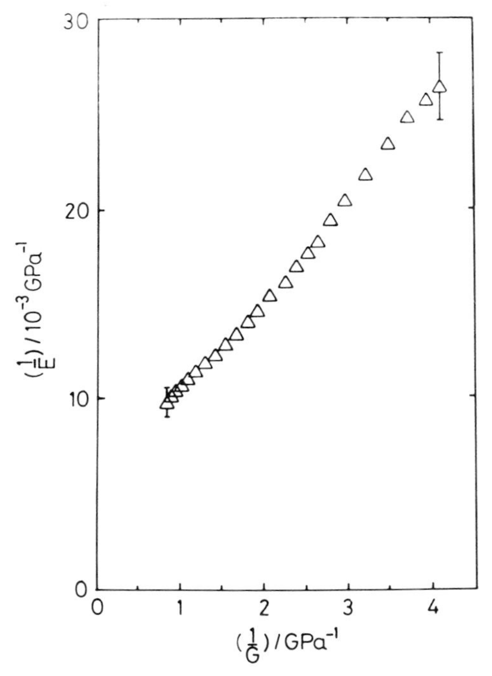


**Figure 10** The aggregate model for COTBP assuming temperature-independent unit modulus.

The second topic where the aggregate model has proved constructive recently concerns the mechanical behaviour of highly oriented polyethylene produced by hydrostatic extrusion of chain-extended polyethylene produced by chain extension (Powell *et al* 1990). Samples of different extrusion ratio were obtained from a range of different molecular weight polyethylenes which had been subjected to a variety of different treatments, in all cases annealing under high pressures, but with significant differences with regard to temperature, pressure and time. Again data for the tensile modulus $E$ and shear modulus $G$ over a wide temperature range can be related by equation (7), in this case with $S_{33} = 1/E_c$ the crystal chain modulus of polyethylene. Figure 11 shows the plotted data for a range of R006-60 polyethylene extrudates of different extrusion ratios, prepared from initial billets with identical pressure annealing conditions. The data certainly confirm the validity of equation (7) to a good approximation. Furthermore, values of $\langle \sin^2\theta \rangle$ are obtained which compare reasonably with those obtained directly from x-ray diffraction, which are shown in figure 12. The important fact about the $\langle \sin^2\theta \rangle$ values is that to a first approximation they conform to the so-called pseudo-affine deformation scheme, where the unique axes of the anisotropic units (in this case extended chain crystallites) rotate toward the extrusion direction like lines marked on the deforming macroscopic polymer. Sir Charles Frank has described this as akin to the rotation of needles in plasticine during plastic deformation.

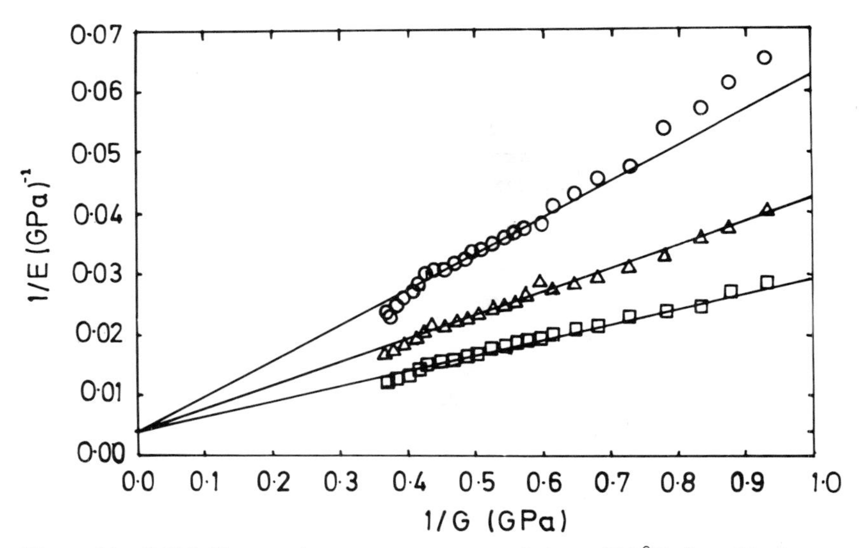


**Figure 11** R006-60 extrudates pressure annealed at 234 °C for 15 min at 450 MPa pressure and subsequently extruded to the extrusion ratio indicated. ○, 5 : 1; △, 7 : 1; □, 10 : 1.

The aggregate model can also be used to gain further insight into the effects of annealing treatment and polymer molecular weight on the mechanical properties of the extrudates. In figure 13 results for four different annealing times are included. In figure 13($a$) the data are offset to enable the four sets of data to be distinguished, whereas in figure 13(*b*) they are not transposed and can be seen to be on the same line. It can be concluded that $\langle \sin^2\theta \rangle$ is independent of annealing time (as confirmed by x-ray diffraction measurements) and that differences in tensile modulus relate to differences in the shear modulus of the orienting units. The indications from structural studies are that these differences arise from segregation effects during crystallisation.

## 4.2 Composite models

Takayanagi (1963) recognised the analogy between a crystalline polymer with a rigid crystalline phase and a flexible amorphous phase and an incompatible blend of two polymers with very different glass transition temperatures. This recognition led to the well-known Takayanagi models which have been of considerable value in understanding both the mechanical behaviour and, as shown above, the measurements of crystal strain in oriented polymers.

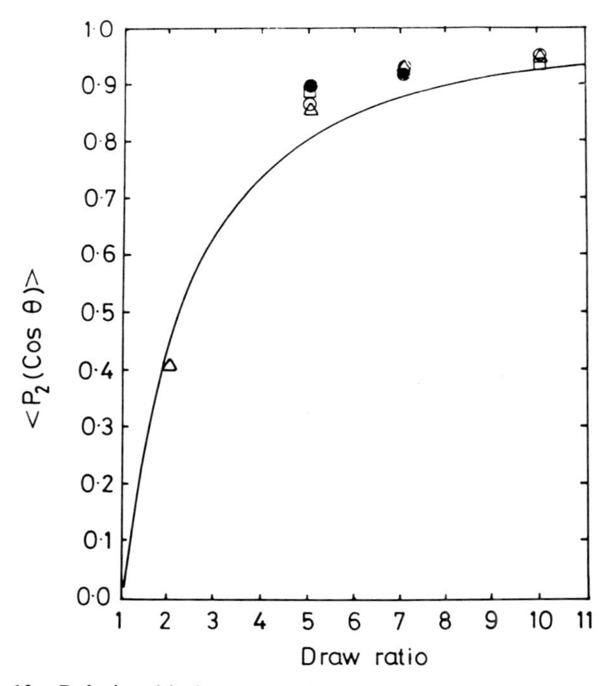


**Figure 12** Relationship between orientation function $\langle P_2(\cos\theta)\rangle$ and extrusion ratio for various extrudates pressure annealed at 230 °C, 450 MPa for times indicated: •, 5 min; △, 15 min; ○, 60 min; ▽, 120 min.

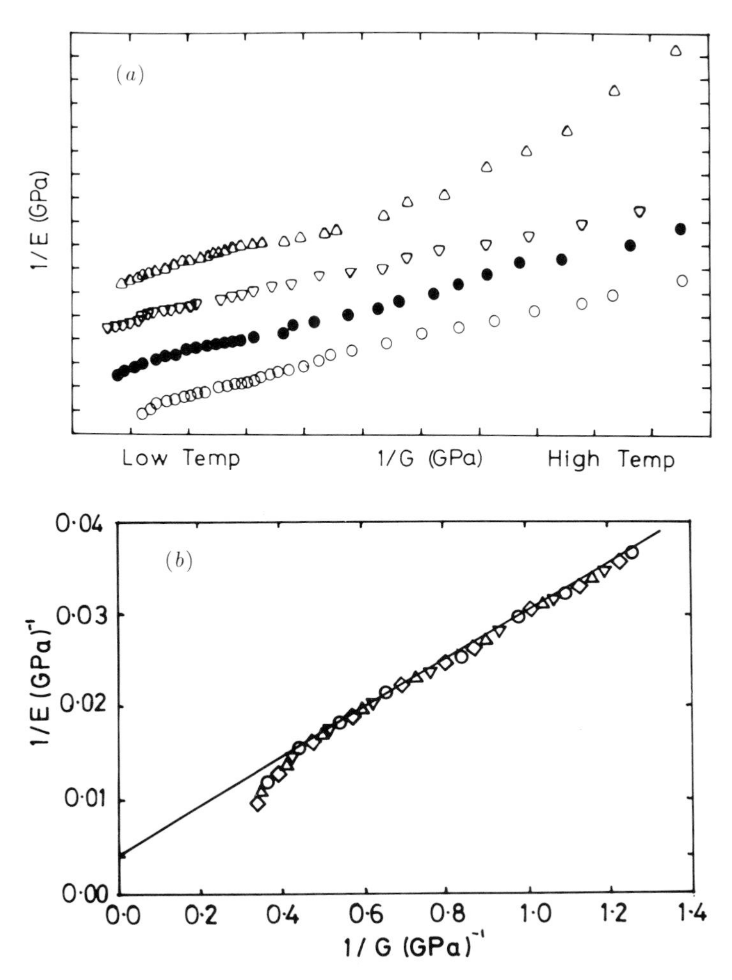


**Figure 13** (*a*),(*b*) Aggregate model parameters for R006-60 extrudates (extrusion ratio 10 : 1) pressure annealed at 234 °C at 450 MPa pressure for the time indicated: △, 5 min; ◇, 15 min; ○, 60 min; ▽, 180 min.

### *4.2.1 Interlamellar shear and chain-axis shear processes*

The severe limitation of the Takayanagi models is that they are one dimensional and therefore cannot describe the true nature of the relationship between stress and strain, which requires us to be able to distinguish between tensile stresses and shear stresses, and between tensile strains and shear strains. The Takayanagi models cannot therefore describe the influence of lamellar *orientation*, although they have been, and are still, of great value in dealing simplistically with a wide range of polymers where the lamellar normals are parallel to the principal orientation direction (the draw direction which usually coincides with the chain-axis direction).

The most definitive study of the effect of lamellar orientation on mechanical anisotropy is the research on specially oriented sheets of low-density polyethylene undertaken at Bristol University by Keller, Ward and their collaborators in the 1960s under the aegis of Sir Charles Frank. Stachurski and Ward (1968) examined the mechanical loss spectra of such sheets, which were prepared following the definitive structural research of Hay and Keller (1966, 1967). The key results are summarised in figure 14. It can be seen that the low-temperature process *P* shows maximum loss when the tensile stress is along the direction of the *c* axis in the '*b–c*' sheet (figure 14(*a*)) or along the *a*-axis direction in the '*a–b*' sheet (figure 14(*c*)) or at 45° to the draw direction in the 'parallel-lamellae' sheet (figure 14(*b*)). This then identifies the process as interlamellar shear. The high-temperature loss process *Q*, on the other hand, shows maximum loss in the '*b–c*' sheet when the tensile stress is at 45° to the *c*-axis direction, or along the draw direction in the 'parallel-lamellae' sheet. This identifies the process as *c*-shear, shear in the *c*-axis direction on planes containing the *c* axis.

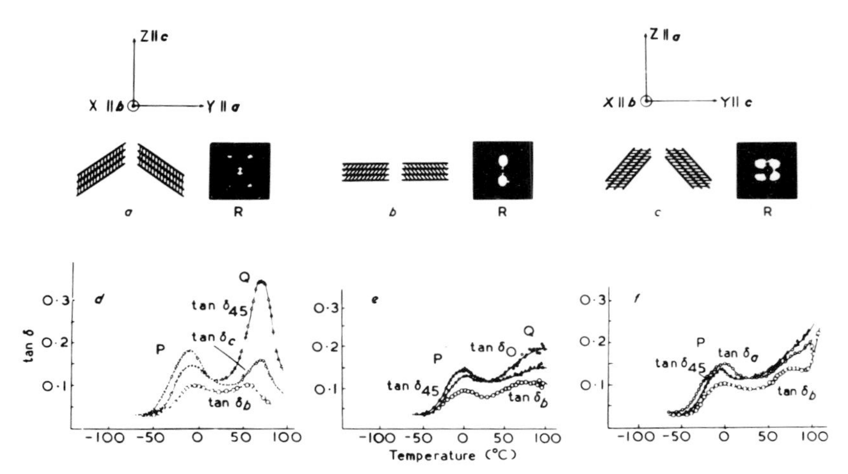


**Figure 14** Schematic structure diagrams and mechanical loss spectra; (*a*) and (*d*): for *b–c* sheet; (*b*) and (*e*): for parallel-lamellae sheet; (*c*) and (*f*): for *a–b* sheet. *P*, interlamellar shear process; *Q*, *c*-shear process; *R*, small-angle x-ray pattern. Reproduced from Ward (1970) by permission of The Institute of Physics (c).

To a first approximation, these results can be explained in terms of the interlamellar shear being simple shear only. A more rigorous analysis (Richardson and Ward 1978) indicates, however, that when the 'parallel-lamellae' sheet (figure 14(*b*)) is subjected to a tensile stress parallel to the draw direction *pure* shear will occur. On the simple model of plank-like lamellae, the compliance $S_{33}$ relates to pure shear of the interlamellar amorphous phase in the $yz$ plane. For total constraint in the $x$ direction (if the $b$ axis of the lamellae is assumed to be of infinite extent) $S_{11}=0$, $S_{23}=-S_{33}$ and $\nu_{13}=0$, $\nu_{23}=1$. The results in table 4 show that this is a good approximation for the parallel-lamellae sheets. Recent research (Humphreys *et al* 1988) has indicated a somewhat similar pattern of anisotropy for oriented sheets of polyvinylidene fluoride (PVDF), as also shown in table 4.

**Table 4** Compliance constants ($GPa^{-1}$) for parallel-lamellae polyethylene and oriented PVDF sheets.

| | Polyethylene | Low draw PVDF |
|---|---|---|
| $S_{33}$ | 9.76 | 0.34 |
| $S_{23}$ | −7.8 | −0.24 |
| $\nu_{23}$ | 0.8 | 0.74 |
| $S_{13}$ | −0.82 | −0.06 |
| $\nu_{13}$ | 0.08 | 0.18 |
| $S_{11}$ | 1.77 | 0.41 |

**Table 5** Shear compliance constants ($\times 10^{-10}\ m^2 N^{-1}$) for parallel-lamellae polyethylene sheet, PET sheet, and wet and dry nylon 6.

| | | | α-form nylon 6 | | γ-form nylon 6 | |
|---|---|---|---|---|---|---|
| | Polyethylene | PET | 0% RH | 65% RH | 0% RH | 65% RH |
| $S_{44}$ | 267 | 97 | 26 | 30.9 | 12.1 | 20.1 |
| $S_{55}$ | 214 | 5.64 | 7.1 | 19.0 | 7.7 | 15.6 |
| $S_{66}$ | 105 | 141 | 19.1 | 6.0 | 6.2 | 7.2 |

Very interesting information can be obtained from direct measurements of the shear compliances of oriented polymer films. Table 5 shows collected data for $S_{44}$, $S_{55}$ and $S_{66}$ for the parallel-lamellae polyethylene, for polyethylene terephthalate (PET) (Lewis and Ward 1980a) and for $\alpha$ and $\gamma$ form nylon sheets (Lewis and Ward 1980b, 1981) whose structures are shown schematically in figure 15.

The parallel-lamellae polyethylene sheets show $S_{44} \sim S_{55} > S_{66}$, confirming that interlamellar shear predominates, but $S_{66}$ is still quite large, indicating that $c$-shear is also occurring. The PET sheet shows remarkable anisotropy. Both $S_{44}$ and $S_{66}$ are large compared with $S_{55}$, because the planar terephthalate units are aligned in the plane of the sheet and can easily shear past each other, whereas $S_{55}$ involves distortion of the plane of the molecules. The results for the nylon sheets are more complicated. Consider first the dry $\alpha$-form results, where $S_{55} < S_{44} \sim S_{66}$. This contrasts directly with the parallel-lamellae polyethylene sheets and suggests that shear parallel to the chain axis predominates ($S_{44}$). We have called this $b$-shear (for the monoclinic structure the chain axis is the $b$ axis of the unit cell). For wet $\alpha$-form on the other hand, $S_{44} \sim S_{55} > S_{66}$, which suggests interlamellar shear.

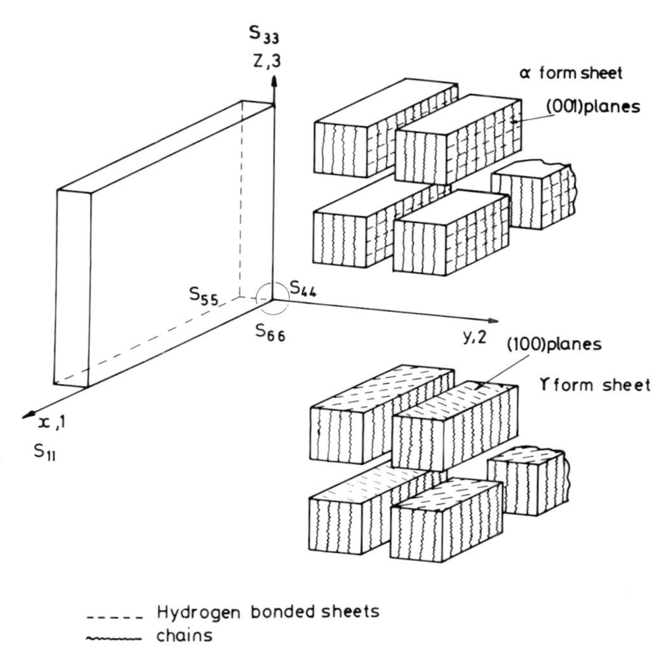


**Figure 15** Schematic diagram showing the morphologies of the $\alpha$- and $\gamma$-form nylon 6 sheets in relation to the principal axes of the sheets. Reproduced from Ward (1983) by permission of John Wiley and Sons Ltd (c).

The shear compliances of the $\gamma$-form sheet are somewhat different. Here the values of $S_{44}$ are smaller than for the $\alpha$-form sheet, which can be attributed to the fact that the hydrogen bonds now form sheets at 60° from the sample plane, making it harder for $b$-shear to occur. Again, however, the wet sheet results indicate $S_{44} \sim S_{55} > S_{66}$, which is consistent with interlamellar shear.

### *4.2.2 High modulus flexible polymers: fibre composite models, intercrystalline bridges, tie molecules*

The discovery of high modulus polyethylenes where the Young's modulus reaches values close to those predicted theoretically required a serious reappraisal of the simple Takayanagi composite model. Two approaches have emerged: (1) the analogy with a short fibre composite and (2) the addition of tie molecules or intercrystalline bridges between adjacent crystal blocks.

The short fibre composite model is more satisfying because it does embrace both tensile and shear deformation. Barham and Arridge (1977) considered that the fibre phase consists of needle-like crystals, which are identified with the macroscopic fibres. A constant proportion ($\sim 0.8$) of this fibre is assumed and the increase in modulus on drawing is attributed to the increase in the aspect ratio of the needle-like crystals. In the ultra drawing process the material 'tapers down' from a draw ratio of $\sim 10$ to a final draw ratio of $\sim 30$. The Young's modulus of an aligned short fibre composite is given by (Cox 1952)

$$E = E_f \nu_f \Phi + E_m \nu_m \tag{8}$$

where $E_f$, $\nu_f$ are the modulus and volume fraction of the fibre phase and $E_m$, $\nu_m$ are the modulus and volume fraction of the matrix phase. $\Phi$ is the shear lag factor which allows for ineffective stress transfer caused by the finite aspect ratio of the fibre. Barham and Arridge showed that the increase in Young's modulus with draw ratio could be quantitatively explained by the increase in $\Phi$ with draw ratio, through the affine transformation of the aspect ratio of the needle-like crystals.

Gibson *et al* (1978) used the same approach, but assumed that the fibre phase consisted of long crystals, i.e. crystalline sequences which link two or more adjacent crystal blocks. The *proportion* of this fibre phase increased with draw ratio due to the increased number of randomly arranged intercrystalline bridges. The probability $f_n$ that $n$ crystal blocks are linked is defined by a single parameter $p$ and

$$f_n = p^{n-1}(1 - p). \tag{9}$$

It can be shown that the Young's modulus at $-50\ ^\circ$C (between the $\alpha$ and $\gamma$ relaxations) is given by

$$E = E_c \chi p(2 - p) \tag{10}$$

where $E_c$ is the crystal chain modulus and $\chi$ is the crystallinity.

The most convincing demonstration of this simple relationship has come from a study of the effects of draw temperature on the modulus/draw ratio relationship (Clements and Ward 1983). The parameter $p$ can be determined from the integral breadth of the 002 meridional reflection and the long period $L$. It was found, as anticipated, that $L$ increased with increased draw temperature. For a given draw ratio it was also found that the integral breadth decreased with increasing draw temperature, i.e. the structure scaled with temperature to maintain a unique relationship between $p$ and draw ratio, and hence the Young's modulus and the draw ratio. The results are illustrated in figure 16 and show a good fit to equation (10).

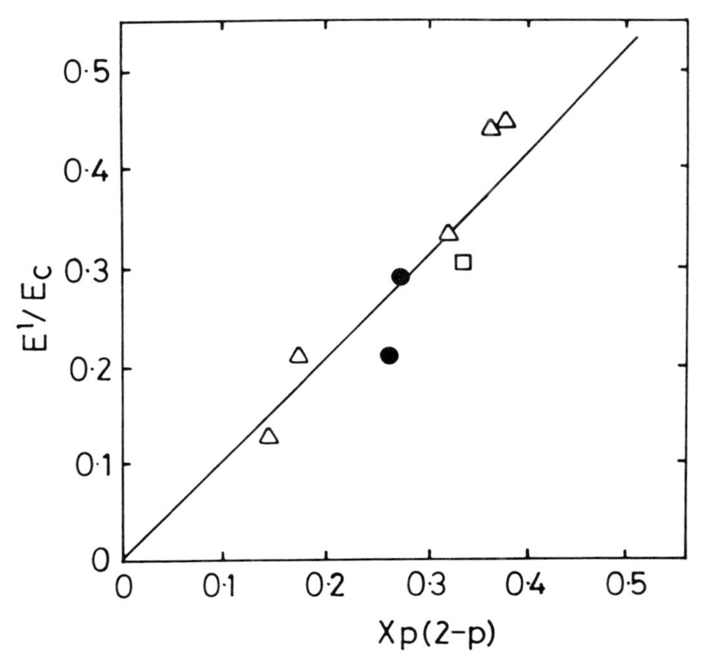


**Figure 16** A comparison of axial low-temperature moduli (measured at $-50\,^{\circ}$C) with the predictions of the 'random inter-crystalline bridge model'. △, Rigidex 50; •, H020-54P, $T_d = 95\,^{\circ}$C; □, H020-54P, $T_d = 115\,^{\circ}$C. Reproduced from Clements and Ward (1983) by permission of Butterworth & Co, Publishers Ltd (c).

It was also shown, in a separate study (Gibson *et al* 1982), that the temperature dependence of the Young's modulus could be quantitatively modelled in terms of the change in shear modulus with temperature. This affects the shear lag factor $\Phi$, which is related to the ratio $G_m/E_f$, where $G_m$ is the shear modulus of the matrix phase and $E_f$ the Young's modulus of the fibre phase. $\Phi$ falls with decreasing $G_m/E_f$, so that the Young's modulus of the oriented material $E$ also falls, in accordance with equation (8).

In the case of other high modulus flexible oriented polymers, such as polypropylene and polyoxymethylene, there is no evidence for intercrystalline bridge or very long crystals. As discussed above it appears that the high modulus of the highly oriented POM samples relates to the presence of a third phase in parallel with the remaining crystalline and amorphous material. In PP, there is more direct evidence for structural changes with draw ratio (Taraiya *et al* 1988). Figure 17 shows the development of a new endothermic peak with increasing draw ratio. This peak shows remarkable able superheating effects and has been tentatively attributed to the presence

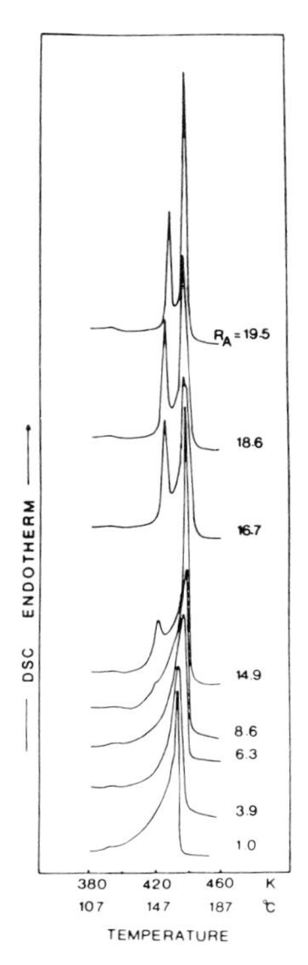


**Figure 17** DSC endotherms of die-drawn polypropylene rods; draw temperature = 110 °C; draw ratio $R_A$ as indicated. Reproduced from Taraiya *et al* (1988) by permission of John Wiley and Sons Inc (c).

superheating effects and has been tentatively attributed to the presence of extended crystals with constrained chains. An increase in the heat of fusion with increasing draw ratio was also observed. Using the parallel–series Takayanagi model (figure 3(*b*)) the parallel component $b$ can be calculated. On the basis that $E_a \ll E_c$ we have

$$E/E_c = b.$$

Figure 18 shows a plot of the first DSC melting peak area (figure 17) as a fractional percentage of the total area of both peaks plotted against the calculated $b$ values. In figure 18 $b$ is identical to $1 - \lambda$. There is an excellent correlation, but it can only be concluded that the reinforcing elements could be either intercrystalline bridges or taut tie molecules, both of which could be expected to contribute to the low-temperature peak. Further evidence for this view has been obtained from annealing experiments where the modulus falls as this peak shifts towards the high-temperature peak and eventually merges with it.

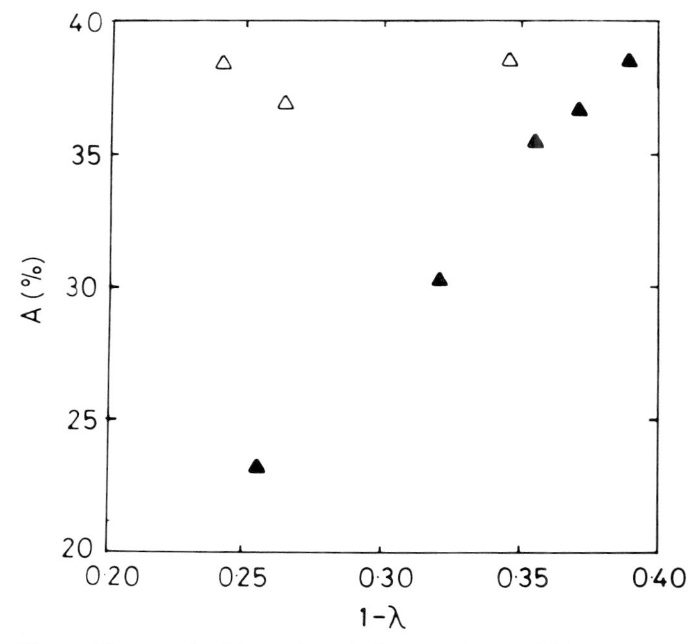


**Figure 18** DSC double-peak endotherms; first melting peak area as a fraction percentage $A$ of the total area of both peaks against the calculated fraction $(1-\lambda)$ of crystal bridges and taut tie molecules. ▲, Unannealed; △, annealed at different temperatures. Reproduced from Taraiya *et al* (1988) by permission of John Wiley and Sons Inc (c).

## 5 CONCLUSIONS

In this review it has been shown that much progress has been made towards an understanding of the anisotropic mechanical behaviour of polymers. There is a wide variety of behaviour, and different approaches appear to be valid for different types of polymer. For amorphous polymers and liquid crystalline polymers, molecular orientation is of primary importance. For crystalline polymers, an understanding of the morphology is vital, and Takayanagi models must only be regarded as a first step.

Much quantitative research remains to be done, perhaps especially with regard to the determination and understanding of shear compliances and lateral compliances, including Poisson's ratios. Viewed objectively, this area of research is one where mechanical and structural studies must advance together and it is only where this has been possible that genuine physical insight has been obtained. It is therefore a pleasure to dedicate this article to Sir Charles Frank who saw so clearly the essential synergism between these different facets of research activity in polymer science and by his example and encouragement played a key role in the development of this area of research.

## REFERENCES


Arridge R G C, Barham P J and Keller A 1977 *J. Polymer Sci. Polymer Phys. Ed.* **15** 389

Barham P J and Arridge R G C 1977 *J Polymer Sci. Polymer Phys. Ed.* **15** 1177

Chiba A, Masagawa A, Hikichi K and Furuichi J 1966 *J. Phys. Soc. Jpn* **21** 1777

Clements J, Jakeways R and Ward I M 1978 *Polymer* **19** 639

Clements J and Ward I M 1983 *Polymer* **24** 27

Cox H L 1952 *Br. J. Appl. Phys.* **3** 72

Davies G R and Ward I M 1988 *High Modulus Polymers: Approaches to Design and Development* ed A E Zachariades and R S Porter (Basel: Marcel Dekker) p 37

Dulmage W J and Contois L E 1958 *J. Polymer Sci.* **28** 275

Fischer E W, Goddar H and Peisczek W J 1971 *J. Polymer Sci.* C **32** 149

Frank F C 1970 *Proc. R. Soc.* A **319** 127

Gibson A G, Davies G R and Ward I M 1978 *Polymer* **19** 683

Gibson A G, Jawad S H, Davies G R and Ward I M 1982 *Polymer* **23** 349

Green D, Unwin A P, Davies G R and Ward I M 1990 *Polymer* **31** 579

Hay I L and Keller A 1966 *J. Mater. Sci.* **1** 41

—— 1967 *J. Mater. Sci.* **2** 538

Holliday L 1975 *Structure and Properties of Oriented Polymers* ed I M Ward (London: Applied Science) p 242

Holliday L and White J W 1971 *Pure Appl. Chem.* **26** 545

Humphreys J, Lewis E L V, Ward I M, Nix E L and McGrath J C 1988 *J. Polymer Sci. Polymer Phys. Ed.* **26** 141

Jungnitz S, Jakeways R and Ward I M 1986 *Polymer* **27** 1651

Kaji K, Shintaku T, Nakamae K and Sakurada I 1974 *J. Polymer Sci. Polymer Phys. Ed.* **12** 1457

Leung W P, Chen F E, Choy C L, Richardson A and Ward I M 1984 *Polymer* **25** 447

Lewis E L V and Ward I M 1980a *J. Mater. Sci.* **15** 2354

—— 1980b *J. Macromol. Sci.* B **18** 1

—— 1981 *J. Macromol. Sci.* B **19** 75

Mizushima S and Shimanouchi T 1949 *J. Am. Chem. Soc.* **71** 1320

Odajima A and Maeda M 1966 *J. Polymer Sci.* C **15** 55

Olf H G and Peterlin A 1964 *J Appl. Phys.* **35** 3108

Powell A K, Craggs G and Ward I M 1990 *J. Mater. Sci.* **25** 3990

Prevorsek D C, Tirpak G A, Harget P J and Reinshussel A C 1974 *J. Macromol. Sci.* B **9** 733

Richardson I D and Ward I M 1978 *J. Polymer Sci. Polymer Phys. Ed.* **16** 667

Sakurada I, Ito T and Nakamae K 1966 *J. Polymer Sci.* C **15** 75

Sakurada I and Kaji K 1970 *J. Polymer Sci.* C **31** 57

Shauffele R F and Shimanouchi T 1967 *J Chem. Phys.* **47** 3605

Sorensen R A, Liau W B, Kesner L and Boyd R H 1988 *Macromol.* **21** 200
Stachurski Z H and Ward I M 1968 *J. Polymer Sci.* **A2** (6) 1817
Ströbl G R and Eckel R 1976 *J. Polymer Sci. Polymer Phys. Ed.* **14** 913
Sugeta H and Miyazawa T 1970 *Polymer. J.* **1** 226
Takayanagi M 1963 *Mem. Fac. Engng. Kyushu Univ.* **23** 41
Taraiya A K, Unwin A P and Ward I M 1988 *J. Polymer Sci. Polymer Phys. Ed.* **26** 817
Tashiro K, Kobayashi M and Tadokoro H 1977 *Macromol.* **10** 413
—— 1978 *Macromol.* **11** 914
Thistlethwaite T, Jakeways R and Ward I M 1988 *Polymer* **29** 61
Troughton M J, Unwin A P, Davies G R and Ward I M 1988 *Polymer* **29** 1389
Ward I M 1962 *Proc. Phys. Soc.* **80** 1176
—— 1970 *Phys. Bull.* **21** 71
—— 1982 *Development in Oriented Polymers* vol 1, ed I M Ward (London: Applied Science)
—— 1983 *Mechanical Properties of Solid Polymers* 2nd edn (Chichester: Wiley) p 246
White J W 1976 *Structural Studies of Macromolecules by Spectroscopic Methods* ed K J Ivin (London: Wiley)
Wu G, Tashiro K, Kobayashi M, Komatsu T and Nakagawa K 1989 *Macromol.* **22** 758

# Bidimensional Compression

K H G Ashbee

> Imagine four old men with beards, facing the same way round a square, each sucking the beard of his neighbour
> F C Frank's verbal description of a bidimensional compression device from his hospital bed in 1983.

The earliest published description of a bidimensional compression apparatus that is now known to me is that reported by Hambly and Roscoe (1969) at the 7th International Conference on Soil Mechanics and Foundation Engineering held in Mexico in 1969. This apparatus comprises four identical platens, each fitted with a guide rod and each having a guide hole, such that the guide rod of any one platen fits into the guide hole of its neighbour, thereby constraining relative motion of the platens to be mutually perpendicular. The four-platen assembly, defining a confined cell of square cross section, is illustrated in figure 1. Taricco (1980), in his radial compression device, used a rigid frame external to the platens in order to confine his platens to move in pairs in a plane, the pairs moving at right angles to each other. His apparatus is illustrated in figure 2. Since the lines of action of the forces applied to Taricco's platens do not pass through the centre of the confined cell, provision has to be made for lateral displacement of those forces as the platens advance.

On one of his visits to Orsay in the early 1980s Charles Frank had seen at first hand Deloche's (Deloche and Samulski 1981) deuteron magnetic resonance experiments with tensioned elastomers. The magnetic fields used in NMR experiments are of the order of 1 T. In such fields the sample is macroscopically oriented with the director parallel on average to the

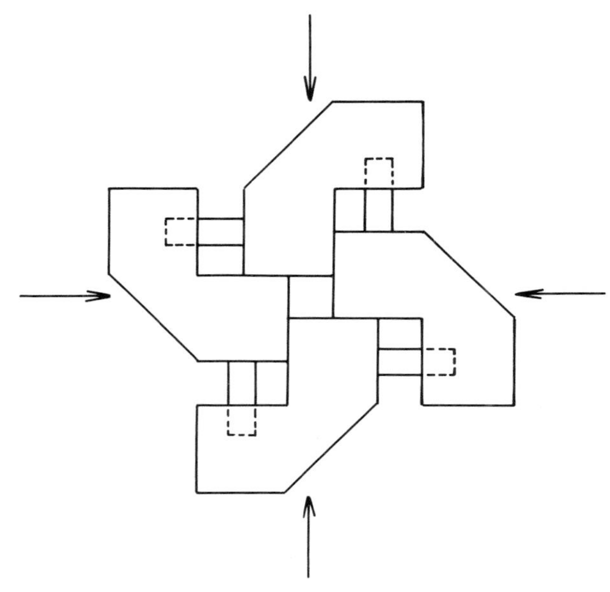

**Figure 1** Perspective view of the Hambly and Roscoe (1969) bidimensional compression device.

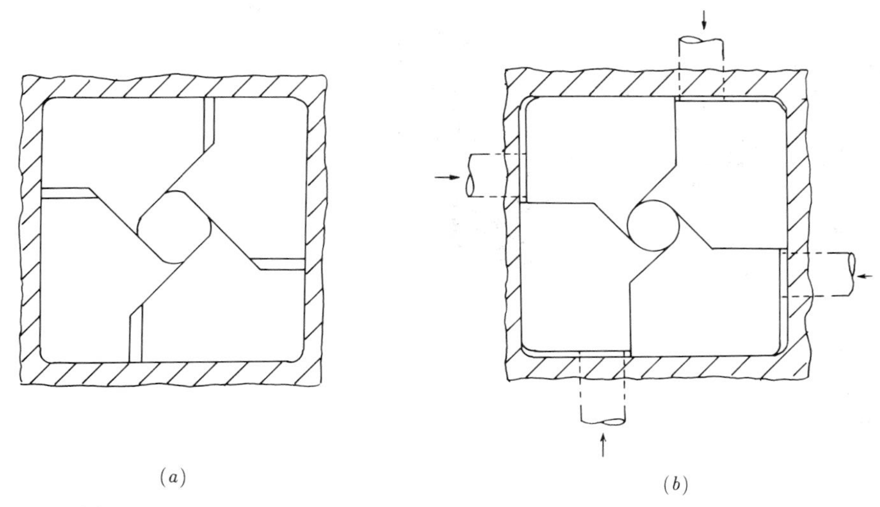


**Figure 2** Taricco's (1980) radial compression moulding apparatus (*a*) fully open, (*b*) fully closed.

magnetic field. The protons naturally present in the molecule usually constitute the NMR probes. However, as the number of protons increases, so does the number of dipolar couplings and the PMR spectrum becomes complex. When protons are replaced by deuterons, a much simpler PMR spectrum is observed. The simplification comes from the fact that the magnetic moment of the deuteron is smaller than that of the proton. Hence their NMR signals are distinct and the dipolar couplings between them are small in comparison with those which exist with protons. Figure 3 ($\lambda = 1$) shows the $^2H$ NMR spectrum for hexadeuteriobenzene ($C_6D_6$) in polyisoprene. When uniaxially tensioned ($\lambda = 1.47$), the single line changes to a quadrupolar doublet. Not only that but, as the extension ratio ($\lambda$) is increased, the splitting ($\Delta\nu$) increases. The occurrence, under tension, of finite values of $\Delta\nu$ is indicative of anisotropic reorientational diffusion of $C_6D_6$ now constrained by the uniaxial field associated with the deformed rubber network. That is, the solvent acquires non-zero average

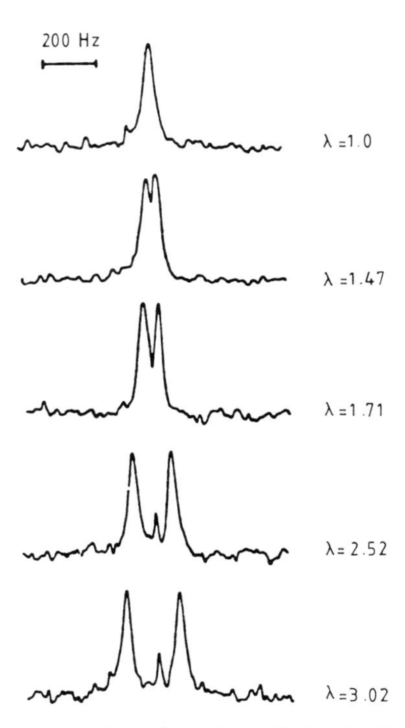


**Figure 3** Development of quadrupolar splitting in the deuterium NMR spectrum of $C_6D_6$ dissolved in polyisoprene during uniaxial extension. The polymer volume fraction is $\frac{1}{2}$. Reprinted with permission from *Macromolecules* **14** (1981) 575–81. Copyright (1981) American Chemical Society.

orientational order as it diffuses through the deformed elastomer. Deloche and co-workers had followed the development of orientational correlations between chain segments with increasing extension ratio right up to the point of mechanical failure, and Charles Frank wondered if Deloche's data could be extended to far greater extension ratios, and hence to far greater degrees of chain orientation, by subjecting the polymer to bidimensional compression instead of to uniaxial tension. The point here is that the material would increase in length in the dimension perpendicular to the plane of bidimensional compression, and the polymer chains would presumably tend to become aligned in that direction but, since both principal stresses are compressive, fracture is unlikely even at, in principle, infinite bidimensional compressive strain. By this time, late 1983, Charles' handwriting had become decidedly shaky, to the point where even his sketches were hard to decipher. All of this was, of course, markedly improved by the surgery that he underwent at the time of my bedside discussions with him of how best to devise a bidimensional compression experiment that Deloche might be able to incorporate within his NMR cavity.

**Figure 4** The rectangular cross-section bidimensional compression deuterium magnetic resonance device fabricated from ebonite.

A photograph of the miniature bidimensional compression device that we came up with is reproduced in figure 4. Harry Young machined it from ebonite using fairly rough drawings that I had developed from the notion that we should use four interfitting G-shaped platens—four old men with beards as Charles had described it. The G-shape of the platens incorporates a leg and a leg-receiving slot; each leg-receiving slot receives and then guides displacement of the leg of an adjacent platen, in much the same way as the guide hole in each of Hambly and Roscoe's platens guides the guide rod of an adjacent platen. The device shown in figure 4 measures one inch in diameter when fully closed. The G-shaped elements are bolted together using nylon screws in pairs so that they can be advanced in such a way as to achieve independent displacements parallel to each of two perpendicular directions.

And so, I set off for the boat train to Paris. Well, we tried very hard to make the experiment work. In fact, we subsequently made two more bidimensional compression rigs for Deloche, both machined from Macor®, a machinable glass ceramic, an insulator because it had to be non-conducting, kindly provided by Dr Ken Chyung of Corning Glass

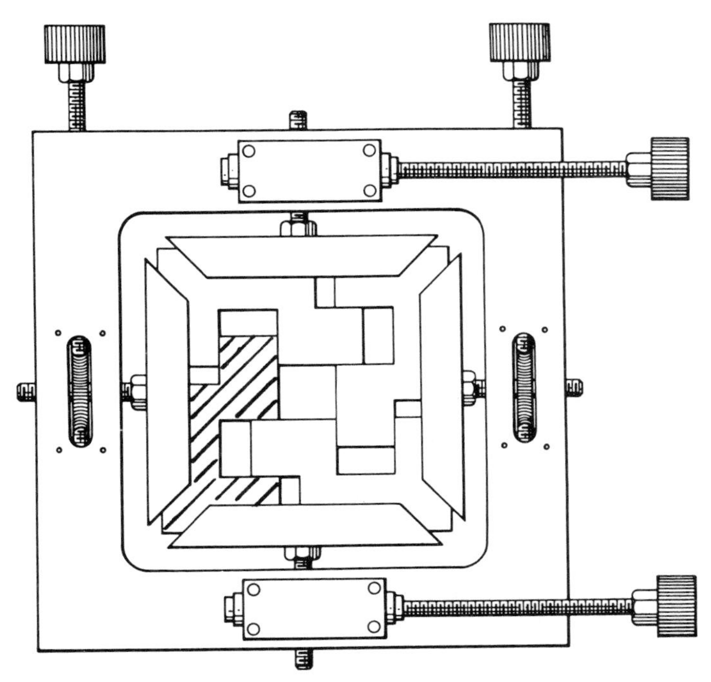

**Figure 5** Plan view of the worm and screw driven bidimensional compression device. To assist visualisation of the geometry of the four interfitting G-shaped platens, one of them is cross-hatched.

Works. One problem with our experiment was that we were unable to sufficiently lubricate the compressive contacts between specimen and platens for meaningful data to be obtained at extension ratios beyond that which Deloche could more easily realise in tension.

So what else could we do with bidimensional compression? In my laboratory at Bristol, Jeff Sargent, who had been studying the problem of constraint at the ill-lubricated platen surfaces using photoelastic methods applied to transparent samples, briefly turned his attention to a bidimensional compression analogue of the Rivlin and Saunders (1952) bidimensional tension experiment that Ian Ward had rigged up in the early days of the MSc Course in the Physics of Materials at Bristol. Jeff had printed a square grid on a one-inch square elastomer sample and proceeded to measure the grid deformations for various combinations of two-dimensional applied compressive stresses. We tried two or three energy functions in an attempt to fit the data to large strain elasticity theory but

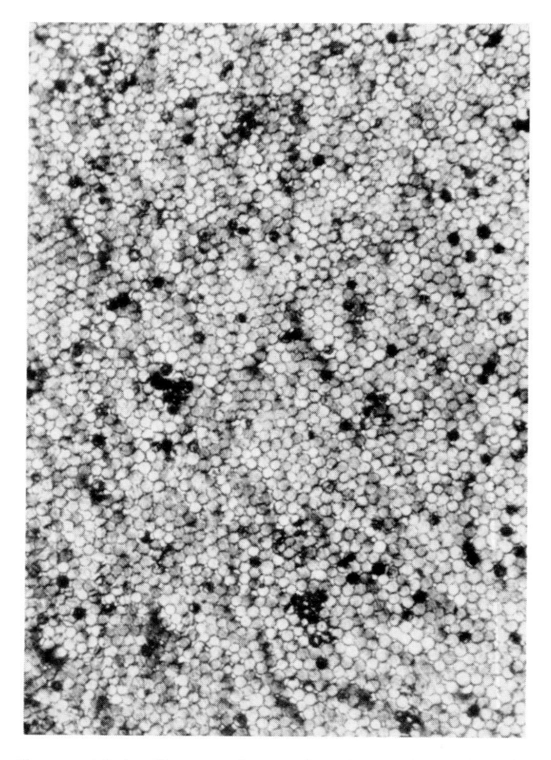

**Figure 6** Super high fibre volume fraction glass fibre/epoxy resin uniaxial composite. Fibre diameter nominally 10 $\mu$m. Fibre volume fraction close to theoretical ($\pi/2\sqrt{3}$). Reprinted by permission of Technomic Publishing Co Inc.

without much success. Consoling ourselves that what we were really doing was investigating the deformation of an elastomer in the 'no mans land' between that described by rubber elasticity and that described by small strain elasticity, we shelved that line of research for a later date.

It was still the early 1980s and, on one of my trips to the USA, I had been struck by the fact that the major suppliers of fibre-reinforced polymer pre-pregs for the manufacture of high-performance laminates had decided that 60% fibre volume fraction was optimum. I knew, from a then recent Bristol tutorial question, set by Mike Berry I think, that the packing fraction of identical straight cylinders is $\pi/2\sqrt{3}$, a little over 90%. So the advanced fibre-reinforced composite materials community was opting for only two-thirds of the fibre content that is available to them. By way of justification for the 60% fibre loading, it was claimed that a higher fibre volume fraction pre-preg with all of its fibres fully wetted by resin is difficult to mass produce by squeezing between rollers. The reasons for this are explained as follows. Consider a uniaxial array of fibres in a mechanically deformable matrix. During application of external pressure to fabricate the composite, any two fibres will resist approaching each other. This is because the matrix material has to flow outwards ever faster as the fibres approach each other, that is the shear rate in the matrix material is required to increase as the fibres become closer together. Thus an effective repulsive force is generated between the two fibres. Now consider a parallel array of fibres well-spaced on a hexagonal grid in a mechanically more plastic matrix. Unidirectional squeezing causes fibres to slide past each other, thereby destroying the hexagonal packing. If, instead of undirectional squeeze, a bidimensional squeeze is applied to the same array, the hexagonal arrangement is stabilised. Squeezing equally along mutually perpendicular axes in the plane at right angles to the fibre direction will not permit separation of any two fibres to allow passage between them of a third and, since the rate of approach of fibres which are close together will be less than that of fibres which are separated by larger distances, it is evident that a bidimensional squeeze will uniformise the packing, and then will encourage the matrix material to drain away in the fibre direction, thereby resulting in a higher overall fibre volume fraction.

Using the apparatus shown in figure 5, we were able to create near ideal fibre packing in composites fabricated from resin matrix pre-pregs (Ashbee 1986). An example, for glass-fibre-reinforced epoxy resin, is shown in figure 6. At this time, Frank and I filed for and were granted a patent on this invention (Frank and Ashbee 1984).

Thus far, we had made only worm and wheel operated small-scale bidimensional compression rigs. In 1985, I spent the Spring semester at Texas A&M University and, within a week of arrival, Dick Schapery afforded me $10 000 with which to build the hydraulically operated version shown in figure 7. This was a 15 000 psi machine in which we simultaneously advanced opposed pistons across the two diagonals of the square cross section of the confined cell. The super high fibre volume fraction composites manufactured with this rig were 6″ lengths by approximately 1″ square cross section, amply large enough to get meaningful data from mechanical tests. Well, as expected, the tensional

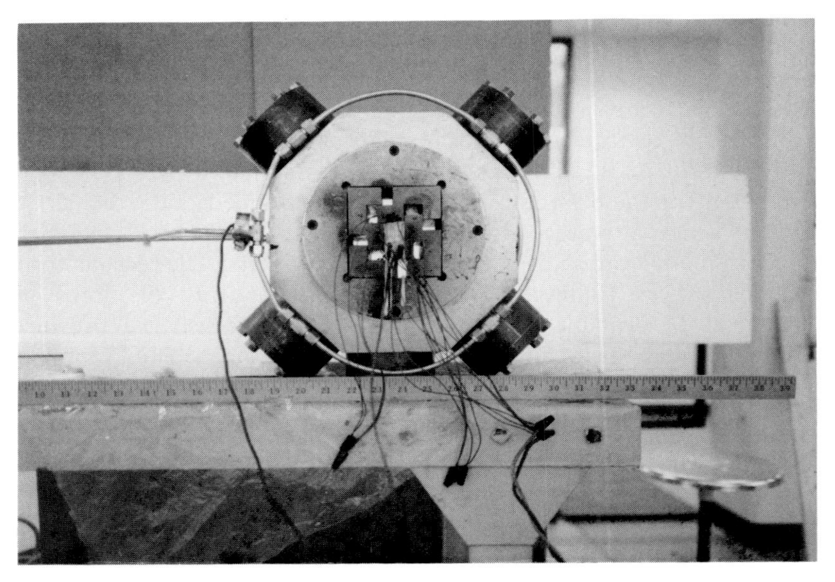

**Figure 7** The hydraulically driven apparatus built at Texas A&M University.

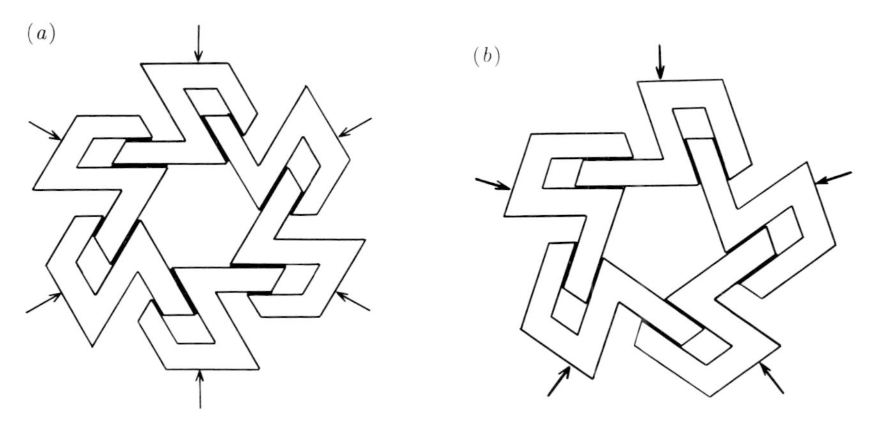


**Figure 8** Six-member (*a*) and five-member (*b*) bidimensional compression methods for fabricating hexagonal and pentagonal panels, respectively, for a radially fibre-reinforced pentagonal dodecahedron.

modulus (actually, flexural modulus since we had difficulty in gripping such strong specimens for testing in tension) measurements were remarkably high, following the law of mixtures relationship between the modulus of matrix and fibre materials that had long since been established for more modest fibre volume fractions. More interesting were the compressive strength data. By filament winding the outside of compression test pieces, fibre buckling, the usual mechanism of axial compressive failure of uniaxial composites, was circumvented and very high compression strengths, of more than 100 ksi for Kevlar/epoxy composites, for example, were realised. These latter results later led me to propose a radially reinforced soccer ball design for fibre-reinforced submersibles. In our patent, we had worked out the theorem for bidimensional compression moulding of bar stock of any geometrical cross section. So we know how to fabricate the hexagonal and pentagonal prisms needed for the soccer ball panels of a pentagonal dodecahedron (refer to figures 8(*a*) and (*b*)).

One other experiment that I tried while working in Schapery's department was to explore the fracture in bidimensional compression of rigid particle/elastomeric matrix composites. I had remembered Orowan's graphical construction (Orowan 1948/49) of the two Griffith failure criteria

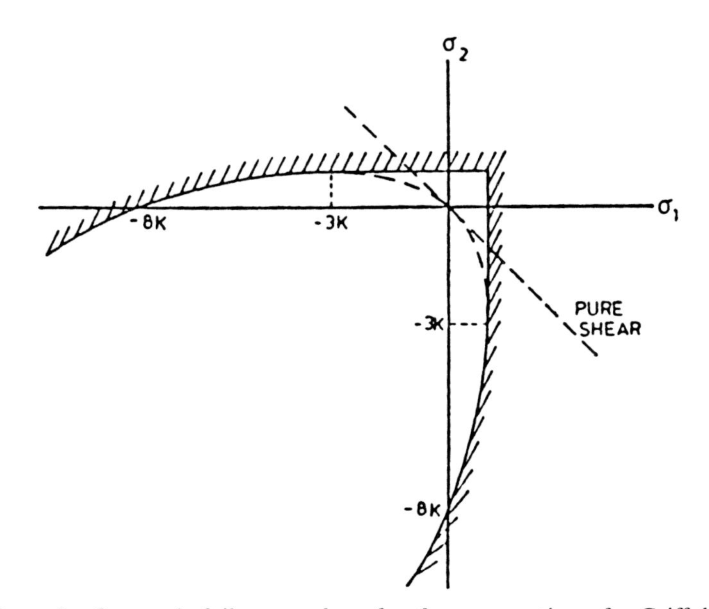


**Figure 9** Orowan's failure envelope for the propagation of a Griffith crack by application of biaxial stress. *K* is the uniaxial tensile strength. Note that the envelope, although open in the bidimensional compression quadrant, predicts failure when both principal stresses are compressive and strongly unequal.

for elastic solids subjected to two-dimensional stress fields. As shown in figure 9, there exists a region within which failure is predicted in the compression–compression quadrant. My test pieces were pencil erasers of the PVC/chalk particle variety and, sure enough, they swelled appreciably (by about 40% in volume, accounted for by the large density of internal cracks) after removal of the bidimensional compressive stresses (figure 10). The resulting very large internal porosity is best seen by stereoscopic pair scanning electron microscopy, an example of which is reproduced in figure 11.

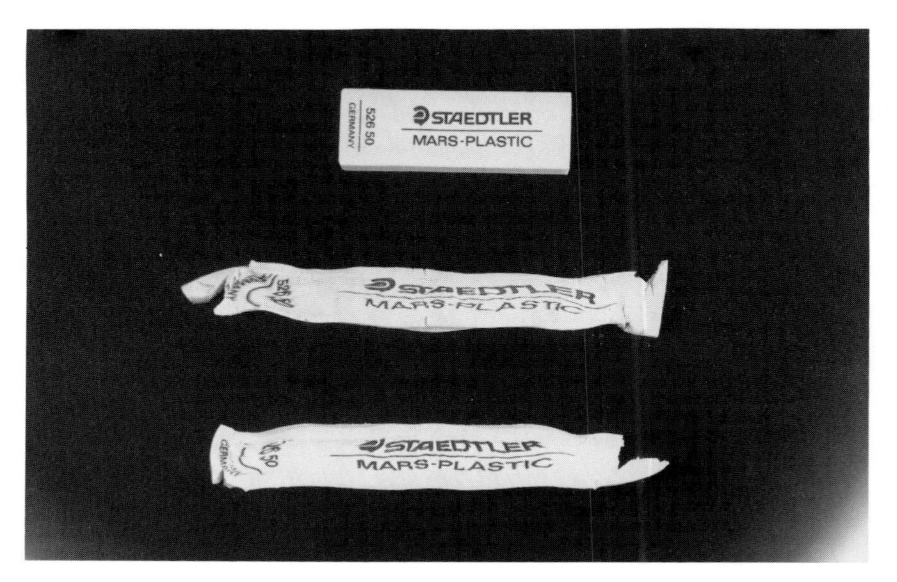


**Figure 10** The volume expansion imparted to chalk particle filled PVC erasers by bidimensional compression at room temperature.

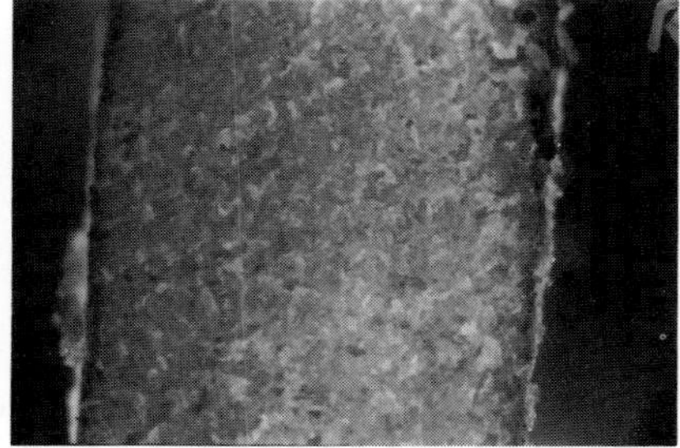

**Figure 11** Stereoscopic pair scanning electron microscope photographs of the porosity introduced into chalk particle filled polyvinylchloride during bidimensional compression.

Soon after my return to Bristol in the summer of 1985, I was offered a Chair of Excellence at the University of Tennessee, Knoxville. Exciting developments were under way in East Tennessee. A so-called Science Alliance programme had been established between UTK and the nearby Oak Ridge National Laboratory; I had spent a year as a visiting scientist at ORNL in 1970–71. David Joy, Bernhard Wunderlich and Jack Weitsman were to be recruited into that programme, all of whom I had known for some years. The timing was right for a move. The Science and Engineering Research Council had finally sealed the demise of the Physics of Materials MSc course at Bristol; Frank, Thompson and Lang had retired, Nye was nearing retirement, Forty, Ward and Hart had long since moved on to higher things and those of us who remained felt that this really was the end of the Charles Frank era. And so it was that I accepted my chair at UTK.

Now a bidimensional squeeze is precisely the deformation needed in polymer engineering to co-mingle continuous fibres with yarn and, in metallurgical engineering, to co-mingle continuous fibres with metal wires. Soon after my arrival in Knoxville, I visited Vinod Sitka at the Oak Ridge National Laboratory and took with me a new duplicate version of the

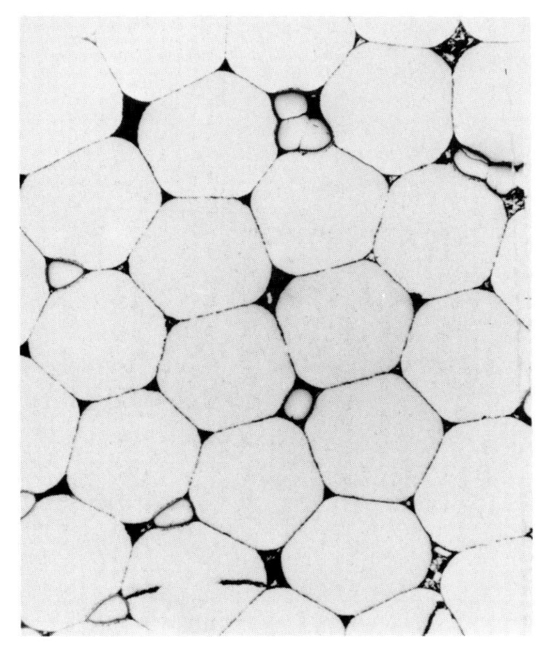

**Figure 12** Nickel aluminide and tungsten wires co-mingled by bidimensional compression for about 3 seconds at approximately 800 °C. 50X. The larger cross-section wires are the $Ni_3Al$.

apparatus that I had built at Texas A&M. C T Lui and co-workers at ORNL had been investigating superplasticity of intermetallic compounds and were, for example, able to superplastically wire draw $Ni_3Al$ on a routine basis. I had a new metallurgical engineering graduate student, Steve Joslin, with me, and Steve mixed a bundle of six inch lengths of $Ni_3Al$ wire with similar lengths of tungsten wire. Vinod had one of his welders encapsulate the mixture of wires inside a stainless steel tube which we then heated to 1000 °C in an argon atmosphere, withdrew from the furnace, and inserted into the confined cell of our bidimensional compression rig. It took Steve Joslin about three seconds to apply the full pressure (about 10 000 psi) available with our hydraulics, during which time we estimated that the temperature of the wires probably dropped a couple of hundred degrees. When cool, we cut the specimen into two and found that the mixture had consolidated. In fact, the wires had welded together about as well as they would have done after several hours at the same temperature in an isostatic press. The microstructure seen in a transverse section through that specimen is reproduced in figure 12. The $Ni_3Al$ and W wires have deformed from circular to space-filling polygonal cross sections. Although a few gaps remain, sufficient consolidation for the macroscopic

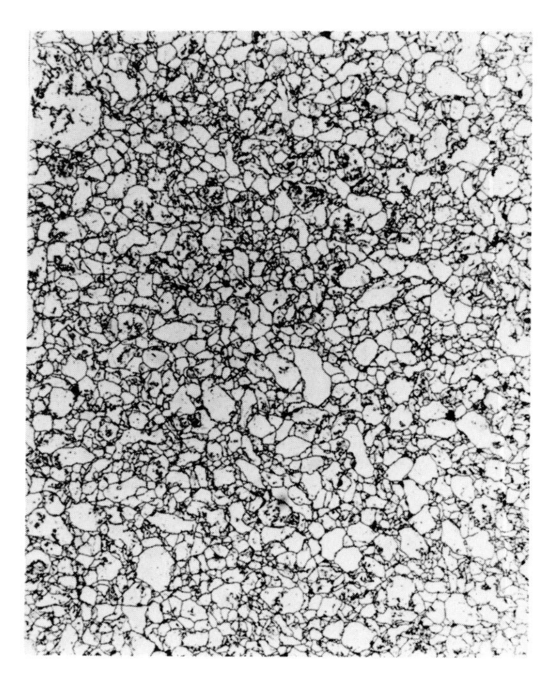

**Figure 13** Compacted aluminium powder encapsulated in an aluminium can and bidimensionally compressed at room temperature. 400X.

hardness to be on par with that for bulk $Ni_3Al$ had been obtained by bidimensional compression for only 3 seconds!

Steve Joslin now had a research project. If, as a consequence of the very large shear stresses present in our bidimensional stress field, mechanisms like grain boundary sliding and dislocation glide can cause metal wires to deform and weld together in times of the order of seconds, it should be possible to similarly fabricate other metal-based materials. In particular, it may be possible to substantially reduce the time at temperature when fabricating fibre-reinforced metals using fibres susceptible to thermal degradation. And so, Steve embarked on his master's project to discover if metal powders, encapsulated inside a metal sheath, can be rapidly consolidated by bidimensional compression.

Figures 13–15 show examples of the microstructures obtained by Joslin with encapsulated 1100 aluminium powder samples kindly provided by The Aluminum Company of America. Compacts having 98% theoretical density can be routinely fabricated at room temperature (figure 13). Figure 14 shows the cross section of a similar compact bidimensionally compressed at 200 °C. The density corresponds to the theoretical density for aluminium. Tensile strength and Charpy impact energy measurements demonstrate that these compacts have about the same strength and toughness as specimens of the same metal fabricated by more conventional

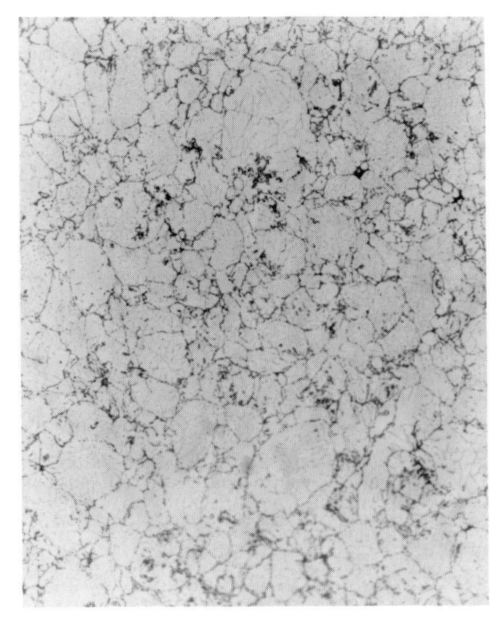

**Figure 14** The same powder compacted at 200 °C. 400X.

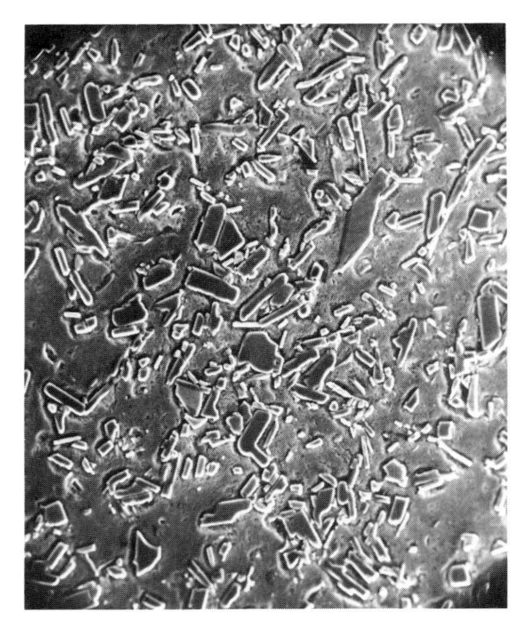

**Figure 15** The uniform distribution of silicon carbide platelets in aluminium that is achieved by bidimensional compaction. 200X.

methods. Figure 15 is typical of the microstructure obtained with mixtures of aluminium powder and silicon carbide single-crystal platelets. Notice the uniform dispersion of the platelets.

Finally, the two-dimensional cider press. More than one person who has seen the Frank–Ashbee bidimensional compression device in action has remarked that surely it has potential applications in food processing. To date, we have only experimented with the extraction of juice from fruit. Traditional cider presses impart a one-dimensional squeeze to the pomage delivered by the cider mill. Describing his 1788 observations on cider making in Herefordshire, William Marshall (1796) writes '...The principle of the press is that of the packing press, and the common napkin press; a screw working with a square frame.' He goes on to remark that 'All fluids endeavour to escape pressure. If it be applied in the perpendicular direction, they endeavour to fly out horizontally. Thus the pomage (a subfluid) is forced toward the outer edges of the cloths; which, in this form of the pile *have no immediate pressure*: an indirect horizontal pressure, arising from the natural law of fluids, being all the compression which the outer sides of the pile receive.' 'Hence, men, who exel in the art, continue to press, so long as a drop can be drawn; and, unsatisfied with this, return the FIRST RESIDUE to the mill, to be REGROUND: not with water, but with some of the first runnings of the liquor; moistening the materials to be ground as occasion requires'.

The virtue of exerting equal bidimensional compression along two axes at right angles is that it preserves the geometry of arrangement of solid fractions (of flesh, rind and kernel in the pomage) whilst decreasing the distances between them. Figures 16(*a*) and (*b*) show the two-dimensional press in operation. The confined cell, within which the pomage is squeezed,

(*a*)

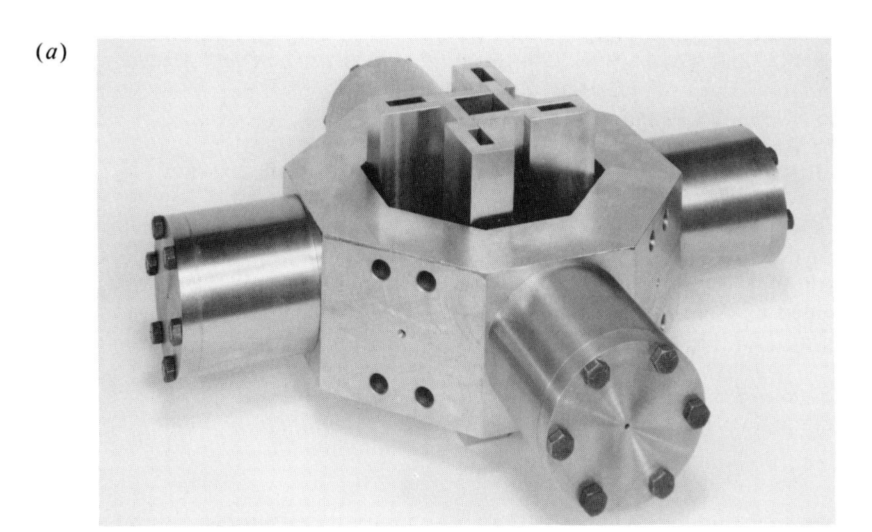

(*b*)

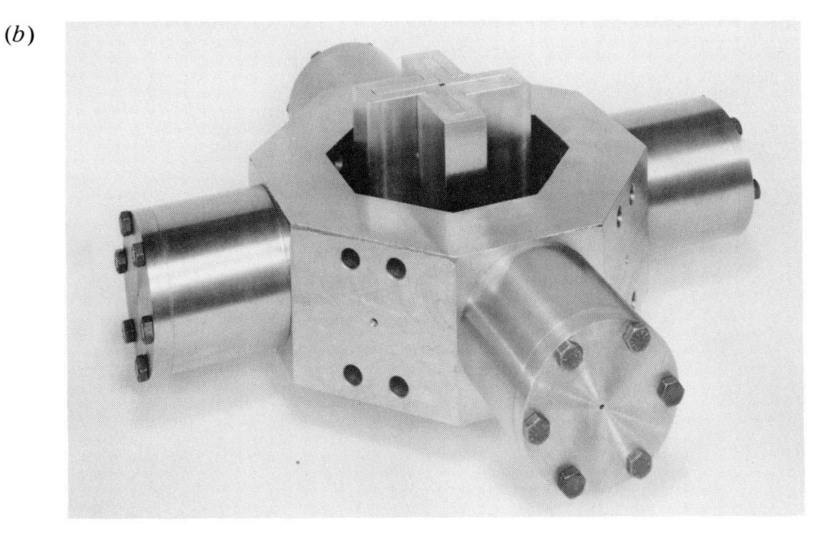

**Figure 16** Demonstrator model of the two-dimensional cider press. Anvils (*a*) fully open, and (*b*) fully closed.

is vertically oriented so as to facilitate introduction of fruit or pomage from an overhead hopper or mill. The two pairs of interfitting G-shaped anvils are advanced by hydraulic pressure. Pressures as high as 10 000 psi can be applied. The unfermented apple juice tastes as good as any that I can purchase here in the mountains of Tennessee.

## REFERENCES

Hambly E C and Roscoe K H 1969 *Proc. 7th Int. Conf. on Soil Mechanics and Foundation Engineering, Mexico* vol 2 pp 173–81

Taricco S 1980 *US Patent no 4,208,174* June 17, 1980

Deloche B and Samulski E T 1981 *Macromol.* **14** 575–81

Rivlin R S and Saunders D W 1952 *Trans. Faraday Soc.* **48** 200–6

Ashbee K H G 1986 *J. Composite Mater.* **20** 114–24

Frank F C and Ashbee K H G 1984 *UK Patent no 2, 159, 456* June 1, 1984

Orowan E 1948/49 *Rep. Prog. Phys.* **12** 185–232

Marshall W 1796 *The Rural Economy of Gloucestershire* vol II (1979 facsimile reprinted by Alan Sutton Publishing Ltd)

# On the Relative Stability of Graphite and Diamond

J Friedel

## 1 INTRODUCTION

The special stability of unsaturated sp bonds as compared with saturated ones is a characteristic of carbon, and is not observed with the other tetravalent elements Si, Ge and Sn. Indeed graphite, with layers of $sp^2(\sigma)$ bonds plus $p(\pi)$ bonds, is more stable than diamond, with $sp^3(\sigma)$ bonds, while the diamond structure is most stable in Si, Ge and Sn. The same difference appears in small aggregates, with $\pi$ bonding for carbon but $\sigma$ or metallic bondings in Si and Ge. Finally, unsaturated molecules are at the root of (carbon) organic chemistry, and indeed of life, while they are practically absent from the chemistry of Si, Ge and Sn (cf Friedel 1987).

This behaviour of C is *a priori* surprising, if one describes the $\sigma$ and $\pi$ bondings in the simplest LCAO terms (figure 1). To first approximation, one can treat the $\sigma(sp^n)$ bonds as independent, with energies

$$E = E_0 \pm t_\sigma$$

where

$$E_0 = \frac{E_s + nE_p}{1+n}$$

and

$$t_\sigma = | \langle iJ | V_i | jJ \rangle |$$

can be assumed to vary little with $n$ ($|iJ\rangle, |jJ\rangle$ are the sp$^n$ hybrids along $\sigma$ bond $J$ from site $i$ to site $j$, figure 2). The $\pi$ band, with its centre of gravity at $E_p$, extends between $E_p \pm pt_\pi$, where $p$ is the number of first neighbours, such that

$$p = n + 1$$

and

$$t_\pi = |\langle p_i^\pi | V_i | p_j^\pi \rangle|.$$

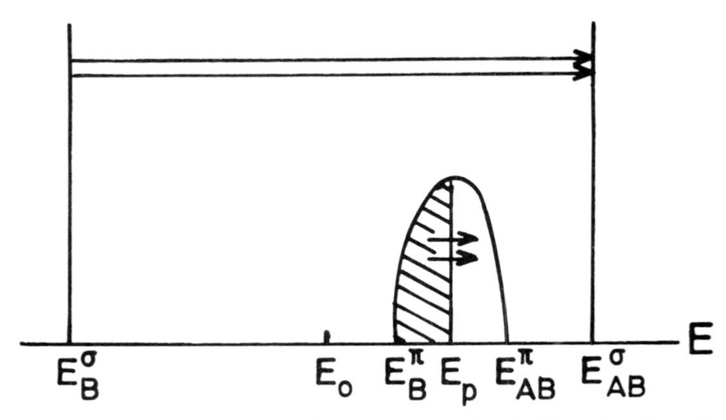


**Figure 1** Relative energy positions of $\sigma$ bonding ($E_B^\sigma$) and antibonding ($E_{AB}^\sigma$) states and of the possible $\pi$ band. Arrows show virtual pair excitations due to correlations.

Now going from the diamond to the graphite structure corresponds to transferring one electron per atom from the bonding part $E_B^\sigma$ of the $\sigma$ bonds to the lower half of the $\pi$ band. In graphite, $p = 3$ and the lower limit of the $\pi$ band is thus at energy

$$E_B^\pi = E_p - 3t_\pi.$$

Even if we neglect the promotion energy

$$E_{sp} = E_p - E_s$$

it is clear that, for carbon, where

$$t_\pi \cong 0.2 t_\sigma \ll t_\sigma$$

one has

$$E_\pi^B > E_\sigma^B$$

and the diamond structure should be more stable, as is indeed observed in Si, Ge and Sn, but not in C.

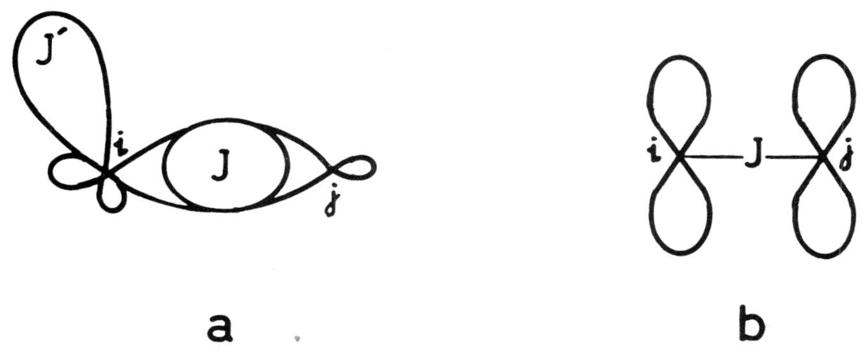


**Figure 2** Definition of (a) $t_\sigma$ and $\delta$; (b) $t_\pi$.

This reasoning is of course very rough. Even within the admittedly over-simplified LCAO approximation, with only nearest-neighbour transfer integrals $t_\sigma, t_\pi$ and quasi-orthogonal atomic functions $|iJ\rangle$ and $|p_i^\pi\rangle$, two corrections must be considered.

First, in the delocalised independent electrons picture, the bonding and antibonding $\sigma$ states are actually broadened into *bands*. This is related, in the LCAO approximation, to the fact that two hybrids $|iJ\rangle, |iJ'\rangle$ which point along different bonds $J, J'$ from the same site $\imath$, are related by a matrix element proportional to the *promotion energy* $E_{\mathrm{sp}}$:

$$\delta = \frac{E_{\mathrm{sp}}}{p} = |\langle iJ|V_i|iJ'\rangle|\,\delta_{JJ'}.$$

In C, Si and Ge, this broadening is less than the $\sigma$ bonding energy $E_{\mathrm{AB}}^\sigma - E_{\mathrm{B}}^\sigma$, and the corresponding valence and conduction bands are separated by a sizable energy gap (figure 3). In the simple LCAO model considered, this broadening is just equal to $E_{\mathrm{sp}}$. The asymmetry of the broad valence band varies with the number $p$ of $\sigma$ bonds; and as we shall see, the ensuing small change in the average electronic energy in the valence band favours somewhat the *diamond* structure. This term cannot therefore explain the stability of graphite.

Second, the valence electrons, if delocalised, are not completely independent but present *correlations*. In a Hubbard model, where the electrons only interact when on the same atom at the same time, the first significant correction in a perturbation study of correlations can be viewed as due to virtual excitations of the valence electrons into empty states, due to collisions through $U$, the average intra-atomic Coulomb interaction. The ensuing stabilising effect is inversely proportional to the energy of these pair excitations. It is therefore larger in $\pi$ bands than in $\sigma$ bonds (figure 1) and favours the *graphite* structure.

The band broadening term due to the promotion energy $E_{\mathrm{sp}}$ can be computed for any ratio of this energy to the $\sigma$ bonding energy (Friedel and

Lannoo 1973, Friedel 1978). Similarly, variational schemes can be set up, in specific cases, to study correlation effects (Stollhoff and Fulde 1977, Olès *et al* 1986). But to be able to compare simply the two effects, we shall restrict ourselves here to a study within *second-order perturbations*, which is at least approximately valid and gives simple results (Friedel 1978).

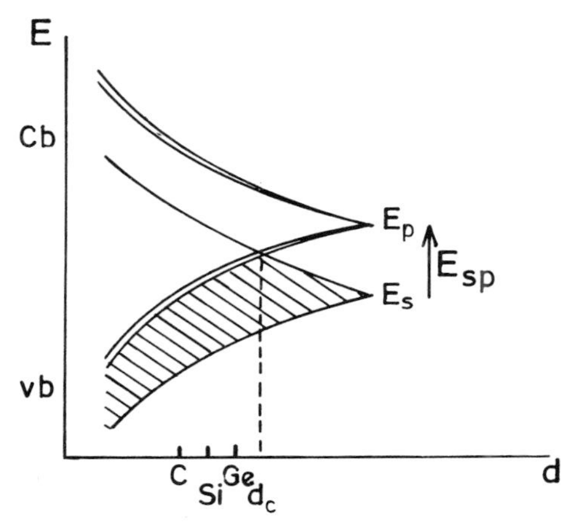


**Figure 3** Broadening of the bonding and antibonding $\sigma$ states into valence and conduction bands, due to a finite promotion energy $E_{\mathrm{sp}}$ (band limits against the interatomic distance $d$).

## 2 BAND BROADENING AND ASYMMETRY DUE TO FINITE SP PROMOTION ENERGY

The independent electron states of a covalent sp$^n$ macroscopic aggregate can be analysed in terms of $|iJ\rangle$ hybrids:

$$|\psi\rangle = \sum_{i,J} a_{iJ} \, | \, iJ\rangle$$

satisfying

$$H \, | \, \psi\rangle = E \, | \, \psi\rangle$$

with

$$H = \sum_{i,j} E_0 \, | \, iJ\rangle \langle iJ \, | - t_\sigma \times \sum_{\substack{i,J \\ j \text{ near to } i}} | \, iJ\rangle \langle jJ \, | - \delta \sum_{i,J,J' \neq J} | \, iJ\rangle \langle iJ' \, | \, .$$

The band broadening due to the $\delta$ term leads in fact to two sp bands which satisfy effective LCAO relations (Leman 1963, Thorpe and Weaire 1971, Friedel and Lannoo 1973):

$$\mathscr{H} | \Psi \rangle = \varepsilon | \Psi \rangle$$

with

$$| \Psi \rangle = \sum_i x_i | i \rangle$$

$$x_i = \sum_J a_{iJ}$$

$$\langle i | j \rangle = \delta_{ij}$$

and

$$\mathscr{H} = -\delta t_\sigma \sum_{i,j \text{ near to } i} | i \rangle \langle j |$$

$$\varepsilon = t_\sigma^2 + p\delta(E_\mathrm{p} - E) - (E_\mathrm{p} - E)^2.$$

The classical result

$$-p\delta t_\sigma \leqslant \varepsilon < p\delta t_\sigma$$

fixes the limits for $E$, i.e.

$$E_\mathrm{s} - t_\sigma \leqslant E < E_\mathrm{p} - t_\sigma \qquad E_\mathrm{s} + t_\sigma < E \leqslant E_\mathrm{p} + t_\sigma$$

in the case $d < d_\mathrm{c}$ which holds for C, Si and Ge (figure 3). These are the broad valence and conduction bands pictured in figure 3. One has to add $n - 1$ flat p bands

$$\bar{E}_\mathrm{p} = E_\mathrm{p} \pm t_\sigma.$$

The average electronic energy in the valence sp band then reads

$$\bar{E}_\mathrm{sp} = \int_\mathrm{occ} n(E)\, E\, \mathrm{d}E$$

where $n(E)$ is the density of states per spin direction. If $\nu(\varepsilon)$ is the density of states of the $\varepsilon$ states, the previous relations lead to

$$\bar{E}_\mathrm{sp} = E_\mathrm{p} - \frac{E_\mathrm{sp}}{2} - \frac{1}{2} \int (4t_\sigma^2 + E_\mathrm{sp}^2)^{1/2} \left(1 - \frac{4\varepsilon}{4t_\sigma^2 + E_\mathrm{sp}^2}\right)^{1/2} \nu(\varepsilon) \mathrm{d}\varepsilon$$

if one uses the fact that

$$n(E)\, \mathrm{d}E = \nu(\varepsilon)\, \mathrm{d}\varepsilon.$$

A development in successive powers of $\delta$ gives

$$\bar{E}_\mathrm{sp} \cong E_\mathrm{p} - \frac{E_\mathrm{sp}}{2} - \frac{1}{2} \left( (4t_\sigma^2 + E_\mathrm{sp}^2)^{1/2} - \frac{2pt_\sigma^2\, \delta^2}{(4t_\sigma^2 + E_\mathrm{sp}^2)^{3/2}} + \dots \right)$$

if one uses the values of the corresponding moments of $\nu(\varepsilon)$:

$$\mathcal{M}_0 = \int \nu(\varepsilon)\, d\varepsilon = 1$$
$$\mathcal{M}_1 = \int \nu(\varepsilon)\varepsilon\, d\varepsilon = 0$$
$$\mathcal{M}_2 = \int \nu(\varepsilon)\varepsilon^2\, d\varepsilon = pt_\sigma^2\, \delta^2.$$

The development of the square root gives finally

$$\bar{E}_{sp} \cong E_p - \frac{E_{sp}}{2} - t_\sigma - \frac{p(p-1)\,\delta^2}{8t_\sigma} + \dots .$$

If there is a $\pi$ band, with $4 - p$ electrons per atom, the average energy of the $\pi$ electrons is approximately

$$\bar{E}_\pi \cong E_p - \left(\frac{3p}{4}\right)^{1/2} t_\pi$$

if one replaces the real $\pi$ band by a rectangular band with the same second moment, a good approximation (Friedel 1978, Treglia *et al* 1982).

The cohesive energy per atom then reads

$$E_c = 2E_s + 2E_p - 2\bar{E}_{sp} - (n-1)\bar{E}_p - (4-p)\bar{E}_\pi.$$

Hence

$$E_c \cong pt_\sigma + (4-p)\left(\frac{3p}{4}\right)^{1/2} t_\pi - E_{sp} + \frac{p(p-1)\,\delta^2}{4t_\sigma}.$$

For the diamond $(p = 4)$ and graphite $(p = 3)$ structures, the corresponding cohesive energies are such that

$$E_c^d - E_0^g \cong t_\sigma - \tfrac{3}{2} t_\pi + \frac{E_{sp}^2}{48t_\sigma}$$

if one assumes the same value for $t_\sigma$ for sp and $sp^2$ bonds. The broadening due to $E_{sp}$ thus leads to a *second*-order term which *stabilises slightly the diamond structure.*

The sign of this term can be understood as follows. The term itself comes from the difference in average energy in the broad valence (sp) band (figure 3). This band has the same limits $E_s - t_\sigma, E_p - t_\sigma$ for $sp^3$ diamond and $sp^2$ graphite. But its *asymmetry* differs. This comes from the fact that the density $\nu(\varepsilon)$ of $\varepsilon$ states is symmetrical, owing to the alternate characters of the diamond and graphite structures, but the corresponding density $n(E)$ of $E$ states is not symmetrical, because of the non-linear relation between $\varepsilon$ and $E$. The $\varepsilon$ band has a total width function only of $E_{sp} = p\delta$, but it has an effective width proportional to

$$M_2^{1/2} = \frac{E_{sp} t_\sigma}{\sqrt{p}}$$

which decreases with increasing $p$. With a parabolic relation between $E$ and

$\varepsilon$, one easily sees that the valence sp band has an *asymmetry with a longer tail towards the lower energies*; and *this asymmetry decreases for increasing values of p*. The sign of the correction in $E_{\rm sp}^2$ follows.

## 3 CORRELATION CORRECTION

This term has been fully discussed previously (Friedel 1978). In a development in successive powers of $U$, the correction to second order in $U$ reads for diamond, in an approximation compatible with those used so far,

$$\Delta E_{\rm c}^{\rm d} \cong -U + \frac{7}{16}\frac{U^2}{t_\sigma}.$$

The denominator in $t_\sigma$ of the $U^2$ term due to correlation comes from the energy of excitation $4t_\sigma$ of a pair of $\sigma$ bonding electrons, in the approximation of figure 1. It is easily seen that the broadening into bands pictured in figure 3 and discussed above does not change the order of magnitude of this estimate, as long as the band broadening $E_{\rm sp}$ is less than the bonding energy $E_{\rm AB}^\sigma - E_{\rm B}^\sigma$ which is the case for C, Si and Ge.

For graphite, one must take into account not only virtual collisions of electrons within $\sigma$ or $\pi$ bands, but collisions between electrons belonging one to the $\sigma$ and the other to the $\pi$ bands. As a result, the $U^2$ term is more complex, and one obtains, within the same approximations,

$$\Delta E_{\rm c}^{\rm g} \cong -U + \frac{U^2}{64t_\sigma}\left\{15 + 24\left[1 + \left(\frac{3}{2}\right)^{1/2}\frac{t_\pi}{t_\sigma}\right]^{-1} + \left[\left(\frac{3}{2}\right)^{1/2}\frac{t_\pi}{t_\sigma}\right]^{-1}\right\}.$$

Hence a contribution to the relative stability of diamond and graphite

$$\Delta E_{\rm c}^{\rm d} - \Delta E_{\rm c}^{\rm g} \cong \frac{U^2}{64t_\sigma}\left\{13 - \left(\frac{2}{3}\right)^{1/2}\frac{t_\sigma}{t_\pi} - 24\left[1 + \left(\frac{3}{2}\right)^{1/2}\frac{t_\pi}{t_\sigma}\right]^{-1}\right\}.$$

This term is *negative* for any likely value of $t_\pi \leqslant 0.5t_\sigma$. It thus stabilises the *graphite* structure.

For carbon, where

$$t_{\rm p} \cong 0.2\, t_\sigma$$

it reads

$$\Delta E_{\rm c}^{\rm d} - \Delta E_{\rm c}^{\rm g} \cong -U^2/6t_\sigma.$$

## 4 DISCUSSION

Table 1 gives values of $U, E_{\rm sp}$ and $t_\sigma$ deduced from experiment (Friedel 1978). Thus

$$U = I + A$$

where $I$ is the first ionisation potential and $A$ the electron affinity.

$$E_{sp} = p\delta$$

the atomic sp promotion energy, and

$$2t_\sigma = \delta + g$$

where $g$ is the measured energy gap between the valence and conduction bands. This is somewhat larger but of the same order of magnitude as the values used by Harrison (1978). From these admittedly very crude estimates, one can conclude the following.

(1) In carbon, the special stability of graphite can only be due to larger Coulomb correlations. This is indeed in keeping with the general feeling of chemists. The very simple approximations used show that the second-order correlation term is of the right sign and right order of magnitude, although its exact value as deduced from table 1 is slightly too small.

**Table 1** Values of parameters (in eV).

| | C | Si | Ge |
|---|---|---|---|
| $U$ | 12 | 8 | 8 |
| $E_{sp}$ | 8.2 | 6.7 | 7 |
| $t_\sigma$ | 6.7 | 4 | 3.9 |

(2) In Si and Ge, the same conclusions would hold for the same ratio of $t_\pi/t_\sigma$ as in C, as $U$ and $t_\sigma$ are reduced in practically the same proportions from one element to another. However, the full difference of cohesion energies $E_c^d + \Delta E_c^d - E_c^g - \Delta E_c^g$ is very sensitive to the ratio of $t_\pi/t_\sigma$; the diamond structure is stabilised by an increase of this ratio, because the reduction in correlation energy due to the $\pi$ band is larger than the increase in stability due to the widening of the $\pi$ band. In Si and Ge, such larger values of $t_\pi/t_\sigma$ are likely; they would explain the stability of the diamond structure by a reduced role of the change in correlations from $\sigma$ to $\pi$ bonding.

(3) The promotion energy $E_{sp}$ is, by definition, smaller than the first ionisation potential $I$. It is thus definitely smaller than the value taken for $U$, as shown in table 1. Then, because of the large numerical factor in its denominator, the correction $E_{sp}^2/48t_\sigma$ due to band broadening is more than ten times smaller than the correlation term $-U^2/6t_\sigma$, and thus negligible in this discussion.

This paper was written in memory of all the happy and stimulating moments Mary and I spent with Charles and Maïta.

**REFERENCES**

Friedel J 1978 *J. Physique* **36** 651, 671
—— 1987 *Elemental and Molecular Clusters* (Springer Series in Material Sciences **6**) ed G Benedeck, T P Martin and G Pacchioni (Berlin: Springer)
Friedel J and Lannoo M 1973 *J. Physique* **34** 115, 483
Harrison W A 1978 *The Physics of the Chemical Bond* (San Francisco: Freeman)
Leman G 1963 *Ann. Phys., Paris* **18** 1
Olès A M, Pfirsch P, Fulde P and Böhm M C 1986 *J. Chem. Phys.* **85** 5183
Stollhoff G and Fulde P 1977 *Z. Phys.* B **26** 257
Thorpe M F and Weaire D 1971 *Phys. Rev.* B **4** 3518
Treglia G, Ducastelle F and Spanjaard D 1982 *J. Physique* **43** 34

# Diamond—A Letter from the Depths

T Evans

At a conference in Oxford in 1966 Charles Frank proposed that the major reason for studying defects in diamonds was to learn what they can tell us about the history of their formation (Frank 1966). He quoted some words from a poem by Ukichiro Nakaya, 'A snowflake is a letter from the sky', and added that

> Diamonds are letters
> Still better worth the reading
> We can visit the sky.

Frank went on to say: 'It is a truism in diamond lore that every diamond is an individual but the perfect diamond is a blank page. The individuality of each diamond lies in its imperfections and these afford our only clue as to what has happened to that crystal since its birth. The surface tells us the last things that happened to it and the interior something about its origin, growth, and perhaps the changes in temperature, pressure and shear stress to which it has been subjected if we can read the record aright. Of all the defects of diamond the most individual, the most unexpected and I suppose the most significant, if their significance can be deciphered, are those associated with the presence of nitrogen.'

In the main part of his paper he proposed a scheme for the history of diamonds, within the limits of the information then available. Knowledge acquired since that time now makes it possible for that history to be revised. It is still true to say that significant clues about the history of diamonds are provided by nitrogenous defects in the diamond lattice, and the first part of this paper will describe the present state of knowledge about them.

However, there are other defects that also give much information; these defects are not in the lattice itself but are non-diamond mineral inclusions that are easily visible with a hand lens. They have been known about for centuries and have depressed the value of gemstones by their presence. Since Frank wrote his article detailed scientific studies have been made of inclusions and other geological factors, adding considerably to knowledge about the conditions under which diamonds grew and their subsequent history. The results of those studies will be described in the second part of the paper.

In the third part, knowledge about the formation of nitrogen defects and information gleaned from the study of mineral inclusions will be combined to suggest a history of natural diamonds within our present limited understanding, and to indicate areas where more research is needed before there can be a clearer reading of the letter from the depths.

## 1 NITROGEN IN DIAMONDS

### 1.1 Classification of diamonds

Early work on the optical absorption and luminescence of natural diamonds showed that many features of the spectra varied widely from sample to sample. On the basis of their optical absorption properties diamonds were divided into two broad groups, types I and II, with the assumption that type II diamonds contained fewer impurities than type I. Sutherland *et al* (1954) and Clark *et al* (1956) divided the absorption features in the ultraviolet, visible and infrared regions into two classes, A and B. The absorption spectra for different diamonds showed wide variations in the intensity of both A and B features, but the characteristic pattern of each stayed the same. A very important advance was made by Kaiser and Bond (1959) who heated diamonds to 2000 °C, transforming them to graphite. The gas content freed by this process was found to contain a large proportion of nitrogen, up to 3000 atomic ppm. They also found a direct correlation between the concentration of nitrogen and the height of an infrared peak at 1282 $cm^{-1}$ (7.8 $\mu$m) which had been classified as one of the A features. This correlation gives a relationship for the nitrogen concentration responsible for the A absorption features:

$$N_A = 33.3\mu_A$$

where $N_A$ is in atomic ppm and $\mu_A$ is the absorption coefficient in $cm^{-1}$ at

1282 cm$^{-1}$ (7.8 μm). Since it was discovered, this relationship has been considered the most reliable measure of the concentration of nitrogen when it is present in the state that produces the A absorption features. However, recent results, to be described later, question the reliability of this relationship.

Since 1959, the role of nitrogen in the absorption characteristics of diamonds has been studied extensively and the A and B absorption features have been examined and identified over a wide spectral range. For the classification of diamonds used in this paper, only absorption features in the infrared region will be considered although it must be remembered that absorption in the visible and the ultraviolet is also involved in the general classification. In the infrared region, between about 2660 and 1670 cm$^{-1}$, there is two-phonon absorption that is intrinsic to the diamond lattice, i.e. all diamonds irrespective of type have this absorption. The Raman energy for diamond is 0.165 eV, equivalent to a wavenumber of 1332 cm$^{-1}$ (7.5 μm). The one-phonon absorption at wavenumbers between 1332 and about 1000 cm$^{-1}$ is forbidden in a perfect lattice (Lax and Burstein 1955) and only appears when induced by lattice defects.

*Type II diamonds* contain less nitrogen than type I and so exhibit more closely the absorption features characteristic of the perfect diamond lattice. There are two subclassifictions: type IIa with a high electrical resistivity, and type IIb, which are semiconductors due to the presence of boron as an impurity. Type IIa diamonds have no significant absorption in the one-phonon region, indicating a concentration of nitrogen that is too low (less than 1 atomic ppm) for detection by infrared absorption measurements.

*Type I diamonds* have a nitrogen content sufficiently large for detection and examination by optical absorption techniques. They may be subdivided into two, the concentration of nitrogen in type Ib being less than in type Ia diamonds. The characteristic of type Ib diamonds is that most of the nitrogen is present as single substitutional atoms. All synthetic diamonds that contain detectable amounts of nitrogen are type Ib. Type Ib natural diamonds are comparatively rare when compared with the number of type Ia diamonds. The one-phonon absorption spectrum for a type Ib diamond containing nitrogen that is present only as single substitutional atoms is shown in figure 1. The height of the 1130 cm$^{-1}$ (8.85 μm) band can be used as a measure of the amount of nitrogen in single substitutional form. About 98% of natural diamonds are type Ia, and contain nitrogen concentrations between about 50 and 3000 atomic ppm. In these specimens nitrogen is present as aggregates and each type of aggregate gives characteristic optical absorption features. An A centre is considered to be two nitrogen atoms on adjacent lattice sites (Davies 1976), and is responsible for the A absorption features. The diamonds used by Kaiser and Bond (1959) in their nitrogen determination contained nitrogen mainly in this state of aggregation. The infrared absorption spectrum in the one-phonon

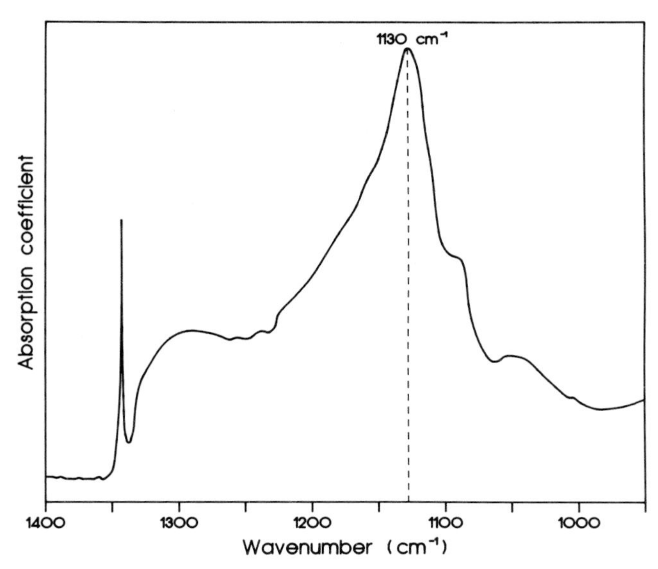


**Figure 1** Infrared absorption spectrum in the one-phonon region, for a type Ib diamond. The Raman wavenumber is at 1332 $cm^{-1}$.

region associated with A aggregates is shown in figure 2. The prominent peak at 1282 $cm^{-1}$ (7.8 $\mu$m) is the one that is used to determine the nitrogen content in the A form. A second type of aggregate is termed the N3 centre and is believed to consist of three nitrogen atoms and a vacancy (Loubser and Wright 1973, Davies *et al* 1978). The presence of N3 centres results in an absorption peak in the visible region at 24 000 $cm^{-1}$ (415 nm). A third type of aggregate is the B centre and it has been suggested that it consists of four nitrogen atoms and a vacancy (Loubser and van Wyk 1981); this centre is responsible for the B absorption features. The infrared absorption spectrum in the one-phonon region for a diamond containing only B centres is shown in figure 3. In type Ia diamonds there are usually, but not always, platelets in the cube planes at concentrations of between $10^{20}$ and $10^{22}$ $m^{-3}$ and ranging in size from about 8 nm to a few micrometres. The presence of these platelets was predicted by Frank (1956) to explain the anomalous x-ray spikes projecting from the Laue spots for certain diamonds (Raman and Nilakantan 1940). The defects can be examined by electron microscopy (Evans and Phaal 1962) and also by infrared absorption measurements as there is an absorption peak at about 1370 $cm^{-1}$ (7.3 $\mu$m) associated with the platelets (Sobolev *et al* 1967, Evans and Rainey 1975, Klyuev *et al* 1977, Sumida and Lang 1988). This peak is at a higher wavenumber than the Raman and is probably due to local mode absorption, presumably at the platelets. The precise wavenumber of the

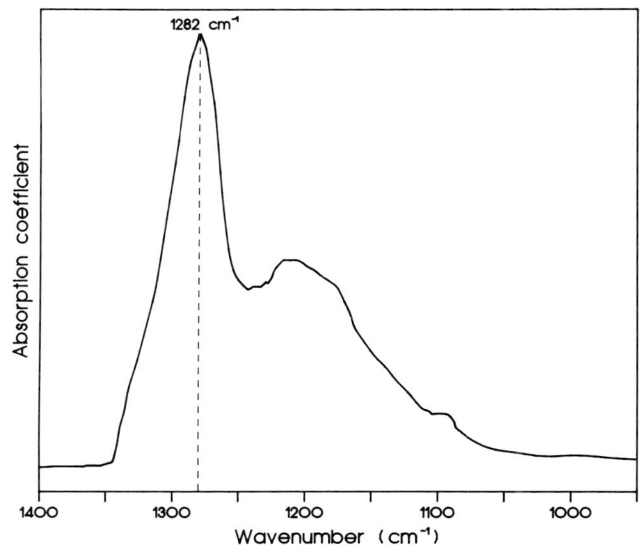


**Figure 2** Infrared absorption spectrum in the one-phonon region for a type IaA diamond; nitrogen present only as A centres.

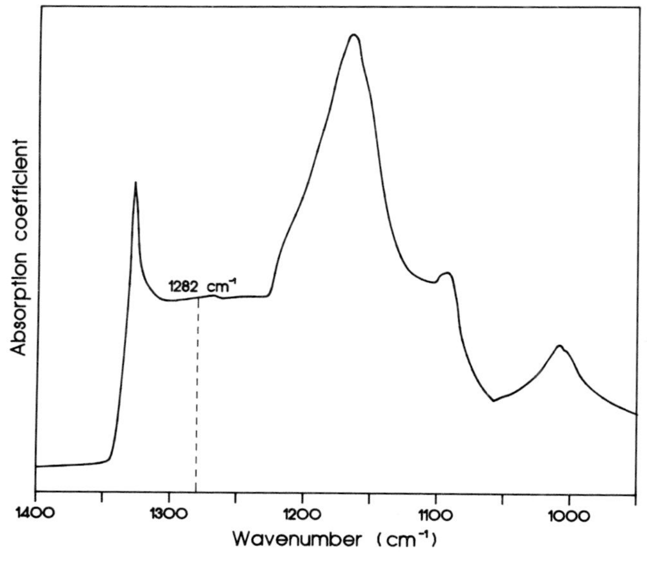


**Figure 3** Infrared absorption spectrum in the one-phonon region for a type IaB diamond; nitrogen present only as B centres.

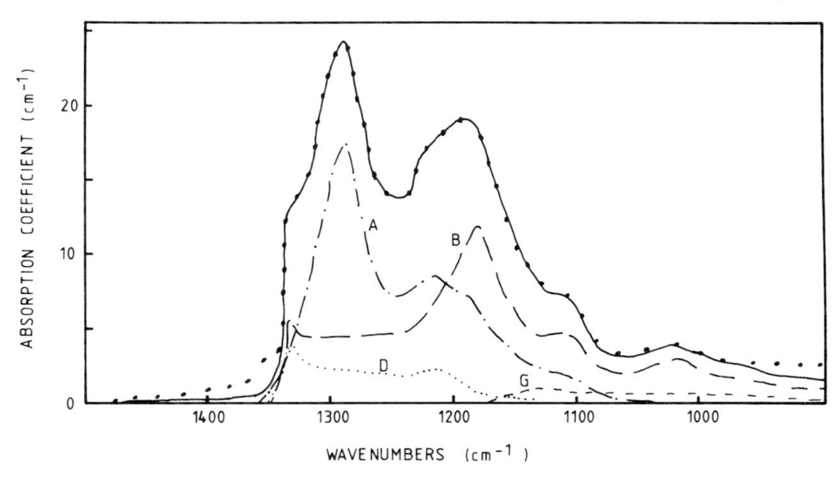


**Figure 4** Infrared absorption spectrum of a type IaA/B diamond showing decomposition into A, B and D spectra with an additional G spectrum at lower wavenumbers. The full curve is the spectrum resulting from the addition of the A, B, D and G spectra; black dots on or near the full curve constitute the measured infrared absorption spectrum.

platelet absorption peak depends upon the size of the platelets: the larger the platelets, the lower the wavenumber (Sobolev *et al* 1967).

In type Ia diamonds these various defects are present in widely differing proportions. A diamond that contains nitrogen only as A aggregates is termed type IaA, only in the B form type IaB, and a mixture of the two, type IaA/B. Thus a type IaA diamond would have the one-phonon absorption spectrum shown in figure 2; a type IaB spectrum would be as in figure 3, and an example of such a spectrum for type IaA/B is shown in figure 4. Clark and Davey (1984a,b) have decomposed the spectra in the one-phonon region for type IaA/B diamonds into six spectra in all. The three main ones are A (figure 2), B (figure 3) and a D component that can be attributed to the platelets (Woods 1986).

### 1.2 Voidites

In addition to defects containing nitrogen that show the characteristic optical absorption features used in classification, there is another defect containing nitrogen, that cannot so far be identified by a particular optical absorption feature; this is called a *voidite* (not an apt term, as will become apparent later). These defects were first reported by Stephenson (1978) and Evans (1978) and were described as dots about 8 nm in diameter, on and within a dislocation loop in a {100} plane. The electron microscope used was of insufficient resolution to show any detail other than to suggest a square outline to the dots when looking along a ⟨100⟩ direction. The defects have subsequently been studied in detail by groups in Bristol, Reading, and Oxford Universities using the high-resolution electron microscopes at Oxford. Figure 5 shows a dislocation loop and voidites within the loop.

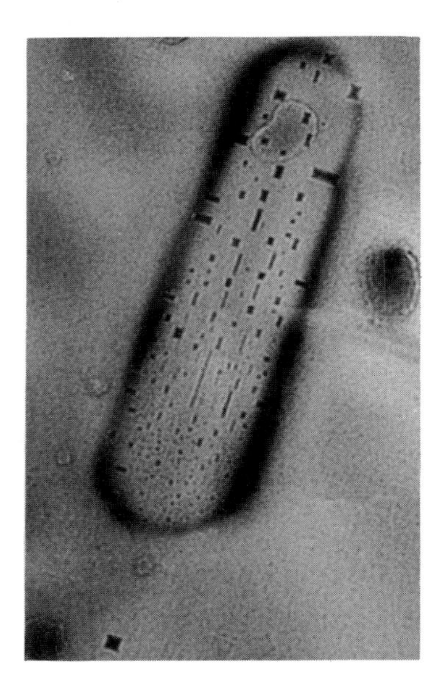

**Figure 5** Dislocation loop containing voidites. The voidites appear square or rectangular as they are bounded by {111} faces and viewed from a [001] direction. The square voidites are regular octahedra and the rectangular ones are elongated octahedra. Note the voidite outside the dislocation loop. (Magnification $2.5 \times 10^5$.)

It has been established that the voidites are octahedra bounded by {111} planes, 1 to 10 nm in diameter (Anstis and Hutchison 1982, Barry 1986). Phase contrast electron microscopy has shown that the scattering density of the voidites is considerably less than that of the surrounding diamond (Hutchison and Bursill 1983). The voidites are usually associated with dislocation loops whose Burgers vector is either $\frac{1}{2}a\langle 100\rangle$ or $a\langle 100\rangle$ (Hirsch *et al* 1986b). Hirsch *et al* also proposed a mechanism by which the dislocation loops and associated voidites are the result of dissociation of platelets. Significant advances were made in the understanding of the nature of these defects by Barry (1986) and Hirsch *et al* (1986a). Barry used the electron microscope to study the contrast under various diffraction conditions and suggested that voidites contained a solid crystalline phase, possible ammonia under pressure. As well as showing that they in fact contained nitrogen, Hirsch *et al* (1986a) obtained an electron diffraction pattern from several voidites in turn which indicated that they contained a crystalline cubic phase with unit cell side of about 0.5 nm. It was suggested that the voidites most probably contained solid ammonia at a pressure of between about 2.2 to 3.5 GPa, although solid nitrogen could not be ruled out. Evans and Woods (1987) could not detect any absorption features in the infrared that could be assigned to ammonia, in diamonds considered to contain sufficient voidites to show such features. Bruley and Brown (1989) using electron energy loss spectroscopy confirmed that the voidites contain a solid phase that consists mainly or entirely of nitrogen. On the assumption that the voidites in fact contain only solid nitrogen, they suggested higher pressures in the voidites than Hirsch *et al* (1986a), namely between 6 and 14 GPa at room temperature, or between 11 and 20 GPa at about 1300 K. Diamonds that contain dislocation loops and voidites are type 1a with a high proportion of B features compared to A features, i.e. they are type 1aA/B diamonds tending towards type 1aB (Burt 1980, Maguire 1983). Not all such diamonds contain these defects and Woods (1986) has suggested that type Ia diamonds that do not contain them are termed regular diamonds, irregular diamonds being those in which there are voidites. The reason for such terminology and the place of voidites in the history of diamond will be considered later.

### 1.3 Inhomogeneous distribution of nitrogen in synthetic and natural diamonds

In section 1.1 the classification of diamonds is based upon the concentration of nitrogen in its various aggregated states; but it cannot be assumed that there is a uniform distribution of nitrogen throughout a particular diamond as both synthetic and natural diamonds sometimes have a markedly inhomogeneous distribution. Also in natural diamonds the ratio of A : B centres and platelet concentration can vary in different parts of the same diamond.

The inhomogeneous distribution of nitrogen is particularly striking in synthetic diamonds (type Ib) as the nitrogen is incorporated as single substitutional atoms which absorb in the blue end of the optical spectrum, thus giving the diamond a yellow or greenish-yellow appearance. Synthetic diamonds grow mainly with a cuboctahedral morphology often modified by minor {110} and {113} facets (Strong and Chrenko 1971) and this morphology can be influenced both by the growth temperature and the type of solvent/catalyst used in the growth process. The concentration of single substitutional nitrogen atoms has been found to be different for the various growth sectors. A thin polished plate with the [110] direction perpendicular to the plate was prepared from a large diamond grown at the De Beers Diamond Research Laboratory using Fe–Ni as the solvent/catalyst (Burns *et al* 1990). Figure 6 is an optical micrograph of the plate and shows the inhomogeneous nitrogen uptake. Infrared absorption spectra were taken of the various growth sectors. Data of Chrenko *et al* (1971) indicate that in synthetic type Ib diamonds, 25 atomic parts per million of nitrogen produces an absorption coefficient of 1 $cm^{-1}$ at 1130 $cm^{-1}$. The results show that the concentrations of the dispersed nitrogen in the growth sectors were

| | | | |
|---|---|---|---|
| $(1\bar{1}1)$ | 99 ppm | $(\bar{1}13)$ | 10 ppm |
| $(\bar{1}11)$ | 100 ppm | $(1\bar{1}3)$ | 7 ppm |
| (001) | 46 ppm | $(1\bar{1}0)$ | 1 ppm. |

As the primary growth morphology for synthetic diamonds is cuboctahedral the ratio of most significance is the (111) : (100) uptake of about 2 : 1, although this ratio varies according to the growth conditions.

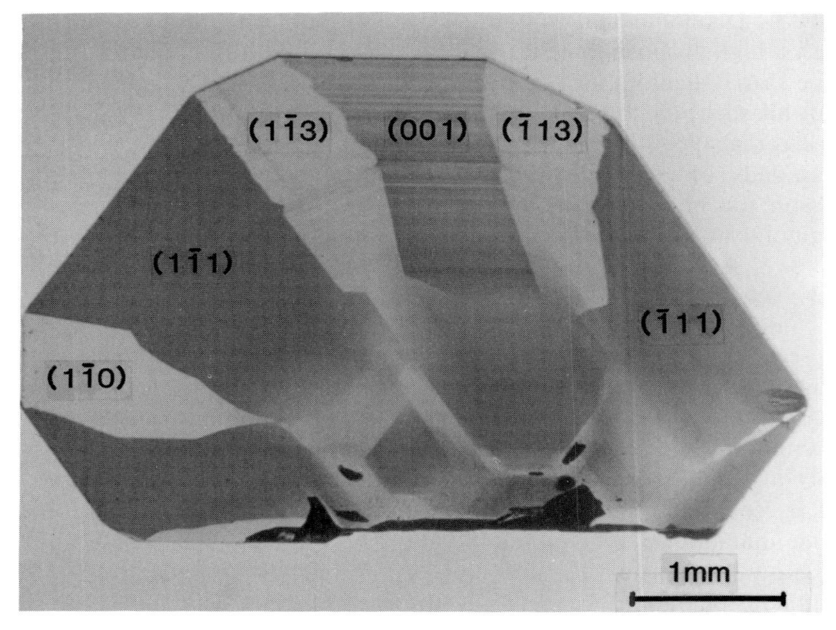


**Figure 6** Optical micrograph of a synthetic diamond slice showing the differential uptake of nitrogen in different growth zones.

Inhomogeneous distribution of nitrogen can also occur in natural diamonds. Hanley *et al* (1977) examined many natural diamonds and found that a number of crystals had a growth stratigraphy that included zones of type II (low nitrogen content) diamond intercalated within zones of the more usual type Ia diamond. This suggests that for diamonds showing these features the surrounding chemical environment varied during the period of growth, resulting in a differential uptake of nitrogen. The same result can be inferred from figure 4 of Evans and Phaal (1962), which is an electron micrograph of an extremely sharp boundary between a platelet-rich region and one with no platelets.

Almost all natural diamonds grew with an octahedral morphology and thus there was no differential uptake of nitrogen due to differing growth sectors. However, certain very rare natural diamonds experienced epochs of growth when as well as the normal octahedral facets they were also bounded by non-facetted surfaces with a mean {100} orientation—'cuboid' surfaces (Frank 1966, Lang 1974). There appears to have been a differential uptake of nitrogen for these two types of surfaces during growth.

### 1.4 Aggregation of nitrogen

It was suggested by Klyuev *et al* (1974) and Evans (1976) that nitrogen in an atomically dispersed state (type Ib) was incorporated into natural diamonds during growth. The type I state was produced subsequently by the slow aggregation of the nitrogen during the diamond's long residence in the high-temperature environment deep within the earth. Chrenko *et al* (1977) were the first to examine this aggregation process in the laboratory, using two synthetic diamonds containing dispersed nitrogen atoms at concentrations of 365 and 237 ppm, respectively. The first was heated at about 1900 °C for a total of 90 min under a stabilising pressure; 80% of the dispersed nitrogen aggregated to form A centres (two nitrogen atoms). The second was heated at about 1700° for 30 min, when 24% of the nitrogen aggregated. No thermocouple was incorporated in the actual experiment as the cell was calibrated separately for temperature. It was suggested that the reaction was second order with an activation energy of 2.6 eV.

This work was followed by Collins (1980) who irradiated synthetic diamonds containing dispersed nitrogen with 2 MeV electrons and then heated at 1500 °C for 400 min without pressure being imposed; about 60% of the dispersed nitrogen aggregated into A centres. When a synthetic diamond containing nitrogen was given the same heat treatment with no prior irradiation, no detectable aggregation took place. Evans and Qi (1982) confirmed this last result. Using the activation energy of 2.6 eV and the kinetics of Chrenko *et al* (1977), sufficient aggregation of dispersed nitrogen atoms into A centres for easy detection would have been expected. The results of Collins (1980) show that considerable enhancement of the aggregation process can be achieved by prior irradiation with electrons of sufficient energy to produce vacancies and interstitial atoms. This was confirmed by Allen and Evans (1981) who irradiated synthetic diamonds with heavy doses of 2 MeV electrons and then heated the specimens at temperatures up to 2200 °C under a stabilising pressure of 8.5 GPa. By this process A centres, B centres, N3 centres and platelets were produced. If no prior irradiation had been used, only aggregation of single nitrogen atoms into A centres would have been detected.

The aggregation of single nitrogen atoms to A centres without prior irradiation was studied in detail by Evans and Qi (1982). The apparatus contained a pressure corrected iridium–tungsten thermocouple that was stable under pressure to 2100 °C but degraded above that temperature. An estimation of temperatures above 2100 °C was done by extrapolation of the linear temperature against power input graph. After confirming that a heat treatment of 400 min at 1500 °C under stabilising pressure also produced no aggregation detectable by infrared absorption measurements, synthetic diamonds containing dispersed nitrogen were heat treated at temperatures between 1700 and 2500 °C, again under stabilising pressures. Table 1 gives a summary of the main results.

**Table 1** Heat treatment of synthetic diamonds containing single substitutional nitrogen atoms.

| Temperature (°C) | Time (h) | | Nitrogen concentration† ($m^{-3}$) | Nitrogen aggregated (%) |
|---|---|---|---|---|
| 1700 | 4 | | Between 2.6 and $5 \times 10^{25}$‡ | ~12 |
| 1700<br>2100 | 4<br>1 | § | Between 2.6 and $5 \times 10^{25}$‡ | ~83 |
| 2200 | 2 | | $2.6 \times 10^{25}$ | ~94 |
| 2400<br>2500 | 4<br>2 | § | $5 \times 10^{25}$ | ~98‖ |

† Using the correlation of Woods *et al* (1990b).
‡ Several specimens treated.
§ Consecutive heat treatments.
‖ B features and platelets produced as well as A features (also weak N3 peak).

Figure 7 shows infrared evidence that 94% of the single nitrogen atoms are converted to A centres by treatment for 2 h at about 2200 °C with a pressure of 9.5 GPa. The dotted curve was decomposed into two spectra; one being the A spectrum as in figure 2, and the other the single nitrogen spectrum as in figure 1. From these the percentage of single nitrogen atoms converted could be calculated. The concentrations in the table were not the actual concentrations throughout a specimen as examination with an optical microscope showed that there was considerable inhomogeneity in the distribution of the nitrogen atoms. The rates of conversion of single nitrogen atoms to A centres were considerably lower than those reported by Chrenko *et al* (1977) under the same conditions. The table shows that reasonable rates of conversion can be achieved by treatments between 1700 and 2200 °C with no evidence for the formation of B centres, N3 centres or platelets. Control experiments indicated that there was no detectable enhanced aggregation rate due to the presence of excess vacancies and interstitials; these might have been produced if plastic deformation had taken place under pressure during the high-temperature treatments. The results also showed that to produce B centres and platelets, temperatures at or above 2400 °C under pressure are required. Figure 8 shows an electron micrograph of a synthetic diamond that had been given the last treatment in table 1. The diffraction conditions were such that platelets in only one set of {100} planes are clearly visible and these have the same type of contrast as found at platelets in natural diamonds. The concentration of platelets was about $10^{21}$ $m^{-3}$ with sizes between 20 and 50 nm. In this experiment a weak absorption peak at 24 000 $cm^{-1}$ (415 nm) was detected, indicating that a small amount of N3 centres had been produced.

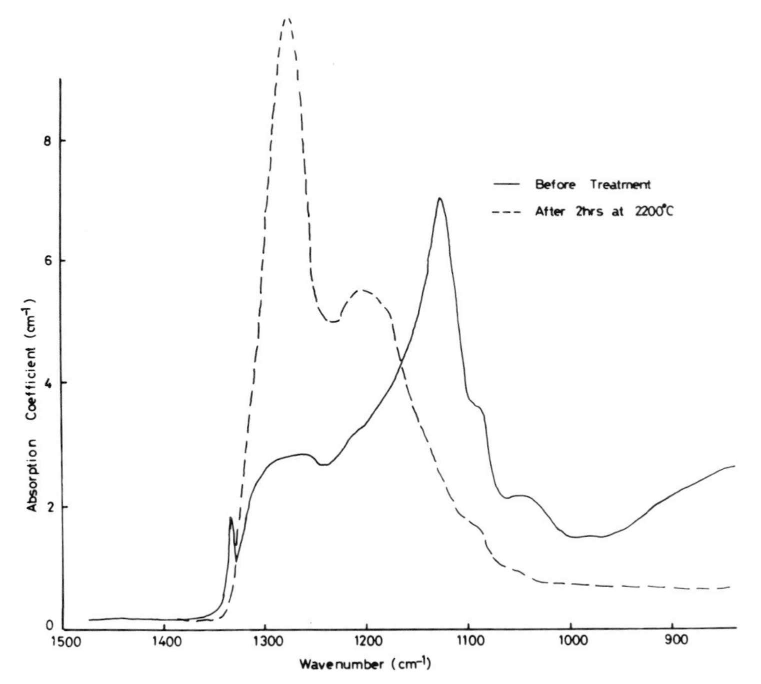


**Figure 7** Infrared absorption spectra in the one-phonon region of a synthetic diamond before and after high-temperature annealing.

Treatments at temperatures above about 2500 °C resulted in synthetic diamonds breaking up into small pieces, due presumably to the weakening effect of metallic inclusions. Therefore natural diamonds with strong A features were chosen for heating to these higher temperatures to study the conversion of A centres to B centres. They were not suitable for study of platelet formation as platelets would have been present prior to treatment. A summary of the results for natural diamonds is given in table 2. As the temperatures were derived by extrapolation of a temperature against power graph, they are not as accurate as in table 1.

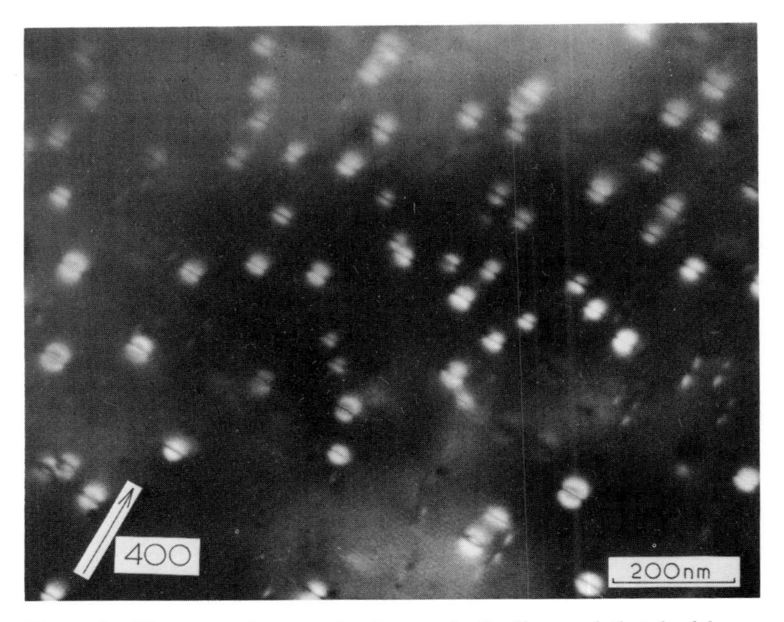


**Figure 8** Electron micrograph of a synthetic diamond that had been heat treated at 2400 and 2500 °C under a stabilising pressure. The platelets are viewed edge-on and the light lobes on either side are regions of high strain due to the displacement outwards of the diamond lattice.

**Table 2** Heat treatment of natural diamonds.

| Temperature (°C) | Time (h) | Nitrogen concentration† ($m^{-3}$) | A centres converted (%) |
|---|---|---|---|
| 2400 | 3 | $1.8 \times 10^{26}$ | ~16 |
| 2600 | 1 ‡ | | |
| 2700 | $\frac{1}{2}$ ‡ | $1.8 \times 10^{26}$ | ~75 |
| 2500 | 3 ‡ | | |

† Using the correlation of Woods *et al* (1990a).
‡ Consecutive heat treatments.

The conclusions drawn from this study of the aggregation of nitrogen are as follows.

(a) The first stage is the aggregation of single nitrogen atoms to form A centres. This obeys second-order kinetics with an activation energy of $5 \pm 0.3$ eV.

(b) The second stage is the formation of B centres and platelets by the aggregation of A centres. The reason for this conclusion is that as the B centres and platelets form, the concentration of A centres diminishes. This does not imply that a platelet is effectively an aggregation of A centres as it is not clear at present whether or not the platelets consist primarily of nitrogen.

No study of the kinetics of the second stage of aggregation was possible because of the inaccurate method of temperature measurement and the instability of the conditions at such high temperatures. Also the range of temperatures at which this process could be studied was too small for reliable kinetic data to be determined.

The N3 centres (three nitrogen atoms and a vacancy) also formed as the B centres and platelets were produced. However, the experimental results suggested that the N3 centres were a by-product of the primary aggregation sequence (single nitrogen atoms to A centres, and from A centres to B centres and platelets).

### 1.5 Regular and irregular diamonds

The types of nitrogen aggregate found in type I natural diamonds may be accounted for by the following sequence of events. During growth of these diamonds the nitrogen was incorporated as single substitutional atoms; aggregation then occurred under the high temperature and pressure conditions found in the Earth's mantle and this aggregation was arrested by the drop in temperature and pressure associated with ejection of the diamonds to the Earth's surface. For type I diamonds there is a wide range in the amount of aggregation that has taken place before being quenched; the extremes being type Ib, which still contain a high proportion of single nitrogen atoms, and type IaB which contain only B centres. It would be expected that the amount of aggregation achieved in a particular diamond prior to quenching would depend upon the concentration of nitrogen, the pressure to which it was subjected, the temperature and the time that it spent at that temperature.

If natural diamonds followed the suggested sequence of aggregation it would be expected that type IaA diamonds would not contain a platelet peak at about 1370 $cm^{-1}$: the platelet peak would become more prominent as the proportion of B centres increased in comparison with A centres. At the completion of this stage a type IaB diamond would have a strong platelet peak with no A centres detectable. Brozel *et al* (1978) and Davies (1981) examined the spectra of a large number of type Ia diamonds and found a certain regularity in plots of the platelet peak as a function of the relative amount of A centres present, i.e. as the amount of A centres diminished when B centres formed, so the platelet peak became more prominent. However, there were exceptions, particularly in diamonds that had a relatively high B centre content compared with the content of A centres; the platelet peak was much lower than would be expected from the aggregation sequence. In the extreme case there were diamonds that were type IaB with no platelet peak at all. Burt (1980) and Maguire (1983) suggested, as a result of extensive experimental work, that the unexpectedly low value of the platelet peak in these diamonds was due to the degradation of some or all of the platelets to form dislocation loops and voidites, thus weakening or removing entirely the platelet peak at about 1370 $cm^{-1}$ (7.3 $\mu$m).

Woods (1986) measured the infrared and visible spectra of a large number of type Ia diamonds. The infrared spectrum in the one-phonon region was decomposed into three spectra, A, B and D, with the platelet peak also being measured. In what he termed *regular* diamonds there was a strict proportionality between the strength of the platelet peak and the strength of the absorption due to the B aggregates. This indicates an epoch of smooth undisturbed aggregation of A aggregates to B aggregates with the concurrent nucleation and growth of platelets. For *irregular* diamonds this proportionality does not hold and such specimens are considered to be those in which a catastrophic degradation of some or all of the platelets has occurred to form dislocation loops and voidites. Almost all the irregular diamonds were at a late stage of the aggregation sequence with a higher B centre content than that of A centres. Woods also found a direct correlation between the platelet peak and the D component of the lattice absorption spectrum for both regular *and* irregular specimens. The D spectrum was attributed to the platelet-stimulated lattice vibrational modes in the one-phonon region. Also amongst regular specimens the strength of the N3 optical absorption was proportional to the platelet peak and hence also to B absorption strengths. The N3 centres were considered to result from minor side reactions occurring during the conversion of A centres to B centres. Figure 9 from Woods (1986) shows the change in shape of the infrared spectra from a pure IaA specimen to the final aggregation stage for regular and irregular diamonds. It is interesting to note that a sequence of this kind does not hold for diamonds from the Argyle mine in Western Australia (Harris and Collins 1985). These diamonds show evidence of extensive plastic deformation that presumably occurred in the upper mantle and may have affected the suggested aggregation sequence.

## 1.6 Estimates of nitrogen concentration

As mentioned previously, Kaiser and Bond (1959) found considerable amounts of nitrogen in diamonds that contained mainly A absorption features. The nitrogen concentration was related to one of the A features, the 1282 $cm^{-1}$ (7.8 $\mu$m) peak, by $N_A = 33.3\mu_A$, where $N_A$ is the nitrogen concentration in atomic ppm and $\mu_A$ is the absorption coefficient in $cm^{-1}$ at 1282 $cm^{-1}$. Lighthowlers and Dean (1964), as well as measuring this absorption coefficient, also analysed several diamonds for nitrogen using gamma-ray activation analysis. Their results were consistent with those of Kaiser and Bond, despite the fact that several of their specimens had pronounced B absorption features. The results of the nitrogen aggregation experiments described in the last section brought to light a number of discrepancies in nitrogen concentrations as estimated by optical absorption measurements.

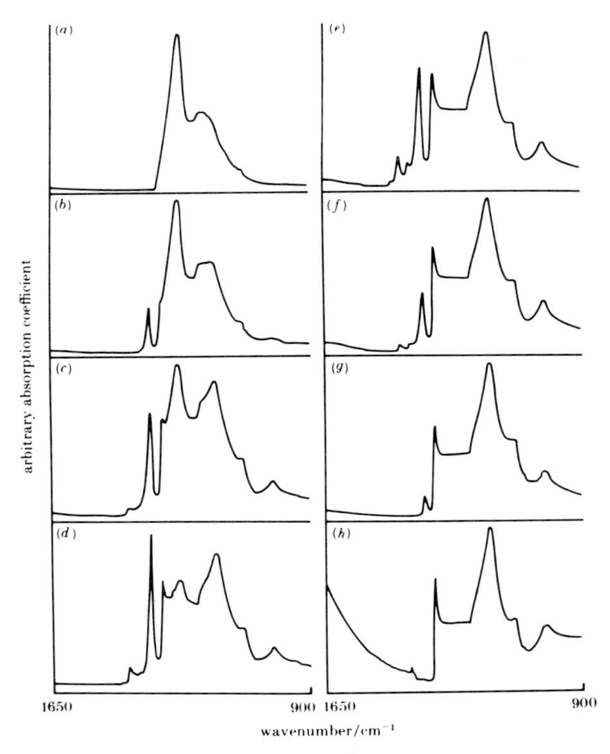


**Figure 9** Infrared absorption spectra from a pure type IaA specimen (*a*), through (*b*)–(*d*), to the case of a regular specimen with only B and D lattice absorptions and a large 1370 $cm^{-1}$ platelet absorption peak (*e*) followed by a departure from regularity (*f*), (*g*) towards a pure type IaB specimen (*h*).

In the first stage of aggregation the single substitutional nitrogen atoms (dispersed nitrogen) form A centres. Chrenko *et al* (1971) measured the concentration of dispersed nitrogen in type Ib by electron paramagnetic resonance (EPR) (Smith *et al* 1959). The absorption coefficient of a band at 1130 $cm^{-1}$ (see figure 1) was found to correlate with the intensity of an EPR signal assigned to dispersed paramagnetic nitrogen (Dyer *et al* 1965). The relationship derived by Chrenko *et al* was

$$N_s = 25\mu_s$$

where $N_s$ is the dispersed nitrogen concentration in atomic ppm and $\mu_s$ is the absorption coefficient in $cm^{-1}$ at 1130 $cm^{-1}$.

In their studies of the aggregation of dispersed nitrogen to form A centres both Chrenko *et al* (1977) and Collins (1980) measured the infrared spectra of their specimens before and after heat treatments. Collins analysed the spectra of both groups into component spectra and attempted to balance the nitrogen content before and after treatment. He used the relation $N_A = 33.3\mu_A$ (Kaiser and Bond 1959) for the concentration of nitrogen in the A centres after treatment, and the relation $N_s = 25\mu_s$ (Chrenko *et al* 1971) for the concentration of dispersed nitrogen before treatment. Using these relations the nitrogen concentration apparently increased during heat treatment by a factor of about 1.5. To resolve this difficulty Collins assumed that the Kaiser and Bond relationship was correct, and deduced that the relation for the dispersed nitrogen was in fact $N_s \approx 45\mu_s$ if the same amount of nitrogen was present before and after treatments.

The relationship between the nitrogen concentration in its dispersed form in type Ib synthetic diamonds and the absorption coefficient at 1130 $cm^{-1}$ has been re-examined in detail by Woods *et al* (1990b). They estimated the nitrogen concentration in two ways, by EPR measurements and by inert gas fusion. The two methods gave good agreement; values for the nitrogen concentration of 22.0 ± 1.1 atomic ppm using EPR, and 20.6 ± 1.5 atomic ppm using gas fusion, for an absorption coefficient of 1 $cm^{-1}$ at 1130 $cm^{-1}$. These values agree tolerably well with the value of 25 atomic ppm given by Chrenko *et al* (1971). Woods *et al* (1990b) propose the relationship

$$N_s = (22 \pm 1.1)\mu_s$$

where again $N_s$ is in atomic ppm and $\mu_s$ the absorption coefficient in $cm^{-1}$ at 1130 $cm^{-1}$. If this value is correct it brings into question the relationship proposed by Kaiser and Bond for the concentration of nitrogen in A centres. Woods *et al* (1990a) measured the amount of nitrogen using the same gas fusion technique as for type Ib diamonds. The new relationship between the nitrogen in A form and the absorption coefficient $\mu_A$ in $cm^{-1}$ at 1282 $cm^{-1}$ is

$$N_A = (17.5 \pm 0.4)\mu_A$$

where $N_A$ is in atomic ppm. This reduces the estimate of nitrogen from 33.3 to 17.5 atomic ppm in A centres giving an absorption coefficient of 1 $cm^{-1}$ at 1282 $cm^{-1}$.

There were also inconsistencies in relating the amount of nitrogen in the B centres to the absorption coefficient at 1282 $cm^{-1}$ (7.8 $\mu$m) of the B spectrum of type IaB diamonds (see figure 3). Davies and Summersgill (1973) deduced that the B defect as well as the A defect was nitrogenous in nature;

they assumed that the absorption coefficient $\mu_B$ at 1282 cm$^{-1}$ for the B spectrum was related in the same way to the nitrogen concentration $N_B$ in the B centres as the relationship for the A centres (Kaiser and Bond 1959). They proposed therefore that $N_B = 33.3\mu_B$. However, photon activation analyses by Sobolev and Lisoivan (1972) on diamonds showing only B absorption indicated a different relationship. Brozel *et al* (1978) suggested the relationship $N_B = 117\mu_B$ from the results of Sobolev and Lisoivan. Also working from the same results, Burgemeister (1980) arrives at a slightly different relationship $N_B = 103\mu_B$. It is clear that although the last two relationships are similar they differ by a factor of between 3 and 4 from the one given by Davies and Summersgill (1973). At the time that this work was done the third component, D, in the one-phonon absorption region for type IA/B diamonds, was not known. As this new component also absorbs at 1282 cm$^{-1}$ errors could result in assigning a relationship between $N_B$ and $\mu_B$. Woods *et al* (1990a) selected pure type IaB diamonds which showed no trace of a platelet peak at about 1370 cm$^{-1}$ thus ensuring that there was no underlying D spectrum admixed with the B spectrum in the one-phonon region. They examined 18 pure type IaB diamonds and related the concentration $N_B$ of the nitrogen in the B centres with the absorption coefficient at 1282 cm$^{-1}$ by the relationship $N_B = (103.8 \pm 2.5)\mu_B$. Woods *et al* realised that the linear relationship between $N_B$ and $\mu_B$ was puzzling, given that such diamonds are irregular and expected to contain voidites which themselves contain nitrogen. This emphasises the difficulty that still remains in partitioning nitrogen between different defects in both regular and irregular type IaA/B diamonds containing various combinations of point defects, platelets and voidites. This difficulty has also been realised and considered by Sumida and Lang (1988).

## 2 GEOLOGICAL FACTORS

### 2.1 Introduction

Diamonds were transported from the upper mantle to the Earth's surface in a magma termed kimberlite; in Western Australia lamproite, a related alkaline rock, also acted as a transporting medium. Study of kimberlite and lamproite is of considerable importance because these rocks carry a wide variety of mantle and crustal fragments to the surface during their eruptions. These *xenoliths* (material in an igneous rock that is not genetically related) include rocks such as peridotites and eclogites that can give direct information about the physical and chemical structure of the mantle. Estimates of eruption ages can be calculated by measuring radiogenic isotopic ratios in minerals found directly within the alkaline volcanic rocks. Mineral inclusions occur in some diamonds and study of these gives information about the chemical environments, temperatures, and sometimes pressures, at the time when the diamonds grew. Some of these inclusions can also help in estimating the age of diamonds.

### 2.2 Eruption ages

The first estimates of eruption ages of kimberlites and lamproites were made by Davis (1977) using U–Pb ratios in the mineral zircon. He made the assumption that, no matter at what time the zircons crystallised prior to eruption, at temperatures of about 1100 °C or more the mineral would lose Pb by diffusion; it could not then accumulate and therefore, there was no radioactive clock. The radioactive clock would operate only when diffusion was restricted by a sudden lowering of temperature, a situation most likely to occur once the kimberlite had erupted to the surface. The problem with the use of zircons in age dating is that there are usually two generations recovered from kimberlite and it is difficult to distinguish between them. One group belongs to an ancient megacryst suite of upper mantle minerals, the other to a group which formed much closer to the age of kimberlite eruption. Thus there is an ambiguity about the initial starting time of the radioactive clock and the U–Pb data suggest that not all the upper mantle Pb may have purged itself from both generations of zircons before eruption. Nevertheless, using zircons Davis (1977) determined the ages of 40 kimberlites from 13 countries. Of the 27 kimberlite sites in Southern Africa, with the exception of those in Lesotho and Swaziland, the eruption ages varied from 55 to 95 Ma (1 Ma = $10^6$ yr). Four sites from Lesotho gave ages of 1500, 1050, 151 and 149 Ma, whilst one from Swaziland gave an eruption age of 650 Ma. Other kimberlites that were examined from Angola, Zaire, Tanzania and Brazil gave eruption ages from about 50 to 190 Ma whilst one site from Melanesia gave the youngest eruption age, of 34 Ma. More recently, Kinny *et al* (1989) dated the Jwaneng kimberlite mine in Botswana at 255 ± 2 Ma using U–Pb in zircon. U–Pb and Pb–Pb methods have also been used in age dating the mineral perovskite from kimberlite concentrates as well as from whole rock kimberlite samples (Kramers and Smith 1983). Perovskite appears to be more suitable than zircon for dating purposes as it does not occur in the upper mantle and is considered to be formed exclusively out of residual kimberlite liquid during eruption. The mantle contains insufficient concentrations of some of the elements necessary to form perovskite prior to eruption and it is only at a late stage that crystals form. This probably accounts for the small crystallite size, 100–200 μm, of the perovskite found in kimberlite. Perovskite ages from De Beers Pool, Kimberley (95 Ma), were in good general agreement with earlier work (Davis 1977). U–Pb whole rock analyses for Jagersfontein and New Elands kimberlite gave ages of 89 ± 4 and 189 ± 75 Ma, respectively. At Premier mine, Transvaal (not considered by Davis (1977)) combined mineral fractions and whole rock work gave an eruption age of 1202 ± 72 Ma (Kramers and Smith 1983).

Smith *et al* (1985) have used the Rb–Sr method on the mineral phlogopite and whole rock samples to determine eruption ages of between about 90 and 160 Ma for 15 South African mines. Similar Rb–Sr measurements on whole rock and phlogopite samples from the Argyle lamproite pipe in Western Australia give an eruption age of 1126 ± 9 Ma.

There appears to be no specific geological period in which diamondiferous kimberlite and lamproite magmas erupted to the surface. In South Africa the age varies from about 55 to about 1200 Ma (Premier mine). In Australia eruptions occurred at 20–25 Ma (Ellendale mine) to over 1100 Ma (Argyle mine). What is of particular interest here is the time of eruption at a specific mine and the estimated age of diamonds from that mine.

### 2.3 Diamond ages

Diamonds are present on the Earth's surface through eruptions of magma from the mantle. The relationship between diamonds and the magma has long been controversial. Essentially the problem is whether diamonds crystallised from the magma and are phenocrysts, or have been picked up and transported to the surface by the passing magma and are therefore xenocrysts. This problem can be resolved by determining the age of diamonds from a specific mine and comparing it with the time the magma erupted to the surface.

Dating diamonds is best achieved by determining the age of syngenetic inclusions in the host diamond. It is necessary to establish that the inclusion morphology was imposed by the surrounding diamond in order to show that both grew at the same time. As inclusions are small, one method of age dating involves examining a composite collection of one type of inclusion in order to get sufficiently accurate results. Diamonds chosen for the study of their inclusions need to be from the same mine and to have grown in the same chemical environment. The growth environment of diamond on the basis of chemical composition is mainly of two types: (1) peridotitic (or ultramafic), and (2) eclogitic.

Peridotite consists mainly of olivine and enstatite with minor components of pyrope garnet and chrome diopside. The inclusions of the peridotitic paragenesis comprise these minerals with, in addition, chromite and sulphides. Eclogite is mainly almandine garnet and omphacitic clinopyroxene and has a quite different chemical composition from peridotite. Eclogitic inclusions, apart from garnet and clinopyroxene, comprise sulphides with minor kyanite, rutile and coesite. It is essential that the inclusions in the composites being studied are entirely from one or other of these two growth parageneses. Geological studies show that the decay of long-lived radioactive elements, such as U, Sm, Rb and K, give rise to small but significant changes in the isotopic compositions of the daughter elements, respectively Pb, Nd, Sr and Ar (Bristow *et al* 1987).

Following earlier work on isotopic ratios from inclusions, Kramers (1979) measured the isotopic compositions of sulphide inclusions from three South African mines and derived very old ages. At both Finsch and De Beers Pool, Kimberley, diamond ages of greater than 2000 Ma were obtained. These results indicate that diamonds are xenocrysts, the kimberlite emplacement having been dated in the Cretaceous period, between 60–140 Ma ago. A further age of about 1200 Ma for diamonds from the Premier Mine, Transvaal, is similar to the emplacement age (see table 3). Uncertainty about the specific paragenesis of the sulphide inclusions has led to some criticism of this work; nevertheless it was an important milestone in diamond age-dating techniques.

Richardson *et al* (1984) selected diamonds with peridotitic garnet inclusions from both the Finsch and De Beers Pool, Kimberley, mines and careful measurements of the Sm–Nd isotopic compositions of composites of these inclusions yielded ancient model ages of between 3200 and 3300 Ma. As the kimberlites from these mines were ejected to the surface about 85–118 Ma ago, the results confirm that diamonds of the peridotitic paragenesis from these mines are xenocrysts in the kimberlites. Richardson (1986) measured the Sm–Nd isotopic compositions of inclusions of eclogitic garnet and clinopyroxene in diamonds from the Premier mine and the Argyle mine (Western Australia); he estimated ages of 1150 and 1580 Ma, respectively, for such diamonds. At Premier the age appears to be comparable, within the errors, to the magma emplacement age, but nitrogen aggregation characteristics indicate that these diamonds are xenocrysts. In 1989, Richardson published Sm–Nd work on eclogitic garnet and clinopyroxene inclusions from the Finsch and Orapa (Botswana) mines, obtaining ages of 1580 ± 50 Ma and 990 ± 50 Ma, respectively. Table 3 shows that for the Finsch mine the growth of diamonds containing peridotitic inclusions and those with eclogitic inclusions were separated by 1500 Ma.

The isotope work described so far has been done on composites of inclusions, i.e. collections of inclusions of the same minerals from the same mine. A significant advance was made by Phillips *et al* (1989) and Burgess *et al* (1989) using $^{40}Ar/^{39}Ar$ laser-probe analyses of *individual* eclogitic clinopyroxene inclusions from diamonds of the Premier mine; they estimated mean ages of 1198 ± 14 and 1185 ± 94 Ma, respectively. These ages agree well with the Sm–Nd work of Richardson (1986) on the same diamond paragenesis. Also in 1989, Smith *et al* determined the ages of relatively large (up to 1 mm) single eclogitic garnet inclusions in large diamonds from Finsch. Although age variations between 1400 and 2400 Ma were obtained, model ages ranging between 1400 and 1700 Ma were considered most valid, which is comparable with Richardson's work (see before) on eclogitic garnet and clinopyroxine composites.

It can be anticipated that the development of isotope measurements of individual inclusions will be of immense benefit in studying the history of diamonds. Such geological information used in conjunction with studies of nitrogen aggregation for the same diamonds will be of particular interest. The most reliable estimates of diamond ages and eruption ages, using our knowledge in 1990, are shown in table 3.

**Table 3** Diamond ages and eruption ages.

| | Method | Environment | Source | Age (Ma) | Eruption age (Ma) |
|---|---|---|---|---|---|
| † | U–Pb | ?(sulphides) | Finsch | >2000 | 118 |
| ‡ | Sm–Nd | P(garnet) | Finsch | 3300 ± 100 | 118 |
| § | Sm–Nd | E(gnt + cpx) | Finsch | 1580 ± 50 | 118 |
| § | Sm–Nd | E(gnt + cpx) | Orapa | 990 ± 50 | 94 |
| ‡ | Sm–Nd | P(garnet) | De Beers Pool | 3300 ± 100 | 85 ± 5 |
| † | U–Pb | ?(sulphides) | De Beers Pool | >2000 | 85 ± 5 |
| ‖ | Sm–Nd | E(gnt + cpx) | Premier | 1150 ± 60 | 1207 ± 72 |
| † | U–Pb | ?(sulphides) | Premier | 1200 (approx.) | 1207 ± 72 |
| ¶ | $Ar^{40}$–$Ar^{39}$ | E(cpx) | Premier | 1198 ± 14 | 1207 ± 72 |
| * | $Ar^{40}$–$Ar^{39}$ | E(cpx) | Premier | 1185 ± 94 | 1207 ± 72 |
| ‖ | Sm–Nd | E(gnt + cpx) | Argyle | 1580 ± 60 | 1126 ± 9 |

† Kramers (1979)
‡ Richardson *et al* (1984)
§ Richardson (1989)
‖ Richardson (1986)
¶ Phillips *et al* (1989)
* Burgess *et al* (1989)

**2.4 Equilibrium temperatures and pressures**

Some of the xenoliths that have been transported from the mantle contain diamonds. Examination of the mineralogy of these nodules can indicate the general environment in which diamonds grew. By combining this mineralogical information with the results obtained with experimental petrology in the laboratory, an estimate of temperature and sometimes pressure of formation of these xenoliths can be made. Caution must be exercised in these calculations as chemical re-equilibrium or the alteration of the constituent minerals often occurs after growth and prior to ejection to the Earth's surface. Also some re-equilibrium may take place during eruption when the temperature and pressure fall. As has been described in section 2.3, there are two growth environments for diamond, peridotitic and eclogitic. With the peridotitic minerals the partition of elements in co-existing crystals has led to the development of geothermometry and geobarometry techniques. For instance, Lindsley and Dixon (1976) considered the relationship between the chemical composition of diopside with that of co-existing enstatite, and were able to obtain a geothermometer for this system. More recently, O'Neill and Wood (1979) considered the partitioning of iron and magnesium between garnet and co-existing olivine and developed a geothermometer for this system. As an example of a geobarometer Macgregor (1974) determined the solubilities of alumina in enstatites for spinel and garnet peridotite compositions and showed that the amount of alumina in enstatite was a function of pressure.

With eclogitic minerals, high-temperature and pressure experiments indicate that there is a partitioning of iron and magnesium between the two principal minerals of this suite, i.e. almandine garnet and omphacitic clinopyroxene. Ellis and Green (1979), for example, used this partitioning for a geothermometer based on the assumption that a particular pressure, usually 5 GPa, is appropriate to diamond formation in the upper mantle. A geobarometer in the eclogitic system involves the extent to which clinopyroxene is in solid solution with garnet. This relationship was first reported by Moore and Gurney (1985) and later determined experimentally over the pressure interval 4.0 to 17.5 GPa by Irifune (1987).

The geothermometry and geobarometry work on diamondifereous xenoliths has now been extended to diamond itself. The minerals of interest are the same as those found in the xenoliths, but in this case they occur as mineral inclusions within the diamond and have been isolated from the surface of the diamond by complete enclosure by the host. Two or more different syngenetic mineral inclusions are required and as they are usually separated by the host diamond there is no chemical exchange between them. To estimate the temperature and pressure of formation of a particular diamond it must be assumed that the non-touching minerals in the diamond were incorporated at almost the same time and that they were both in equilibrium with the same chemical environment before confinement. For some diamonds, growth appears to have been a discontinuous process that resulted in growth zones of varying impurity and defect content. In such diamonds individual inclusions may have been incorporated at different times which would be a cause of error in estimating temperature and pressure. With these limitations in mind, by analysing at least ten inclusion pairs in diamonds from a particular source and growth environment, peridotitic or eclogitic, an estimate of temperature and sometimes pressure at the time of diamond formation can be made.

Temperature and pressure estimates from inclusions in diamond are contained in table 139 of a review paper by Meyer (1987). The table shows that for the diamonds from specific African and Siberian mines the temperature range is from about 900 to 1300 °C with a pressure range spanning 45–65 kbars (4.5–6.5 GPa). This pressure range indicates that these diamonds originated at depths of between about 140 to 200 km in the upper mantle. Recent results reported at the Fourth International Kimberlite Conference in 1986 (published 1989) indicate that this comparatively narrow range of depths may be considerably widened in the future. For example, Moore and Gurney (1989) have found mineral inclusions which suggest that the diamonds containing them came from depths down to 450 km or even deeper. Thus these diamonds from a particular locality in South Africa provide a unique source of sampling of the deep upper mantle. These recent results also suggest that some diamonds experienced temperatures considerably greater than the 1300 °C that had previously been assumed to be about the upper limit for growth and equilibration.

## 3 DISCUSSION

From the limited geophysical and geochemical data that is available, it appears that most diamonds grew at depths of between about 140 to 200 km in the upper mantle and then spent a considerable time there at temperatures between 900 and 1300 °C, before being collected and transported very rapidly to the surface by a rising magma. Because this work is so recent, mostly in the 1980s, significant advances in the understanding of the geological factors can be expected in the 1990s. Evidence is already emerging that some diamonds grew at much deeper regions, down to 450 km or more, at correspondingly higher temperatures.

It is instructive to examine how the aggregation of nitrogen in diamond fits into this proposed history of diamonds. A reasonable proposition is that, as with synthetic diamonds, nitrogen was incorporated into a growing natural diamond as single substitutional atoms; these subsequently aggregated over long periods of time in the mantle before the process was stopped by the drop in temperature and pressure associated with rapid eruption to the surface of the Earth. Natural diamonds with the least amount of aggregation are type Ib diamonds. Using the results of the aggregation work it can be shown that this type would have experienced annealing times of about 3000 Ma if at 750 °C to a much shorter time, 20 yr, at 1300 °C.

Specimens that have advanced to a later stage of aggregation are the type IaA diamonds. Again using the kinetic equations derived in aggregation studies, it was suggested that to achieve the amount of aggregation associated with a type IaA diamond would require about 1000 Ma at 1000 °C with correspondingly much shorter times at higher temperatures, e.g. 0.2 Ma at 1300 °C.

The next stage of aggregation, with the formation of B centres and platelets at the expense of A centres, was experienced by regular type IaA/B diamonds. No experimental results are available from which the kinetics for this stage can be derived, although there is a measurement of the rate of aggregation at about 2600 °C. Reliable measurements of such rates over the necessary temperature range, say 2600–3000 °C, must await advances in high-temperature/pressure technology before the kinetics can be determined with any degree of confidence. However, an indirect way of getting kinetic information was attempted by Evans and Harris (1989). Three type IaA/B diamonds were chosen which had been examined in detail in order to assign each specimen with a temperature which it had experienced for a known time in the upper mantle. The infrared absorption spectra of the three diamonds were decomposed to give the nitrogen concentrations and the different stages that they had reached in the aggregation process. It was assumed that second-order kinetics was appropriate for this aggregation of A centres to form B centres and platelets, and an activation energy of about 6.5 eV was derived to explain the amounts of aggregation in the three diamonds. It can be shown that if first-order kinetics is assumed then an activation energy of about 7 eV is appropriate. It is probable that the actual kinetics in fact lie between first and second order with an activation energy that is between 6.5 and 7 eV.

The satisfactory feature of the proposed nitrogen aggregation sequence and its associated kinetic conditions is that, in spite of the obvious inaccuracies, the amount of aggregation found in *regular* natural diamonds could occur, within a reasonable geological time, only at temperatures between about 800 and 1300 °C; this is approximately the temperature range proposed by the geological work on inclusions and xenolith nodules. If temperatures were much less, the required amounts of aggregation would not be achieved at times within the age of the Earth, 4600 Ma. If temperatures were much higher no type Ib and IaA diamonds would have been produced if they were originally xenocrysts in the rising magma. It seems reasonable to propose that at a depth of about 200 km the rising magma picked up regular diamonds that had reached the most advanced stage of aggregation, due to annealing at about 1300 °C. Diamonds at a less advanced stage, types IaA/B, were picked up at shallower depths, until at about 150 km, type IaA diamonds that had annealed at about 1000 °C for geological periods of time were collected. At still shallower depths type Ib diamonds would have been collected and joined the transportation of the other types of diamond to the surface. This sequence applies only to *regular* natural diamonds that experienced the aggregation sequence

Single nitrogen atoms → A centres → B centres<br>
↘ platelets.

This proposal that diamonds of different types were collected by the magma from different depths could be tested by examination of inclusions in sets of diamonds of a particular type and the associated xenoliths so as to assign annealing temperatures and times.

There is an ambiguity about the temperatures that have been assigned to diamonds from particular mines, summarised by Meyer (1987). Each of these temperatures has been determined by examination of the inclusions, either peridotitic or eclogitic, from more than ten diamonds that contained pairs of inclusions. Thus in principle the assigned temperature could be some average from all the types of diamond. In practice the large majority of diamonds containing nitrogen are of type IaA/B with most of these having an A/B ratio greater than 1, for the infrared absorption peak at 1282 $cm^{-1}$ (7.8 $\mu$m), as in figure 9(*a*), (*b*) and (*c*). Thus it could be expected that the assigned temperature for diamonds from a particular mine would be heavily biased towards diamonds of this type. Advances in experimental techniques will make it possible to assign temperatures, together with times spent at these temperatures, to *individual* diamonds of all types; this would put into context the temperatures so far derived by examination of inclusions.

Results from the nitrogen aggregation work are of use to geologists as they set limits on the physical conditions that can be proposed to explain geological results. An instance is the observation that it has not been possible experimentally to differentiate within experimental errors between the age of diamonds from the Premier mine and eruption of the kimberlite. This might suggest that diamonds from this mine were originally phenocrysts rather than xenocrysts in the rising magma. The diamonds recovered from this mine show normal amounts of nitrogen aggregation. Examination of table 3 shows that the error limits derived for the values of eruption ages and diamond ages for the Premier mine allow reasonably for times of up to 20 Ma between the two. The temperature determined for diamonds from this mine is rather high at about 1250 to 1270 °C (Meyer 1987). It can be estimated that, using the proposed kinetic parameters for this temperature range, for times of 20 Ma, would result in the required amount of aggregation to form type IaA/B with an A/B ratio greater than 1. Thus the aggregation work suggests that diamonds from the Premier mine were originally xenocrysts in the rising magma in common with diamonds from other mines.

So far only regular diamonds have been considered and for irregular diamonds other factors need to be taken into account. The most obvious irregular diamonds are those that contain dislocation loops together with voidites that are sometimes associated with the loops. Much information has been derived using sophisticated analytical techniques about the nature of voidites. They certainly contain a solid phase consisting mainly if not entirely of nitrogen under high pressure. It has not been possible to produce them in the laboratory despite considerable efforts and it may be that to do so temperatures even higher than 2800 °C with a suitable stabilising pressure would be necessary. Thus not enough information is available to fit voidite formation and platelet dissociation into a geological context. It is tempting to speculate that diamonds containing these defects came from greater depths than regular diamonds and experienced correspondingly higher temperature and pressure conditions. It may be that new results suggesting that some diamonds originated at depths down to 450 km or greater will eventually be linked to irregular diamonds that contain voidites.

Other irregular diamonds are those from some Australian mines. These diamonds seem to have more defects than diamonds from other localities and Humble (1988) has suggested that considerable plastic deformation and dislocation climb has occurred during the time that they spent in the upper mantle. It is not surprising that these diamonds do not appear to have followed the nitrogen aggregation sequence in the same way as regular diamonds. Plastic deformation results in enhanced point defect concentrations as well as a high density of dislocations. This implies that nitrogen diffusion can be affected in two ways, by enhancement due to excess point defects and by pipe diffusion along the dislocations. Before an explanation for the nature of these Australian diamonds is forthcoming, much more experimental work needs to be done on the distribution of nitrogenous and other defects as well as further studies of geological factors.

It is clear that considerable advances have been made since Frank's original proposition that diamonds are letters from the depths. Much of the language in which the letter is written has been translated and perhaps the first page is now understood; further pages are still there to be read and their complex sentences analysed, before the full beauty and meaning of the letter can be appreciated.

## ACKNOWLEDGMENTS

My grateful thanks are due to Dr J W Harris for his help in the section on geological factors and to Drs C M Welbourn and G S Woods for access to results of their work prior to publication.

## REFERENCES

Allen B P and Evans T 1981 *Proc. R. Soc.* A **375** 93–104
Anstis G R and Hutchison J L 1982 *Proc. 8th Int. Conf. Electron Microscopy, Hamburg* **2** 923–4
Barry J C 1986 *Ultramicroscopy* **20** 169–76
Bristow J W, Allsop H L, Smith C B and Skinner E M W 1987 *Nucl. Active, Int. J. of Atomic Energy Corp. S. Afr.* **36** 21–6
Brozel M R, Evans T and Stephenson R F 1978 *Proc. R. Soc.* A **361** 107–27
Bruley J and Brown L M 1989 *Phil. Mag.* **59** 247–61
Burgemeister E A 1980 *J. Phys. C: Solid State Phys.* **13** L963–8

Burgess R, Turner G, Laurenzi M and Harris J W 1989 *Earth Planet. Sci. Lett.* **94** 22–8
Burns R C, Cvetkovic V, Dodge C N, Evans D J F, Rooney M L T, Spear P M and Welbourn C M 1990 *J. Crystal Growth* **104** 257–79
Burt K A H 1980 *PhD thesis* University of Reading
Chrenko R M, Strong H M and Tuft R E 1971 *Phil. Mag.* **23** 313–18
Chrenko R M, Tuft R E and Strong H M 1977 *Nature* **270** 141–4
Clark C D and Davey S T 1984a *J. Phys. C: Solid State Phys.* **17** 1127–40
—— 1984b *J. Phys. C: Solid State Phys.* **17** L399–403
Clark C D, Ditchburn R W and Dyer H B 1956 *Proc. R. Soc.* A **234** 363–81
Collins A T 1980 *J. Phys. C: Solid State Phys.* **13** 2641–50
Davies G 1976 *J. Phys. C: Solid State Phys.* **9** L537–42
—— 1981 *Nature* **290** 40–1
Davies G and Summersgill I 1973 *Diamond Research* (London: Industrial Diamond Information Bureau) pp 6–15
Davies G, Welbourn C M and Loubser J H N 1978 *Diamond Research* (Ascot: De Beers Industrial Diamond Division (Pty) Ltd) pp 23–30
Davis G L 1977 *Extended Abstracts 2nd Int. Kimb. Conf., Santa Fe*
Dyer H B, Raal F A, du Preez L and Loubser J H N 1965 *Phil. Mag.* **11** 763–74
Ellis D J and Green D H 1979 *Contrib. Mineral. Petrol.* **71** 13–22
Evans T 1976 *Contemp. Phys.* **17** 45–70
—— 1978 *Diamond Research* (Ascot: De Beers Industrial Diamond Division (Pty) Ltd) pp 17–22
Evans T and Harris J W 1989 *Proc. 4th Int. Kimb. Conf., Perth* **2** 1001–6
Evans T and Phaal C 1962 *Proc. R. Soc.* A **270** 538–52
Evans T and Qi Z 1982 *Proc. R. Soc.* A **381** 159–78
Evans T and Rainey P 1975 *Proc. R. Soc.* A **344** 111–30
Evans T and Woods G S 1987 *Phil. Mag. Lett.* **55** 295–9
Frank F C 1956 *Proc. R. Soc.* A **237** 168–74
—— 1966 *Proc. Int. Industrial Diamond Conf. Oxford* (London: Industrial Diamond Information Bureau) pp 119–35
Hanley P L, Kiflawi I and Lang A R 1977 *Phil. Trans. R. Soc.* **284** 329–68
Harris J W and Collins A T 1965 *Industr. Diam. Rev.* **45** 128–30
Hirsch P B, Hutchison J L and Titchmarsh J 1986a *Phil. Mag.* **54** L49–54
Hirsch P B, Pirouz P and Barry J C 1986b *Proc. R. Soc.* A **407** 239–58
Hutchison J L and Bursill L A 1983 *J. Microsc.* **131** 63–6
Humble P 1988 *Diamond Conf. Cambridge* unpublished
Irifune T 1987 *Phys. Earth Planet. Interiors* **45** 324–36
Kaiser W and Bond W L 1959 *Phys. Rev.* **115** 857–63
Kinny P D, Compston W, Bristow J W and Williams I S 1989 *Proc. 4th Int. Kimb. Conf., Perth* **2** 833–42
Klyuev Y A, Naletov A M, Nepsha V I, Apishina N I and Bulygina T I 1977 *Fiz. Tverd. Tela* (*Leningrad*) **19** 14–19 (Engl. transl. 1977 *Sov. Phys.–Solid State* **19** 7–10)
Klyuev Y A, Nepsha V I and Naletov A M 1974 *Fiz. Tverd. Tela* (*Leningrad*) **16** 3259–67 (Engl. transl. 1975 *Sov. Phys.–Solid State* **16** 2118–26)
Kramers J D 1979 *Earth Planet. Sci. Lett.* **42** 58–70
Kramers J D and Smith C B 1983 *Isotope Geosci.* **1** 23–38

Lang A R 1974 *Proc. R. Soc.* A **340** 233–48
Lax M and Burstein E 1955 *Phys. Rev.* **97** 39–52
Lighthowlers E C and Dean P J 1964 *Diamond Research* (London: Industrial Diamond Information Bureau) pp 21–5
Lindsley D H and Dixon S A 1976 *Am. J. Sci.* **276** 1285–301
Loubser J H N and van Wyk J A 1981 *Diamond Conference, Reading* unpublished
Loubser J H N and Wright A C J 1973 *Diamond Research* (Ascot: De Beers Industrial Diamond Division (Pty) Ltd) pp 16–20
MacGregor I D 1974 *Am. Mineral.* **59** 110–19
Maguire J 1983 *PhD thesis* University of Reading
Meyer H O A 1987 *Mantle Xenoliths* ed P H Nixon (New York: Wiley) pp 501–25
Moore R O and Gurney J J 1985 *Nature* **318** 553–5
—— 1989 *Proc. 4th Int. Kimb. Conf., Perth* **2** 1029–41
O'Neill H St C and Wood B J 1979 *Contrib. Mineral. Petrol.* **70** 59–70
Phillips D, Onstott T C and Harris J W 1989 *Nature* **340** 460–2
Raman C V and Nilakantan P 1940 *Proc. Indian Acad. Sci.* A **11** 389–97
Richardson S H 1986 *Nature* **322** 623–6
—— 1989 *Extended Abstracts of Workshop on Diamonds* (Washington, DC: Carnegie Institute) pp 87–90
Richardson S H, Gurney J J, Erlank A J and Harris J W 1984 *Nature* **310** 198–202
Smith C B, Gurney J J, Harris J W, Otter M L, Kirkley M B and Jagontz E 1989 *Extended Abstracts of Workshop on Diamonds* (Washington, DC: Carnegie Institute) pp 102–4
Smith C B, Gurney J J, Skinner E M W, Clement C R and Ebrahim N 1985 *Trans. Geol. Soc. S. Afr.* **88** 267–80
Smith W V, Sorokin P P, Gelles I L and Lasher V I 1959 *Phys. Rev.* **115** 1546–52
Sobolev E V and Lisoivan V I 1972 *Dokl. Akad. Nauk SSSR* **204** 88–91 (Engl. transl. 1972 *Sov. Phys.–Crystallogr.* **17** 425–7)
Sobolev E V, Lisoivan V I and Lenskaya S V 1967 *Dokl. Akad. Nauk SSSR* **175** 582–5 (Engl. transl. 1968 *Sov. Phys.–Crystallogr.* **12** 665–8)
Stephenson R F 1978 *PhD thesis* University of Reading
Strong H M and Chrenko R M 1971 *J. Phys. Chem.* **75** 1838–43
Sumida N and Lang A R 1988 *Proc. R. Soc.* A **419** 235–57
Sutherland G B B M, Blackwell D E and Simeral W G 1954 *Nature* **174** 901–4
Woods G S 1986 *Proc. R. Soc.* A **407** 219–38
Woods G S, Purser G C, Mtimkulu A S S and Collins A T 1990a *J. Phys. Chem. Solids* **51** 1191–7
Woods G S, van Wyk J A and Collins A T 1990b *Phil. Mag.* in press

# Dislocations and Solid Earth Geophysics

J P Poirier

## 1 INTRODUCTION

In addition to his commitment to the physics of materials, Sir Charles Frank has for a long time taken an active interest in solid Earth geophysics and has made important contributions in that field. Let us only recall here that he introduced the concept of dilatancy in relation to seismic sources (Frank 1965) and suggested the so-called 'Frank–Griggs mechanism' (Griggs 1967), still at the basis of recent theories of the hydrolytic weakening of quartz.

In a paper entitled 'The Earth as a ceramic body' (Frank 1969), we can find these words of advice: 'It is very high time for geophysicists to give up treating the Earth's mantle as a Newtonian fluid of very great viscosity and try assigning to it possible real mechanical properties.'

The Earth's mantle extends from the bottom of the crust, some 30 km deep, down to the top of the liquid core, at a depth of 2900 km, where the pressure reaches 1.35 Mbar and where the temperature is probably as high as 3000 K. Its mineralogical composition as a function of depth is inferred from the knowledge of seismic velocities and of the density profile, in turn extracted from the travel times of seismic waves; it is constrained by reasonable assumptions about the global composition of the Earth. Although there still is some controversy about the detailed chemical composition and mineralogy of various regions, there is general agreement about the mantle being mostly constituted of ferromagnesian silicates.

There is also no very strong reason to depart from the idea that the mantle is, to first order, chemically homogeneous from top to bottom, the seismic discontinuities encountered in depth being only due to pressure-driven phase transitions in the constitutive minerals.

The principal minerals in the upper mantle, down to 400 km, are olivine $(Mg,Fe)_2SiO_4$ and pyroxenes $(Mg,Fe)SiO_3$, with some aluminous silicate phases. In the 'transition zone', below 400 km and down to 670 km, olivine transforms to a phase with the same composition but with spinel structure ($\gamma$-spinel or ringwoodite) and pyroxenes transform to garnet (incorporating the aluminous component) and ilmenite structure. Below 670 km, in the lower mantle, the pyroxenes have been entirely converted to the perovskite structure, with $SiO_6$ octahedra instead of the $SiO_4$ tetrahedra of the lower pressure phases, and spinel disproportionates to perovskite and magnesiowüstite (Mg,Fe)O:

$$(Mg,Fe)_2SiO_4 \rightarrow (Mg,Fe)SiO_3 + (Mg,Fe)O.$$

The silicate perovskite $(Mg,Fe)SiO_3$ is therefore the principal mineral in the lower mantle that occupies about 70% of the volume of the Earth.

'Assigning possible real mechanical properties to the mantle' would indeed be highly desirable, since it would constrain the convection and thermal models of the Earth. However, the high-pressure minerals of the mantle can be synthesised only in very small quantities in diamond-anvil cells or multi-anvil presses and, although they are metastable at ambient pressure, they revert to low-pressure phases by heating. It is therefore impossible to perform high-temperature creep experiments *in situ* at high pressure or on synthesised samples at ambient pressure. Even though there is little hope in the near future of being able to determine the viscosity of the mantle minerals, let alone that of the mantle itself (Poirier 1988), we may nevertheless speculate on their 'possible real mechanical properties' by investigating the stability of the dislocations that can exist in the relevant mineral structures and the possible consequences on the ease of slip or climb. The speculations can be checked on minerals accessible to experimentation such as olivine and by observation of dislocations by transmission electron microscopy (TEM) in synthetic high-pressure phases or analogue phases with the same structures.

In the following discussion we will use 'the assumption that the energy per unit length of a dislocation line is proportional to the square of the Burgers vector' (Frank and Nicholas 1953) to discuss the possible dislocations and their stability in the principal mineral structures relevant to solid Earth geophysics: olivine, spinel and perovskite.

## 2 OLIVINE AND SPINEL

Let us consider the magnesian end-member of the ferromagnesian olivine series: forsterite, $Mg_2SiO_4$. The orthorhombic crystalline edifice of isolated

$SiO_4$ tetrahedra and $Mg^{2+}$ cations can be described as a slightly distorted hexagonal close-packed lattice of oxygen ions where one eighth of the tetrahedral sites are occupied by Si ions and one half of the octahedral sites are occupied by Mg ions. The $c/a$ ratio of the HCP oxygen lattice is equal to 1.59, smaller than the ideal close-packing ratio.

At high temperatures the documented slip systems of natural olivines are: (010)[100] and (001)[100] corresponding to slip on prism planes of first order $\{10\bar{1}0\}$ and of second order $\{11\bar{2}0\}$, respectively, in the $\boldsymbol{c}$ direction of the HCP oxygen lattice.

A simple geometrical analysis of the dislocations in the hard-sphere model (Poirier 1975) shows that [100] screw dislocations could split in the three prism planes of first order according to the following schemes:

$$\begin{aligned}
&(010) \qquad [100] \rightarrow [\tfrac{1}{6}, \tfrac{1}{36}, \tfrac{1}{4}] + [\tfrac{2}{3}, 0, 0] + [\tfrac{1}{6}, \tfrac{1}{36}, \tfrac{1}{4}] \\
&(0\bar{1}1) \qquad [100] \rightarrow [\tfrac{1}{6}, \tfrac{1}{9}, \tfrac{1}{6}] + [\tfrac{2}{3}, 0, 0] + [\tfrac{1}{6}, \tfrac{1}{9}, \tfrac{1}{6}] \\
&(011) \qquad [100] \rightarrow [\tfrac{1}{6}, \tfrac{5}{36}, \tfrac{1}{12}] + [\tfrac{2}{3}, 0, 0] + [\tfrac{1}{6}, \tfrac{5}{36}, \tfrac{1}{12}].
\end{aligned} \tag{1}$$

In all three cases the sum of the squares of the Burgers vectors of the partial dislocations is smaller than the square of the Burgers vector [100] of the unsplit dislocation (Frank's criterion). The splitting is therefore geometrically possible, although whether the dislocations are widely extended or not depends on the unknown stacking fault energy. This simple analysis suffices to interpret the documented macroscopic 'pencil glide' $\{0kl\}$ [100] as due to cross slip of the [100] screw dislocations on various prism planes of first order of the HCP oxygen lattice. In this hard-sphere splitting scheme, the Burgers vectors of the outer partials lie somewhat outside the glide planes (e.g. $[\tfrac{1}{6}, \tfrac{1}{36}, \tfrac{1}{4}]$ does not lie in the (010) plane), and this is suggestive of a slight dilatation associated with the stacking fault that would make recombination of the partials and cross slip easier under high pressure (Poirier 1975), as is the case for halite NaCl (Fontaine and Haasen 1969, Aladag *et al* 1970).

In the same way, [100] screw dislocations could split on the (001) glide plane (prism plane of second order) of the HCP oxygen lattice, that does not cut the Si–O bonds, according to the scheme (Poirier and Vergobbi 1978)

$$(001) \qquad [100] \rightarrow [\tfrac{1}{6}, \tfrac{1}{9}, \tfrac{1}{6}] + [\tfrac{2}{3}, 0, 0] + [\tfrac{1}{6}, \tfrac{1}{9}, \tfrac{1}{6}]. \tag{2}$$

Sessile configurations can also be envisaged that would account for the observation of straight screw dislocations.

Cross slip of the [100] dislocations onto planes in zone with [100] suggests the possibility that high-temperature creep might be cross-slip-controlled, with a stress-dependent activation energy. This mechanism has been shown to be compatible with the experimental results of Kohlstedt and Goetze (1974) and it is not impossible that it might be active, concurrently with climb-controlled power-law creep, in the convecting upper mantle (Poirier and Vergobbi 1978).

The (100) [001] slip system corresponding to HCP basal slip in the ***a*** direction is only observed at low temperature and/or high-strain rates, e.g. in shocked meteorites (Madon and Poirier 1983, Madon *et al* 1989).

Dislocations with Burgers vector ***a*** in the basal plane of the HCP lattice can dissociate according to the classical scheme

$$(100) \qquad [001] \rightarrow \tfrac{1}{2}[001] + \tfrac{1}{2}[001] \tag{3}$$

or:

$$[001] \rightarrow \tfrac{1}{12}[013] + \tfrac{1}{12}[0\bar{1}3] + \tfrac{1}{12}[013] + \tfrac{1}{12}[0\bar{1}3]. \tag{4}$$

The stacking faults in the oxygen lattice have a FCC structure appropriate for spinel, however after the first partial moves through the crystal, the cations find themselves in sites with the wrong coordination, and they must therefore jump to correct sites as the dislocation moves; this process is known as synchroshear (Kronberg 1957) and it must take place for slip to occur. However, it is probably not an easy process, as witnessed by the very straight [001] screw dislocations observed in shocked olivine (Madon and Poirier 1983) that obviously have a low mobility, implying a high Peierls force, consistent with the fact that (100)[001] slip occurs mostly at high strain rates.

The fact that the stacking fault has the spinel structure led Poirier (1981) to propose that the transition olivine–spinel might occur by invasion of the crystal of olivine by stacking faults whose energy would be lowered as the stability domain of spinel is approached, much in the way hexagonal cobalt goes to FCC at high temperature. This led to some controversy: Poirier's hypothesis was supported by experiments in diamond-anvil cells (Lacam *et al* 1980, Boland and Liu 1983), although Furnish and Bassett (1983) showed that synchroshear did not occur and that the cations did not immediately fall into place after shearing of the oxygen lattice; the hypothesis, however, was contradicted by experiments in large-volume apparatus which did not produce the expected topotactic relations between spinel and olivine and were more consistent with a nucleation and growth mechanism than with a shearing one (Vaughan *et al* 1982, Boland and Liebermann 1983). It is now proposed that the non-hydrostaticity of the stress regime has a dominant influence on the mechanism (Burnley and Green 1989), the deviatoric stress in the diamond-anvil cell favouring the shearing mechanism whereas the large-volume apparatus would be more hydrostatic. Another reason may also be that the transformation in the diamond-anvil cell usually occurs far from equilibrium, the olivine being brought to a state well into the spinel stability field before heating; when the laser beam is turned on, the temperature rapidly rises allowing the transformation to proceed with a very large driving force, which favours the shearing mechanism, faster than nucleation and growth; in a large-volume apparatus, on the other hand, the transformation usually takes place near the Clapeyron, allowing more time for diffusion.

In $\gamma$-spinel, the (111) plane corresponds to the (100) plane of olivine and slip occurs by glide of $\frac{1}{2}[1\bar{1}0]$ dislocations that can dissociate according to the scheme:

$$(111) \qquad \tfrac{1}{2}[1\bar{1}0] \rightarrow \tfrac{1}{4}[1\bar{1}0] + \tfrac{1}{4}[1\bar{1}0] \qquad (5)$$

with a low-energy stacking fault (Ashworth and Barber 1977), or

$$(111) \qquad \tfrac{1}{2}[1\bar{1}0] \rightarrow \tfrac{1}{12}[\bar{1}2\bar{1}] + \tfrac{1}{12}[\bar{2}11] + \tfrac{1}{12}[\bar{1}2\bar{1}] + \tfrac{1}{12}[\bar{2}11]. \qquad (6)$$

The stacking fault has the structure of olivine and slip cannot occur without synchroshear as pointed out by Hornstra (1961). We therefore expect to see straight dislocations, which is indeed what is found in TEM in ringwoodite produced by shock in meteorites: Madon and Poirier (1983) observed very straight $60^\circ$ $\frac{1}{2}[1\bar{1}0]$ dislocations, as well as sessile prismatic Frank loops containing a stacking fault with displacement vector $\frac{1}{2}[111]$. Abundant planar defects were also observed on [110] planes with displacement vector $R = \frac{1}{4}[1\bar{1}0]$ or equivalently $R = \frac{1}{4}[11\bar{2}]$. These defects, typical of spinels, can be analysed as antiphase boundaries for cations, twins or stacking faults of low energy preserving the oxygen sublattice and affecting only the cations (Price and Putnis 1979, Madon and Poirier 1980, 1983). These faults are the same as would appear in the dissociation of the $\frac{1}{2}[1\bar{1}0]$ dislocations on their climb planes as seen in $MgAl_2O_4$ spinels (Doukhan *et al* 1979). Climb splitting, although not visible in TEM, may also affect the core structure leading to a high Peierls stress: $\gamma$-spinel, a dominant material in the upper mantle transition zone, is probably hard to deform by $(111)\frac{1}{2}[1\bar{1}0]$ slip.

## 3 SILICATE PEROVSKITE

According to all Earth models, the silicate perovskite of chemical formula close to $(Mg_{0.9}Fe_{0.1})SiO_3$ is the dominant material in the lower mantle, from a depth of 670 km to the core–mantle boundary at 2900 km; it probably occupies at least 80% of the volume of the lower mantle, it is therefore necessarily connected and it follows that whether it is more or less creep resistant than the other main constituent (Mg,Fe)O, the silicate perovskite must control the viscosity of the lower mantle.

The perovskite structure $ABO_3$ is one of the most frequently encountered in crystals. The ideal structure consists of a three-dimensional framework of corner-linked $BO_6$ octahedra, all of whose dodecahedral interstices are filled with A cations. The unit cell is cubic, with eight A cations at the corners, six oxygen anions at the centres of the cube faces and one B cation at the centre of the cube.

For electrical neutrality, the sum of the cation vacancies must be equal to six in oxide perovskites and we therefore have 1–5 perovskites (e.g. $KTaO_3$, $KNbO_3$), 2–4 perovskites (e.g. $BaTiO_3$, $CaTiO_3$, $CaGeO_3$,

$MgSiO_3$) and 3–3 perovskites (e.g. $GdAlO_3$). Depending on the ionic sizes, the structure can be distorted in various ways: in some cases, the B cation is off-centre in the $BO_6$ octahedra, thus causing the perovskite to be ferroelectric and piezoelectric (e.g. $BaTiO_3$, $KNbO_3$), in other cases, the $BO_6$ octahedra are tilted, thus distorting the dodecahedral sites and giving an orthorhombic lattice at room temperature (e.g. $CaTiO_3$, $CaGeO_3$, $MgSiO_3$).

As temperature increases, the distorted perovskites go through a series of displacive phase transitions toward more and more symmetrical structures until they become cubic at high enough temperatures. Pressure has usually the opposite effect, at least up to a certain point: the density is initially more easily increased by further tilting of the relatively less compressible octahedra than by uniformly compressing the structure.

Notwithstanding the fact that the fine structure of the dislocation cores and the stacking fault energies must depend on the type of distortion, it is worthwhile, as a first-order approximation, to investigate the geometrically possible slip systems and dislocation dissociation schemes. The speculations will be checked against recent experimental results of high-temperature creep experiments and TEM examination of various perovskites stable at ambient pressure.

The A cations and oxygen anions are usually of comparable size and, taken together, form a FCC lattice. Due to the presence of the B cation at the centre of the cell, the usual Burgers vector $\frac{1}{2}\langle 110\rangle$ of the FCC lattice is replaced by $\langle 110\rangle$; the shortest Burgers vectors are therefore $\langle 100\rangle$ and $\langle 110\rangle$. The potential slip systems are of the type: $(001)[100]$, $(110)[1\bar{1}0]$ and $(111)[1\bar{1}0]$.

(i) Slip on $(001)[100]$ has been documented at high temperature in $BaTiO_3$ (Beauchesne and Poirier 1989), $KTaO_3$ and $KNbO_3$ (Beauchesne and Poirier 1990), as well as in fluoride perovskites $KZnF_3$ (Poirier *et al* 1983) and $RbCaF_3$ (Beauchesne and Poirier 1990). A possible splitting is

$$[100] \rightarrow \tfrac{1}{2}[100] + \tfrac{1}{2}[100]. \tag{7}$$

In $KTaO_3$ and $KNbO_3$ the [100] dislocations observed in TEM are mostly edge, less mobile than the screw dislocations; this may be due to a possible splitting in the climb plane (010):

$$[100] \rightarrow \tfrac{1}{2}[101] + \tfrac{1}{2}[10\bar{1}]. \tag{8}$$

Climb on {100} planes is much easier in $KNbO_3$ than in $KTaO_3$ and dislocations can be seen expanding in their climb planes.

(ii) Slip on $(110)[1\bar{1}0]$ also is present in $BaTiO_3$, $KTaO_3$ and $KNbO_3$ perovskites deformed at high temperatures, with the possible dissociation

$$[1\bar{1}0] \rightarrow \tfrac{1}{2}[1\bar{1}0] + \tfrac{1}{2}[1\bar{1}0]. \tag{9}$$

Whether the dissociation is seen or not in TEM depends of course on the stacking fault energy; it is not seen in the above mentioned perovskites but it has been observed in $CaTiO_3$, on grown-in dislocations (Doukhan and Doukhan 1986), and in $CaGeO_3$ deformed at 1000 °C (Wang *et al* 1989). The width of the split dislocations can be measured in $CaGeO_3$ perovskite and yields a value for the stacking fault energy of 35 $mJ\,m^{-2}$ (Wang *et al* 1989). The $[1\bar{1}0]$ dislocations in $CaGeO_3$, split according to (9), react in the (110) plane with [001] dislocations split according to (7) to give octagonal extended nodes with $\frac{1}{2}[1\bar{1}1]$ displacement vector.

In split screw dislocations, the $\frac{1}{2}[1\bar{1}0]$ partial can potentially dissociate on {111} planes in zone with the line according to the usual Shockley scheme with formation of a sessile stair rod dislocation at the intersection of the faults (Poirier *et al* 1983):

$$[1\bar{1}0] \rightarrow \tfrac{1}{6}[1\bar{2}1] + \tfrac{1}{3}[2\bar{1}0] + \tfrac{1}{6}[1\bar{2}\bar{1}]. \tag{10}$$

This reaction can account for the extremely straight dislocations aligned in $[1\bar{1}0]$ directions seen in $KTaO_3$ (Beauchesne and Poirier 1990).

(iii) Slip on systems of the type $(111)[1\bar{1}0]$ have never been documented in any of the investigated perovskites. This may be due to the fact that the Shockley splitting corresponding to glide in the {111} planes may simultaneously occur on two {111} planes as in (10), thus blocking the screw dislocations.

Despite the interest of such geometrical considerations on dislocations in ideal cubic perovskites and their relative success in predicting slip systems, they fall short of allowing the prediction of high-temperature creep mechanisms, probably because the core structures, hence the values of the Peierls stresses and of the stacking fault energies of the faults corresponding to various splittings, as well as the ability to climb, depend on the interatomic potentials responsible for the distortions at lower temperatures.

Recent experiments on cubic $KTaO_3$ and on $BaTiO_3$ and $KNbO_3$, cubic at high temperatures but orthorhombic and ferroelectric at room temperature (Poirier and Beauchesne 1989, 1990) have demonstrated the limits of the analogy: whereas $BaTiO_3$ and $KNbO_3$ deform by power-law creep with similar rheological equations but different dislocation structures, $KTaO_3$ and $KZnF_3$, cubic at all temperatures, deform by Newtonian Harper–Dorn creep.

More experiments on perovskites with the same distortion as $MgSiO_3$ (e.g. $CaTiO_3$) are needed before one can 'assign possible real mechanical properties' to the lower mantle.

## REFERENCES

Aladag E, Davis L A and Gordon R B 1970 *Phil. Mag.* **21** 469–78
Ashworth J R and Barber D J 1977 *Phil. Trans. R. Soc.* A **286** 493–506

Beauchesne S and Poirier J P 1989 *Phys. Earth Planet. Interiors* **55** 187–99
—— 1990 *Phys. Earth Planet. Interiors* **61** 182–98
Boland J N and Liebermann R C 1983 *Geophys. Res. Lett.* **10** 87–90
Boland J N and Liu L 1983 *Nature* **303** 233–5
Burnley P C and Green H W 1989 *Nature* **338** 753–6
Doukhan J C and Doukhan N 1986 *Phys. Chem. Minerals* **13** 403–10
Doukhan N, Duclos R and Escaig B 1979 *J. Physique* **40** 381–7
Fontaine G and Haasen P 1969 *phys. status solidi* **31** K61–70
Frank F C 1965 *Rev. Geophys.* **3** 485–503
—— 1969 *Chemical and Mechanical Behaviour of Inorganic Materials* ed A W Searcy, D V Ragone and U Colombo (New York: Wiley) pp 697–707
Frank F C and Nicholas J F 1953 *Phil. Mag.* **44** 1213–35
Furnish M D and Bassett W A 1983 *J. Geophys. Res.* **88** 10333–41
Griggs D 1967 *Geophys. J. R. Astron. Soc.* **14** 19–31
Hornstra J 1961 *J. Phys. Chem. Solids.* **15** 311–23
Kohlstedt D L and Goetze C 1974 *J. Geophys. Res.* **79** 2045–51
Kronberg M L 1957 *Acta Metall.* **5** 507–24
Lacam A, Madon M and Poirier J P 1980 *Nature* **289** 155–7
Madon M, Guyot F, Peyronneau J and Poirier J P 1989 *Phys. Chem. Minerals* **16** 320–30
Madon M and Poirier J P 1980 *Science* **207** 66–8
—— 1983 *Phys. Earth Planet. Interiors* **33** 31–44
Poirier J P 1975 *J. Geophys. Res.* **80** 4059–61
—— 1981 *Anelasticity in the Earth* (Geodynamics Series **4**) (Washington, DC: American Geophysical Union) pp 113–17
—— 1988 *The Physics of the Planets* ed S K Runcorn (New York: Wiley) pp 161–71
Poirier J P, Peyronneau J, Gesland J Y and Brébec G 1983 *Phys. Earth Planet. Interiors* **32** 273–87
Poirier J P and Vergobbi B 1978 *Phys. Earth Planet. Interiors* **16** 370–8
Price G D and Putnis A 1979 *Nature* **280** 217–18
Vaughan P J, Green H W and Coe R S 1982 *Nature* **298** 357–8
Wang Y, Poirier J P and Liebermann R C 1989 *Phys. Chem. Minerals* **16** 630–3

# The Evolution of the Terrestrial Planets

D C Tozer

## 1 INTRODUCTION

The evolution of the Earth must rank as one of the longest running topics about which it is still possible to have major differences of opinion even about its salient features. To primitive men who happened to live within sight of a fitfully active volcano, a belief that the Earth's interior is hot and can thereby influence the surface appearance would have been natural long before anyone might have convinced them of the idea of the Earth as a planet. Speculation about the Earth's interior, particularly in response to such spectacular and dangerous manifestations of planetary change, has always been rife among natural philosophers and theologians, but a discernible scientific approach to the subject of the Earth's heat had to await a cultural shift away from catastrophism and thoughts of divine intervention. It began with the growing suspicion among seventeenth and eighteenth century geologists that immensely longer periods of time than previously thought possible would be needed to create the far more ubiquitous non-volcanic scenery with known physical processes. Heat flow studies were soon an active participant in this trend, it being recognised well before there was any clear idea of how to scale laboratory observations that the cooling of the Earth might give a clue as to how old the Earth was. This development was to culminate in Lord Kelvin's famous controversy with geologists who had overreacted to the earlier few thousand year estimates of the Earth's age by introducing the doctrine of uniformitarianism—a vague term still used by geologists but then literally taken to mean that the Earth was a steady-state perpetual motion machine. Kelvin's neglect of a then unknown radiogenic heating made his estimates of the Earth's age too short, but the real lesson to be learnt from this controversy is that effects one would still find impossibly difficult to measure directly in a laboratory experiment—in this case a heating rate less than $10^{-6}$ K yr$^{-1}$—can be important in a planetary context. The difficulties in comprehending the relevant physics of planetary heat transfer are still with us in a more subtle form, many past surprises and mistakes being attributable to a naive belief that its solution will ultimately prove to be just a greatly magnified version of a situation one could in principle create in a laboratory experiment. However, the self-gravitation of matter in planetary amounts restricts heat transfer solutions in ways and produces novel effects which now appear crucial to an understanding of the observed dynamical activity—or lack of it, in terrestrial planets. Although the physics of the situation produced by self-gravitation in planetary and stellar mass objects is very different I think there are some interesting parallels between the study of planetary dynamics now and the development of stellar evolution theory early in this century. Motivated to a great extent by views on the antiquity of the Earth, it will be recalled that people like Eddington were trying to persuade laboratory scientists that an unknown process of energy generation, made self-regulating through its strong temperature dependence and the effectively negative total specific heat of such large self-gravitating masses, balanced the observed heat loss and defined an evolutionary path for stellar structure for billions of years. The part of this story worth emphasising is that the strong non-linearity in the problem, which made laboratory scaling impossible, introduced a self-regulatory behaviour which gave Eddington the confidence to make important predictions before virtually anything was understood about the presumed energy generation. For terrestrial planets it is an analogous self-regulation of the heat loss process which is emerging as the key factor defining the general course of their physical and chemical evolution. Perhaps the most notable difference about the stellar and planetary problems is that fundamental physics had to be invented before Eddington's position was seen to be tenable, whereas it is the empirically well known rapidly increasing 'creep' deformation of nominally 'solid' materials with temperature which lies behind the constraint imposed by self-regulation on the internal states of terrestrial planets for long periods.

Thermal convection was mentioned as a cause of superficial deformation over a century ago, but it was to remain just one among several competing hypotheses about tectonic change for many years because the problem of deformation was not seen in a planetary perspective and little attention was given to it as a planetary heat transfer theory. Something of that attitude survives in the myth that plate tectonics is a theory distinct from heat transfer theory. Attitudes to planetary heat transfer were and, in many cases, still are controlled by loose statements in elementary textbooks that heat conduction and convection are different modes of heat transfer with mutually exclusive and easily discernible domains of validity. The two processes are much more intimately connected, conduction being an underlying cause of movement in 'free' convection problems and heat conduction theory being perhaps most accurately described as a truncated version of convection theory one might apply if one knew beforehand that material velocities $v$, throughout a system were $\ll \varkappa/L$ ($\varkappa$ is a representative thermal diffusivity and $L$ is the system's length scale). This amounts to a very stringent condition for a planet sized object ($v \ll 10^{-5}$ m yr$^{-1}$) and no one could ever have justified a conduction approximation to planetary heat transfer on observational grounds. In practice, it rested on the assumption that the change of rheology with temperature was essentially a qualitative one of transition from elastic to viscous behaviour in shear at a 'melting' temperature $T_M$—certainly an obvious simplification of much that had long been known about rock deformation, and one that was to make the observation of shear wave transmission look far more decisive than it is in defining the form of planetary heat transfer theory (see below).

The early attempts to evaluate the consequences of convection for the thermal state of the Earth's interior were non-controversial because they accepted as a starting point what had become an article of faith among heat transfer theorists by the 1950s: temperatures of erupting 'magma' verified conduction theory estimates of internal temperature to ~ 100 km depths. Any thoughts petrologists may have had that magmatism did not always originate at such great depths, or qualms about the representativeness of a material that had somehow reached the surface, were not allowed to disturb the notion of magmatism as leakage from a layer of 'partially molten' silicate phases. The identification of a minor feature in the distributions of speed and attenuation of seismic S waves at sub-oceanic depths ~ 100 km (rather inappropriately named 'the low velocity layer' (LVL) because the S wave velocity is even lower above it) was taken as confirmation of the conduction theorist's view. However, that confirmation was only convincing to those who saw silicate 'melting' as the only cause of S wave attenuation. What should not have fooled anyone was that the agreement between conduction theory estimates of temperature at ~ 100 km depths and magma temperatures was an entirely contrived one. Reconciling an active magmatism with the apparent 'solidity' of Earth material to great depth† has always been a delicate issue for those who see *in situ* planetary material either as heat conducting elastic 'solid' or viscous 'liquid'—one

† Extrapolation of the typical near surface temperature gradient downward leads to wholesale melting at only a 50–100 km depth (~ 1% of the Earth's radius).

usually has no magma at all or embarrassing amounts of it. The effect of this overly simplistic view of the temperature dependence of a material's rheology was to make planetary heat transfer theory look an elaborate exercise in juggling with distributions of radiogenic heating, thermal conductivity, 'initial' temperature, etc, which conduction theory had made look so important in fixing internal temperature values at the present time. Faced with the fact that internal temperature is *not* a directly observed quantity but the only output from a conduction theory calculation, rules of the game and doctrine had been invented to keep speculation within 'sensible' bounds and to give conduction theory the appearance of a subject concerned with detail and precision. It was to this end that magma temperatures had long been used as a constraint on temperature estimates for $\sim 100$ km depths because that was the depth range in which 'sensible' spherically or near spherically symmetric conduction models generally showed greatest likelihood of 'melting'. Repeated use of this doctrine, particularly by people claiming an 'empirical approach to geothermometry', did nothing but harden opinion and prolong its influence among many who have been otherwise so keen to proclaim their rejection of 'static' Earth models. The use plate tectonicians have made of seismological data to define the thickness of 'rigid plates' is a graphic illustration of how entrenched and totally muddled the attempt to characterise the rheological temperature dependence of planetary material as an elastic to viscous transition has now become.

An inevitable consequence of these attempts to fit magma temperatures to spherically symmetric models is that one could only offer *ad hoc* arguments for active vulcanism's (visible magmatism) very non-random occurrences over the Earth's surface. Similarly, it looks impossible to give any explanation for vulcanism's temporal characteristics so long as one invokes a diffusive theory of heat transfer and sees direct radiogenic heating as the only way of generating local temperature maxima. These problems convinced me in the early 1960s that the whole question of the Earth's present internal state and how it had evolved was being seriously misconceived and that it would be interesting to think it through afresh with a qualitatively different description of a temperature-dependent rheology. It was easy to envisage models that ascribed quasi-elastic and inelastic behaviour to planetary material in the same rather than different temperature ranges and that this 'viscoelastic' behaviour could be made temperature dependent to express 'melting' in different terms. I rather naively thought that such a model would be judged against the earlier Hookean models by how well it could be quantified to account for a wider set of Earth observations than earthquake travel times. Such modest aims were soon in conflict with experimental geophysicists who saw material properties in objective terms rather than as the way continuum mechanical theories are designed to represent a system's observed behaviour. Instead of realising that was the way the concept of material properties was always used in laboratory experiment—it is of great importance to see they are *never* directly observed (see below)—there were assertions of what the rheology of *in situ* Earth material really was, based on creep measurements of 'good' specimens of the mineral olivine $(Mg,Fe)_2SiO_4$ that had frequently figured in petrological interpretations of the Hookean Earth models. I had suggested (Tozer 1965) that creep measurements on this refractory mineral could only serve as a plausible upper bound to the creep resistance of *in situ* planetary material, but beneath the experimentalists' great concern with olivine properties there was really a reluctance to face up to a lesson a broader programme of experimentation or even the existing literature would have given: there is no useful connection between a rock's creep resistance and its quasi-elasticity. The velocity of seismic waves through such composite materials is controlled by the quasi-elasticity and density of its major phases, but it is well known that their creep resistance can be controlled by possibly quite minor low 'melting' point phases: the highly refractory nature of the phases apparently controlling seismic velocities made this situation look quite likely. This made it look logical not to try to build directly on the earlier Hookean model interpretations, but to explore the possibility of interpreting a wider set of data than earthquake travel times, normal mode frequencies, etc, with an initially wide choice of temperature-dependent viscoelastic planetary models. Restricting that choice by comparison with planetary observation of their consequences promised to give petrological insight one might never get from the Hookean models. Seismology was to be seen as part rather than an arbiter of this enterprise—one could even discern the chance of this approach explaining why the Earth is still a seismically active planet! I was disappointed when some of the first results raised by this approach, e.g. the possibility of a plate-like convective velocity field at the Earth's surface, were promptly taken out of their 'viscoelastic' context and crassly forced into the old elastic 'solid'/viscous 'liquid' conception of rheology's temperature dependence. It was not surprising this step led 'plate tectonicians' into difficulties understanding what they called 'the forces driving the rigid plates'—their misapprehension of the rheological situation was just as naive and internally inconsistent as the earlier opposition to continental drift on the grounds seismologists had proved the Earth was 'elastic'.

The formalism of continuum mechanics can easily be misconstrued as implying that the properties of a system are those of its parts juxtaposed —the basis of experimental geophysics and source of the frustration that so much of the planetary material is inaccessible for such study. Without wishing to deny much would be learnt if planets could be taken apart, I think any longing to do so on grounds of definitive property measurement fails to appreciate the non-local, relative nature of material properties—in brief the abstract nature of explanations based on continuum mechanics.

The scale of visible planetary deformation makes it clearer than in most laboratory experiments that relevant properties are not localised—structural geologists have long recognised that the deformability of the Earth's surface rocks is controlled by the size of enormous defects ('faults', etc) introduced by previous deformation rather than any 'intrinsic' properties one might measure in a laboratory. In other words, no clear distinction can be drawn between the cause of deformation and the material properties controlling its presence, as was possible with earlier 'static' planetary models. In particular, I would emphasise there is no guarantee of being able to predict a pattern of current superficial planetary deformation without appealing to one in the past—what could easily appear to be begging the question for the many who still see the existence of active deformation rather than its absence as the only thing requiring an explanation. I cannot claim to have sorted out the complicated technical and methodological issues raised by deformation-induced deformability (DID), but to refer to superficial material as constituting 'rigid plates' is not only a misreading of this situation but an obstacle to understanding how it has arisen. The plate picture was originally just a way of conveying the sense of a prediction that any existing surface velocity field arising from heat transfer in terrestrial planets would show deformation at any particular time largely confined to narrow zones—because a DID seemed a *sine qua non* of its existence. It was offered as the missing rationale of continental drifters' efforts to match widely separated structures of the geological *status quo* with quasi-rigid-body translations.

It is unrealistic to expect any theory to predict a current surface velocity field pattern (number and shape of 'plates') if only because 'initial' conditions are not available; cf the weather forecasting problem. Theoreticians have not always emphasised this indeterminacy in planetary convection problems—splendid computer graphics of unobservable internal velocity fields merely raise false hopes about what a theory can achieve and deflect attention away from the fact that nothing substantial, i.e. not easily changed, has been predicted. To anyone puzzled that there are still† surface movements at the present stage of Earth history, I can only suggest they are probably ignoring or grossly underestimating how big the weakening effect of a DID could be in the context of the Earth's superficial rock deformation. Talking of terrestrial planets generally, how long their surface rocks share the few centimetres per year convective velocities of their interiors as the near surface temperature gradient supported by

† Perhaps 'still' is the wrong word because some geologists appear to think large-scale movement may be only a feature of the most recent 10–20% of the Earth's history. If really good positive evidence could be offered for that view it would seriously weaken the idea of tectonics as a heat transfer phenomenon. For heat transfer theorists, the existence of an active surface deformation makes an extremely strong case for its existence in all the previous planetary history.

radiogenic heating declines, depends on how long a DID can effectively compensate for a negligible thermally induced 'creep' deformability at ever increasing depths. It can be expressed in this way because a buckling instability of surface material behaving quasi-elastically even on the $10^8$ year timescale of the convective process (what I define as 'lithospheric' material) will not occur at smaller stresses than its failure by 'faulting' (Jeffreys 1976). Note: where a 'veneer' of weak sediments forms the visible surface, its buckling or 'folding' might well mark the sites of a DID in the underlying rocks. If water can persist on a planetary surface, it prolongs the duration of plate-like surface movements by increasing the maximum depth to which a DID can be induced and activated by the stresses convective heat transfer imposes. While seismology gives no direct sign of the magnitude of this weakening, reflection seismology (COCORP and BIRPS projects), probing to depths of a few tens of kilometres using artificial seismic sources, has shown that 'fault' structures necessary for superficial movement exist through the critical depth range. Also noteworthy was the Russian discovery by deep drilling of water circulation at $\sim 10$ km depths. More ambiguously, attempts to infer an electrical conductivity from natural magnetic field variations suggest an extensive network of groundwater channels at such depths.

While the complications introduced by a DID only directly affect the deformation of a small superficial fraction of any terrestrial planet's material, it is not true to say it doesn't have evolutionary implications for a much bigger volume of planetary material. 'Dynamic' planetary models, the term I use to describe models where the material properties defining a process are also to varying degrees under its control, are non-linear and do not allow the kind of separation of problems by region that was possible with earlier 'static' planetary models. It is obvious, for example, that temperature-dependent properties take values dependent on the thermal insulation provided by shallower material, though this insulation is not simply a function of that material's properties if it is convecting. A further example of such 'global' behaviour is the way significant chemical evolution of deep terrestrial planet material seems to depend on whether their superficial material participates in the convective circulation. An observationally inferred spatial association of much of Earth's current vulcanism (and seismicity) with the descent of cool, recent sea-floor material is the most vivid, seemingly paradoxical evidence we have for this suspected association. Seismicity, particularly a deep seismicity, provides a good example of how material properties used to represent behaviour at a laboratory scale can mislead one as to what is relevant for a planetary scale problem. Even for shallow seismicity, where its association with the activation and/or generation of 'faults' has long been agreed, there must be few people with any knowledge of the complexity of earthquake occurrence who believe prediction can be based on some locally definable criterion, e.g. local shear stress exceeding some critical value. Efforts to explain deep seismicity in terms of a locally defined 'strength' look even more implausible because there is neither field evidence nor theoretical argument for shear stress ever remotely approaching the values laboratory experiment suggests would be required to 'break' Earth material also subjected to the pressure prevailing at many deep earthquake foci. Disregarding any argument there might be about the concept of an intrinsic 'breaking' stress to describe even the laboratory observations, the assumed invariance of a 'strength' used by such arguments ultimately rests on whether a continuum mechanical theory interpreting laboratory observations has scaling laws. While never strictly true when the chosen material properties show variation with the local thermodynamic state, approximations of such property data which do have scaling laws so often prove adequate for the modest scaling factors contemplated in numerous model engineering and scientific applications as to confuse the question of their objectivity and applicability to planetary scale dynamical problems. Analysis shows that as the length scale of a system is increased, any impression one might have of an invariant 'breaking' stress is eventually vitiated by the heating effect of irrecoverable 'creep' deformation occurring at ever smaller shear stress values. If one assumes a 'creep' deformation rate at any finite shear stress and that it increases in a characteristic exponential way with temperature, one can show that a discontinuity occurs in the steady deformation rate/shear stress relationship for the system as a whole at a stress value which tends to zero with increasing system size. Such discontinuities are associated with transitions of a smooth internal strain rate field to one showing (intensely heated) slip 'zones'. For well-annealed, initially uniform systems of laboratory and engineering size, such transitions occur at stress values well in excess of those at which dislocations are mobilised or cracks propagate and are therefore of purely academic interest in determining the onset of a 'catastrophic' rate of deformation in these cases. Although, one can readily estimate such a transition could occur within the range of shear stress values $0-10^8$ Pa associated with self-regulating convective states of the terrestrial planets, this result is purely suggestive for viscoelastic planetary models because one then also needs to take into account a pre-existing inhomogeneity of the viscoelastic properties and storage of shear strain energy throughout the convecting material. Detailed numerical arguments, still at a primitive stage of development, are required to decide whether, where and when such shear transitions and the associated concentration of deformational heating might be a recurring feature of the heat loss process and, in particular, whether they are currently manifest as a seismicity or 'primary' magmatism. I say 'primary' magmatism because a 'secondary' magmatism may eventually be possible as a result of a chemical differentiation of the convecting material—a cumulative effect of the 'primary' magmatism. Note: this 'secondary' magmatism has its origin in direct radiogenic rather than deformational heating and consequently its appearances are so rare on a human timescale as not to be classed as 'active' magmatism.

Everyone may agree long-term structural changes occur at a rate fixed by the rate at which planetary material can be 'magmatised'—only in such low-viscosity conditions can large-scale chemical differentiation occur, but the fact that an 'active' magmatism can now only be attributed to an alternating storage and release of shear strain energy throughout convecting planetary material makes for a definite break with past thinking on the cause of endogenous terrestrial planet evolution. In heat conduction (and the simplest convection models) magmatism could only be understood as the result of direct radiogenic heating, but now one specifically requires the existence of an extremely unmagmatic state (an effective viscosity $\gg 10^{15}$ times larger than that assigned to 'magma') throughout very large volumes of planetary material in order that there be enough stored strain energy for such material to be episodically and locally 'magmatised'. Although most planetologists are now committed to a view involving large-scale planetary movements, this particular break with past heat transfer theory still seems too total for many to make. There is certainly a growing number of geophysicists who recognise my original point (Tozer 1967) that thermal conditions inside terrestrial planets are largely determined by the temperature dependence of their material's creep resistance (effective viscosity), but the sticking point always seems to be the 'absurdly low' horizontally averaged (HA) temperatures I expect to be associated with these extremely viscous self-regulating convective states (Tozer 1972). As already mentioned, the key to resolving this controversy is to realise one has no direct knowledge of the Earth's internal HA temperatures and that magma only rises to the surface because its temperature is so much higher than the average value. The convective regulation of HA viscosities $\gtrsim 10^{20}$ Pa s throughout these temperature-dependent viscoelastic models of terrestrial planets automatically leads to a mean strain energy density $\sim 1$ J m$^{-3}$ if their short-period quasi-elastic response is made seismologically acceptable. This store of energy, a quantity of the same order as the radiogenic heating over $\sim 100$ years, is potentially available for a slip zone magmatisation of the material and perhaps the concomitant radiation of a pulse of strain waves in a time of the order of the transit time of shear waves across the convective flow length scale ($\sim 10^6$ m). One can show that such an instability of the convective flow would result in a global rate of magmatisation $\sim 10$ km$^3$ yr$^{-1}$ and a level of seismic activity averaging $\sim 10^{18}$ J yr$^{-1}$ (taken over decades) in a body like the Earth—just the kind of behaviour needed to give planetary heat transfer a plate tectonic signature†. This kind of 'stick/slip' behaviour is very distinctive of temperature-dependent viscoelasticity, but quite impossible to explain with the previous elastic

† Much more involved argument is required to demonstrate that this potential behaviour would be realised in the Earth but not, for example in the Moon.

solid/viscous liquid view of a rheological temperature dependence. Abandoning untested views on HA internal temperature produced by this earlier rheological outlook is a very small price to pay for reconciling and explaining the salient characteristics of magmatism and seismicity.

## 2 FORMULATION OF PLANETARY EVOLUTION PROBLEMS

I have written the extended introduction because it is very difficult to give a straightforward account of a problem whose essence is a non-linearity and feedback connecting it with diverse planetary phenomena. One has to get well inside the problem before seeing there is no direct clash with observation—merely that some 'facts' have to be reinterpreted.

Without access to 'initial' conditions for planetary evolution problems there is no option but to treat observed planetary states as one would an isolated weather pattern—essentially unpredictable in all detail but hopefully exhibiting 'climatic' type regularities. Theoretical analysis can only hope to demonstrate features common to a vast ensemble of possible planetary states from which the observed one(s) is imagined to be drawn. Convective heat transfer in most laboratory and engineering situations is achieved with a velocity pattern that is unpredictable and unsteady, but it does nevertheless exhibit well defined 'ensemble' features which are often adequately explained by quite simple arguments. Perhaps the most familiar example of an extremely well defined ensemble behaviour of convection emerging from a temperature variation of properties is the near constancy and homogeneity of HA temperatures throughout a so-called 'boiling' liquid for a wide range of geometries and heating rates. Here, the theoretical complication of a heat transfer problem introduced by strong temperature variation of an effective expansion coefficient and specific heat capacity actually makes it possible to summarise the situation with a 'boiling' temperature for all practical purposes independent of the mode of heating or its previous history. In terrestrial planet heat transfer problems, an analogous behaviour comes from the rapid decline of an effective viscosity with temperature†. The methods used to describe 'ensemble' behaviour in the latter case are described in Tozer (1972). They have been successfully tested in laboratory experiments using waxes as a convecting material (Davy 1981).

† Analogous, because as far as convective heat transfer rates are concerned what matters is an effective Rayleigh number value Ra ($=g\alpha L^3\Delta T/\varkappa\nu$). Whether a big increase in Ra comes, as in 'boiling' from a large increase in $\alpha$, the expansion coefficient, or a large decrease in $\nu$, the kinematic viscosity, the effect is similar in creating conditions largely independent of a heating rate and uncertainties in the other material properties.

There has been much discussion in the geophysical literature of whether the rheology of *in situ* planetary material is linear with the stress: more than 30 years ago it went under the general heading of 'finite strength' arguments, but more recently the concern has been with whether the creep rate increases continuously but faster than linearly with the stress. While very frequently observed in laboratory 'creep' rate studies, I take the view this is a relatively small effect at the relevant stress values compared with temperature effects and that it is academic and premature to consider it until questions of composition and internal temperature have been far more closely answered—or some planetary phenomenon is found whose interpretation demands its introduction. This comes back to my point that material properties are not observed but assigned to a system. Theoretical seismology would have got nowhere if the subject had been constantly beset by experimentalists claiming materials never behave like a Hookean elastic medium. There is, however, a crucial choice to be made—what kind of temperature-dependent viscoelastic model should one try to fit to planetary observations? Viscoelastic shear models come in two general classes depending on whether they show continuous deformation under constantly sustained stress, i.e. have a finite 'asymptotic' viscosity $\eta$. My approach is based on the assumption *in situ* planetary material anneals to an isotropic state. This annealing assumption is sufficient to prove (Truesdell and Noll 1965) that a material does not have a 'finite strength' and has short and long timescale responses to 'small' stress resembling a Hookean elasticity (shear modulus $\mu$) and an 'asymptotic' Newtonian viscosity, $\eta$. A Maxwellian viscoelasticity is the simplest of an infinite variety of models for annealable materials' low-stress rheological behaviour. Clearly their $\mu$ values could be identified with what seismologists previously took as the Earth's Hookean behaviour in shear, though now that identification cannot be entirely independent of views on their long-term response (represented by $\eta$). For example, the absence of detectable shear wave transmission in the Earth's core can now be taken to mean $\mu = 0$ (as was formerly obligatory) or, more plausibly, that the Maxwell stress relaxation time $\tau(=\eta/\mu)$ is there very much shorter than the period of seismic waves. Strictly speaking, it is already clear that an isotropic Maxwellian behaviour cannot entirely account for seismic wave propagation or attenuation in the outer parts of the present Earth, but one must not forget that one is trying to develop 'dynamic' planetary models in which a property like anisotropy is seen as a product of planetary evolution rather than something given *ab initio*. It is perhaps dangerous to prejudge what more elaborate form future linear viscoelastic models might take, but in my opinion the really big and interesting changes in outlook come with this change from Hookean to Maxwellian models. As hinted above, even this 'basic' viscoelastic behaviour gets augmented by a deformation due to cracking and fault slip where or if shear stresses exceed an ambient pressure—indeed that was the original reason for saying any convective velocity field involving terrestrial planet surfaces would have a plate-like character. While such non-linear rheological behaviour may be most visible to the field geologist, that visibility is ultimately attributable to Earth material also having a highly temperature-dependent irrecoverable deformability at low shear stresses.

In approaching the question of a plausible form of temperature dependence for the Maxwellian viscoelasticity, the big conceptual change is that 'solid' and 'liquid' lose their earlier mutually exclusive meanings and a 'melting' or 'solidus' temperature, $T_M$, merely assumes the status of a parameter describing the functional dependence of $\eta$ on temperature. In other words, a $T_M$ value is empirically and purely conventionally chosen to be the minimum temperature value at which the finiteness of $\eta$ is apparent even to the most casual observer. More careful observation at lower temperatures suggests a Maxwellian asymptotic viscosity model in which the temperature dependence is given the form $\eta = \eta_0 \exp(gT_M/T)$. $g \sim 20$ and $\eta_0 \sim 10^5$ Pa s would be plausible central values to choose if one knew one was dealing with a single-phase planetary material but an $\eta(T)$ for terrestrial planet rocks can be presumed to be some appropriated weighted function of $\eta(T)$ values for its constituent phases. How big a role do perhaps quite minor phases with lower $T_M$ values play in controlling the overall $\eta(T)$ of *in situ* planetary material? My views on this point are less restricted than those of colleagues who look to seismology to give an answer. Mine are more influenced by thoughts on whether terrestrial planet material could have been rid through demonstrable processes of all rheological influence from phases like $H_2O$ and $CO_2$ (see next section). Notice that, while such low $T_M$ phases make virtually no difference to the HA values of an overall $\eta(T)$ regulated by convection, they can have a big effect in lengthening the period of time that surface material will be part of a convective circulation.

Given the uncertainty about the effect of low $T_M$ components on overall $\eta(T)$ values, I shall ignore the usually relatively small effects of pressure on it. These are in any case roughly cancelled in their effect on convection by the rise in temperature that occurs along 'adiabats'. Note: 'adiabats' define a more satisfactory reference state to use for planetary scale convection problems than an isothermal state, since questions of convective stability may hinge on whether the heat flow supports a radial temperature gradient $> (\partial T/\partial r)_s = (-g\alpha T/C_p)$, and the tendency of any convection is to create, apart from thermal boundary layers, a homentropic state in a system. This neglect of pressure effects may not be justified in the larger terrestrial planets, Earth and Venus, where pressures appear large enough to induce first-order phase changes in major mineral phases. If the volume change and latent heat associated with a phase change raises the effective expansion coefficient/specific heat ratio, $\alpha/C_p$, by $> 10$, the increased adiabatic gradient in the phase transition region may well exceed the value associated with a conduction of the radiogenic heat through this region. In these circumstances the phase transition region will serve as a barrier to the convective mixing of the material above and below it, thereby facilitating the possible evolution of a compositional difference. While there is very considerable geochemical interest in whether this has happened throughout the Earth's history as the result of major phase transitions occurring at 400–700 km depths, this question is still rather secondary to the task of explaining how magmatism needed to explain the development of chemical inhomogeneity could have arisen in a planet where $\eta(T)$ values are so tightly controlled by a convective heat loss process.

## 3 PLANETARY FORMATION

Terrestrial planet material, along with a somewhat larger mass of similar material now thought to form the central regions of the Jovian planets, is generally believed to have been expelled to interstellar space by supernova explosion(s) more than $4.5 \times 10^9$ years ago. There have been many accounts of how this material was eventually reconstituted to form planets, but that story is now restricted in some important respects by astronomical and meteorite studies. Infrared astronomy and the discovery of isotopic anomalies in refractory meteorite inclusions indicative of short-lived radioactivities, e.g. $^{26}Al$, half life $= 7 \times 10^5$ years, both support the idea of a rapid cooling ($< 10^4$ years?) to temperatures at which the most refractory compounds form and condense to grains. Isotopic anomalies would arise because such refractory grains cooled much faster than the ambient well mixed gas. The cool grains served as nuclei for the condensation of volatiles like $H_2O$ and $CO_2$. While some of these 'condensible' volatiles might chemically bond to the refractory grains, most would form a coating of 'excess' volatiles. Particle growth and aggregation to macroscopic bodies would have been stopped by the declining density of this expanding supernova remnant, but with a temperature of only $\sim 30$ K even microscopic grains would survive the unknown but possibly multimillion year period before further growth was promoted by a rise in their number density. Astronomers have evidence of coated grains in the interstellar medium, but perhaps the most convincing evidence that this intimate mixture of refractories and condensible volatiles was a stage in the growth of all planets comes from comets. The Giotto mission to Halley's comet revealed its nucleus as refractory material mixed with a volatile fraction that was 80% $H_2O$, and the rest largely CO and $CO_2$. In a mass ratio that justifies the description 'dirty snowball', cometary material may be visualised as the fraction of a collapsing dust/gas cloud which was captured in orbits so far from the centre of collapse that it has not yet had time (or ever will) to form larger and chemically differentiated objects.

The observed behaviour of cometary material when forced to travel through the inner Solar System shows what would have quickly happened to 'excess' volatiles orbiting within a few astronomical units of a proto Sun. Uncertainty about the luminosity of such an object at the relevant time makes it impossible to decide to what heliocentric distance grains would have lost any excess volatile coating, but the thermal stability of several hydrate and carbonate phases makes it very improbable that a complete separation of the condensible volatiles and refractories would be achieved even at a Mercurian heliocentric distance (0.46 AU). It should not be forgotten that at this fine dust stage of accumulation heliocentric orbits would have been rounded and made coplanar by collisions with the pervading and much more dense gas of 'non-condensible' volatiles, $H_2$ and He. It is also possible the orbits would have been significantly perturbed by non-gravitational forces that could be chemically selective. An orbital evolution due to differences in the density and solar radiation reflectivity of metallic and silicate dust would produce a systematic change in mean dust particle composition with orbital radius (Harris and Tozer 1967)—the cause of the differing zero-pressure densities of the terrestrial planets? Note: the different densities of the Earth and the Moon, like the departures from coplanarity of the existing planetary orbits and scatter of planetary spin axis directions, are thought to date from a much later, near terminal stage of planetary material accumulation (see below).

Several studies to determine the effect of accumulation on the thermal and chemical evolution of the planetary material envisage 'embryo' planets growing by the impact of relatively tiny particles on their surfaces. Depending on the degree of 'rounding' of the particle orbits, the impact kinetic energy is mostly acquired from the embryo's gravitational attraction and is assumed to be dissipated at the site of impact. However, this picture has given a false impression of how readily this dissipated energy and any volatiles it may have liberated would escape to space. There are no dynamical reasons for believing in the kind of selective growth implied by 'embryo' planets. Leaving aside any fragmentation by individual collisions, the net accumulation process is more accurately described as a free for all; every object grows at the expense of its companions (Tozer 1978). The immediate tendency of such a process was to create well mixed and, in a statistical sense, uniformly heated accumulations—most of the heating comes from the large collisions that form a terminal stage of any such mutual growth process. In toto, this heating would be $\sim 2G\rho R^2$ per unit mass for an accumulation of density $\rho$ and radius $R$—an amount just sufficient to 'melt' an object like the Moon, or in the case of the Earth, raise the mean temperature $\sim 30\,000$ K!

This marginal sufficiency of accumulational heating to 'melt' the Moon was a decisive point when Apollo project studies indicated lunar material had been extensively differentiated well over $4 \times 10^9$ years ago. Supporters of the 'embryo' accretion picture were forced to consider accumulational timescales $\sim 10^3$ years and/or short-lived radioactivities as the source of the necessary heating. However, apart from any time such radioactivity had to decay in interstellar space, many studies have indicated that a much longer timescale ($\sim 10^7$ years) is necessary for terrestrial planet material already in heliocentric orbits to form planets. The 'free for all' or mutual accumulation picture is quite easily able to reconcile the early lunar heating with this timescale, because the accumulational heating is mainly introduced by large collisions that also bury it to great depths. An extreme version of the mutual growth picture, increasingly supported for quite separate reasons, is that the whole Earth and Moon are themselves fragments of a glancing collision between objects that were both considerably larger than the present Moon. Notice that fragmentation at such a late stage of accumulation introduces the possibility of lunar material having been more strongly heated by accumulation than is indicated by the present lunar value of $2G\rho R^2$ per unit mass—useful in explaining a relative depletion (compared with Earth) of even only moderately volatile elements in superficial lunar material, but still a very ineffective way of devolatilising the Moon as a whole (see below).

A feature of early planetary accumulation shown by all theoretical studies is a strong dependence of its rate on heliocentric distance. While one might expect the refractory grains orbiting at $>5$ AU distances to retain much, if not all, their original coating of 'excess' volatiles, this growth rate dependence would soon produce a planetesimal population whose mean size at any one time decreased with increasing heliocentric distance. However, this trend would eventually have been upset and the gross characteristics of the future planetary system settled by their capture of the pervasive and much more massive 'non-condensible' gas fraction of the circumsolar material. Capture probabilities for the $H_2$ and He depend exponentially on the surface temperature and radius of a planetesimal so that one must expect their capture to be entirely controlled by only a few objects reaching the necessary size. While one cannot be very precise about the heliocentric distance of such objects, it is reasonable to think the capture probability was highest for the largest proto Saturnian and Jovian objects orbiting at 5–10 AU heliocentric distances. These few objects would have acted as 'cold traps' of increasing effectiveness depriving protoplanets at smaller and greater heliocentric distances of the chance to pursue a similar path. This behaviour not only guaranteed the eventual existence of largely refractory terrestrial planets, but would also explain the somewhat higher density of Uranus and Neptune *vis à vis* Saturn and Jupiter. I now concentrate on terrestrial planet accumulation, which may even have been substantially complete before 'non-condensible' gas collection got under way. Even at Saturnian heliocentric distances a protoplanet would probably have to reach a few thousand kilometre radius before 'triggering' a 'runaway' capture of $H_2$ and He.

As a well mixed polyphase protoplanet grew, the amount of self-gravitational potential energy per unit mass that would be dissipated in a particular rearrangement of its phases by density also increased. This energy is proportional to a parameter $P = f(1-f)\Delta\rho G R^2$, where $f$ is a particular mass fraction and $\Delta\rho$ its density contrast with the remaining $(1-f)$. The absolute rate of any such rearrangement depends on a viscosity which rapidly declines (due to the accumulational heating) as the radius of the protoplanet, $R$, increases. $P$ values for all conceivable binary separations give a measure of the order in which they are likely to be realised—one differentiation process serves as a mixing process for one of lower $P$ value. For typical terrestrial planet material, the $P$ value associated with a metal/silicate separation is by far the largest, and in this case there is a real possibility of the separation becoming self-sustaining once initiated (Tozer 1978). Separation would probably be initiated by a major collision producing a protoplanet little more than $\sim 10^3$ km in radius. While such arguments account for the evidence of metal/silicate separations in the smaller terrestrial planets, it has been traditional to discuss formation of the Earth's core as if it would not commence until Earth material had reached its present degree of accumulation ($R \sim 6.4 \times 10^3$ km). It is, in fact, far more probable that all existing terrestrial planet core/shell structures, let alone the Earth's, are the outcome of many separations and amalgamations of core/shell structures in several generations of their respective protoplanets. Note that early separations of metal and silicate fractions are an essential part of the theory attributing the differing densities of the Earth and the Moon to a collision between objects already possessing a core/shell structure. One should also notice the pressure/temperature conditions of these early core/shell separations are very different from those subsequently prevailing near their interface in the larger terrestrial planets. There is seismological and laboratory evidence to suggest this has promoted some limited remixing of core and shell materials (through chemical reaction) in these cases.

## 4 POST ACCUMULATION EVOLUTION

In every case of terrestrial planetary material accumulation it now seems virtually certain that peak values of average temperature exceeded the $T_M$ values of its most refractory phases. Even if there were silicate liquid immiscibility, the repeated and vigorous mixing action of movements associated with amalgamations of the cores and shells of colliding protoplanets would largely cancel any tendency to differentiate the shell material during the main phase of accumulation. However, the eventual decline in frequency and scale of these collisions gave increasing opportunity for convective loss of the accumulation/differentiation heat to control the internal situation. At the very earliest stages of this cooling it is reasonable to talk of effective Rayleigh numbers $\gg 10^{20}$, thermal convection velocities even in superficial shell material $\gtrsim 1\ \mathrm{cm\,s^{-1}}$ and planetary cooling rates $\sim 1\,^{\circ}\mathrm{C\,yr^{-1}}$. Changes of the HA internal temperature with radius would have been confined to thermal boundary layers (TBL) of negligible thickness attached to the external surface and shell/core interface. Although HA temperatures at the external surface would have fallen below $T_{\mathrm{M}}$ values for some phases within thousands of years of any collision, 'solidification' could not proceed inward to any significant degree in these large objects until 'solidification' proceeding from near the base of the shell material had reduced the effective Rayleigh number of the situation by a factor $> 10^{12}$. Since this upward 'solidification', predicted many years ago from a comparison of the radial gradients of adiabats and $T_{\mathrm{M}}$ values, facilitates the loss of latent heat, I estimate shell material solidification would have been substantially complete within $10^{6}$ years of the last really major collisions. How completely the core material also 'solidified' at this time is considered later.

Until at least the 1960s, this upward 'solidification' of shell material was regarded as a good argument for a $T_{\mathrm{M}}(r)$ distribution being taken as an 'initial' temperature condition for heat conduction calculations. However, with $\eta(T)$ values representing 'solid' state creep rising to no more than $\sim 10^{14}$ Pa s just below a melting temperature, an effective Rayleigh number for sub-solidus cooling initially exceeded $10^{12}$. Radiogenic heat would have negligible effect in slowing the rate of planetary cooling until HA $\eta(T)$ values below the external surface TBL approached $10^{20}$ Pa s. A quasi steady state, ensuring a very close balance of the global heat production and loss rates, would exist within a few hundred million years of accumulation effectively ending, i.e. within 10% of the present planetary ages. Assuming there was no chemical differentiation of shell material by magmatism that produced big changes in the $\eta(T)$ dependence at different depths, the subsequent thermal evolution for $> 10^{10}$ years can be very simply described. There is a progressive thickening of the external surface TBL (due to the decline of radiogenic heat production) but virtually no change in HA temperature at greater depths where the $\eta(T)$ dependence fixes its value.

As this radiogenically sustainable convective state (RSCS) was attained, the HA temperature increment across the TBL at the core/shell interface would decline to a value supportable by the radiogenic heat production in core material. Using the radiogenic heat production of iron meteorites as a guide, one expects this TBL temperature difference to be $< 100\,^{\circ}\mathrm{C}$ even in the case of the largest terrestrial planet cores. Since the regulation of HA $\eta(T)$ values $\sim 10^{20}$ Pa s in shell material could not plausibly be associated with HA temperatures more than $\sim 60\%$ of the $T_{\mathrm{M}}$ values of its most refractory phases, one can show that convection in shell material has the effect of imposing an upper limit $\sim 1200$ K on the core temperatures associated with a RSCS of the smaller terrestrial planets. This result, obtained before the advent of planetary missions, made it clear that any magnetic fields now existing in the vicinity of the smaller terrestrial planets could not be due to magnetohydrodynamic action in a 'liquid' iron/nickel core. Although one might plausibly assume the 'solidification' of these cores $>4 \times 10^9$ years ago was delayed by the presence of a core component like sulphur which forms a eutectic with iron, this was not a large enough effect to deter one from predicting that any surviving magnetism of the Moon, Mars and Mercury could only be due to a remanence of their near surface rocks. This was a real possibility on the grounds that the vigorous core convection which would have attended adjustment to a RSCS would coincide in time with a downward movement of a Curie-point isothermal surface in the shallow shell material. One therefore saw the detection of a magnetic field as a way of resolving whether the cessation of magnetic-field generation predated or post dated the involvement of sub-Curie-point superficial material in the shell convective circulation.

Small magnetic fields were subsequently discovered around the Moon, Mars and Mercury which could be attributed to remanence, though it remains one of the more tractable experiments for future missions to check it does not have a time variation proving otherwise. Experiments on lunar rocks indicate a magnetising field $\sim 4 \times 10^9$ years ago comparable in magnitude to the Earth's present surface magnetic field. Since we know any metallic lunar core occupies a much smaller fraction of the interior than does the Earth's, such a large magnetising field is only likely to have originated in it if convective adjustment to a RSCS of the Moon was still in progress at that time. It is more surprising that a lunar magnetising field does not appear to have entirely disappeared until $\sim 10^9$ years after lunar accumulation effectively ended. This could be interpreted as evidence of an interplanetary field (see also below).

The unknown effect of pressures $>100$ GPa on the $T_M$ values of possible core materials in the Earth and Venus makes prediction of the duration of magnetic-field generation by dynamo action in these objects much more uncertain. However, as the techniques and range of high-pressure experimentation have expanded, it has become increasingly apparent that these planetary cores would also have frozen before a RSCS was established $\sim 4 \times 10^9$ years ago if they were Fe/Ni alloys. This may be the reason for the quite unexpected discovery that Venus now has no magnetic field. The absence of even a small field due to magnetic remanence is attributable to the high ($\sim 460$ °C) surface temperature and probably the continuance of surface participation in shell convective movements long after a core field disappeared. However, it may be indirectly pointing to misapprehensions about the cause of the Earth's magnetic field. At the moment, one could argue that the abundance of the alloying element that appears increasingly necessary to keep the Earth's outer core 'liquid' is sufficiently less abundant in a Venusian core for any 'liquid' region there surviving the adjustment to a RSCS to be too small for sustainable dynamo action. However, if it becomes possible to show that the Venusian core is similar to the Earth's in its liquidity, it will reopen the question of what now drives core movements. It has never been very convincing that a core radioactivity is enough and some theorists, unaware of how quickly and effectively planets adjust to a RSCS, still look to a 'solidification' of Earth's inner core as a source of density differences driving core movements. In my view, it would then be much more plausible to attribute the magnetic difference of the two planets to the very different rates at which their cores are being stirred by their respective precessional motions. It is also possible that the duration of a lunar magnetic field generation was increased by the energy fed in by large lunar precessional movements induced by the Earth's gravitational field when the two objects were much closer and synchronous lunar rotation was still to be established.

The prompt cessation of convective surface movements in the case of small terrestrial planets, indicated by the density of cratering and survival of a remanent magnetism, immediately prompts the question of what keeps the Earth so geologically active so long after its accumulation. While it may be confidently expected that all volatile/refractory compounds were dissociated by any terrestrial planet's accumulational heating, one can convincingly show from the enormous ratios of thermal diffusivity to any molecular species' diffusivity in condensed matter that only a tiny proportion of a global volatile inventory could be lost solely on that account. Perhaps the most persuasive evidence that all terrestrial planets contain volatiles is that even the 'excess' and relatively highly mobile volatile species $^3$He is still being liberated from the Earth (Craig and Lupton 1976)—the terrestrial planet material most strongly heated by its accumulation and the one whose subsequent convective movements appear to have offered the best opportunity for escape. A further major obstacle to the loss of chemically active species like $H_2O$ and $CO_2$ from terrestrial planets is that their action in reducing the overall creep resistance of shell material lowers the HA temperatures associated with a RSCS to values at which substantial rehydration and recarbonation takes place. An interesting feature of the rehydration is that it is likely to be more complete in near surface material, due to a reaction (creating amphibole from a basaltic fraction) which proceeds at lower temperatures the higher the pressure (see, for example, Wyllie 1971). I would predict a free (intergranular) water concentration at depths corresponding to $\sim 2.5$ GPa pressures to be a very persistent feature of shell material, since if convection were to bring it up to levels at which escape through a network of cracks might be a possibility, it would already be bound chemically. I have suggested this is a much better interpretation of the Earth's LVL $\sim 100$ km beneath its ocean basins, now that the regulation of extremely large HA viscosities can be seen to make the previous idea of a 'partially molten' silicate layer untenable. I would also use the discovery of a similar low '*Q*' (high-absorption) zone of seismic waves at corresponding pressures in the Moon (~600 km depths) as evidence of the phenomenon's widespread occurrence. It is also noteworthy in the lunar case that a seismicity originates at such depths having a well defined monthly periodicity—a phenomenon that seems to link it unequivocally to the tidal flexing of the Moon as it pursues its elliptical orbit about Earth. Since such tidal stresses are $\sim 10^{-5}$ of the lithostatic pressures at the seismic sources, one can only suppose an intergranular fluid carries virtually all the force acting between rock masses. Note: deep lunar seismic events are extremely small by terrestrial earthquake standards and any similar behaviour in the Earth's LVL would be entirely masked by the higher seismic noise level.

The barrier to escape volatiles create for themselves by chemical recombination with refractory phases will only be surmounted on any important scale if, and only for so long as, a RSCS involves the generation and upward migration of 'magmatised' shell material to levels at which its (uncombined) volatile content exsolves and makes its own escape channels. However, this picture only crudely conveys one facet of what is a delicately balanced and, in general, highly complicated dynamical feedback problem controlling the chemical evolution and external surface appearance of terrestrial planets. The feedback arises because liberated volatiles remaining on a planetary surface can, by lubricating fault slippage and the development of a DID in the cool near surface material, facilitate their reintroduction to a planet's interior. The delicate balance of the problem is indicated by numerical studies of shear instability in temperature-dependent viscoelastic materials, which suggest that magma producing 'stick/slip' instabilities only occur if a RSCS involves superficial material subduction to the interior. These interlocking feedback loops are not understood in any quantitative detail, though one can readily see that their effect in prolonging tectonic activity and magmatism is confined to those planets whose gravitation is strong enough to prevent volatile loss to interplanetary space. It has often been thought that a smaller heat flow/unit area across the smaller terrestrial planet surfaces, creating a larger thickness of 'lithospheric' material, is the reason their endogenous tectonic and magmatic histories were so short-lived or non-existent compared with the Earth's. However, other things being equal, this heat flow effect is largely cancelled by a smaller radial pressure gradient that allows cracking and fault slippage to occur at the correspondingly greater depths required for surface material to participate in convective movements. It seems much more likely that it was their inability to liberate their internal water through an endogenous magmatic activity or to retain as a lubricant of lithospheric fault movements any that might have been released by earlier accumulational heating which shortened, even to extinction, the epoch of plate like surface movements associated with their RSCS.

Geologists have long debated the influence of free water in promoting rock deformation in specific regional contexts, but I want to discuss the case for giving it a much greater significance. The thickness of the Earth's lithospheric material which has to be faulted if subduction is to occur is certainly much less than the ~100 km depth to the LVL. However, given the difficulty of promoting a compressive failure in lithospheric material only ten kilometres thick if its faults were dry (Sibson 1974), I have suggested (Tozer 1985) water may have been indispensible in sustaining geotectonic activity for the last one to two billion years. If I am right in thinking magma-producing shear instabilities in a convective flow require the participation of lithospheric material in the circulation, a significant part of the present crust/mantle division of shell material owes its existence to the presence of water. The expression of this crust/mantle differentiation in the broad topographic relief we call continents probably has its origin in a curious feature of shell material differentiation suggested by phase equilibrium studies: under the pressure/temperature conditions in which the various shell material fractions exist after separation, they are all less dense than the starting material. The density contrasts created by differentiation are much greater than the thermally induced density differences ($\sim 10^{-1}$ kg m$^{-3}$) which drive movements in a RSCS and play a major role in keeping the various differentiated fractions together. This differentiation according to density also tends to sort the shell material phases according to their creep resistance—lower melting point silicate phases and volatiles becoming concentrated above refractory phases that form a region of higher creep resistance than the starting metal. Seismologists have sensed differences between sub-continental and sub-oceanic material extending to several hundred kilometre depths and I have cited the persistence of such differences throughout a history of continental drift as excellent evidence that the rheology of undifferentiated shell material is not controlled by its most refractory phases. A major increase in creep resistance must have occurred when such continental 'roots' were formed if one is to explain why they have not become horizontally displaced from the low melting point phases forming continental crust. Such a major rise in creep resistance is also interesting in providing a way of circumventing the regulation of extremely large HA viscosities by the convective process. I see a 'secondary' magmatisation by direct radiogenic heating of undifferentiated material existing below such roots as the mechanism producing the rare but very large 'flood' basalt magmatism within continental areas, and possibly the kimberlite intrusion phenomenon.

I want to end with some comments on how the forthcoming Magellan mapping of Venus may help to reveal the crucial role of water in the evolution of terrestrial planet shell material. Venus is a planet so similar to the

Earth in its global characteristics as to make one wonder why its evolution has produced such a different superficial environment (see, for example, Kaula 1990). Low-resolution mapping has already revealed a surface with a significantly smaller fraction of high continental areas than the Earth's. There is no topography comparable to the Earth's mid-ocean ridge system that would indicate current surface involvement in convective movements. The density of cratering revealed by a higher resolution but limited area Russian survey also suggests a surface of much greater average age than the Earth's. What makes this relative surface immobility all the more remarkable is the relatively high ($\sim$730 K) surface temperature compared with the Earth's (288 K)—an interesting example of how even liberated volatiles may profoundly change the $\eta(T)$ values of superficial material. However, explanations of the absence of any ridge type topography being due to very much smaller $\eta(T)$ values in surface material are immediately refuted by the multi-kilometre 'continental' style topography which orbiting spacecraft tracking has shown to be less isostatically compensated than the Earth's continents. The existence of definite continental areas on Venus would also seem to be against Kaula's idea that the surface of Venus does not move because it is entirely 'choked' by a buoyant layer of crustal material. If these arguments for Venusian surface immobility can be sustained after the Magellan mapping, I believe it will be very difficult to avoid concluding that the lubricating effect of free water on the deformability of the Earth's lithosphere is greater than the effect a $> 400\,^{\circ}$C temperature increase would have if it were dry.

The present Venusian atmosphere is virtually devoid of water in any form ($< 10^{-5}$ of that in the Earth's oceans), but a much higher deuterium/hydrogen ratio suggests this water is the residue of a much larger amount. It seems generally agreed the great contrast in present atmospheric compositions and states stems from the $\sim$two-fold ratio of solar radiation intensities falling on their liberated volatiles. One visualises the two planets evolving along closely similar paths, perhaps for only a small fraction of their present ages, until the increasing mass of liberated volatiles (and solar luminosity) triggered a 'runaway greenhouse' effect with the Venusian oceans. The largely aqueous atmosphere may have initially produced even higher surface temperatures than the present value, but it was susceptible to photo dissociation at great heights. The hydrogen escaped to interplanetary space, the deuterium somewhat less rapidly, and the free oxygen would have disappeared through combination with surface rocks and such atmospheric components as ammonia or hydrocarbons. $CO_2$ previously dissolved in the Venusian oceans and returned to the planetary interior as carbonate sediments was henceforth a gas eventually able to take over from water the role of principal greenhouse gas.

The timing of this runaway greenhouse effect with the Venusian oceanic water is most uncertain. If it did happen $> 4 \times 10^{9}$ years ago with the volatiles liberated by accumulational heating rather than by subsequent endogenous magmatism, and when the solar luminosity is thought to have been only ~60% of its present value, one could argue the luminosity has only to rise by another 20% for a similar 'runaway' greenhouse effect to be triggered with the Earth's oceans. If the 'lubricant' effect of water in the Earth's surface rocks is as important as I surmise, this development will advance by several billion years the date at which seismicity, vulcanism and surface movements would have stopped in any case due to the diminishing rate of radiogenic heating.

## 5 CONCLUSIONS

Some theoretical ideas, predictions and observations have been discussed which seem to be most decisive in determining why terrestrial-planet-like objects might exist and what their surface appearance and internal states might be after a few billion years. I have had to be very selective in my discussion of a problem with many recognised facets—and no doubt many others we have yet to discover. For some of the omissions I would readily apologise, but planetary studies being what they are, I have also thought it salutary to correct any impression there might be that a successful theory of planetary evolution should, indeed ought to, predict much of what is most immediately visible about observed terrestrial planet states.

The plate tectonic way of picturing the current spatial pattern of changes in the Earth's state has certainly managed to convey the point that one may have to theorise on a global scale before finding a satisfactory explanation of even a minor local geological phenomenon. However, it has been far less successful, indeed positively unhelpful, in putting the observed state in an appropriate temporal perspective. Its constant attempts to impose kinematic constraints and supposed laws on the way change occurs not only gives an undeserved prominence and weight to the observed state and its changes, but conveys a spurious air of determinacy and precision where none can be shown to exist. Without precise and complete knowledge of an earlier planetary state, the best one can ever hope to do is predict a 'climate' of planetary change and further hope to see it exhibited by the observed planetary states.

I have suggested that the dominant factor defining the 'climate' of terrestrial planet changes for many billions of years is a convective regulation of these bodies' highly temperature-dependent rheology. Convective heat transfer quickly removed the heat dissipated within these objects in the course of their accumulation and differentiation of a metallic core/silicate shell structure. The marked temperature dependence of the convective loss resulted in a very close balance being set up between radiogenic heat production and loss within a few hundred million years of accumulation ending. From that time for $\sim 10^{10}$ years, terrestrial planet shell states are characterised by average velocities of a few centimetres per year and values of an average viscosity $\sim 10^{20}$ Pa s below a thermal boundary layer that thickens as the radiogenic heating diminishes. Average temperatures in this sub-thermal boundary region are almost entirely under the control of the viscosity dependence on temperature. This has stimulated interest in a possible volatile content of terrestrial planets, which could profoundly alter this dependence if present in relatively small quantities. My interest in this matter is not so much for the effect it might have in depressing internal temperatures well below previously estimated values (average temperature is not an observable), but for the effect volatiles can have in determining whether and for how long surface material shares the few centimetres per year convective velocities of the deeper shell material. I present reasons for believing all terrestrial planets contain rheologically important amounts of water, and that the continuance of endogenous geological activity on the Earth is in various ways due to its effect in promoting crack propagation and fault slippage in its near surface rocks.

The self-regulating convective states of terrestrial planet shells are so viscous as to make it necessary to consider the fact that one is dealing with a temperature-dependent viscoelastic material that stores shear strain energy throughout the flow—unfortunately, not a situation one can ever simulate in a laboratory scale experiment or one whose distinctive consequences, involving an enormous range of length scales, can be easily evaluated by numerical techniques. It is entirely due to this storage of strain energy that convective heat transfer on a planetary scale can sometimes be conveniently delimited for short-lived humans by a pattern of seismic and volcanic activity. For the first time we appear to have a theory of planetary heat transfer which not only reconciles the paradox of a 'solid' Earth that erupts 'liquid' rock but makes the association a necessary one.

## REFERENCES

Craig H and Lupton J E 1976 *Earth Planet. Sci. Lett.* **31** 369
Davy B W 1981 *PhD thesis* University of Newcastle-Upon-Tyne
Harris P G and Tozer D C 1967 *Nature* **215** 449
Jeffreys H 1976 *The Earth* 6th edn (Cambridge: Cambridge University Press) p 439
Kaula W M 1990 *Science* **247** 1191
Sibson R H 1974 *Nature* **249** 542
Tozer D C 1965 *Phil. Trans. R. Soc.* **258** 252
—— 1967 *The Earth's Mantle* ed T F Gaskell (New York: Academic) p 325
—— 1972 *Phys. Earth Planet. Interiors* **6** 182
—— 1978 *The Origin of the Solar System* ed S F Dermott (New York: Wiley) p 433
—— 1985 *Geophys. Surv.* **7** 213
Truesdell C and Noll W 1965 *The Non Linear Field Theories of Mechanics* (Handbuch der Physik **111/3**) (Berlin: Springer) p 113
Wyllie P J 1971 *The Dynamic Earth* (New York: Wiley) p 175

# Geophysical Excursions

S K Runcorn

Sir Charles Frank made a valuable and most timely contribution to geoscience by introducing to the community some of the ideas of solid state physics, especially those bearing on the mechanics of the Earth. At that time, for most geophysicists and geologists, Sir Harold Jeffreys' great treatise *The Earth* was the starting point. It presented a model of the Earth based on pre-Maxwellian physics: Newtonian mechanics, Fourier's law of the transport of heat by conduction and the Hookean laws of elasticity. Up to a point it was most successful. The observations of seismology could all be explained on the basis of classical elasticity and by an Earth model, the properties of which depended on radius only (except for modifications due to the equatorial bulge). That a theory making such economical use of physical principles was so successful was a remarkable triumph. This Earth model, in which the density as a fraction of radius could be derived and thus important information concerning the composition of the Earth's interior deduced, seemed remarkably secure. In fact even today the seismological data can be explained on the basis of classical elasticity with a single exception (reconciling Earth models from free oscillation and body wave data) and only recently have departures from radial symmetry been found.

It is characteristic of the laws of classical physics that they apply both on the microscopic and the large, even cosmic, scale and alike on short and vast timescales. Newtonian mechanics has been successful in discussing the motions of double stars and even galaxies and a major part of motions on the atomic scale. The gas laws are applicable to the interior of stars and to ordinary laboratory apparatus. It is not therefore entirely surprising that when Jeffreys sought to discuss departures from perfect elasticity in solid bodies of the Solar System he searched for a law which would explain equally well all the phenomena he recognised to require for their explanation departure from perfect elasticity. These were the damping of shear waves in the Earth's mantle, the damping of the Chandler wobble, the synchronous rotation of satellites in the Solar System and the existence at the present time of departures of the Earth and Moon's gravitational fields in their low harmonics from their hydrostatic figures. The great disparities in the timescales of these phenomena and the wide differences in physicochemical characteristics of solid bodies in the Solar System makes it rather amazing that Jeffreys could find such a relation. 'Lomnitz's law', although based on even at that time crude experiments, did go some way to explaining the data. But the processes of solid state physics result in laws which are qualitatively different over different time and length scales. Thus the solid state physicist would have regarded Jeffreys' search as quixotic. The most general considerations therefore at once explain why the Earth's mantle possesses finite strength only in the outer layer, called the lithosphere. Above a temperature of about half the melting point, creep becomes the dominant mechanical property on the geological timescale. The lithosphere is below the threshold temperature. The mantle behaves according to a creep law over the timescales of geology. The Boltzmann factor causes the boundary between lithosphere and flowing mantle to be quite sharp. Solid state theory therefore has given the foundation of physics necessary to reconcile the evidence for mobility of the Earth with seismology. For an impasse had been reached when quantitative evidence, initially from the palaeomagnetism of continental rocks, gave support to Wegener's theory of continental drift. Such great horizontal displacements of the lithosphere could not be accommodated in the Earth model that had successfully explained the data of the seismologists. Thus at this crucial point in the development of geoscience Sir Charles Frank's contribution (1965, 1966) was seminal.

It is natural that anyone involved with the modern theory of solids should attempt to understand the earthquake mechanism and Sir Charles was no exception. In his paper of 1965 he suggests that the Osborne–Reynolds dilatancy phenomenon might be a key to solving the fundamental problem: fracture involving the opening of cracks is impossible at the modest depths where the pressure much exceeds the shear stress. A granular mass, such as a rock, when sheared tends to expand as the close-packed state is disturbed, but if there is fluid between the grains the increase in volume greatly reduces its pressure and the contact pressure between the grains increases as does the frictional resistance to deformation. This is called dilatancy hardening and, as Frank beautifully describes, accounts for the firmness of wet sand. This paper was written at the Institute of Geophysics and Planetary Physics at the University of California, San Diego and one can well imagine a barefoot Sir Charles deriving much pleasure in demonstrating these phenomena to the graduate students on the La Jolla beach.

Frank sought to show how dilatancy hardening can result in long-term instability in which localised yield occurs. He supposed that the yield becomes catastrophic. Before this happens the pressure of the fluid changes. The earthquake mechanism still remains obscure, and this idea has been of importance in the development of research on earthquake prediction. The marked change in the level of water in wells and changes in the telluric currents and in Earth resistivity measurements have all been reported prior to an earthquake and have been suggested as possible observations which might prove a key to earthquake prediction. The phenomena of dilatancy certainly shows how these possible precursors could result from slow deformation of rocks near the surface prior to the earthquake. Thus Charles' work was a considerable stimulus to the large amount of research effort which has gone into earthquake prediction. That there has been some waning of interest in recent years reflects the complexity of the phenomena involved in the earthquake mechanism and the corresponding disappointment that there is no promising method in sight. The paper, however, has been an important stimulus to this most important endeavour.

Charles was among the first to propose that the plates moved by gravity sliding, by supposing that the plates on the flanks of the oceanic rises would move away from the rifts where magma rose by partial differentiation in the upper mantle. He argued that the analogy with the flow of glaciers was helpful: new ice is added at the mountain top just as new material is added at the ocean rise. Like all analogies this one is not complete; the gravitational field in which the plate and the glacier move differ significantly. Others have also proposed gravity sliding away from the ridge—a ridge push—as the mechanism of plate motions, e.g. Hales, but Frank's account is very lucid and carefully considers possible difficulties with the process that escaped the attention of others (Frank 1972, 1973).

This theory of a 'ridge push' originated in the success of plate tectonics in bringing together within a common description the theory of the relative movement of continents, the concepts of sea-floor spreading and transform faults and the palaeomagnetic and seismic evidence for them. Its elegant geometrical features concentrated the attention of geoscientists on the properties of the lithosphere and the plates. It was natural that their boundaries, the ocean rises and the trenches, active seismically and volcanically, should be thought of as the sources of driving force for plate motions. Thus the push from the ridge and the pull towards the sinking lithospheric edges in the trenches became widely accepted as theories of plate motions. To this necessary stage in the yet unsolved problem of the mechanism of plate tectonics Charles' paper was a valuable contribution. However, it does not face the essential problem of what dynamical processes in the mantle have their expression in the phenomena of continental drift and sea-floor spreading. What initiated the rifts or the trenches? One hesitates to draw the analogy with the doctor treating a serious malfunction of an internal organ by prescribing an emollient ointment to the skin eruption produced. The driving force of plate tectonics must be solid state convection in the mantle—Nature's way of turning heat or potential energy into motion. I had so argued in 1962 but mantle convection was anathama in geoscience. The educative influence of Sir Charles in expanding the view of geophysicists to appreciate that creep is the fundamental long-term mechanical property of the silicate mantle was invaluable. The primary evidence now for solid state convection in the mantles of terrestrial planets is the departure of the low harmonics of their gravitational fields from hydrostatic values: the surface manifestations—plate movements, or the still obscure features of Venus' surface—depends on the properties of the lithosphere, its thickness, finite strength, etc. Thus these great horizontal movements—a major part of geology—turn out to be a secondary effect of the fundamental process in the mantle of solid state convection!

In spite of all the work done in modelling convection, which has only recently progressed to solving problems more nearly relevant to the situation in the Earth's mantle, the way in which convection is coupled to the plates and provides the driving forces still remains obscure. The gravity sliding on which Charles worked may, therefore, still turn out to be an important part of the mechanism.

Perhaps nowhere is there a better example of Charles' approach to science than in a short letter entitled 'Curvature of Island Arcs' (Frank 1968). This is a simple, elegant, geometrical explanation of what to the geologist and geographer is a commonplace observation—the curvature of island arcs. A ping-pong ball (analogous to the inextensible but flexible lithosphere) is bent inwards through an angle $\theta$: he proves that this occurs only on a circle of radius of curvature $\frac{1}{2}\theta$. Asserting that deep focus earthquakes show that the lithospheric slab or ocean floor moves downwards into the mantle at an angle of roughly 45°, he measures the Kamchatka–Kuribe, Alaska Aleutian, Sumatra–Java arcs and the less obvious arcuate Himalayas and Andes at 20°–22°: *quod erat demonstrandum*!

## REFERENCES

Jeffreys H 1962 *The Earth* 4th edn (Cambridge: Cambridge University Press)
Frank F C 1965 *Rev. Geophys.* **3** 485–503
—— 1966 *Rev. Geophys.* **4** 405–8
—— 1968 *Nature* **220** 360
—— 1972 *Flow and Fracture in Rock, AGU Monograph 16* ed H C Heard, I Y Borg, N L Carter and C B Raleigh pp 285–92
—— 1973 *Phil. Trans. R. Soc.* A **274** 351–4
Runcorn S K 1962 *Nature* **193** 311–14

# Index